QUANTITATIVE PHARMACEUTICAL CHEMISTRY

QUANTITATIVE PHARMACEUTICAL CHEMISTRY

GLENN L. JENKINS, Ph.D., L.L.D.
Professor of Medicinal Chemistry and Dean Emeritus
Purdue University School of Pharmacy and Pharmacal Sciences

ADELBERT M. KNEVEL, Ph.D.
Professor of Medicinal Chemistry
Purdue University School of Pharmacy and Pharmacal Sciences

FRANK E. DiGANGI, Ph.D.
Professor of Pharmaceutical Chemistry
University of Minnesola College of Pharmacy

SIXTH EDITION

CBS Publishers & Distributors Pvt. Ltd.

New Delhi • Bengaluru • Chennai • Kochi • Kolkata • Mumbai
Hyderabad • Uttarakhand • Nagpur • Patna • Pune • Jharkhand

ISBN: 978-81-239-1585-2

First Indian Reprint: 2008

Published by **Satish Kumar Jain** and produced by **Varun Jain** for
CBS Publishers & Distributors Pvt. Ltd.,
4819/XI Prahlad Street, 24 Ansari Road, Daryaganj, New Delhi - 110002
delhi@cbspd.com, cbspubs@airtelmail.in • www.cbspd.com
Ph.: 23289259, 23266861, 23266867 • Fax: 011-23243014

Corporate Office: 204 FIE, Industrial Area, Patparganj, Delhi - 110 092
Ph: 49344934 • Fax: 011-49344935
E-mail: publishing@cbspd.com • publicity@cbspd.com

Branches:
- ***Bengaluru:*** 2975, 17th Cross, K.R. Road, Bansankari 2nd Stage, Bengaluru - 70 • Ph: +91-80-26771678/79 • Fax: +91-80-26771680 E-mail: cbsbng@gmail.com, bangalore@cbspd.com
- ***Chennai:*** No. 7, Subbaraya Street, Shenoy Nagar, Chennai - 600030 Ph: +91-44-26681266, 26680620 • Fax: +91-44-42032115 E-mail: chennai@cbspd.com
- ***Kochi:*** Ashana House, 39/1904, A.M. Thomas Road, Valanjambalam, Ernakulum, Kochi • Ph: +91-484-4059061-65 Fax: +91-484-4059065 • E-mail: cochin@cbspd.com
- ***Kolkata:*** 6-B, Ground Floor, Rameshwar Shaw Road, Kolkata - 700014 Ph: +91-33-22891126/7/8 • E-mail: kolkata@cbspd.com
- ***Mumbai:*** 83-C, Dr. E. Moses Road, Worli, Mumbai - 400018 Ph: +91-9833017933, 022-24902340/41 • E-mail: mumbai@cbspd.com

Representatives:

- Hyderabad: 0-9885175004
- Patna: 0-9334159340
- Jharkhand: 0-9811541605
- Nagpur: 0-9021734563
- Pune: 0-9623451994
- Uttarakhand: 0-9716462459

Printed at:
Neekunj Print Process,

PREFACE

The present trend in medical science makes pharmacy an increasingly important health profession. Large numbers of new drug products have been introduced, and the quality control of these products has assumed greater importance in industrial and governmental laboratories. Progress has been so rapid that frequent editions of a textbook in the area of quantitative pharmaceutical analysis become necessary. The fact that previous editions of "Quantitative Pharmaceutical Chemistry" have been received favorably as a valuable reference by analysts and as an effective textbook by teachers and students of pharmacy makes this new sixth edition possible.

Since the publication of the fifth edition of this book, numerous changes have been made in the official methods of analysis of pharmaceutical materials in The United States Pharmacopeia XVII and in The National Formulary XII. The detailed procedures and discussions have been changed or rewritten in the present edition to conform to the revised official methods.

The object of this book is threefold: first, to furnish students of pharmacy with a systematic course covering the quantitative chemical and physical methods official in The United States Pharmacopeia and The National Formulary through the selection and explanation of typical procedures; second, to classify the large number of official procedures according to type methods of analysis for effective and convenient instruction; and third, to present some of the generally applicable, nonofficial methods of analysis that are widely used in pharmacy and with which all students pursuing the profession should be familiar. In seeking to attain all three objectives the theory and practice of analytical chemistry as applied in quantitative pharmaceutical procedures have been correlated. Secondary objectives have been to adhere closely to unity of subject matter, to place emphasis on the accuracy and proper technique required in good analytical and pharmaceutical practice, to present type methods in an orderly sequence leading from relatively simple to complex procedures, and to avoid expansion of the book to unwieldy size by inclusion of detail of subject matter that is readily available in general references.

The book is intended for use in didactic instruction and also as a laboratory manual. Its use as a text should be supplemented by lecture, recitation, and laboratory instruction. It is obviously impracticable to include detailed explanations of all the quantitative determinations in The United States Pharmacopeia and The National Formulary. Typical analyses illustrating all the various methods have therefore been selected and explained in considerable detail, whereas those determinations requiring the same or closely similar procedures are indicated in tables or otherwise. The instructor may select other determinations from the Pharmacopeia or The National Formulary and assign to the students the task of applying the theory and explanation of the procedure involved as given in the text.

The content of the book is arranged in three parts. Part 1 treats of general methods of gravimetric and volumetric analysis, Part 2 contains the special methods of pharmaceutical analysis, and Part 3 treats of physicochemical and instrumental methods. Part 1 includes suitable material for a one-semester course. Parts 2 and 3 may be combined to constitute the work of the second semester.

All chapters of the book have been thoroughly revised. New and additional questions and problems have been added to permit greater selection for assignment. New chapters on nonaqueous titrimetric analysis, precipitation and complexation methods of analysis, potentiometry, and polarography have been added. The chapter on chromatography has been expanded to include sections on thin-layer and gas chromatography. The qualitative and quantitative aspects of infrared spectrophotometry have been added to the chapter on spectrometry. New illustrations and much new material pertinent to instrumental methods have been introduced. The theory, type procedures, and tables have been revised in accord with the official requirements.

The authors are indebted to Mrs. Theodora A. Andrews, Pharmacy Librarian with the rank of Associate Professor, Purdue University, who revised the compilation of General References. The useful references to other textbooks and treatises on general and special methods of analytical procedure have been revised and expanded.

The authors wish to acknowledge their appreciation of the courtesy extended by the following companies, which granted permission to use illustrations from their catalogues and books: American Pharmaceutical Association; United States Pharmacopeial Convention; McGraw-Hill Book Company; Beckman Instruments Inc.; E. H. Sargent and Co.; Bausch & Lomb; The Torsion Balance Co.; Fisher Scientific Co.; Mettler Instrument Corp.; Warner-Chilcott Lab. Instruments Div.

The wide popularity that the earlier editions of this book have enjoyed among drug analysts, students, and teachers has been gratifying.

It is hoped that the new edition will similarly commend itself to a still larger circle. To the many kind friends who have so generously aided by means of helpful suggestions, the authors wish to express their thanks.

GLENN L. JENKINS
ADELBERT M. KNEVEL
FRANK E. DIGANGI

CONTENTS

PART 3
Physicochemical and Instrumental Methods Used in Official Pharmaceutical Analyses

INTRODUCTION

Quantitative pharmaceutical analysis may be defined as the application of the procedures of quantitative analytical chemistry to the analysis of the materials used in pharmacy and, especially, to the determination of the purity and quality of the drugs and chemicals official in The United States Pharmacopeia and in The National Formulary. A complete chemical analysis of any substance requires the establishment of the identity of its component parts by qualitative analysis and the determination of the proportion in which these components are present by the processes of quantitative analysis.

Quantitative chemical methods are commonly classified as gravimetric, volumetric, gasometric, and instrumental. It is impractical, however, to adhere strictly to this classification in presenting the subject matter of quantitative pharmaceutical analysis, since certain phases of the work, such as alkaloidal assaying and the assay of essential oils, have been developed to a point where they constitute a distinct subject matter in which the principles involved and the technique required are the same whether the procedure is gravimetric or volumetric.

Quantitative pharmaceutical analysis may be appropriately subdivided according to the following procedures, each of which requires a special technique:

Gravimetric analysis is the separation by extraction, precipitation, or other means of the constituent to be determined either in the natural state or in the form of a definite compound the composition of which is known to the analyst, and the weighing of the resulting product.

Volumetric analysis is the determination of the volume of a solution of known concentration required to react with a given amount of the substance to be analyzed.

Gasometric analysis is the measurement of the volume of a liberated gas or the decrease in volume of a mixture of gases when a suitable reagent is used to remove one of the gases present.

Special methods are those which require a distinct type of technique, such as alkaloidal assaying.

Physicochemical methods of analysis are based on some specific physical or chemical property or properties of the substance being analyzed. In many assays the measurement of these properties requires the utilization of an *instrumental* technique. Those physicochemical methods which commonly employ instrumental techniques as applied in quantitative pharmaceutical chemistry are *spectrophotometric, electrometric, thermal conductimetric, refractometric,* and *polarimetric.* They constitute some of the most valuable methods used in analytical procedures. Since each of them requires special treatment, they will be discussed separately.

The official assay methods serve as an exact measure of the purity of a substance only when the results are considered in conjunction with the qualitative tests. Thus, in the assay of zinc oxide, the determination of the purity of the oxide by assay must follow qualitative tests for other metals which if present would be estimated as zinc oxide. A complete analysis, therefore, requires the qualitative estimation.

The theoretical considerations applicable in any given analysis are inherently a part of the analytical procedure. They should be mastered before a determination is made, and they should be carefully applied throughout each step of the procedure.

Those who desire a more comprehensive treatment of special aspects of analytical procedure will find of value the texts and reference books listed under General References.

GENERAL REFERENCES

Textbooks

BECKETT, A. H., and J. B. STENLAKE, "Practical Pharmaceutical Chemistry: Quantitative Analysis," Oxford University Press, New York, 1962.

BENEDETTI-PICHLER, A. A., "Essentials of Quantitative Analysis," The Ronald Press Company, New York, 1956.

BLAEDEL, W. J., and V. W. MELOCHE, "Elementary Quantitative Analysis," 2d ed., Harper & Row, Publishers, Incorporated, New York, 1964.

BROWN, G. H., and E. M. SALLEE, "Quantitative Chemistry," Prentice-Hall, Inc., Englewood Cliffs, N.J., 1963.

BRUMBLAY, R. U., "Quantitative Analysis," Barnes & Noble, Inc., New York, 1960.

CLARK, E. R., "Tutorial Questions in Quantitative Analysis," University of London Press, Ltd., London, 1964.

DAY, R. L., and A. L. UNDERWOOD, "Quantitative Analysis," Prentice-Hall, Inc., Englewood Cliffs, N.J., 1958.

ERDEY, L., "Gravimetric Analysis," vol. I, The Macmillan Company, New York, 1963.

FISCHER, R. B., "Quantitative Chemical Analysis," W. B. Saunders Company, Philadelphia, 1961.

FURMAN, N. H., F. J. WELCHER, et al., "Standard Methods of Chemical Analysis," vols. I to III, D. Van Nostrand Company, Inc., Princeton, N.J., 1962, 1963, and 1966.

HAMILTON, L. F., and S. SIMPSON, "Quantitative Chemical Analysis," 12th ed., The Macmillan Company, New York, 1964.

KOLTHOFF, I. M., and E. B. SANDELL, "Textbook of Quantitative Inorganic Analysis," 3d ed., The Macmillan Company, New York, 1952.

MERRITT, P. E., "Quantitative Analysis," D. C. Heath and Company, Boston, 1964.

PIERCE, W. C., E. L. HAENISCH, and D. T. SAWYER, "Quantitative Analysis," 4th ed., John Wiley & Sons, Inc., New York, 1958.

RIEMAN, W., J. D. NEUSS, and B. NAIMAN, "Quantitative Analysis," 3d ed., McGraw-Hill Book Company, New York, 1951.

SKOOG, D. A., and D. M. WEST, "Fundamentals of Analytical Chemistry," Holt, Rinehart and Winston, Inc., New York, 1963.

TALBOT, H. P., "Quantitative Chemical Analysis," 12th ed., The Macmillan Company, New York, 1964.

VOGEL, A. I., "Textbook of Quantitative Inorganic Analysis," 3d ed., John Wiley & Sons, Inc., New York, 1961.

WASER, J., "Quantitative Chemistry; A Laboratory Text," rev. ed., W. A. Benjamin, Inc., New York, 1964.

WILLARD, H. H., N. H. FURMAN, and E. K. BACON, "A Short Course in Quantitative Analysis," 2d ed., D. Van Nostrand Company, Inc., Princeton, N.J., 1957.

WILLARD, H. H., N. H. FURMAN, and C. E. BRICKER, "Elements of Quantitative Analysis," 4th ed., D. Van Nostrand Company, Inc., Princeton, N.J., 1956.

General References

American Association of Clinical Chemists (Miriam Reiner, editor-in-chief), "Standard Methods of Clinical Chemistry," vol. I, Academic Press Inc., New York, 1953.

BATES, R. G., "Determination of pH: Theory and Practice," John Wiley & Sons, Inc., New York, 1964.

BELCHER, R., and C. L. WILSON, "New Methods in Analytical Chemistry," 2d ed., Reinhold Publishing Corporation, New York, 1964.

BUTLER, J. N., "Ionic Equilibrium: A Mathematical Approach," Addison-Wesley Publishing Company, Reading, Mass., 1964.

FREISER, H., and Q. FERNANDO, "Ionic Equilibria in Analytical Chemistry," John Wiley & Sons, Inc., New York, 1963.

GLICK, D., "Methods of Biochemical Analysis," vols. I to XIV, Interscience Publishers, Inc., New York, 1954–1966.

HIGUCHI, T., and BROCKMANN-HANSSEN, E., "Pharmaceutical Analysis," Interscience Publishers, Inc., New York, 1961.

KLINE, G. M., "Analytical Chemistry of Polymers," parts 1 to 3, Interscience Publishers, Inc., New York, 1959–1962.

LAITINEN, H. A., "Chemical Analysis," McGraw-Hill Book Company, New York, 1960.

LAMBIE, D. A., "Techniques for the Use of Radioisotopes in Analysis," D. Van Nostrand Company, Inc., Princeton, N.J., 1964.

MEITES, L., "Handbook of Analytical Chemistry," McGraw-Hill Book Company, New York, 1963.

MITCHELL, J., and D. M. SMITH, "Aquametry," John Wiley & Sons, Inc., New York. Second edition in preparation.

MITCHELL, J., et al., "Organic Analysis," vols. I to IV, Interscience Publishers, Inc., New York, 1953–1960. Additional volumes in preparation.

NALIMOV, V. V., "Application of Mathematical Statistics to Chemical Analysis," Addison-Wesley Publishing Company, Inc., Reading, Mass., 1962.

NIEUWENBERG, C. J. VAN, and J. W. L. VAN LIGEN, "Quantitative Chemical Micro-analysis," American Elsevier Publishing Company, New York, 1963.

OSTROUMOV, E. A., "Application of Organic Bases in Analytical Chemistry," Pergamon Press, London, 1963.

REILLY, C. N., "Advances in Analytical Chemistry and Instrumentation," vols. I to IV, Interscience Publishers, Inc., New York, 1960–1964.

SANDELL, E. B., "Colorimetric Determination of Traces of Metals," 3d ed., Interscience Publishers, Inc., New York, 1959.

SNELL, F. D., and F. M. BIFFEN, "Commercial Methods of Analysis," rev. ed., Chemical Publishing Company, Inc., New York, 1964.

Society for Analytical Chemistry, "The Determination of Sterols," London, 1964.

STEYERMARK, A., "Quantitative Organic Microanalysis," 2d ed., Academic Press Inc., New York, 1961.

STROUTS, C. N. R., J. H. GILFILLAN, and H. N. WILSON, "Analytical Chemistry: The Working Tools," vols. I and II, Oxford University Press, New York, 1955.

TAYLOR, D., "The Measurement of Radio Isotopes," 2d ed., John Wiley & Sons, Inc., Englewood Cliffs, N.J., 1958.

WALTON, H. F., "Principles and Methods of Chemical Analysis," 2d ed., Prentice-Hall, Inc., Englewood Cliffs, N.J., 1964.

WILSON, C. L., and D. W. WILSON, "Comprehensive Analytical Chemistry," vols. I and II, American Elsevier Publishing Company, New York, 1959–1964.

ZWEIG, G., "Analytical Methods of Pesticides, Plant Growth Regulators, and Food Additives," vol. I to IV, Academic Press Inc., New York, 1963–1964.

Calculations

ANDERSON, H. V., "Chemical Calculations," 6th ed., McGraw-Hill Book Company, New York, 1955.

ARENSON, S. B., "How to Solve Problems in Quantitative Analysis," Thomas Y. Crowell Company, New York, 1959.

BENSON, S. W., "Chemical Calculations," 2d ed., John Wiley & Sons, Inc., New York, 1963.

DANIELS, F., "Mathematical Preparation for Physical Chemistry," McGraw-Hill Book Company, New York, 1928.

HAMILTON, L. F., and S. G. SIMPSON, "Calculations of Analytical Chemistry," 6th ed., McGraw-Hill Book Company, New York, 1960.

WEST, P. W., "Calculations of Quantitative Analysis," The Macmillan Company, New York, 1948.

Drug Analysis

American Pharmaceutical Association, "The National Formulary," 12th ed., Mack Publishing Company, Easton, Pa., 1965.

HIGUCHI, T., and E. BROCKMANN-HANSSEN, "Pharmaceutical Analysis," Interscience Publishers, Inc., New York, 1961.

The United States Pharmacopeial Convention, Inc., "The Pharmacopeia of the United States of America," 17th rev., Mack Publishing Company, Easton, Pa., 1965.

Food Analysis

BLOCK, R. J., and D. BOLLING, "Amino Acid Composition of Proteins and Foods: Analytical Methods and Results," 2d ed., Charles C Thomas, Publisher, Springfield, Ill., 1951.

JACOBS, M. B., "The Chemical Analysis of Foods and Food Products," 3d ed., D. Van Nostrand Company, Inc., Princeton, N.J., 1958.

JOSLYN, M. A., "Methods in Food Analysis Applied to Plant Products," 2d ed., Academic Press Inc., New York. In preparation.

PEARSON, D., "The Chemical Analysis of Foods," 5th ed., J. and A. Churchill, Ltd., London, 1962.

Food and Drug Analysis

HORWITZ, W., "Official Methods of Analysis of the Association of Official Agricultural Chemists," 9th ed., Association of Official Agricultural Chemists, Washington, 1960.

Physical and Chemical Data

BERL, W. G., "Physical Methods in Chemical Analysis," vols. I and II, 2d rev. ed., Academic Press Inc., New York, 1960 and 1963.

HODGMAN, C. D., "Handbook of Chemistry and Physics," 45th ed., The Chemical Rubber Company, Cleveland, 1964.

"International Critical Tables," McGraw-Hill Book Company, New York, 1930.

LANGE, N. A., "Handbook of Chemistry," 10th ed., McGraw-Hill Book Company, New York, 1961.

LATIMER, W. M., and J. H. HILDEBRAND, "Reference Book of Inorganic Chemistry," 3d ed., The Macmillan Company, New York, 1952.

LINKE, W. F., "Seidell's Solubilities," 4th ed., vol. I, D. Van Nostrand Company, Inc., Princeton, N.J., 1958. Volumes II and III in preparation.

ROSIN, J., "Reagent Chemicals and Standards," 4th ed., D. Van Nostrand Company, Inc., Princeton, N.J., 1961.

TIMMERMANS, J., "Physico-Chemical Constants of Pure Organic Compounds," vols. I and II, American Elsevier Publishing Company, New York, 1950 and 1963.

Spectrometry

BEAVER, G. H. et al., "Molecular Spectroscopy: Methods and Applications in Chemistry," The Macmillan Company, New York, 1962.

BRAND, J. C. E., and G. EGLINTON, "Applications of Spectroscopy to Organic Chemistry," Oldbourne Press, London, 1965.

COLTHUP, N., L. A. DALY, and S. E. WIBERLEY, "Introduction to Infrared and Raman Spectroscopy," Academic Press Inc., New York, 1964.

ELWELL, W. T., and J. A. F. GIDLEY, "Atomic-absorption Spectrophotometry," The Macmillan Company, New York, 1962.

HERRMANN, R., and C. T. J. ALKEMADE, "Chemical Analysis by Flame Photometry," 2d rev. ed., Interscience Publishers, Inc., New York, 1963.

HUMMEL, D., "Identification and Analysis of Surface-active Agents by Infrared and Chemical Methods," vols. I and II, Interscience Publishers, Inc., New York, 1964.

MELOAN, C. E., "Elementary Infrared Spectroscopy," The Macmillan Company, New York, 1963.

POTTS, W. J., Jr., "Chemical Infrared Spectroscopy," vol. 1, John Wiley & Sons, Inc., New York, 1963.

RAO, C. N. R., "Chemical Applications of Infrared Spectroscopy," Academic Press Inc., New York, 1963.

SQUIRREL, D. C. M., "Automatic Methods in Volumetric Analysis," D. Van Nostrand Company, Inc., Princeton, N.J., 1964.

UDENFRIEND, S., "Fluorescence Assay in Biology and Medicine," Academic Press Inc., New York, 1962.

WHITE, R. G., "Handbook of Industrial Infrared Analysis," Plenum Press, New York, 1964.

pH and Hydrogen Ion Concentration

BATES, R. G., "Determination of pH, Theory and Practice," John Wiley & Sons, Inc., New York, 1964.

PHILLIPS, J. P., "Automatic Titrators," Academic Press Inc., New York, 1959.

Polarography

BREYER, B., and H. H. BAUER, "Alternating Current Polarography and Tensammetry," Interscience Publishers, Inc., New York, 1963.

BREZINA, M., and P. ZUMAN, "Polarography in Medicine, Biochemistry, and Pharmacy," Interscience Publishers, Inc., New York, 1958.

LEVENSON, L. L., "Introduction to Electroanalysis," Butterworth, Inc., Washington, 1964.

ZUMAN, P., "Organic Polarographic Analysis," Pergamon Press, New York, 1964.

Chromatography

LEDERER, M., "Chromatography," 2d ed., American Elsevier Publishing Company, New York, 1957.

NEHER, R., "Steroid Chromatography," American Elsevier Publishing Company, New York, 1964.

SMITH, I., "Chromatographic and Electrophoretic Techniques," 2d ed., vols. I and II, Interscience Publishers, Inc., New York, 1960.

WEISSBERGER, A., "Fundamentals of Chromatography," Interscience Publishers, Inc., New York, 1957.

Thin Layer Chromatography

BOBBITT, J. M., "Thin Layer Chromatography," Reinhold Publishing Corporation, New York, 1963.

MARINI-BETTOLO, G. B., "Thin-layer Chromatography," American Elsevier Publishing Company, New York, 1964.

RANDERATH, K., "Thin-Layer Chromatography," translated by D. D. Libman, Academic Press Inc., New York, 1963.

STAHL, E., "Thin-Layer Chromatography: A Laboratory Handbook," Academic Press Inc., New York, 1965.

Gas Chromatography

BAYER, E., "Gas Chromatography," American Elsevier Publishing Company, New York, 1961.

BURCHFIELD, H. P., and E. E. STORRS, "Biochemical Applications of Gas Chromatography," Academic Press Inc., New York, 1962.

DAL NOGARE, S., and R. S. JUVET, "Gas-Liquid Chromatography," Interscience Publishers, Inc., New York, 1962.

KEULEMANS, A. I. M., "Gas Chromatography," 2d ed., Reinhold Publishing Corporation, New York, 1959.

PURNELL, H., "Gas Chromatography," John Wiley & Sons, Inc., New York, 1962.

SIGNEUR, A. V., "Guide to Gas Chromatography Literature through Early 1963," Plenum Press, New York, 1963.

SZYMANSKI, H. A., "Biomedical Applications of Gas Chromatography," Plenum Press, New York, 1964.

WOTIZ, H. H., and S. J. CLARK, "Gas Chromatography in the Analysis of Steroids," Plenum Press, New York, 1965.

Special Analytical Methods

ALIMARIN, I. P., and M. N. PETRIKOVA, "Inorganic Ultramicroanalysis," The Macmillan Company, New York, 1964.

ASHWORTH, M. R. F., "Titrimetric Organic Analysis," part 1, Direct Methods, Interscience Publishers, Inc., New York, 1964. Part 2, Indirect Methods, in preparation.

BENEDETTI-PICHLER, A. A., "Identification of Materials via Physical Properties, Chemical Tests, and Microscopy," Academic Press Inc., New York, 1964.

BERGMEYER, H. U., "Methods of Enzymatic Analysis," Academic Press Inc., New York, 1963.

DELAHAY, P., "Instrumental Analysis," The Macmillan Company, New York, 1957.

DELAHAY, P., "New Instrumental Methods in Electrochemistry: Theory, Instrumentation and Applications to Analytical and Physical Chemistry," Interscience Publishers, Inc., New York, 1954.

DUVAL, C., "Inorganic Thermogravimetric Analysis," 2d rev. ed., American Elsevier Publishing Company, New York, 1963.

EWING, G. W., "Instrumental Methods of Chemical Analysis," 2d ed., McGraw-Hill Book Company, New York, 1960.

FLASCHKA, H., "EDTA Titrations: An Introduction to the Theory and Practice," Pergamon Press, New York, 1959.

GILLAM, A. E., and E. S. STERN, "An Introduction to Electronic Absorption Spectroscopy in Organic Chemistry," St Martin's Press, Inc., New York, 1954.

GUNTHER, F. A., and R. C. BLINN, "Analysis of Insecticides and Acaricides," Interscience Publishers, Inc., New York, 1955.

HEADRIDGE, J. B., "Photometric Titrations," Pergamon Press, 1961.

JACOBS, M. B., "The Analytical Chemistry of Industrial Poisons, Hazards, and Solvents," 2d ed., Interscience Publishers, Inc., New York, 1949.

JACOBS, M. B., and L. SCHEFLAN, "Chemical Analysis of Industrial Solvents," Interscience Publishers, Inc., New York, 1953.

JAMES, A. T., and L. J. MORRIS, "New Biochemical Separations," D. Van Nostrand Company, Inc., Princeton, N.J., 1964.

KOLTHOFF, I. M., and J. J. LINGANE, "Polarography," 2d ed., vols. I and II Interscience Publishers, Inc., New York, 1952.

KOLTHOFF, I. M., and V. A. STENGER, "Volumetric Analysis," 2d ed., vol. III Interscience Publishers, Inc., New York, 1957.

KUCHARSKY, J., and L. SARAFIK, "Titrations in Non-aqueous Solvents," American Elsevier Publishing Company, New York, 1964.

LINGANE, J. J., "Electroanalytical Chemistry," 2d ed., Interscience Publishers, Inc., New York, 1958.

MACNEVIN, W. M., "Analytical Balance: Care and Use," McGraw-Hill Book Company, New York, 1951.

MEITES, L., "Polarographic Techniques," 2d ed., Interscience Publishers, Inc., New York, 1965.

MITCHELL, J., I. M. KOLTHOFF, E. S. PROSKAUER, and A. WEISSBERGER, "Organic Analysis," vols. I to IV, Interscience Publishers, Inc., New York, 1953–1960.

MORRISON, G. H., and H. FREISER, "Solvent Extraction in Analytical Chemistry," John Wiley & Sons, Inc., New York, 1957.

MULLER, O. H., "Polarographic Method of Analysis," 2d ed., Chemical Education Publishing Company, Easton, Pa., 1951.

RINGBOM, A., "Complexation in Analytical Chemistry," Interscience Publishers, Inc., New York, 1963.

SANDELL, E. B., "Colorimetric Determination of Traces of Metals," 3d rev. ed., Interscience Publishers, Inc., New York, 1959.

SCHWARZENBACH, G., "Complexometric Titrations," Interscience Publishers, Inc., New York, 1957.

SNELL, F. D., and C. T. SNELL, "Colorimetric Methods of Analysis," 3d ed., vols. I to IV, D. Van Nostrand Company, Inc., Princeton, N.J., 1948–1954.

WILLARD, H. H., L. L. MERRITT, and J. A. DEAN, "Instrumental Methods of Analysis," 4th ed., D. Van Nostrand Company, Inc., Princeton, N.J., 1965.

PART 1

GENERAL METHODS USED IN OFFICIAL PHARMACEUTICAL ANALYSES

The theoretical considerations and the procedures involved in gravimetric, volumetric, and gasometric methods of analysis are the same in quantitative pharmaceutical analysis as in quantitative chemical analysis. In quantitative pharmaceutical analysis, however, the general theories of analytical chemistry are applied to the procedures used in the analysis of the drugs, chemicals, and medicinal preparations employed in pharmacy; chiefly to those materials official in The United States Pharmacopeia (U.S.P.) and The National Formulary (N.F.).

Chapter 1

REMARKS AND GENERAL DIRECTIONS

Success as an Analyst. To be successful as an analyst, the student must realize that analytical chemistry is not simple routine procedure. Manipulative skill acquired by experience with the ability to follow directions under the supervision of a skilled analyst may enable one to carry out successfully certain analytical procedures. An important objective in studying quantitative pharmaceutical chemistry, however, is the acquisition of a fundamental knowledge of the theory as well as a practical ability in application of analytical methods.

The successful analyst must know what reactions are taking place during an analysis and be able to understand and apply the theory upon which the method is dependent. The analyst must acquire skill of technique, patience, neatness, and accuracy. The fact should be borne in mind that analysis is carried out not for the sake of analysis but as a means of controlling the purity and strength of substances used as medicaments. Above all, to be successful, the analyst must not only use common sense in the laboratory but think throughout each step of the procedure.

Accuracy and Honesty. It is absolutely essential that all quantitative work be performed without even a slight loss of material or gain of extraneous matter. Every precaution should be taken to prevent such loss or gain. All solutions and precipitates should be covered to protect them from dust, whenever possible, and every particle of sample, solution, or precipitate must be regarded as indispensable to the ultimate success of the analysis.

Absolute integrity is demanded of every quantitative analyst. The operator is the only person who is familiar with the entire history of an analysis and who knows whether it is worthy of full confidence. When confidence is lost for any definitely established reason in the course of an analysis, operations should be discontinued at once, and a new start made. At least two determinations of any analytical value are required, and the results should agree closely. Failure to obtain duplicate values which check is conclusive evidence that at least one value is wrong and that neither value is dependable.

AGREEMENT IN ANALYTICAL DETERMINATIONS

Percent of constituent in sample	Difference in percent between two determinations
50 to 100	0.20 to 0.25
25 to 50	0.10 to 0.20
10 to 25	0.07 to 0.10
5 to 10	0.05 to 0.07
1 to 5	0.03 to 0.05
0.1 to 1	0.03

Notebooks. All notebooks should be of the type designated by the instructor. A small bound notebook which will slide under the balance while weighing has proved most satisfactory in the writer's experience.

The following information relative to each determination should be included:

Date
Object or title
Experimental data
Reactions
Calculations
Results
Remarks

This information should be arranged in a neat, systematic manner so that anyone familiar with quantitative analysis may understand it readily.

The date is very important in all scientific work. Numerous legal decisions have been lost through the failure of the analyst to affix the date at the time of analysis. The object or title should be briefly stated and should identify the sample and the analytical value measured. The experimental data should include a complete record of *all* weighings and measurements set down at the time they are made. The practice of recording data on loose pieces of paper or in a scrap notebook and copying the original entries into a second notebook should not be tolerated. These data may be conveniently placed and followed by the calculations on the left-hand page of the notebook, leaving the right-hand page for a concise, orderly write-up of the entire experiment. All reactions should be given using structural formulas where possible. The use of logarithms for all calculations is highly recommended for accuracy and economy of time. Results should always be reported as percentages unless otherwise directed. Remarks should include any unexpected development during the analysis, explanation of error, and comments which seem pertinent to the technique or theory involved in the analytical method.

Economy of Time. Economical utilization of laboratory hours is best achieved through a preliminary study of the work to be done, followed by a plan for its prompt execution. There is no such thing as "speed" in quantitative work. Economy of time is achieved not by haste, but by planning the work so that two or more operations may be carried out at one time and especially by working so accurately that it will be unnecessary to repeat a determination. The following suggestions will save time and increase accuracy: (1) Keep the desk scrupulously clean at all times. (2) Clean all apparatus immediately after using it and put it in the desk. (3) Label all solutions, filtrates, and precipitates systematically throughout the analytical procedure. (4) Keep two or more operations going at one time; thus, while cooling a precipitate in the desiccator, make a weighing, or while washing one precipitate, ignite another. (5) Utilize all time between operations in making calculations and writing up experiments.

Cleaning Solution. A good cleaning solution can be prepared by dissolving 200 Gm of sodium dichromate in 100 ml of water and then adding 1,500 ml of sulfuric acid slowly and with stirring.

The solution should be stored in a thick-walled, glass-stoppered bottle and used repeatedly for cleaning glass and porcelain ware until it deposits a large amount of solid material or turns green.

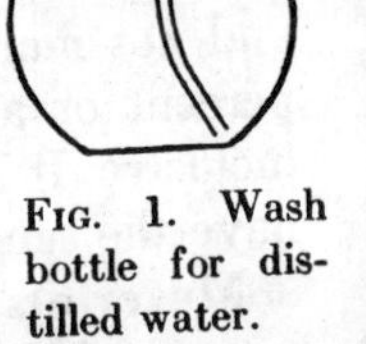

FIG. 1. Wash bottle for distilled water.

Wash Bottles. Wash bottles for use in quantitative work usually consist of rubber-stoppered, flat-bottomed flasks of 500 to 1,000 ml capacity, fitted with smoothly bent tubes and a jet made flexible by means of a rubber joint. The jet should be drawn out so that it will deliver a thin, even stream of liquid. Prepare two wash bottles of the type illustrated in Fig. 1, one of 1,000 ml and one of 500 ml capacity, for cold and hot water. The necks of wash bottles for use with hot water should be wrapped with asbestos paper or twine or other suitable material.

To bore a hole through a rubber stopper, dip the end of the borer in a strong solution of sodium hydroxide or in glycerin before starting to bore. Boil all rubber parts in dilute sodium hydroxide for from 2 to 3 min and wash them thoroughly with distilled water. Fire-polish the ends of all glass tubing by holding them in the blue flame of a burner until the sharp edges have become rounded.

Wash bottles made of polyethylene are also suitable for quantitative analysis. These bottles are inexpensive and commercially available, and they provide a flow of washing solution which ranges from a full stream to a single drop.

"Policeman." The so-called "policeman" is a piece of soft rubber tubing fitted to the end of a glass rod. The rod should be about 20 cm long with a diameter between 5 and 7 mm, and it should have plain annealed ends. A piece of rubber tubing about 3 cm long should be fitted tightly over, and project about 2 cm beyond, the rod. The policeman is used to loosen from the walls of the containing vessel adhering particles that are not removable by a stream of water from the wash bottle. A policeman should never be used as a stirring rod or allowed to remain in analytical solutions.

Reagents. The reagents used in quantitative analysis must be pure. Reagents which conform to the requirements of the United States Pharmacopeia and of the National Formulary should be used in all assays which have for their purpose the establishment of the strength or quality of official substances. Those reagents which meet the specifications of the American Chemical Society as well as those sold by reputable manufacturers as "analyzed" or "reagent" grade will usually be found to meet all official requirements.

Purity and Strength Requirements. The purity and strength of chemicals and drugs of the U.S.P. and of the N.F. are usually expressed in terms of *percent.* The official requirements of purity and strength pertain only when the official methods of assay are employed. When the percent strength or purity is given in round numbers, it is understood that the figures represent exactly the minimum or maximum requirement, e.g., the requirement that hydrochloric acid contain "not less than 35 percent and not more than 38 percent of HCl" permits a product of 35 or 38 percent or any intermediate strength; in other words, the figures are inclusive. It is established by rule that the figures 35 and 38, respectively, have the same significance as the figures 35.00 and 38.00.

Materials Required. The statement of materials required in the exercises is intended as a guide for the laboratory instructor in the preparation of an adequate supply for the class. The quantities indicated are for each sample. It is possible to give only approximate quantities for materials used as solvents, for washing, for acidifying, and so forth. All reagent materials should meet the official requirements of purity and strength.

SAMPLING

The first important consideration in an analysis is the securing of a representative sample, a consideration too frequently ignored by analysts. Fixed rules cannot be laid down in sampling, for much depends on the nature of the material and the quantity from which a representative sample is to be taken. Whenever possible, a sample from each package should be examined separately. When a single package is sampled, the contents should be mixed or samples should be taken from different parts of the container and mixed, and then a portion of this mixture should be taken for analysis. Bulk quantities of chemicals and drugs may be sampled by the methods given in the U.S.P. for sampling vegetable drugs:

a. It is recommended that gross samples of vegetable drugs in which the component parts are 1 cm. or less in any dimension, and all powdered or ground drugs, be taken by means of a sampler which removes a core from the top to the bottom of the container, not less than two cores being taken in opposite directions; and that when the total weight of the drug to be sampled is less than 100 kg. (220 pounds) at least 250 Gm. shall constitute an *official sample.* When the total weight of the drug to be sampled is in excess of 100 kg., several samples shall be taken by the above method, and mixed and quartered, two of the diagonal quarters being rejected, the remaining two quarters being combined and carefully mixed, and again subjected to a quartering process in the same manner until two of the quarters weigh at least 250 Gm., which latter quarters shall constitute an *official sample.*

b. It is recommended that gross samples of vegetable drugs in which the component parts are over 1 cm. in any dimension be taken by hand. When the total weight of the drug to be sampled is less than 100 kg., at least 500 Gm. shall constitute an *official sample*, and this shall be taken from different parts of the container or containers. When the total weight of the drug to be sampled is in excess of 100 kg., several samples shall be taken by the above method and mixed and quartered, two of the diagonal quarters being rejected, and the remaining two quarters being combined and carefully mixed, and again subjected to a quartering process in the same manner until two of the quarters weigh not less than 500 Gm., which latter quarters shall constitute an *official sample.*

c. When the total weight of a drug to be sampled is less than 10 kg., it is recommended that the above methods be followed but that somewhat smaller quantities be withdrawn, and in no case should the final *official sample* weigh less than 125 Gm.

d. Alternative to withdrawing of *official samples* according to methods *a, b,* and *c*, the *official sample* may consist of the total amount of a direct purchase made by officials charged with enforcement of a federal, state, or municipal Food and Drugs Act.

Students' samples are supplied ready for analysis as a rule. Samples which require drying should be spread on a large watch glass, thoroughly mixed, and dried in an oven at the specified temperature. They should then be transferred to a weighing bottle and kept in the desiccator.

The size of the sample taken for analysis volumetrically should be such that errors in weighing will not be significant. Further, the sample should be of such size that it will require from 20 to 40 ml of titrant to be delivered from a buret. This volume of titrant will also minimize any errors in buret readings.

CALCULATION OF RESULTS AND ERRORS

Source and Nature of Errors. It is only in rare cases that the numerical value of an experimental result can be directly determined. As a rule, it is necessary to calculate the result from the different measurements and observations that have been made. Duplicate results that check very closely are not an assurance of accuracy, as is often taken for granted by the beginner in quantitative work, since errors inherent in the method and apparatus used may be repeated for each sample. Generally, results that agree closely when obtained by two different methods of analysis are a good indication of the reliability of the methods.

The results of repeated analyses or measurements will fail to agree, in general, when made by the same analyst or by different analysts to the full precision of which the method or instrument is capable. The discrepancies in the results are caused by various sources of error to which all experimental data are subject. The errors are of two types, viz., indeterminate or accidental errors, which manifest themselves by slight variations in a series of observations made by the same observer under identical conditions. They result from causes difficult to detect, such as differences in the judgment and skill of the analyst. Generally, indeterminate errors are intangible, and their elimination by the analyst is impossible. Determinate or constant errors are of such nature that they recur in a constant manner in each of a series of determinations. Consequently, it is possible partially to determine their value and reduce their effect on the final result. They arise from causes such as (1) personal errors made by the individual analyst, e.g., inability to judge color changes sharply, resulting in habitual reading of end points in titration too late; (2) errors of method caused by faulty procedure, e.g., incorrect sampling, contamination of precipitates, and improper selection of indicators; (3) apparatus errors due to poor construction or calibration, e.g., inaccuracy in the calibration of burets or pipets, inequality in the length of the arms of the balance, and incorrect weights. Errors of this type are usually detectable and so may be eliminated to a large extent.

Indeterminate Errors. The random distribution of indeterminate errors of various magnitudes is such that a graph showing the relationship between the magnitude of an error (abscissa) and the frequency of its occurrence (ordinate) has the shape of a normal frequency distribution curve or probability curve, sometimes called the *curve of error* (Fig. 2).

An examination of this curve shows that (1) very large errors are unlikely to occur often; (2) small errors occur with greater frequency than large ones; and (3) positive and negative errors of the same numerical magnitude are equally probable since the curve is symmetrical with respect to the Y axis. Consequently, it is apparent that the best value to select for a series of observations is their arithmetical mean.

Average Deviation from the Arithmetical Mean. In order to measure the magnitude of indeterminate or accidental errors, use is made of the theorem proved by the method of least squares, which states that an arithmetical mean computed from n equally trustworthy observations is $\sqrt{n}$ times as trustworthy as any single observation. If the indeterminate error in any one observation is represented by d, the indeterminate error of the arithmetical mean is $d/\sqrt{n}$; in other words, the uncertainty of the arithmetical mean is inversely proportional to the square root of the number of observations. The deviation of a single measurement from the arithmetical mean of a series of similar measurements can readily be determined, and from this the average deviation of a single determination and the average deviation of the mean value can be calculated, as in the following example.

In the standardization of hydrochloric acid by titration of samples of pure sodium carbonate, the buret readings 30.25, 30.22, 30.26, 30.27, and 30.23 ml were obtained. The arithmetical mean of these numbers is 30.246.

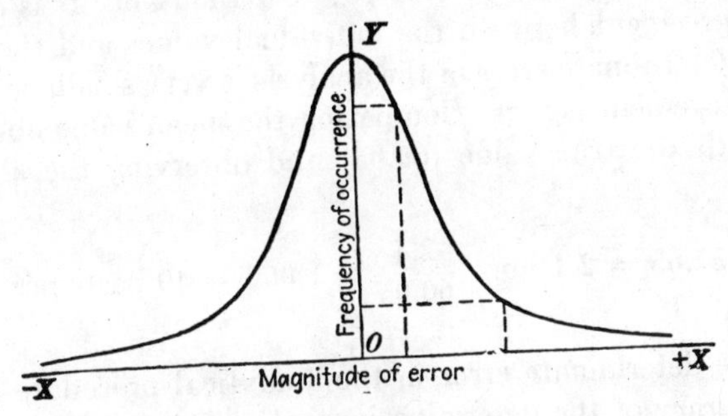

FIG. 2. Probability curve.

The deviations of the individual readings from the mean, neglecting sign, are 0.004, 0.026, 0.014, 0.024, and 0.016. The sum of these five numbers is 0.084, and the average deviation of any of the individual values from the mean value is

$$\frac{0.084}{5} = 0.017 \quad \text{or} \quad \frac{0.017}{30.246} \times 1{,}000 = 0.56 \text{ part per thousand}$$

The average deviation of the mean value, however, is considerably less and is equal to

$$\frac{0.017}{\sqrt{5}} = 0.008 \quad \text{or} \quad \frac{0.008}{30.246} \times 1{,}000 = 0.26 \text{ part per thousand}$$

From any one of the individual values and the average deviation of an individual value from the mean, a range of values which probably includes the true value can be calculated; e.g.,

$$30.25 \pm 0.017 = 30.23 \text{ to } 30.27$$

These values have been rounded off to the nearest 0.01 ml. Because of the greater precision of the mean value, the range of values which probably includes the true value and which can be calculated from the mean and the average deviation of the mean, is much smaller:

$$30.246 \pm 0.008 = 30.24 \text{ to } 30.25$$

Sometimes individual determinations of a given value will agree closely because the indeterminate error is small, but the mean value calculated from them will not be a satisfactory estimate of the true value because the determinate error is large. For example, a sample known to contain 60.05 percent chlorine as chloride was assayed for chlorine and the following values were obtained: 57.63, 57.68, 57.66, and 57.63. The mean calculated from these values is 57.65, and the average deviation of the individual values is only 0.02 or 0.35 part per thousand. It is obvious from the close agreement between the individual values and the mean value that the indeterminate error of the analysis is very small, or the *precision* of the measurement is high. Comparing the mean value obtained in the analysis with the true value (60.05) and observing the absolute error calculated,

$$60.05 - 57.65 = 2.4 \quad \text{or} \quad \frac{2.4}{60.05} \times 1{,}000 = 40 \text{ parts per thousand}$$

A very large determinate error in the analytical procedure is apparent, and the *accuracy* of the determination is low.

Rejection of a Result. Sometimes in a series of determinations of an analytical value, one of the results obtained will differ greatly from the others although no mistake was known to have been made in the work. The question of retaining or rejecting the apparently aberrant result in computation of the arithmetic mean then arises. If at least four determinations have been made, the result in which the error appears to be large may be rejected when the deviation of the questionable value from the arithmetic mean of the other values is four times the average deviation of the other values from the mean. For example, in a series of determinations of the NaCl content of a sample, the following values (percentages) were obtained: 95.72, 95.81, 95.83, 95.92, and 96.18. The arithmetic mean of the first four values is 95.82 and the average deviation of the individual values from the mean is 0.055. The deviation of the last value (96.18) from the mean of the other four values (95.82) is 0.36 or $0.36/0.055 = > 6.5$ times the average deviation of the other values from the mean. Consequently, the last value should be rejected, since it has been shown that the deviation of a result which is four or more times the average deviation is due to a mistake in all except about five cases in a thousand.

Significant Figures. Significant figures are the digits that when placed in order give the value to a number, e.g., in the quantity represented by the number 236, the digits 2, 3, and 6 are significant figures. Zeros are employed in some cases to locate the decimal point; in other cases, they may be significant figures. Thus in the number 20.5, the zero is a significant figure, but in the number 0.0125 the zeros are not significant but show the order of magnitude of the other digits. In the quantity 1.0000 Gm, the zeros are significant, since they indicate that the weighing has been made to $\frac{1}{10}$ mg and that only the last zero is uncertain.

Every measurement involves some error or can be carried out only with a certain degree of accuracy. Consequently, it is evident that the number expressing the value of a measurement can be only approximate. Irrespective of the number of figures written as the result of a calculation, the accuracy of a value beyond that limited by the errors of the measurements cannot be increased. On the other hand, if too few figures are written down, the resulting value may be much less accurate than the data permit. The proper manner of expressing a result is to retain or use such a number of figures that all, except the last, are known with certainty, and that although the last figure is uncertain the error is not greater than ± 5 in the following place. For example: If a buret can be read to 0.01 ml, and 32.5 instead of 32.50 is written as the result of a measurement, the mistake of writing too few figures is committed, since the number 32.5 indicates that the true value lies between 32.45 and 32.55 and the apparent error is five times greater than the error of measurement. Again if, in recording the result of a series of buret readings, the mean of the values 32.52, 32.54, 32.54, and 32.55 ml is written as 32.538, it would be a mistake of retaining too many figures, since it would indicate that the error of measurement is only about ± 0.0005 ml. The number should be rounded off, therefore, to 32.54 ml. For purposes of further calculation, all the figures should be retained; i.e., the number 32.538 should be used.

In general, for analytical work, the operations of addition and subtraction should give as the final result no more decimal places than the least number of decimal places entering into the calculation. This is illustrated by the following example: In the addition $12.4 + 121.502 + 3.6653 = 137.5673$, the maximum apparent error in the result is ± 0.05, which is the greatest apparent error in the number 12.4. Therefore, a derived error of ± 5 exists in the second decimal place, figure 6, of the result, and the result should be written 137.6, but if the result is used for further calculations, it is better to retain one figure more, i.e., to employ 137.57.

In multiplication or division, as a rule, retain one more significant figure, when available, in the result than there are in the number having the least number of significant figures. For example: The product of $11.32 \times 12.2 \times 0.0321 = 4.4331384$. If the last figure in each number is uncertain, the relative error in the number 12.2 will be about 5 in 1,200, or 0.4 percent. Consequently, it would be incorrect to write the result as 4.4331384, for this result has a derived error of 0.4 percent, or of about 2 units in the second place of the decimals. All figures following this are therefore meaningless and should be discarded, the result being written 4.43.

Logarithms and Their Use. The solution of problems and the calculation of the results of analyses from data are greatly facilitated by the use of logarithms. It is advisable for students to make calculations by the use of logarithms in order to save time and avoid tedious calculations by longhand.

A logarithm is an exponent that must be applied to a fixed positive number other than 0 or 1 to produce any given number. The fixed number is called the base. In the common, or Briggsian, system of logarithms, the base is 10. A logarithm consists of two parts, the characteristic, or integral part and the mantissa, or fractional part. The characteristic of a number greater than 1 is always one less than the number of digits to the left of the decimal point, i.e., where n equals the number of digits to the left of the decimal point, the characteristic is represented by a number equal to $n - 1$. For example, the characteristic of 10,000 or 10^4 is 4, of 100 or 10^2 is 2, and of 1 or 10^0 is 0. The characteristic may be either positive or negative. When there are no digits to the left of the decimal point, the characteristic is negative or minus, and when there are no zeros before the significant figures to the right of the decimal point, the characteristic is -1 or $9 - 10$. When there are no digits to the left of the decimal point and n zeros to the right of the decimal point before the significant figures, the characteristic is $(9 - n) - 10$, e.g., the characteristic of 0.1 is $\bar{1}$ or $9 - 10$, of 0.01 is $\bar{2}$ or $8 - 10$, and of 0.000001 is $\bar{6}$ or $4 - 10$.

The mantissas of numbers are furnished by the table of logarithms (see Appendix). The examples given below illustrate the use of the tables.

Example 1. Find the logarithm of 9,465. From the explanation given, it is evident that the characteristic is 3. To find the mantissa, locate 94 in the logarithm table in the column of Natural Numbers and then move directly to the right to the column under 6 at the top of the page. The mantissa given there is 9759. This is the mantissa for 9,460. To find the mantissa for 9,465, move farther to the right to the columns headed Proportional Parts where, under the 5 column, the figure 2 is given. This figure 0002 when added to 9759 shows how much more the logarithm of 9,465 is than that of 9,460. Consequently, the logarithm of 9,465 is

$$3.9759 + 0.0002 = 3.9761$$

Example 2. Find the logarithm of 0.0001058. Since the first significant digit is four places to the right of the decimal point, the characteristic is $6 - 10$. Find the mantissa as in example 1 by following the horizontal line opposite 10 to the vertical column headed by 5 and add to this the proportional part under the column headed 8. The mantissa is then found to be $0212 + 0033 = 0245$, and the logarithm is written $6.0245 - 10$.

Example 3. Multiplying and dividing with logarithms.

To multiply, add the logarithms of the numbers together. To divide subtract one logarithm from the other. Study the following example:

$$\frac{24.46 \times 0.2970}{1{,}000 \times 0.003741}$$

Multiply by adding the logarithms:

log 24.46	=	1.3885
log 0.2970	=	9.4728 − 10
log (24.46 × 0.2970) is the sum of the logs	=	10.8613 − 10
log 1,000	=	3.0000
log 0.003741	=	7.5730 − 10
log (1,000 × 0.003741) is the sum of the logs	=	10.5730 − 10

Divide by subtracting the logarithms:

$$\begin{array}{r} 10.8613 - 10 \\ 10.5730 - 10 \\ \hline 0.2883 \end{array}$$

The result of these operations is the logarithm of the number sought. To find the natural number corresponding to the logarithm 0.2883, turn to the table of antilogarithms (Appendix), move down the column headed Logarithms until 28 is found, move horizontally to the column headed 8 where the number 1941 is found, and then proceed to the Proportional Parts column under 3 where the number 0001 is found; add this to 1941; the sum is 1942. The characteristic is 0. Therefore the one digit is to the left of the decimal point and the required number is 1.942.

Questions and Problems

1. Rewrite each of the following quantities, underscoring the significant figures: (*a*) 1.1200 Gm weighed on the analytical balance; (*b*) 20.459 ml from a buret reading; (*c*) 0.0062 Gm from a weighing on the analytical balance.

2. Make the following calculations retaining only the significant figures:

a. 2.3 + 100.56 + 4.273 − 20.005 =
b. 100 × 0.010 × 1,000 ÷ 10 =
c. (20 × 1.0250) − (10 × 1.2000) ÷ 1.200 =

3. A sample known to be 98.50 percent pure yielded results on analysis in duplicate of 98.30 and 98.12 percent, respectively. What is the error in parts per thousand of each determination?

4. The result of an analysis for percent purity employing the usual apparatus and weighing to the fourth decimal place was reported as 86.3584323 percent. How should the result be reported?

5. The following results were obtained in the determination of chloride ion in a sample of pure potassium chloride: 35.26, 34.98, 35.00, and 35.07 percent. Calculate (*a*) the mean deviation; (*b*) the percent deviation; (*c*) the mean deviation in parts per thousand; (*d*) the mean error in percent.

Should any of the results be rejected? Why?

6. Find the common logarithms of the following numbers: 100, 65, 8,542, 0.221 0.00018.

7. Find the numbers (antilogarithms) that correspond to each of the following logarithms: 3.5448, 0.8250, 0.1260, 2.3979, 1.6990, 8.4321–10.

8. Using logarithms, multiply (*a*) 21,500 × 0.000332; (*b*) 0.0648 × 0.0008426; (*c*) (42.16 − 1.85)(36.20 + 12.82)(0.5444 × 0.1112).

9. Using logarithms, divide (*a*) 0.01648 by 0.9472; (*b*) 20.04 by 140.80; (*c*) 1,020 by 12.64.

10. Look up the logarithm to five places of 2 and 3; it is advisable to fix these values firmly in your mind. Using these values, and without further reference to tables, calculate the logarithms of the following numbers:

4	1.0	111	25	$\frac{1}{4}$	0.9
5	1.2	120	27	$\frac{1}{3}$	0.09
6	1.333	133	30	$\frac{2}{3}$	0.009
8	1.5	144	32	$\frac{5}{18}$	1.0
9	1.6	150	36	$\frac{1}{8}$	0.1
10	1.8	160	40	$\frac{2}{15}$	0.01
12	2.0	180	45	$\frac{9}{32}$	0.001
15	2.22	192	50	$\frac{25}{64}$	0.000001
16	2.4	288	54	$\frac{16}{81}$	0.125
18	2.5	333	60	$\frac{27}{125}$	0.375
20	2.7	480	64	$\frac{125}{25}$	0.875

11. Using the slide rule, solve the following:

a. 27 × 32   *b.* 272 × 328   *c.* 275 × 327

d. 271 × 3.25   *e.* 0.279 × 3,290   *f.* 27.4 × 0.0326

g. 875 × 24.6   *h.* 0.933 × 1.926   *i.* 0.00453 × 0.00345

j. 777 × 853 × 269   *k.* 12.41 × 0.567 × 9.05   *l.* 45/9.3

m. 4.5/9.3   *n.* 0.45/9.3   *o.* 875/1264

p. 125/55   *q.* 0.01565/0.1655   *r.* $\dfrac{377 \times 14.25}{75.3}$

s. $\dfrac{0.981 \times 181.9}{652}$ *t.* $\dfrac{0.0874 \times 0.247}{0.00875}$ *u.* $\dfrac{99.9 \times 88.8}{77.7 \times 66.6}$

v. $\dfrac{1.012 \times 0.001023}{0.0056 \times 0.1049}$ *w.* $\dfrac{1{,}003 \times 0.963}{803 \times 0.005}$ *x.* $\dfrac{\dfrac{6.05}{1.03} \times 2.07}{91 \times \dfrac{0.0001}{0.0009}}$

y. $\dfrac{283 \times 1.411 \times 964 \times 0.333 \times 50.2}{0.032 \times 385 \times 7.34 \times 90.0 \times .431}$

12. What is the approximate difference in parts per hundred between the following pairs of figures:

a. 100 101	*b.* 99 101	*c.* 200 201	*d.* 202 203	*e.* 4,001 4,006
f. 51 50	*g.* 51.0 51.5	*h.* 50 52	*i.* 25 24	*j.* 10 11
k. 155 154	*l.* 15.5 15.4	*m.* 1.55 1.56	*n.* 0.155 0.156	*o.* 0.0155 0.0154
p. 813 805	*q.* 19.8 20.1	*r.* 2.54 2.49	*s.* 0.0999 0.1004	*t.* 87,654 87,599
u. 0.000032 0.0000325	*v.* 0.000091 0.000092	*w.* 0.0056781 0.0056805	*x.* 1.00001 0.98032	*y.* 3,579,028 3,578,799

GENERAL OPERATIONS

Crucibles. A variety of crucibles are used in quantitative analysis. Those most commonly employed are made of high-grade porcelain. They withstand high temperatures and are suitable for use in the ignition of most drugs and precipitates. They are not suitable for fusions because the glaze is attacked by the flux, especially if the flux is basic. Other crucibles consist of fused silica, alundum, nickel, and platinum.

The Gooch Filtration Crucible. The Gooch filtration crucible (Fig. 3) is designed for the separation of precipitates by suction filtration. It has a perforated bottom upon which is bedded a mat of asbestos, thus making it possible to collect, wash, dry, and weigh a precipitate in the same crucible.

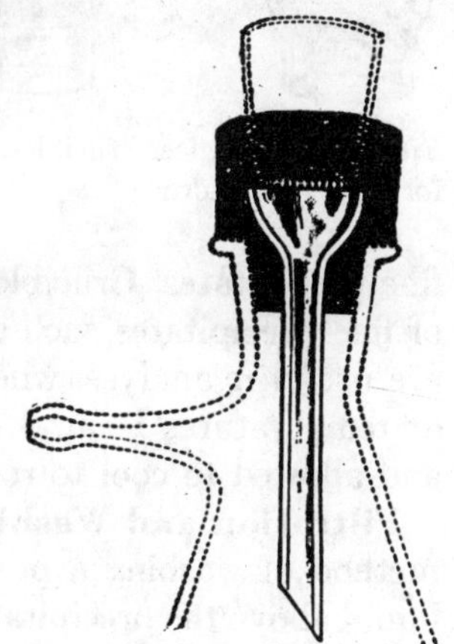

FIG. 3. Diagram of Gooch crucible assembly for filtration.

Preparation and Use of Gooch Crucible. Obtain about 0.5 Gm of asbestos fiber (washed with hydrochloric acid) and shake it thoroughly with 100 ml of distilled water in a suitable flask. Allow the asbestos to settle, and decant most of the water containing the fine fibers. Add about 100 ml of water and agitate the mixture again. Fit the Gooch funnel containing the crucible in the mouth of a filter flask connected to a suction pump. Pour the suspended asbestos mixture into the crucible in small successive portions, applying suction gently after the addition of each portion until a smooth mat not over 1.5 mm thick is obtained. Upon holding the crucible to a strong light, the holes in the bottom should be just perceptible and the mat of asbestos should appear uniform. Cover the asbestos with a perforated porcelain disk (filter plate) and add a very thin layer of asbestos. Wash the filter mat with 200 to 300 ml of water or until no trace of asbestos fiber passes into the washing. Place the prepared crucible on a watch glass and dry it for 1 hr in an oven at 120 to 130°. Allow the crucible to cool in a desiccator, then weigh. Repeat until the weight of the prepared crucible is constant within 0.0002 Gm for two successive weighings.

To use the Gooch crucible, place it in the funnel and apply moderate suction. Pour the liquid from the precipitate into the crucible, wash the precipitate in the beaker by decantation, transfer the precipitate to the crucible, and complete the washing. The rod used to guide the liquid into the crucible should never be allowed to come in contact with the asbestos mat, since it may break the thin film above the filter plate. The filtrate should be examined for traces of asbestos fiber, and if such are found, it should be repassed through the filter.

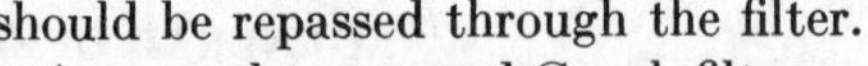

FIG. 4. Fritted-glass crucibles, low form and high form.

A properly prepared Gooch filter may be used for a number of successive determinations where the precipitates are coarse-grained and crystalline.

Fritted-glass Crucibles. Fritted-glass crucibles (Fig. 4) have fused-in sintered or fritted-glass bottoms and are supplied in various porosities such as "medium" and "fine." Crucibles of medium porosity are suitable for the filtration of flocculent and moderately fine precipitates. Crucibles of fine porosity are suitable for the filtration of fine precipitates such as mercuric or bismuth sulfides. These crucibles are useful in analyses where the residue is to be dried to constant weight at temperatures below 150°. The crucibles should be heated in an oven and allowed to cool to room temperature in a desiccator.

Filtration and Washing Precipitates. The most usual and efficient method of washing a precipitate is by the process of decantation. The liquid above the precipitate is poured, using a stirring rod as a guide, onto a prepared filter, some of the liquid and most of the precipitate being allowed to remain in the container. A given amount of wash water is then poured on the precipitate and stirred vigorously. The precipitate is allowed to settle, and the super-natant liquid is decanted as before. This process is repeated three or four times. The precipitate is then washed onto the filter, using a "policeman" if necessary to dislodge any particles that adhere to the container wall. The filter containing the precipitate is washed by causing a fine stream of liquid from the wash bottle to fall upon the filter, always washing from the top edge down, until the filter is about three-fourths full. The filter is allowed to drain until practically empty, then the rinsing is repeated. After about five such rinsings, 1 or 2 ml of the final filtrate should be tested for completeness of washing.

It should be borne in mind that it is better to wash a precipitate a comparatively large number of times with small portions of liquid than only a few times with large portions and that each portion of wash liquid should be removed as completely as practicable before the next portion is added. Never fill the filter more than to within ¼ in. of its upper edge. The size of the filter should be governed by the amount of precipitate, not by the amount of liquid to be filtered.

Colloidal and Fine-grained Precipitates. Colloidal precipitates, some of them gelatinous, some of them finely divided, are given by the most insoluble substances, such as sulfides, hydroxides, elementary sulfur, etc. Although insoluble, they have the property of being converted by water into colloidal dispersions, which pass through the filter. These substances can usually be coagulated and rendered filterable by boiling them after the addition of an electrolyte, such as NH_4NO_3, NH_4Cl, or HCl. It is in part for this reason that prepared wash liquors, rather than pure water, are used for washing certain of these precipitates.

All crystallization is preceded by supersaturation. If supersaturation is high, crystallization will be spontaneous and the crystals formed will be so fine that they will pass through the filter. If the precipitating agent is added slowly with constant stirring, high local supersaturation will be avoided, and the crystals formed will grow in size to such an extent that they will be retained by the filter. By using different concentrations, it is possible to get the same salt crystallized out in crystals varying in size from full-faced crystals down to the colloidal form.

Evaporation of Liquids. The evaporation of liquids is best carried out in porcelain dishes constructed for that purpose, which expose a large surface of liquid to the air. Evaporations should, with few exceptions, be carried out on the water bath to avoid danger of loss of material due to spattering or bumping and to avoid decomposition. The dish should be covered by an inverted funnel supported above it or with a watch glass elevated above the rim of the dish by means of a glass triangle to prevent the entry of foreign matter into the liquid. Large quantities of liquid are best evaporated in a comparatively small dish by adding fresh portions of the original liquid from time to time to replace that lost by evaporation. This helps to eliminate the source of error involved in the frequent transfer of liquids.

Transfer of Liquids. As few transfers of liquid as possible should be made, since each is a possible source of error. When transfers are necessary, they must be made quantitatively, washing the original container with successive small portions of wash liquid. This procedure is the most efficient, since it most thoroughly removes the material and keeps the volume of liquid within workable limits.

When transferring liquids from one vessel to another, a guide rod should always be used. It should be held tightly against the lip of the container.

Drying and Ignition of Samples and Precipitates. Many substances are directed to be dried under specified conditions before analysis to correct for absorbed moisture. Precipitates frequently must be dried previous to ignition. Numerous types of drying ovens are marketed that are suitable for this purpose, the electrically heated ones being best suited to constant temperature control. Samples should be dried at the specified temperature on a watch glass or in a suitable container. Precipitates may be dried directly on a filter by placing the funnel containing both directly

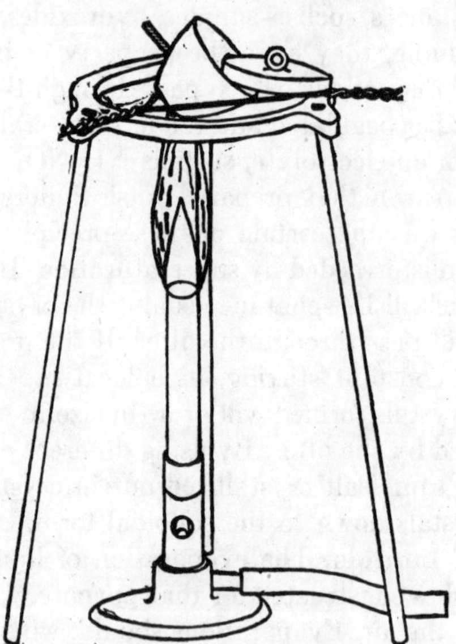

FIG. 5. Position of crucible above flame during ignition.

in a drying oven adjusted to a temperature of 90 to 100° unless otherwise specified. The funnel should be covered with a sheet of filter paper held in place by crimping its edges over the rim of the funnel.

Precipitates may frequently be ignited without previous drying. To do this, the precipitate is folded within the filter paper, and the whole placed in a vertical crucible. The crucible is covered and heated gently by intermittent application of the flame until all the moisture is driven off and the paper is well charred. The cover is then removed, and the crucible placed in an inclined position on the triangle (Fig. 5). The full flame is then turned on and ignition continued until all carbon has been burned away; the crucible is rotated slightly from time to time to expose fresh portions of precipitate. The crucible containing the ignited residue is allowed to cool in a desiccator and weighed. It is then ignited a second time and weighed. If the successive weighings do not agree within 0.2 mg when a porcelain crucible is used, it should be ignited and weighed a third time. These operations should be repeated until constant weight is attained.

Constant Weight. The official methods often direct that a drug be dried to constant weight. The term *dried to constant weight* means that two consecutive weighings do not differ by more than 0.5 mg per Gm of substance taken for the determination, the second weighing following an additional hour of drying. The term *constant weight* is used more often in a different sense in analytical procedures. If the exact weight of an object, e.g., a crucible, is desired, the object is heated at the specified temperature, cooled, and weighed. The object is reheated, cooled, and again weighed. If the two weights agree within 0.2 mg, constant weight is said to have been reached. If the change in weight is more than 0.2 mg, the process of heating, cooling, and weighing is repeated until constant weight is obtained. An object is said to be at constant weight when two consecutive weighings after heating and cooling do not differ by more than 0.2 mg.

Use of Desiccators. Desiccators are a special form of glass vessel, rendered airtight by means of ground contact surfaces, used to maintain a dry atmosphere for objects that might be affected by moisture or carbon dioxide. If the contact surfaces are not airtight, they may be made so by coating them with a thin layer of grease prepared by melting together equal parts of beeswax and petrolatum. The desiccator is usually divided into two compartments separated by a perforated porcelain plate. The lower compartment is charged with a dehydrating agent, one of the most common of which is anhydrous calcium chloride, but concentrated sulfuric acid, solid potassium hydroxide, phosphorus pentoxide, silica gel, magnesium perchlorate, etc., are sometimes used. The porcelain plate is usually fitted with holes for crucibles. The desiccator should always be kept closed when one is not transferring crucibles into or from it.

THE ANALYTICAL BALANCE

A knowledge of the principles and correct use of the analytical balance is essential in all quantitative chemistry. The diagram (Fig. 6) shows the various parts of such a balance with the proper name of each part. Other balances on the market differ slightly in construction, but the parts are essentially the same for all. The analytical balance is a delicate instrument of precision which must be handled with extreme care to maintain its fine adjustment and accuracy. Each student, under the direct supervision of an instructor, should learn the name and function of each part.

The following brief comments will serve to assist the student better to understand the function and care of the balance: The balance beam is supported on a center knife-edge or fulcrum, usually of agate, which rests upon an agate plate. Two pans intended to support the masses to be compared are suspended vertically from each end of the beam by stirrups. Each stirrup has an agate bearing which rests on a knife-edge. The balance beam is notched and graduated in tenths of a gram so that a rider can be placed in any notch by use of a rider rod. Forcible or sudden lowering of the beam upon the knife-edges dulls them and produces furrows in the bearings, which decrease the sensitivity of the balance. A pan arrest is provided, which when released by manipulation of the knob, leaves the pans suspended from the stirrups. A long pointer (or indicator), which

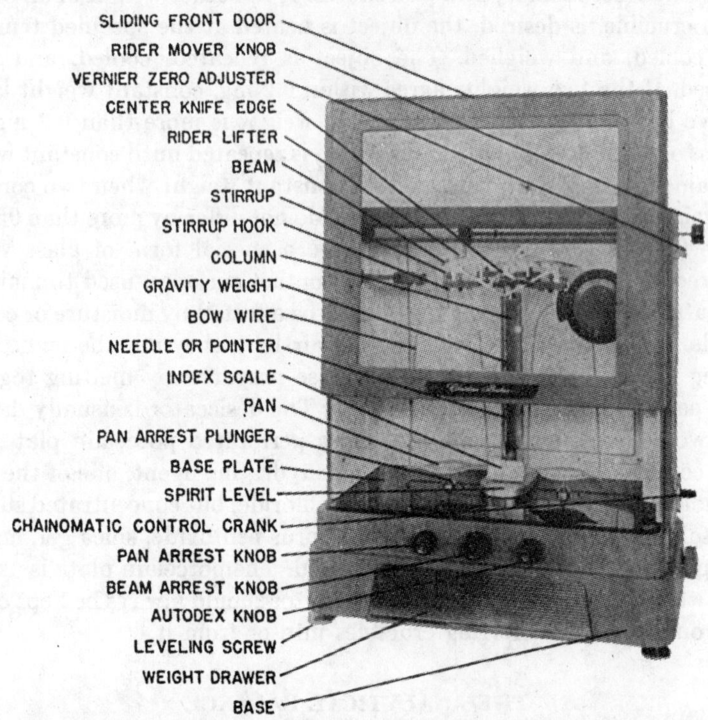

Fig. 6. Typical analytical balance.

magnifies the horizontal displacement of the beam as it moves before the index plate, is attached to the beam in order to render small movements of the beam perceptible.

The introduction of single-pan, substitution-type analytical balances (Fig. 7) has shortened considerably the time formerly spent on weighings in quantitative analysis. A constant weight is built into the analytical balance, and the object is weighed by displacing its equivalent weight from the balance constant weight. Thus, the substitution weighing in effect eliminates the need of determining and applying sensitivity corrections to weighings, since the load on the balance beam is always constant.

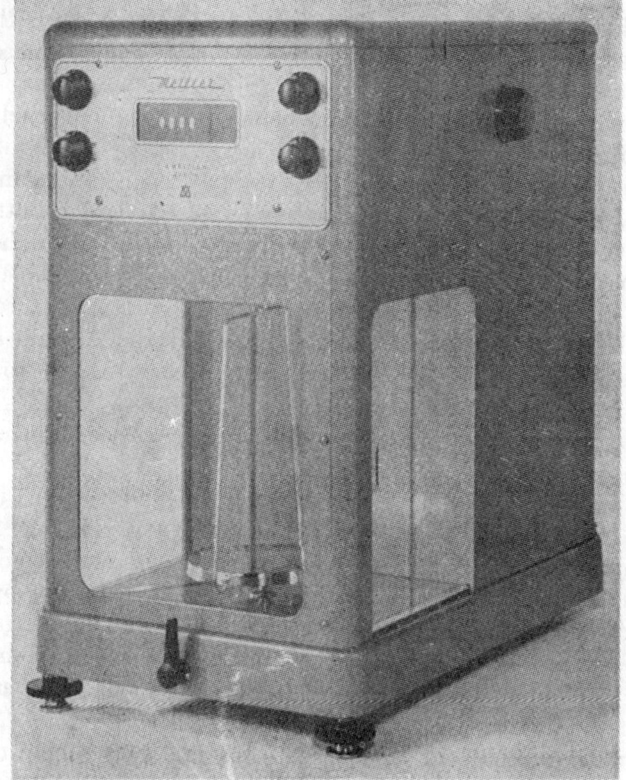

FIG. 7. Single-pan analytical balance. (*Courtesy of Mettler Balance Company*)

Rules for the Use and Care of the Balance:

1. Use the balance assigned by the instructor.
2. Adopt an attitude of personal responsibility for the condition of your balance; the carelessness of one student may render inaccurate the work of all who use the same balance.
3. Brush the pans and floor of the balance with a camel's-hair brush before starting to weigh any substance.
4. Test the adjustment of the balance before each weighing. The balance is properly adjusted only if the following conditions are fulfilled: (*a*) the balance is level as shown by the spirit level; (*b*) the mechanism for releasing and arresting the beam works freely and smoothly; (*c*) the pan arrests just touch the pans when the beam is lowered; (*d*) the pointer rests at zero when the beam is either released or arrested while the pans are supported; and (*e*) the pointer swings approximately equal distances on either side of zero on the index plate when the beam is set in motion

without a load on the pans or with an equal load on each pan. Allowance may be made for a variation of one division on the index scale without adjustment.

5. Do not attempt to adjust the balance. Call an instructor if it requires adjustment.

6. No sample should ever be placed directly upon the balance pan. Solids may be weighed on a watch glass or in a weighing bottle, and liquids must always be weighed in a tightly stoppered bottle. The use of sheets of paper in place of a proper container is not permissible.

7. Hot objects must be allowed to cool to room temperature before they are placed inside the balance case, since currents of warm air tend to buoy up one arm of the balance and also to cause that arm to expand in length.

8. Material spilled on the balance pans or floor must be brushed up at once.

9. The balance door must be closed when the final weighing is made.

10. The beam should be set in motion by gently lowering it so that the pans are supported by the pan arrests; the pan arrests are then carefully lowered, and if the beam does not swing, it may be set in motion by means of the rider or by gently fanning one of the pans. Long swings of the pointer are not necessary. The pointer should swing through an amplitude of not less than three divisions or more than six divisions to either side of the zero of the index scale.

11. All weights should be handled with forceps; they should never be touched with the fingers. Place heavy weights in the center of the pan to prevent oscillation.

12. Record all weights directly into your notebook upon completion of a weighing. Before removing weights, check the balance by arresting and releasing the beam, and be sure that correct balance has been attained. Remember that a slight error in weighing may render worthless all further analysis of a sample.

13. Before leaving the balance, stop the motion of the beam by means of the pan arrests, raise the beam off the agate knife-edges, be sure the balance is clean, and close the door of the balance case.

Sensitivity of the Analytical Balance. The sensitivity varies inversely with the weight of the beam. The lighter the beam, therefore, the greater the sensitivity. The sensitivity varies with the length of the beam. The sensitivity varies directly with the time period of oscillation. The sensitivity varies inversely with the load on the pans. The sensitivity of a balance may be varied by moving the small gravity weight on the pointer upward or downward, thus changing the position of the center of gravity of the beam. The efficient operation and sensitivity of a balance are dependent upon minimum friction between the agate knife-edges and bearings.

Exercise 1

Object. Determination of the Rest Point of the Unloaded Balance.

Procedure. Dust the pans and floor of the balance with a camel's-hair brush. Close the balance door. Gently release the beam and pan arrests, avoiding setting the beam in motion. Cause the beam to swing so that the pointer moves past not more than twelve or less than six divisions of the index scale by carefully touching one arm of the beam with the rider. Allow the pointer to complete two oscillations without taking a reading. Then take three readings of the limit of pointer oscillation on one side of the index scale and two on the other, counting first left then right. Calculate the average reading for each side and the difference between the averages. Divide the difference by 2; the result gives the rest point of the unloaded balance.

The rest point is then computed as follows:

Left	Right
5.2	
5.1	4.6
5.0	4.4
3)15.3	2)9.0
5.1	4.5 average

$5.1 - 4.5 = 0.6$ difference

$\frac{0.6}{2} = 0.3$ divisions to the left

The rest point is 0.3 of a scale division to the left. Make at least three such rest point determinations. The results should check within 0.2 of a scale division.

The rest point of the unloaded balance varies from day to day and should be determined each time the balance is used.

Alternate methods of making readings on the index scale might be recommended by the instructor. For example, many analytical chemists prefer to designate the extreme left scale division as zero and the successive divisions to the right as 10, 20, 30, . . . , 200, thus making the division at the center of the scale 100. Such a procedure eliminates the necessity of recording decimal points. It should be noted that the average readings of the two limits of the period of oscillation are added together and the sum is divided by 2 in calculation of the rest point, when the scale is read according to this alternate procedure.

Exercise 2

Object. Determination of the Sensitivity of a Balance.

Procedure. The sensitivity of a balance is the displacement of the rest point produced by a weight of 1 mg. Determine the rest point of the unloaded balance. Place the rider on the 1-mg division of the beam scale and determine the new rest point on the index scale.

The difference between the two rest points gives the sensitivity of the balance in terms of index scale divisions. Assume that the difference in the rest points is 2; then 1 mg causes a displacement of two scale divisions. The sensitivity of the balance is two divisions per milligram, or 0.5 mg per division. Since the index scale can be read to one-tenth of a division and there are 20 one-tenth divisions, the balance is said to be sensitive to 0.05 mg, i.e., the smallest weight that can cause a readable deflection is 0.05 mg.

Determine the sensitivity of the balance with loads of 10 Gm on each pan and with 20 Gm on each pan. Tabulate your results.

WEIGHTS

The analytical weights used in the assays of official compounds should meet the tolerances of the National Bureau of Standards for Class S if

FIG. 8. Set of analytical weights. (*Courtesy of Fisher Scientific Co.*)

they are to be used without corrections, or Class S-1 if used with corrections. When substances are to be "accurately weighed" for an assay, error is limited to 0.1 percent or less. For example, a quantity of 500 mg is to be weighed to the nearest 0.5 mg.

Analytical weights may be purchased in graduated sets; a set in which the largest weight is 50 Gm and the smallest is 5 mg is satisfactory for most analytical work. The 1-Gm and larger weights are usually made of brass and plated with gold or coated with lacquer to prevent corrosion. The fractional weights are usually made of platinum, tantalum, or aluminum. Figure 8 illustrates a set of analytical weights. Chainomatic balances do not require weights less than 1 Gm, since the smaller denominations can be had directly on the beam or the chainomatic scale.

Exercise 3

Object. Calibration of a Set of Weights.

Procedure. Clean the set of weights to be calibrated by wiping them thoroughly with a clean cloth and brush the weight box with a stiff brush. Mark duplicate weights

CALIBRATION CHART

Left	Right	Rider reading	Actual weight	Aliquot part	Actual correction
0.005	Rider	+0.0050	0.0050	0.0050	0.0
0.010	0.005	+0.0052	0.0102	0.0100	+0.0002
0.010	0.01	+0.0001	0.0103	0.0100	+0.0003
0.020	0.01 + 0.01	−0.0002	0.0203	0.0200	+0.0003
0.050	0.02 + 0.01 + etc.	+0.0042	0.0500	0.0501	−0.0001
0.100	0.05 + 0.02 + etc.	+0.0045	0.1003	0.1002	+0.0001
0.100	0.05 + 0.02 + etc.	+0.0044	0.1002	0.1002	0.0000
0.200	0.100 + 0.100	+0.0008	0.2013	0.2004	+0.0009
0.500	0.200 + etc.	+0.0044	0.5020	0.5010	+0.0010
1.0	0.500 + etc.	+0.0040	1.0036	1.0019	+0.0017
1.0	0.500 + etc.	+0.0024	1.0020	1.0019	+0.0001
1.0	0.500 + etc.	+0.0034	1.0030	1.0019	+0.0011
2.0	1 + 1	−0.0014	2.0042	2.0038	+0.0004
5.0	2 + 1 + 1 + 1	0.0000	5.0128	5.0096	+0.0032
10.0	5 + 2 + 1 + 1 + 1	−0.0064	10.0192	10.0192	0.0000
10.0	5 + 2 + 1 + 1 + 1	−0.0052	10.0204	10.0192	+0.0012
20.0	10 + 10	−0.0024	20.0372	20.0384	−0.0012
50.0	20 + 10 + 10 + etc.	−0.0032	50.0992	50.0960	+0.0032

by punching one or two dots on them so that they may be readily distinguished. Determine the rest point of the unloaded balance. Place the 5-mg weight on the left pan and the rider on the 5-mg position on the right arm of the beam scale and determine the rest point again. Transfer the 5-mg weight to the right-hand pan and place the 10-mg weight on the left-hand pan and secure perfect balance by means of the rider. Proceed in like manner with the remaining weights. Always place the unknown weights on the left pan and the known weights on the right pan. Tabulate the data according to the accompanying chart.

The information secured in the first four columns is all that is required for a calibration of the weights in terms of relative units. This method of calibration is sufficiently accurate for all ordinary quantitative work. The calculation of the actual correction is found by taking the 10-Gm weight in terms of the preliminary standard and finding the aliquot part of this weight represented by each of the other weights. In our table the actual weight of the 10-Gm weight in terms of the preliminary standard was found to be 10.0192; the aliquot part of the 5-Gm weight will then be 10.0192/2 = 5.0096; the aliquot part of the 2-Gm weight will be 10.0192/5 = 2.0038; and the aliquot part of the 1-Gm weight, 10.0192/10 = 1.0019; etc. The aliquot part of 10.0192 is represented in the fifth column as calculated above. In the sixth column are given the actual corrections as plus or minus depending on whether the actual weight is greater or less than the aliquot part.

The corrections in tenths of a milligram should be tabulated on a card, and the card fitted tightly into the cover of the weight box.

Example of the Tabulation of Weight Corrections

Weight, Gm	Correction 0.1 mg	Weight, Gm	Correction 0.1 mg
0.005	0	1.0	+17
0.010	+ 2	1.0	+ 1
0.010	+ 3	1.0	+11
0.020	+ 3	2.0	+ 4
0.050	− 1	5.0	+32
0.100	+ 1	10.0	0
0.100	0	10.0	+12
0.200	+ 9	20.0	−12
0.500	+10	50.0	+32

Questions and Problems

1. Why must the qualitative tests precede the assay to obtain accurate data relative to the purity of an official substance?
2. What is a Gooch crucible? Why should asbestos fiber washed with hydrochloric acid be used in preparing this crucible for use in filtration?
3. What is a desiccator? What two agents are used most commonly to dehydrate official samples before assay?
4. What factors affect the sensitivity of a balance?
5. Why are agate knife-edges and agate supports used in balances?
6. Why should the rest point of the unloaded balance be determined before each weighing?
7. Explain how the sensitivity of a balance may be determined with a load of 10 Gm on each pan.

Chapter 2

GRAVIMETRIC ANALYSIS

THEORY

The student should be familiar with the modern chemical theories which apply in quantitative analysis from the study of general chemistry and qualitative analysis. In quantitative analysis, theoretical considerations must be understood and applied in order to know the full explanation of what takes place in an analysis. A brief review of the fundamentals of the various theories with their quantitative application will be given in conjunction with those procedures to which they apply.

The reactions of quantitative analysis take place in accordance with the established laws and theories of chemistry, i.e., theory of ionization, law of mass action, reversible reactions, solubility product principle, and common-ion effect.

Theory of Ionization. The theory of ionization is based on the following postulates:

1. Solutions of electrolytes are dispersions of ions. Strong electrolytes are ionized even when not dissolved. Weak electrolytes undergo ionization on dissolving; thus, when hydrogen chloride gas dissolves in water, the hydrogen chloride molecules dissociate or ionize to form hydrogen ions and chloride ions so that, from one molecule of hydrogen chloride, two ions are formed: HCl gas$\rightleftharpoons H^+ + Cl^-$. The ions thus formed are solvated by interaction with solvent molecules. In water a hydrogen ion, generally represented for convenience as H^+, is hydrated and is more exactly represented as a hydronium ion or H_3O^+.

2. Some of the ions are charged positively (cations), others negatively (anions); the sum of positive and negative charges always being equal, the solution remains electrically neutral. The hydrogen ions are charged positively, and the chloride ions negatively, the sum of positively charged hydrogen ions being equal to the sum of negatively charged chloride ions; the solution of hydrogen chloride gas remains electrically neutral.

3. The properties of the ions into which the electrolyte dissociates differ from those of the original substance. The hydrogen and chloride ions have properties different from those of the hydrogen chloride molecules, and also from those of hydrogen and chlorine molecules, respectively.

4. The ionization of a weak electrolyte results in an equilibrium because

ionization of the molecules of the weak electrolyte is reversible. The composition of the equilibrium mixture depends on the electrolyte, the solvent, and the concentration; thus, electrolytes differ in the extent to which they ionize; some solvents, such as ether and benzene, produce no ionization; in solvents which produce ionization, the greater the dilution the greater the extent of ionization of the electrolyte.

A certain state of equilibrium exists between the ions and undissociated molecules for every degree of dilution. The equilibrium may be represented as follows:

$$HC_2H_3O_2 \rightleftharpoons H^+ + C_2H_3O_2^-$$

when acetic acid is dissolved in water. Since the extent of dissociation is dependent on dilution, the greater the dilution of the solution the more acetic acid molecules will dissociate into ions until, at infinite dilution, it may be regarded as completely ionized. In accordance with the law of mass action, at any dilution, the ratio of the product of the molar concentrations (moles per liter) of hydrogen ions, $[H^+]$, and acetate ions, $[C_2H_3O_2^-]$, to that of un-ionized acetic acid molecules, $[HC_2H_3O_2]$, is constant, provided temperature is constant.

$$\frac{[H^+][C_2H_3O_2^-]}{[HC_2H_3O_2]} = K$$

Law of Mass Action and Reversible Reactions. Most of the reactions involved in quantitative analysis are of the reversible type. Under certain conditions they may be made to continue to completion, but under other conditions, they may attain equilibrium before completion, resulting in erroneous data pertaining to the substance being analyzed. It is therefore very important to understand what conditions must be satisfied to make the reaction go forward to completion, so that it will be of value in quantitative work. Three general conditions tend to prevent reversal and lead to completion of a reaction: (1) the formation of an insoluble gas; (2) the formation of a sparingly soluble solid; (3) the formation of very slightly ionized molecules.

According to the law of mass action, the rate of a reaction is proportional to the product of the molecular concentrations of the reacting substances. Since the rate of a reaction depends upon the concentration of every substance taking part in the reaction, the point of equilibrium will depend upon the concentration of each of the components of the two opposing reactions. For example, in the reaction, $KNO_3 + H_2SO_4 \rightleftharpoons KHSO_4 + HNO_3$, the rate of the reaction of potassium nitrate with sulfuric acid is expressed by the equation

$$\text{Rate} = [KNO_3] \times [H_2SO_4] \times k$$

where k is a constant which corrects for all factors that affect the rate other than concentration. In the opposing reaction,

$$\text{Rate} = [KHSO_4] \times [HNO_3] \times k_1$$

At equilibrium the rates of the two reactions are equal.

$$[KNO_3] \times [H_2SO_4] \times k = [KHSO_4] \times [HNO_3] \times k_1$$

and
$$\frac{[KHSO_4] \times [HNO_3]}{[KNO_3] \times [H_2SO_4]} = \frac{k}{k_1} = K$$

Since k_1 and k are constants, their quotient K is a constant termed the *equilibrium constant.*

At a definite temperature the equilibrium constant is a fixed value for any given reaction irrespective of the concentration of the substances present. If, therefore, the concentration of sulfuric acid is increased, all the other concentrations must change, the concentration of KNO_3, must become less, and that of both $KHSO_4$ and HNO_3 greater to maintain the equilibrium constant, with the result that equilibrium is forced to shift toward the right. In quantitative analysis, an excess of one component is frequently added to cause the reaction to go as nearly to completion as possible.

Solubility Product Principle. The solubility product principle is an application of the law of mass action to equilibriums that generalizes the behavior of difficultly soluble salts in their saturated solutions. Whenever precipitation occurs or is prevented, or solution is effected, the conditions to which this principle applies are involved. The principle may be stated as follows: *The product of the concentration of the constituent ions in a saturated solution of a difficultly soluble salt for any given temperature is practically a constant, each concentration being raised to a power equal to the relative number of ions supplied by one molecule of the salt upon dissociating.* For example, a difficultly soluble salt $AmBn$ upon dissociating would furnish a relative number of m cations and n anions:

$$AmBn \rightleftharpoons mA^+ + nB^-$$

Using C to designate concentration, the solubility product would be

$$(C_{A+})^m \times (C_{B-})^n = \text{solubility product } AmBn$$

The solubility products of some of the more difficultly soluble salts dealt with in pharmaceutical analysis are given in Table 1.

When a number of different kinds of ions are present in the same solution the greatest concentration that any one of them can attain is determined by the others. Thus, no great concentration of Ag^+ can be present in a solution in the presence of Cl^-, for the two ions unite to form a precipitate of the difficultly soluble salt, silver chloride. Silver chloride is slightly soluble in water, the solubility being about 0.00001 mole (1.5 mg) per liter. When this concentration is exceeded, a precipitate of AgCl forms that is in equilibrium with the dissolved silver chloride. When the equilibrium is established, the supernatant liquid is a saturated solution, and the rate at which the solid goes into solution is exactly equal to the rate at which the dissolved salt precipitates. When this condition exists, the following scheme of equilibriums is set up:

$$\underset{\text{Solid ppt}}{AgCl} \rightleftharpoons \underset{\text{Dissolved un-ionized}}{AgCl} \rightleftharpoons \underset{\text{Dissolved ionized}}{Ag^+ + Cl^-}$$

The ionization equilibrium expressed according to theory is

$$\frac{[Ag^+] \times [Cl^-] \text{ ionized}}{[AgCl] \text{ un-ionized}} = K = \text{ionization constant}$$

Since the solution is saturated with AgCl at a given temperature and the concentration of un-ionized [AgCl] remains constant, it follows that the product $K \times$ [AgCl] also remains constant and that *in a saturated solution of a difficultly soluble salt, the product of the molecular concentrations of its ions is constant.*

TABLE 1. SOLUBILITY PRODUCTS OF SOME IMPORTANT SALTS

Substance	Temperature, °C	Ions involved	Solubility product
Aluminum hydroxide	25	$Al^{3+} + 3OH^-$	1×10^{-33}
Barium carbonate	25	$Ba^{++} + CO_3^=$	8.1×10^{-9}
Barium sulfate	25	$Ba^{++} + SO_4^=$	1.1×10^{-10}
Calcium carbonate	25	$Ca^{++} + CO_3^=$	8.7×10^{-9}
Calcium oxalate	25	$Ca^{++} + C_2O_4^=$	2.6×10^{-9}
Calcium sulfate	25	$Ca^{++} + SO_4^=$	1.9×10^{-4}
Lead carbonate	18	$Pb^{++} + CO_3^=$	4.0×10^{-14}
Lead sulfate	18	$Pb^{++} + SO_4^=$	1.1×10^{-8}
Magnesium ammonium phosphate	25	$Mg^{++} + NH_4^+ + PO_4^{3-}$	2.5×10^{-13}
Magnesium hydroxide	18	$Mg^{++} + 2OH^-$	1.5×10^{-11}
Magnesium oxalate	18	$Mg^{++} + C_2O_4^=$	8.8×10^{-5}
Mercuric sulfide	25	$Hg^{++} + S^=$	1×10^{-50}
Mercurous chloride	25	$Hg_2^{++} + 2Cl^-$	1.5×10^{-18}
Silver bromide	25	$Ag^+ + Br^-$	7.7×10^{-13}
Silver chloride	25	$Ag^+ + Cl^-$	1.5×10^{-10}
Silver iodide	25	$Ag^+ + I^-$	1.5×10^{-16}
Silver thiocyanate	25	$Ag^+ + SCN^-$	1.2×10^{-12}

In the calculation of the solubility product, it is customary to express the concentration in terms of moles per liter, i.e., molar concentrations. A saturated solution of silver chloride contains 1.76×10^{-3} Gm (0.00176 Gm) per liter. The molecular weight of silver chloride is 143.32. In terms of molarity, therefore, the solution contains

$$\frac{1.76 \times 10^{-3}}{143.32} = 1.23 \times 10^{-5} \text{ mole per liter}$$

of silver chloride. At this dilution the dissolved silver chloride may be assumed to be completely ionized, $AgCl \longrightarrow Ag^+ + Cl^-$, so that each mole of silver chloride furnishes 1 mole of silver ions and 1 mole of chloride ions. The solubility product (S.P.) is

$$[Ag^+] \times [Cl^-] = (1.23 \times 10^{-5})(1.23 \times 10^{-5}) = 1.5 \times 10^{-10}$$

If the solubility of a compound is known, the solubility product may be calculated; e.g., a saturated solution of silver iodide at 25° contains about 0.00288 mg of AgI per liter. The molar solubility is equal to the solubility in grams divided by the gram-molecular weight, or

$$\frac{288 \times 10^{-8}}{234.8} = 1.23 \times 10^{-8} \text{ mole per liter}$$

Assuming 100 percent ionization at this high dilution, the concentrations of Ag^+ and of I^- each equal 1.23×10^{-8}, since each molecule of AgI forms one Ag^+ and one I^- ion. The solubility product is equal to the product of the concentrations of the ions:

$$[Ag^+] \times [I^-] = (1.23 \times 10^{-8})(1.23 \times 10^{-8}) = 1.5 \times 10^{-16}$$

From the above calculation of solubility product, it can be predicted that silver iodide will precipitate if $[Ag^+] \times [I^-]$ becomes greater than 1.5×10^{-16}; that in a supersaturated solution, $[Ag^+] \times [I^-]$ becomes greater than 1.5×10^{-16}; that in a saturated solution $[Ag^+] \times [I^-]$ will be equal to 1.5×10^{-16}; and that to dissolve precipitated AgI, $[Ag^+] \times [I^-]$ must be less that 1.5×10^{-16}.

In the above illustration, the solubility product of silver iodide was calculated from solubility data. Conversely, if the solubility product is known, the solubility of a compound may be calculated. In the table, the solubility product of silver chloride is given as 1.5×10^{-10} at 25°. How much silver chloride will dissolve in 100 ml of water at the above temperature?

$$\text{S.P.}_{AgCl} = [Ag^+] \times [Cl^-] = 1.5 \times 10^{-10}$$

If $[Ag^+] = x$, then $[Cl^-] = x$, since $[Ag^+] = [Cl^-]$.

$$x^2 = 1.5 \times 10^{-10}$$
$$x = 1.22 \times 10^{-5}$$

x being expressed in terms of moles per liter. Since solubility of AgCl is equal to 1.22×10^{-5} mole per liter, the solubility in grams per 100 ml is equal to $\frac{1.22 \times 10^{-5}}{10}$ times the gram-molecular weight of AgCl or $1.22 \times 10^{-6} \times 143.32 = 0.000175$ Gm of AgCl soluble in 100 ml of water.

The solubility product is thus seen to be an ultimate value attained by the ionic product when equilibrium has been established between the undissolved solid and the difficultly soluble salt in solution. If the product of the concentration of any pair of ions in solution is made to exceed in value the solubility product of the compound formed by their union, precipitation of the compound will take place until the product of the ionic concentration is exactly equal to the solubility product value; when the product of the ionic concentrations is made less than the solubility product value, the compound formed by their union will dissolve until the product of ionic concentrations is equal to the solubility product value.

Common-ion Effect. The equilibrium constant does not change no matter what the concentration of the reacting substances may be. The relative concentration of the reacting substances may change, but there is no change in the equilibrium constant. Thus, when a solution of silver nitrate is added to a solution of sodium chloride, the chloride ion is momentarily present in a concentration such that its ionic product with the silver ion exceeds the solubility product of silver chloride, and the insoluble silver chloride is precipitated:

$$Ag^+ + Cl^- \longrightarrow AgCl\downarrow$$

When an equivalent amount of silver nitrate has been added, and the system has acquired equilibrium, the concentration of silver ions will be exactly equal to the concentration of chloride ions.

If to the supernatant liquid, which is a saturated solution of silver chloride, a small amount of a soluble silver salt or a soluble chloride be added, a slight further precipitation will take place. It follows from the application of the equilibrium representing the ionization constant,

$$\frac{[Ag^+] \times [Cl^-]}{[AgCl]} = K$$

that if the concentration of silver ion is increased by the addition of a soluble silver salt, the concentration of chloride ion must decrease and, conversely, that if the concentration of chloride ion is increased by adding a soluble chloride, the concentration of silver ion must decrease since their product remains constant. This decrease in the concentration of the ions in either case can be accomplished by the union of silver and chloride ions to form insoluble silver chloride, forcing the reaction toward completion.

The common-ion effect is used frequently in gravimetric pharmaceutical analysis to drive reactions toward completion. Other examples of common-ion effect will be discussed in conjunction with the assays in which they occur.

Questions and Problems

1. What are the postulates upon which the theory of ionization is based?

2. What is a reversible reaction?

3. What conditions tend to destroy equilibrium and cause a reaction to go to completion?

4. Define and illustrate the solubility product principle.

5. If 0.124 mg of barium sulfate dissolves in 50 ml of water at 25°, express the solubility in terms of molar solubility.

6. From the solubility of $BaSO_4$ given in problem 5, calculate its solubility product assuming 100 percent ionization.

7. The solubility product of calcium sulfate is approximately 2×10^{-4}. Calculate the amount of calcium sulfate soluble in 500 ml of water. Assume 100 percent ionization.

8. Illustrate how common-ion effect may be utilized to secure quantitative precipitation.

9. A saturated solution of silver chromate at 20° contains 32 mg per liter. Calculate the molar solubility and the solubility product of the salt.

GRAVIMETRIC METHODS

Gravimetric analysis, the measurement of the weight of a substance in a sample or calculation of the weight of a substance in a sample from the weight of a chemically equivalent amount of some other substance, can be accomplished in various ways. Frequently in pharmaceutical quantitative analysis, the substance to be measured gravimetrically is separated from other substances comprising the sample by physical methods, purified, and weighed without chemical change. For example, the quantity of amobarbital in amobarbital tablets is determined by extracting, purifying further by complete removal of solvent, and weighing the amobarbital in a sample of the powdered tablets. When the other components of the sample are such that separation by physical means of the substance being measured is not possible or convenient, chemical reaction is employed to convert the substance to a chemically equivalent amount of some other substance which can be separated, purified, and weighed. Using the known weight relationship between the substance being measured and the substance which is ultimately weighed, the weight of the former is readily calculated from that of a chemically equivalent amount of the latter.

In the general reaction

$$A + B \longrightarrow C + D$$

the weight of either reactant, A or B, can be calculated from the weight of either product, C or D. For example, in the reaction

$$NaCl + AgNO_3 \longrightarrow AgCl + NaNO_3$$

58.44 Gm of NaCl is required to produce 143.32 Gm of AgCl, or, for each gram of AgCl produced, the sample must have contained

$$\frac{58.44}{143.32} \text{ (1 Gm of AgCl)} = 0.4078 \text{ Gm of NaCl}$$

A chemically equivalent amount of some product can be obtained from a reactant, i.e., the component of the sample being assayed, by any one of various methods: (1) it may be precipitated from solution; (2) it may be a decomposition product resulting from ignition of a compound; (3) it may be deposited on an electrode by electrolysis, etc. The first two methods are considered in this chapter.

Determination of Chlorine in a Soluble Chloride. The weight of chlorine present in a soluble chloride can be determined gravimetrically by calculation from data obtained by precipitation and weighing of a chemically equivalent amount of silver chloride. A solution of a sample of a soluble chloride is acidified with nitric acid and treated with excess silver nitrate solution. The silver chloride which precipitates is collected on a filter, washed, dried, and weighed. Other substances that form insoluble silver salts must be absent from the sample.

Although the chlorine in official soluble chlorides is generally determined by a titrimetric procedure (see Exercise 38, Chap. 7), the gravimetric procedure constitutes a classic example of gravimetric analytical technique with which every student should be familiar.

Exercise 4

Object. Assay of Sodium Chloride.

Materials Requires. 0.25 Gm of sodium chloride
5 percent silver nitrate solution
Diluted nitric acid
Asbestos fiber

Procedure. 1. Accurately weigh two samples of 0.2 to 0.3 Gm of sodium chloride.

This may be done by weighing the weighing bottle and contents accurately to 0.1 mg, without touching it, entering the weight in the notebook. Hold the weighing bottle directly above a 400-ml beaker, labeled 1, remove the stopper, and carefully transfer 0.2 to 0.3 Gm of the sample to the beaker. Replace the stopper and again weigh the weighing bottle to 0.1 mg, entering the weight in the notebook. The first weight less the second is the weight of sample taken. In like manner weigh a second 0.2- to 0.3-Gm sample into beaker 2, entering the weights as before. Treat each sample as follows:

2. Completely dissolve the sample in 100 to 125 ml of distilled water. Add about 1 ml of diluted nitric acid slowly with stirring. Test the solution by touching a strip of blue litmus paper with the moist stirring rod. If it is not acid in reaction, add suffi-

cient acid to make it so. Measure out 5 ml in excess of the amount of silver nitrate solution theoretically required to precipitate all of the chlorine as silver chloride. Add the silver nitrate solution in small successive portions, stirring constantly with a suitable glass rod. Cover the beaker with a watch glass, and, with occasional stirring, gradually heat the mixture to the boiling point. Turn down the flame and digest the mixture, without boiling, for 10 min. This treatment coagulates the precipitate, causes it to settle, and leaves the supernatant liquid clear. Add 1 or 2 drops of silver nitrate solution to the hot supernatant liquid to test for complete precipitation. If a precipitate forms, add 5 ml more silver nitrate solution, stir, allow to settle, and test as before. Set the beakers aside away from direct sunlight to allow the precipitate to settle.

The solution is acidified with nitric acid to prevent the precipitation of other substances insoluble in water but soluble in nitric acid, such as carbonates, oxides, and phosphates. The acid also helps to coagulate any colloidal silver chloride. Too great an excess of nitric acid must not be added, since it exerts some solvent action on silver halides. The acidified solution should not be heated until the silver nitrate has been added, since chlorine may be liberated and lost, thus:

$$NaCl + HNO_3 \longrightarrow HCl + NaNO_3$$
$$6HCl + 2HNO_3 \longrightarrow 3Cl_2\uparrow + 4H_2O + 2NO$$

The silver nitrate is added in excess to drive the reaction to completion through common-ion effect, assuring complete precipitation:

$$AgNO_3 + NaCl \longrightarrow AgCl\downarrow + NaNO_3$$

The precipitation must be performed away from strong light, since silver chloride is decomposed by sunlight with loss of chlorine, while in diffused light the error from this source is negligible.

The precipitate is digested near the boiling point to coagulate colloidal silver chloride which, if present, would pass through the filter.

3. During intervals in the above procedure, carefully clean and prepare two Gooch crucibles for filtration as previously described. After the crucibles have been properly prepared, heated to constant weight, and weighed, fit them into the suction flask. Be sure that the crucibles can be identified as 1 and 2 by a suitable mark. Decant most of the supernatant liquid into the Gooch crucible, using gentle suction to hasten filtration, guiding the liquid into the crucible by means of a glass rod held against the lip of the beaker. Wash the precipitate three times with 15-ml portions of 0.01 *N* nitric acid.

In pouring the clear liquid through the filter, retain as much of the precipitate as possible in the beaker. When the third wash solution has been added to the precipitate, transfer the precipitate into the Gooch crucible using a stream of wash solution from the wash bottle and a "policeman" to dislodge all particles that adhere to the sides of the beaker and guide rod. Wash the precipitate on the filter with several small portions of the wash solution, allowing each portion to run through before adding another. Continue washing the precipitate in this manner until 1 to 2 ml of filtrate gives no test for silver ion upon the addition of a few drops of dilute hydrochloric acid solution. When tested in this way, the filtrate should give no greater turbidity than a portion of the wash solution being used tested in the same manner.

4. When the filtrate is shown by the test to be free of silver nitrate, wash the precipitate twice with 5-ml portions of distilled water to remove most of the nitric acid retained by the precipitate from the former wash solution. Apply strong suction to remove as much liquid from the precipitate as possible. Dry the crucible to constant weight at 110 to 120° in an air bath or electric oven until successive weighings differ by no more than 0.0002 Gm. Allow the crucibles to cool in a desiccator before weighing. The weight of the crucible and precipitate minus the weight of the crucible gives the weight of silver chloride obtained from the sample.

The precipitate is washed to remove soluble salts present in the solution, chiefly sodium nitrate and excess silver nitrate. Washing with 0.01 *N* nitric acid prevents loss of AgCl through its return to colloidal condition (peptization).

Silver chloride is appreciably volatile when ignited, so it should be dried at a relatively low temperature.

The reaction involved in the calculation of the amount of silver nitrate solution to be used and of the purity of the sample is

$$\underset{169.87}{AgNO_3} + \underset{58.44}{NaCl} \longrightarrow \underset{143.32}{AgCl} + \underset{84.99}{NaNO_3}$$

The equation shows that 1 molecule of sodium chloride reacts with 1 molecule of silver nitrate. From the molecular weights NaCl = 58.44 and $AgNO_3$ = 169.87; therefore, the theoretical amount of silver nitrate solution required to precipitate all the chloride ion as silver chloride can be calculated as follows:

To precipitate the chlorine from 1 Gm of NaCl, 169.87/58.44 = 2.9067 Gm of $AgNO_3$ is required. Since, from the data, 0.3160 Gm of NaCl was used, then 0.3160 × 2.9067, or 0.9185 Gm of $AgNO_3$ would be required if the sample were pure NaCl. Since the reagent silver nitrate is a 5 percent solution, each milliliter must contain 0.05 Gm of $AgNO_3$, and the amount of solution that is required is 0.9185/0.05 = 18.37 ml of 5 percent solution.

The weight of the sample and the weight of silver chloride obtained from it being known, the percentage purity of the sample or the weight of chlorine in the sample may readily be calculated:

$$58.44/143.32 = 0.4078 \text{ Gm NaCl equivalent to 1 Gm AgCl}$$

Since 1 Gm of AgCl is equivalent to 0.4078 Gm of NaCl, then 0.7732 Gm of AgCl found is equivalent to 0.4078 × 0.7732 = 0.3153 Gm of NaCl. The percentage purity of the sample in terms of NaCl is therefore 0.3153/0.3160 × 100 = 99.78 percent.

The 0.4078 value used in the above calculation is a *gravimetric factor* or a chemical factor by use of which the weight of NaCl in the sample proportionate to the weight of AgCl formed can be readily calculated. The significance of the gravimetric factor should always be borne in mind: viz., the weight of NaCl in the sample is 0.4078 times the weight of the AgCl formed, or each gram of AgCl formed is equivalent to 0.4078 Gm of NaCl in the sample.

Such a gravimetric factor is included in the official gravimetric assay procedures as outlined in the monographs of the U.S.P. and N.F. and facilitates interpretation of the experimental data. For example, the data of gravimetric procedures are generally interpreted by use of the following formula:

$$\frac{W \times E \times 100}{S} = \text{percent}$$

where W = weight of the product of a chemical reaction involving the substance being assayed

E = gravimetric factor

S = weight of the sample

Using the data given above, the amount of NaCl in 100 Gm of sample, i.e., the percent of NaCl in the sample, is calculated as follows:

$$\frac{0.7732 \times 0.4078 \times 100}{0.3160} = 99.78 \text{ percent}$$

Sample Data for Gravimetric Determinations

Sample No.________ Date________

Determination of Sodium Chloride

Procedure:—Sodium chloride sample precipitated and weighed as silver chloride

Weight	Sample No. 1	Sample No. 2
Container + sample	14.3580	14.0420
Container	14.0420	13.7315
Sample	0.3160	0.3105
Crucible + AgCl	19.0205	18.6409
	19.0201	18.6403
	19.0200	18.6402
Crucible	18.2470	17.8827
	18.2468	17.8827
Silver chloride	0.7732	0.7575
Sodium chloride	0.3153	0.3089
Percent of sodium chloride	99.78	99.64
Average percent	99.71	

Initials of Analyst________

To calculate the quantity of chlorine present as chloride in the sample, the gravimetric factor

$$\frac{Cl}{AgCl} = \frac{35.45}{143.32} = 0.2473 \text{ Gm Cl per gram AgCl}$$

is employed. A similar factor for $CaCl_2$, $CaCl_2 \longrightarrow 2AgCl$, would be

$$\frac{CaCl_2}{2AgCl} = \frac{110.99}{2(143.32)} = 0.3872 \text{ Gm } CaCl_2 \text{ per gram AgCl}$$

or for $CaCl_2{\cdot}6H_2O$,

$$\frac{CaCl_2{\cdot}6H_2O}{2AgCl} = \frac{219.11}{2(143.32)} = 0.7644 \text{ Gm } CaCl_2{\cdot}6H_2O \text{ per gram AgCl}$$

The gravimetric method for determination of chlorine in a soluble chloride can also be applied in the assay of most soluble bromides, iodides, cyanides, thiocyanates, etc., as well as in the standardization of hydrochloric acid solutions and the estimation of silver compounds. Organic halogen compounds usually require preliminary treatment, e.g., oxidation in solution with alkaline permanganate solution, as in the assay of Camphorated Parachlorophenol, or heating with an alkaline flux, as in the assay of bromine in Merbromin, before determination of the halogen by the gravimetric method is possible. The equivalent amount of halide produced by decomposition of the organic compound can be measured gravimetrically as above described, or by an alternate precipitation titrimetric procedure, described in Exercise 38, Chap. 7. The volumetric precipitation method has superseded the gravimetric method because it can be carried out more rapidly.

Questions and Problems

1. Why should the solution be acidified with nitric acid?
2. Why add silver nitrate in excess in the precipitation of the chloride?
3. Why is the precipitate washed with water slightly acidified with nitric acid?
4. Why is it best to wash the precipitate by decantation before it is transferred to the Gooch crucible?
5. Why should this determination be carried out away from direct sunlight?
6. Name several official substances other than sodium chloride that may be assayed by an analogous method.
7. What substances, if present, would interfere with the determination of the purity of a soluble chloride?
8. A 0.3056-Gm sample of a soluble chloride was analyzed gravimetrically for Cl, and 0.7265 Gm of AgCl was obtained. Calculate the Cl and the NaCl contents of the sample, expressing each in percent weight to weight.
9. What volume of 10 percent *W/V* silver nitrate solution would be necessary to precipitate completely the Cl of 1 Gm of 50 percent *W/W* zinc chloride solution?
10. When is a crucible considered to have been heated to constant weight?

11. A sample of Merbromin lost 4.5 percent of its weight on drying. The bromine content of the dried sample was 19.5 percent. What would be the bromine content of the undried material? Why is this dual control of purity necessary?

12. Prepare an outline of steps that would probably be necessary in a procedure for the gravimetric analysis of silver.

13. Define and illustrate the terms: gravimetric factor and chemical equivalent.

14. If the solubility of silver chloride in water is 0.00015 Gm per 100 ml what is its solubility in 0.01 *M* hydrochloric acid?

15. What volume of concentrated hydrochloric acid, sp gr 1.19, 36 percent *W/W*, would be withdrawn from a weighing pipet so that the sample would yield 0.8500 Gm of AgCl when treated with silver nitrate in the usual manner?

16. A precipitate of AgCl weighing 0.2432 Gm was obtained from a 25-ml sample of a hydrochloric acid solution. Calculate (*a*) the HCl content of the solution in percent *W/V*; (*b*) the molarity of the solution; (*c*) the weight of $Ba(OH)_2.8H_2O$ that would be required to neutralize 50 ml of the solution.

17. A 0.5463-Gm sample of the silver salt of a monobasic organic acid yielded 0.4690 Gm of AgCl. Calculate (*a*) the Ag content of the sample in percent *W/W*; (*b*) the molecular weight of the organic acid.

18. If a 0.4600-Gm sample of sodium iodide yielded 0.7200 Gm of AgI, (*a*) what was the weight of iodine in the sample, (*b*) what was the iodine content of the sample in percent *W/W*; (*c*) what weight of lead iodide would be formed from 0.5 Gm of the sample?

19. Calculate (*a*) the molarity; (*b*) the HBr content in percent *W/V*; (*c*) the volume of 10 percent *W/V* sodium hydroxide equivalent to 1 ml of a hydrobromic acid solution, 50 ml of which yielded 0.2342 Gm of AgBr.

20. What would be (*a*) the Br content of a 5-Gm sample; (*b*) the Br content in percent *W/W*; (*c*) the KBr content in percent *W/W* of a sample of potassium bromide, 0.3546 Gm of which yields 0.4749 Gm of AgBr?

21. A 0.6102-Gm sample of potassium chlorate was reduced to chloride, and the reaction mixture diluted to 50.00 ml with water. A 10.00-ml sample of the resulting solution yielded 0.1425 Gm of AgCl. Calculate (*a*) the $KClO_3$ content of the sample in percent *W/W*; (*b*) the weight of Cl formed in the reduction.

22. A 0.5000-Gm sample of potassium iodate was reduced, and the iodide formed converted to silver iodide. When the silver iodide was heated in a stream of chlorine, an equivalent amount of silver chloride was formed and weighed 0.2030 Gm less than the silver iodide. Calculate the KIO_3 content of the sample in percent *W/W*.

23. The AgCl obtained from a 0.6500-Gm sample of sodium chlorate was reduced to 0.6050 Gm of metallic silver. Calculate the $NaClO_3$ content of the sample in percent *W/W*.

24. A sample of mixed calcium and silver nitrates weighs 0.3880 Gm; the silver and calcium are precipitated as mixed oxalates, weighing 0.2800 Gm; the oxalates are subsequently taken up in nitric acid and electrolyzed, 0.1079 Gm of metallic silver being collected. Another sample of the mixed nitrates weighing 0.5820 Gm is treated with excess sodium chloride, giving a precipitate weighing 0.2150 Gm. Calculate the percent of $AgNO_3$ and the percent of $Ca(NO_3)_2$ in the sample. What part of the mixed oxalates is $Ag_2C_2O_4$?

25. The phosphorus in a sample weighing 0.50 Gm is converted to silver phosphate, Ag_3PO_4; this is then converted into AgCl, of which 0.9550 Gm is formed. Calculate the percent of phosphorus.

26. A sample consisting only of pure KCl and pure KI weighs 0.4811 Gm; it is quantitatively converted into KCl and now weighs 0.2982 Gm. Calculate how much KI is present in the original sample.

27. How much NaBr, percent *W/W*, is in a mixture of pure sodium bromide and pure potassium bromide containing 73.4 percent Br?

28. A 1.2500-Gm sample of calcium chloride and calcium bromide yielded a mixture of AgCl and AgBr weighing 2.3740 Gm, and from this, 2.2400 Gm of AgCl was obtained by heating the mixture in a stream of chlorine. What weight of calcium oxalate, $CaC_2O_4 \cdot H_2O$, could be obtained from 0.5 Gm of the original mixture?

Determination of Sulfate Ion in a Soluble Sulfate. The sulfate ion in a soluble sulfate may be determined gravimetrically by precipitation as barium sulfate, the precipitate being collected, dried, ignited, and weighed.

Exercise 5

Object. Assay of Sodium Sulfate.

Materials Required. 1 Gm of sodium sulfate
Barium chloride T.S. (12 percent)

Procedure. "Weigh accurately about 400 mg. of sodium sulfate, previously dried at 105° for 4 hours, dissolve it in 200 ml. of water, and add 1 ml. of hydrochloric acid. Heat to boiling, and gradually add, in small portions at a time and while constantly stirring, an excess of hot barium chloride T.S. (about 8 ml.). Heat the mixture for 1 hour on a steam bath, collect the precipitate of barium sulfate on a tared Gooch filter, wash it until free from chloride, dry, ignite, and weigh. The weight of the barium sulfate so obtained, multiplied by 0.6086, indicates its equivalent of Na_2SO_4."

The solution of the sulfate is acidified with hydrochloric acid to prevent the precipitation of carbonates, etc., which are soluble in hydrochloric acid solution. The acid also increases the solubility of barium sulfate slightly and therefore promotes the growth of large crystals, since small crystals with greater surface area per unit of weight dissolve more rapidly than large ones. The larger crystals with small surface area exposed grow more rapidly than they dissolve. The solution is heated to boiling to expel dissolved CO_2. An excess of hot barium chloride solution is added to ensure complete precipitation and to decrease the solubility of barium sulfate by common-ion effect. The mixture is digested on a steam bath for 1 hr to allow the larger crystals to grow at the expense of the smaller ones.

The barium sulfate is directed to be washed on a filter to remove excess $BaCl_2$ and the NaCl formed, but the washing process may be carried out more rapidly by washing once by decantation. The precipitate is then transferred quantitatively onto an ashless filter by the same technique used in Exercise 4 and washed free of chlorides as shown by tests made with silver nitrate solution on 2-ml portions of the filtrate. The filtrate is always acidified with a drop of nitric acid in carrying out the test for chlorides. Sometimes fine-grained crystalline precipitates will pass through the filter. This may be observed by imparting a rotatory motion to the filtrate, causing any crystals which passed through to gather in the center of the beaker. If such are detected, refilter.

Barium sulfate exhibits the property of "dragging down" some soluble salts when it is precipitated. This property is known as *coprecipitation.* The occluded or coprecipitated salt cannot be removed by ordinary washing. The error from this source is reduced largely by precipitation from hot dilute solutions and by adding the precipitating reagent slowly with continuous stirring. The phenomenon of coprecipitation is particularly likely to occur when barium sulfate is precipitated in the presence of chlorates, nitrates, and the salts of iron, aluminum, chromium, and other trivalent elements.

The precipitate may be dried and ignited by the wet method (see Drying and Ignition of Samples and Precipitates, under General Operations, Chap. 1). The equation

$$\underset{142.04}{Na_2SO_4} + BaCl_2 \cdot 2H_2O \longrightarrow \underset{233.40}{BaSO_4}\downarrow + 2NaCl + 2H_2O$$

expresses the stoichiometric relationship on which the calculations are based:

$$\frac{142.04}{233.40} = 0.6086 \text{ Gm of } Na_2SO_4 \text{ equivalent to 1 Gm of } BaSO_4$$

To calculate the quantity of sulfur in the sample, the gravimetric factor would be

$$\frac{S}{BaSO_4} = \frac{32.06}{233.40} = 0.1374$$

The gravimetric determination of sulfate ion as barium sulfate is used in pharmaceutical analysis. It may be applied to the estimation of metals that form insoluble sulfates such as barium, calcium, etc. The amount of CaO in magnesium carbonate and magnesium oxide is determined by precipitation of an equivalent amount of $CaSO_4$ under conditions such that $MgSO_4$ remains in solution. Sulfur, free and combined, may be oxidized to sulfate and subsequently precipitated as barium sulfate. The combustion of sulfobromophthalein sodium in an *Oxygen Flask Combustion* with oxidizing agents before the gravimetric determination as carried out in the assay of sodium sulfate is illustrative of a procedure frequently used for organic sulfur compounds. Other compounds such as thiosulfates, sulfites, persulfates, etc., may be estimated by the application of similar procedures.

Questions and Problems

1. Why is the solution of soluble sulfate acidified with hydrochloric acid?

2. Why is it necessary to digest the precipitate of barium sulfate previous to filtration?

3. Why should the precipitation of sulfates be carried out in hot, dilute solution?

4. An unknown sample of a soluble sulfate weighing 1.8000 Gm yielded 0.9000 Gm of $BaSO_4$. Calculate the percentage of sulfur in the unknown.

5. Calculate the equivalent amounts of each and the percentage purity if the unknown in problem 4 were $MgSO_4$, K_2SO_4, $Al_2(SO_4)_3$, or H_2SO_4.

6. Write the equations representing the reactions that take place in the assay of sulfur ointment.

7. How much $BaSO_4$ would be formed from 1.2000 Gm of $Na_2SO_4 \cdot 10H_2O$?

8. What is the percent of SO_4 in a sample of K_2SO_4 if 0.5000 Gm yields upon reaction with $BaCl_2$ a precipitate weighing 0.5850 Gm?

9. Calculate the weight of $BaSO_4$ that would be obtained from 200 mg of a 96 percent pure sample of methylthiouracil.

10. Calculate the percent of S and of metal in each of the following:

a. Na_2SO_4 *b.* $Na_2SO_4 \cdot 10H_2O$ *c.* $NaHSO_4$

d. $CaSO_4$ *e.* $BaSO_4$ *f.* $PbSO_4$

11. What weight of $BaCl_2 \cdot 2H_2O$ is required to precipitate the S in 1 Gm of potassium ferric alum, $K_2SO_4 \cdot Fe_2(SO_4)_3$?

12. What volume of barium chloride T.S. would be used in the gravimetric analysis of a 0.4500-Gm sample of sodium sulfate so that 5 percent excess of the theoretical amount would be present in the reaction mixture? The sodium sulfate is assumed to be 100 percent Na_2SO_4.

13. How many grams of barium chloride are necessary to precipitate all the sulfate ion in:

a. 1.4205 Gm of sodium sulfate?
b. 0.4356 Gm of potassium sulfate with $10H_2O$?

14. How many grams of $NaSO_4$ are necessary to precipitate quantitatively as sulfate:

a. The Ba in 0.4167 Gm of $BaCl_2$?
b. The Ca in 0.2220 Gm of $CaCl_2$?
c. The Pb in 0.2781 Gm of $PbCl_2$?

Determination of the Mercury Content of a Mercuric Salt. The mercury is precipitated as sulfide, washed, dried, and weighed.

Exercise 6

Object. Assay of Mercury Bichloride.

Materials Required. 0.5 Gm of mercuric chloride
1 ml of hydrochloric acid
About 70 ml of alcohol
About 40 ml of carbon tetrachloride

Procedure. 1. "Dry about 500 mg. of Mercury Bichloride over phosphorus pentoxide for 4 hours, weigh accurately, and dissolve it in 300 ml. of warm water, to which 1 ml. of hydrochloric acid has been added. Pass hydrogen sulfide through the solution until the precipitate of mercuric sulfide readily settles, leaving a clear, supernatant liquid."

The hydrogen sulfide gas used in the precipitation should be washed by bubbling it through a gas-washing bottle containing water. The glass delivery tube which passes to the bottom of the precipitation beaker should be attached to the gas-washing bottle by means of a rubber tube in such a manner that it can be removed readily and held over the precipitation beaker while washing any adhering precipitate into the beaker.

The hydrogen sulfide reacts with the mercuric chloride to form a double salt of mercuric sulfide and mercuric chloride at first, and then this double salt is decomposed with the formation of black mercuric sulfide as more hydrogen sulfide is passed into the mixture. Since mercuric sulfide is insoluble in acids, hydrogen sulfide may be employed as the precipitating agent and hydrochloric acid may be added to suppress the concentration of sulfide ion so that metals which form sulfides more soluble than mercuric sulfide and which may be present as impurities will not be precipitated.

2. "Collect the precipitate on a counterpoised filter or in a tared filter crucible, wash it well with cold water, and finally with 3 portions of about 10 ml. each of alcohol. Then, without suction, wash the precipitate with 10 ml. portions of carbon tetrachloride until after evaporating about 1 ml. of the last filtrate no visible residue remains. Remove the adhering carbon tetrachloride by washing the precipitate with several portions of 10 ml. each of alcohol, and, after drying in the air, transfer to an oven, and dry at 105° for 2 hours. The weight of mercuric sulfide, multiplied by 1.167, indicates its equivalent in $HgCl_2$."

The precipitate is washed with water to remove soluble sulfides, then with alcohol to remove the water. The carbon tetrachloride, which is added later and is immiscible with water but miscible with alcohol, can permeate the precipitate and dissolve any sulfur formed by the oxidation of hydrogen sulfide. The precipitate is washed with carbon tetrachloride without suction to allow time for the sulfur, which is slowly soluble, to dissolve. After washing the precipitate free of carbon tetrachloride with alcohol, it is dried at 105° because mercuric sulfide is volatile at high temperatures (above 400°).

The reaction that occurs may be represented in its simplest form by the following equation:

$$\underset{271.50}{HgCl_2} + H_2S \longrightarrow \underset{232.65}{HgS} + 2HCl$$

The same procedure can be adapted for the assay of other mercury salts, antimony salts, and, after a preliminary oxidative degradation using acid permanganate, for the assay of such organomercury compounds as Merbromin.

Questions and Problems

1. Show how the gravimetric factors for each of the following are derived: *(a)* mercury bichloride; *(b)* ammoniated mercury; *(c)* mercury in Merbromin.

2. The presence of hydrochloric acid makes possible the gravimetric analysis of mercury bichloride even in the presence of other ions such as nickel, which react with hydrogen sulfide to form insoluble sulfides. Explain, using equations for the reversible reactions involved and the corresponding expressions for calculation of equilibrium constants.

3. Refer to the official monographs in which the assay procedures of ammoniated mercury ointment and Merbromin are described. Outline and explain the steps in the procedures, preliminary to precipitation of the mercuric sulfide.

4. What weight of each of the following should be taken as sample, assuming 100 percent purity, so that 0.5000 Gm of HgS would be obtained in a gravimetric analysis: (*a*) mercury bichloride; (*b*) Merbromin; (*c*) mercury bichloride, large poison tablets?

5. Calculate the $HgCl_2$ content of a sample of mercury bichloride, 0.2551 Gm of which yields 0.1815 Gm of HgS.

6. What weight of AgCl could be obtained from a sample of mercury bichloride capable of yielding 0.3750 Gm of HgS?

Determination of the Aluminum Oxide Content of Dihydroxyaluminum Aminoacetate. The aluminum dihydroxyaminoacetate is precipitated as aluminum hydroxide, ignited to the oxide, and weighed.

Exercise 7

Object. Assay of Dihydroxyaluminum Aminoacetate.

Materials Required. 1 Gm of Dihydroxyaluminum Aminoacetate
5 ml of nitric acid
7 ml of dilute sulfuric acid (1 in 2)
10 ml of hydrochloric acid
Stronger ammonia water
Solution of ammonium chloride (1 in 50)

Procedure. "Transfer about 1 Gm. of Dihydroxyaluminum Aminoacetate, previously dried at 130° to constant weight and accurately weighed, to a platinum dish, add 5 ml. nitric acid and 2 ml. dilute sulfuric acid (1 in 2), evaporate nearly to dryness, and ignite, preferably in a muffle furnace at about 1500°F. Cool, and add to the residue 10 ml. of hydrochloric acid and 5 ml. of dilute sulfuric acid (1 in 2), and transfer to a 600-ml beaker. Dilute the mixture to 400 ml., filter, if necessary, to obtain a clear solution, add a few drops of methyl red T.S., and sufficient stronger ammonia water to change the color of the solution to a distinct yellow. Boil the mixture for 1 or 2 minutes, filter it once through paper, and wash the precipitate thoroughly with a hot solution of ammonium chloride (1 in 50). Transfer the filter containing the moist precipitate to previously ignited and tared crucible, heat to dryness, and ignite, preferably in a muffle furnace at 1800°F., to constant weight. The ignited residue obtained is Al_2O_3."

The ammonium chloride and ammonium sulfate formed upon the addition of stronger ammonia water to the acid solution of the residue, aluminum oxide, has the effects of (1) buffering the ammonia water added and so maintaining optimal pH conditions (pH 7) for the precipitation; (2) preventing the formation of the colloidal form of aluminum hydroxide; and (3) preventing the precipitation of hydroxides of other metals, such as magnesium, minimizing their coprecipitation effects.

The slight excess of ammonia water ensured by the indicator's turning color precipitates the aluminum hydroxide. The ammonia water should be freshly distilled, since upon standing in glass containers it dissolves the silica of glass, the presence of which leads to high results, the silica being precipitated and retained in the precipitate. The ammonia water should be added in very slight excess, since aluminum hydroxide is slightly soluble in strong solutions of ammonia. The mixture is boiled for a short time to convert any colloidal aluminum hydroxide into large particles so as to secure a coarse-grained precipitate. The precipitate is washed with ammonium chloride solution so as to maintain the particle size. A water washing would cause the precipitate to peptize and pass through the filter.

Ignition temperatures above 1200°F ensure a dry, nonhygroscopic residue of aluminum oxide.

$$2Al^{3+} + 6NH_4OH \longrightarrow 2Al(OH)_3\downarrow + 6NH_4^+$$
$$2Al(OH)_3 \longrightarrow Al_2O_3 + 3H_2O$$

No gravimetric factor is used in the calculation of the results of the analysis, since the purity of the sample is based on the sample's yielding a certain percent residue which is aluminum oxide.

$$\frac{Al_2O_3}{2C_2H_6AlNO_4} = \frac{101.96}{270.12} = 37.7 \text{ percent}$$

The theoretical amount of aluminum oxide obtained from dihydroxyaluminum aminoacetate is 37.7 percent, and the official requirement calls for a content of between 35.5 and 38.5 percent of aluminum oxide.

Determination of the Phosphate in Sodium Phosphate Solution. The phosphate is precipitated as magnesium ammonium phosphate, washed, dried, ignited to the pyrophosphate, and weighed.

Exercise 8

Object. Assay of Sodium Phosphate Solution.

Materials Required. 5 ml of Sodium Phosphate Solution
5 ml of hydrochloric acid
15 ml of magnesia mixture T.S.
40 ml of ammonia T.S.
Dilute ammonia water (1 in 3)

Procedure. 1. "Pipet 5 ml. of Sodium Phosphate Solution into a 100-ml volumetric flask, and dilute to volume with water. Mix, and pipet 10 ml. of the diluted solution into a beaker. Dilute to 200 ml. with water, add 5 ml. of hydrochloric acid, an excess (15 ml.) of magnesia mixture T.S., and 40 ml. of ammonia T.S. Mix, and allow to stand overnight."

Magnesia mixture T.S. is a preparation containing magnesium ions in a buffered ammonium chloride–ammonium hydroxide solution. The addition to a solution containing phosphate ions will form a precipitate of magnesium ammonium phosphate if the pH of the reaction is controlled, i.e., through buffering and common-ion effect.

When sodium phosphate is dissolved in water, the phosphoric acid is ionized in three stages, which may be represented by the following equations:

$$H_3PO_4 \rightleftharpoons H^+ + H_2PO_4^-$$
$$H_2PO_4^- \rightleftharpoons H^+ + HPO_4^=$$
$$HPO_4^= \rightleftharpoons H^+ + PO_4^{3-}$$

The degree of ionization becomes less for each successive stage. The relative amounts of PO_4^{3-}, $HPO_4^=$, $H_2PO_4^-$, and H_3PO_4 in equilibrium with one another depend on the hydrogen ion concentration of the solution. In a basic solution, the PO_4^{3-} and $HPO_4^=$ ions predominate. Each of the above species of phosphate ion which exists in the solution of the sample reacts with the Mg^{++} ions contained in magnesia mixture to form the corresponding magnesium salt; all these are more or less insoluble in water. Thus there may be formed $Mg_3(PO_4)_2$, $MgHPO_4$, and $Mg(H_2PO_4)_2$. The addition of a solution of NH_4OH represses the concentration of $H_2PO_4^-$ and prevents the formation of $Mg(H_2PO_4)_2$. In the presence of the ammonium ion, the salt is precipitated as $MgNH_4PO_4 \cdot 6H_2O$. If the solution is too basic, $Mg(OH)_2$ may be precipitated. The precipitation of both $Mg_3(PO_4)_2$ and $Mg(OH)_2$ is prevented by the presence of ammonium salts which, through common-ion effect, act as a buffer and prevent the hydroxyl ion concentration from becoming too great. The hexahydrate, $MgNH_4PO_4 \cdot 6H_2O$, forms in relatively coarse crystals on standing for 3 hr or more at room temperatures. At temperatures above 60°, the monohydrate is formed.

2. "Collect the precipitate in a tared, previously ignited and weighed filtering crucible, and wash it with a mixture of 1 volume of ammonia T.S. and 3 volumes of water until free from chloride. Add a few crystals of ammonium nitrate or a few drops of a saturated solution of ammonium nitrate to the crucible, and dry at 105°. Ignite the precipitate carefully, starting with a very low heat and gradually increasing to bright redness until constant weight is obtained. The weight of the magnesium pyrophosphate so obtained, multiplied by 2.409, indicates its content of $Na_2HPO_4 \cdot 7H_2O$."

The precipitate is washed with aqueous ammonia solution until free of chlorides. In testing the filtrate for chlorides, acidify with nitric acid and add silver nitrate T.S. Upon ignition the magnesium ammonium phosphate is converted into the pyrophosphate as follows:

$$2MgNH_4PO_4 \cdot 6H_2O \longrightarrow Mg_2P_2O_7 + 2NH_3 + 13H_2O$$

For the purpose of making calculations, the equation may be represented as

$$\underset{2(268.07)}{2Na_2HPO_4 \cdot 7H_2O} = 2MgNH_4PO_4 \cdot 6H_2O = \underset{222.57}{Mg_2P_2O_7}$$

From this stoichiometric relationship, it is evident that each gram of $Mg_2P_2O_7$ is equivalent to 2(268.07)/222.57 = 2.409 Gm of $Na_2HPO_4 \cdot 7H_2O$.

The same precipitate, $MgNH_4PO_4 \cdot 6H_2O$, results when magnesium ions are treated with a soluble dibasic ammonium phosphate. The magnesium pyrophosphate obtained on ignition is equivalent to the magnesium ammonium phosphate. The application of this procedure for the determination of citric acid in raspberry juice is indicated in Table 2. In this assay the citric acid reacts with the insoluble magnesium oxide, forming soluble magnesium citrate, which is removed by filtration from the excess magnesium oxide, and then the filtrate is treated with dibasic ammonium phosphate, as the precipitant, to yield the magnesium ammonium phosphate. The relationship between the citric acid and the magnesium pyrophosphate is shown.

$$\underset{4(210.14)}{4H_3C_6H_5O_7 \cdot H_2O} = 2Mg_3(C_6H_5O_7)_2 = 6MgNH_4PO_4 \cdot 6H_2O = \underset{3(222.57)}{3Mg_2P_2O_7}$$

$$\frac{4(210.14)}{3(222.57)} = 0.1259 \text{ (gravimetric factor)}$$

Each gram of magnesium pyrophosphate obtained is then equivalent to 0.1259 Gm of citric acid monohydrate.

Magnesium Citrate Solution and Magnesium Hydroxide were formerly assayed by conversion to equivalent amounts of $Mg_2P_2O_7$, as described above, and are now assayed by a procedure involving the use of an organic precipitant. The high specificity of organic precipitants, especially when the pH of the reaction mixture is controlled, makes possible qualitative and quantitative analyses of certain ions even in the presence of other ions that interfere in analytical procedures in which inorganic precipitating reagents are used. Certain organic precipitants should be familiar from a study of qualitative analysis; e.g., dimethylglyoxime, cupferron, etc. The reagent, 8-hydroxyquinoline or oxine, selectively forms a very insoluble precipitate with Mg^{++}, $Mg(C_9H_6ON)_2 \cdot 2H_2O$, in a mixture buffered with ammonium chloride and stronger ammonia T.S. This reaction is the basis of the procedure for the quantitative analysis of zinc undecylenate in Compound Undecylenic Acid Ointment. The zinc ion behaves similarly to the magnesium.

Some bismuth preparations are assayed by conversion to an equivalent amount of Bi_2O_3. Bismuth Subnitrate Tablets are assayed for their bismuth oxide content by precipitation, etc., of an equivalent amount of bismuth phosphate, $BiPO_4$; these tablets are listed in Table 2.

Certain official gold salts are readily assayed because of the ease of reduction of gold in combination to an equivalent amount of elemental gold. The following are assayed gravimetrically in this way: gold sodium thiosulfate, gold sodium thiomalate, gold chloride reagent, aurothioglucose, and some preparations of these compounds.

There are some miscellaneous gravimetric analyses in which official compounds are converted to insoluble residues that are dried and weighed, e.g., potassium warfarin (K_2SO_4); formamide (NH_4Cl); ammonium

TABLE 2. SUBSTANCES ASSAYED GRAVIMETRICALLY

Substance	Weighed as	Gravimetric factor	Official requirement, percent*
As Silver Halide			
N.F.			
Camphorated parachlorophenol	AgCl	0.8970	C_6H_5ClO = 33 to 37
Merbromin	AgBr	0.4255	Br = 18 to 21.3
Merbromin solution	AgBr	0.4255	Br = 18 to 21.3
Merbromin solution, surgical	AgBr	0.4255	Br = 18 to 21.3
As Sulfate			
U.S.P.			
Magnesium oxide	$CaSO_4$	0.4119	CaO = n.m.t. 1.5
Sodium lauryl sulfate	$BaSO_4$	0.6086	Na_2SO_4 = n.m.t. 8
Sodium sulfobromophthalein	$BaSO_4$	0.1374	S = 7.4 to 8.2
Sulfur, precipitated	$BaSO_4$	0.1374	S = 99.5
Sulfur ointment	$BaSO_4$	0.1374	S = 9.5 to 10.5
N.F.			
Ichthammol	$BaSO_4$	0.5661	$(NH_4)_2SO_4$ = 8
Ichthammol	$BaSO_4$	0.1374	S = 10
Magnesium carbonate	$CaSO_4$	0.4119	CaO = n.m.t. 0.6
Magnesium hydroxide	$CaSO_4$	0.4119	CaO = n.m.t. 1
Methylthiouracil	$BaSO_4$	0.1374	S = 22.2 to 22.8
Sodium cyclamate and sodium saccharin solution	$BaSO_4$	1.033	$C_7H_4NNaO_3S{\cdot}2H_2O$ = 0.54 to 0.66 *W/V*
Sodium cyclamate and sodium saccharin tablets	$BaSO_4$	1.033	$C_7H_4NNaO_3S{\cdot}2H_2O$ = 90 to 110 l.a.
Sodium sulfate	$BaSO_4$	0.6086	Na_2SO_4 = 99
Sulfur, sublimed	$BaSO_4$	0.1374	S = 99.5
As Mercuric Sulfide			
U.S.P.			
Meralluride	HgS	1.929	$C_9H_{16}HgN_2O_6$ = 94
Meralluride injection	HgS	1.929	$C_9H_{16}HgN_2O_6$ = 95 to 105
Mercury ointment, ammoniated	HgS	1.083	$HgNH_2Cl$ = 4.5 to 5.5
Mercury ophthalmic ointment, ammoniated	HgS	1.083	$HgNH_2Cl$ = 2.7 to 3.3
N.F.			
Chlormerodrin	HgS	1.578	$C_5H_{11}ClHgN_2O_2$ = 98.8 to 100.7
Chlormerodrin tablets	HgS	1.578	$C_5H_{11}ClHgN_2\ O_2$ = 90.0 to 110
Merbromin	HgS	0.8622	Hg = 24 to 26.7
Merbromin solution	HgS	0.8622	Hg = 24 to 26.7
Merbromin solution, surgical	HgS	0.8622	Hg = 24 to 26.7
Mercury bichloride	HgS	1.167	$HgCl_2$ = 99.5
Mercury bichloride large poison tablets	HgS	1.167	$HgCl_2$ = 0.42 to 0.52 Gm/tablet

TABLE 2. SUBSTANCES ASSAYED GRAVIMETRICALLY (*Continued*)

Substance	Weighed as	Gravimetric factor	Official requirement, percent*
As Metallic Oxides			
U.S.P.			
Aluminum monostearate.	Al_2O_3	......	Al_2O_3 = 14.5 to 16
Bismuth subcarbonate...	Bi_2O_3	......	Bi_2O_3 = 90
N.F.			
Aluminum aspirin.......	Al_2O_3	......	Al_2O_3 = 12 to 17
Bismuth magma........	Bi_2O_3	......	Bi_2O_3 = 5.2 to 5.8
Bismuth subnitrate......	Bi_2O_3	......	Bi_2O_3 = 79
Dihydroxyaluminum aminoacetate...........	Al_2O_3	......	Al_2O_3 = 35.5 to 38.5
Dihydroxyaluminum aminoacetate magma.....	Al_2O_3	......	Al_2O_3 = 28.5 to 35
Dihydroxyaluminum aminoacetate tablets.....	Al_2O_3	......	Al_2O_3 = 28.5 to 35
Zinc eugenol cement.....	ZnO	......	ZnO = 0.68 to 0.72 Gm/Gm
Zinc eugenol paste and salicylic acid..........	ZnO	......	ZnO = 23.5 to 25.5
As Magnesium Pyrophosphate			
U.S.P.			
Raspberry juice.........	$Mg_2P_2O_7$	1.259	$C_6H_8O_7 \cdot H_2O$ = 1.5
N.F.			
Sodium phosphate solution..................	$Mg_2P_2O_7$	2.409	$Na_2HPO_4 \cdot 7H_2O$ = 71 to 79
As Metallic Oxines			
N.F.			
Magnesium citrate solution..................	$C_{18}H_{12}Mg \cdot N_2O_2 \cdot 2H_2O$	4.624	MgO = 1.55 to 1.9
Magnesium hydroxide....	$C_{18}H_{12}Mg \cdot N_2O_2 \cdot 2H_2O$	0.1673	$Mg(OH)_2$ = 95
Milk of magnesia tablets.............	$C_{18}H_{12}Mg \cdot N_2O_2 \cdot 2H_2O$	0.1673	$Mg(OH)_2$ = 93 to 107 l.a.
Undecylenic acid ointment, compound......	$C_{18}H_{12}Zn \cdot N_2O_2$	1.221	$C_{22}H_{38}O_4Zn$ = 18 to 22
As Free Gold			
U.S.P.			
Aurothioglucose.........	Au	1.9911	$C_6H_{11}AuO_5S$ = 95 to 105
Gold chloride (R)......	Au	......	Au = 49
Gold sodium thiomalate..	Au	2.072	$C_4H_3AuNa_2O_4S \cdot H_2O$ = 93.3 to 101.5
Gold sodium thiomalate injection............	Au	2.072	$C_4H_3AuNa_2O_4S \cdot H_2O$ = 95 to 105 l.a.
N.F.			
Gold sodium thiosulfate..	Au	2.672	$Na_3Au(S_2O_3)_2 \cdot 2H_2O$ = 97.9 to 100.6
Gold sodium thiosulfate, sterile...............	Au	2.672	$Na_3Au(S_2O_3)_2 \cdot 2H_2O$ = 97.9 to 100.6

TABLE 2. SUBSTANCES ASSAYED GRAVIMETRICALLY (*Continued*)

Substance	Weighed as	Gravimetric factor	Official requirement, percent*
As Lead Molybdate			
U.S.P.			
Ammonium molybdate (R)	$PbMoO_4$	0.3920	MoO_3 = 81 to 83
Molybdic acid (R)	$PbMoO_4$	0.3920	MoO_3 = 85
Sodium molybdate (R)	$PbMoO_4$	0.3920	MoO_3 = 57 to 59
N.F.			
Sodium molybdate (R)	$PbMoO_4$	0.3920	MoO_3 = 57 to 59.5
As Miscellaneous Precipitates			
U.S.P.			
Calcium carbonate (R)	$CaCO_3$	......	$CaCO_3$ = 99.5
Formamide (R)	NH_4Cl	0.8420	$HCONH_2$ = 99.3 to 100.2
Sulfurated potash	CuO	0.4030	S = 12.8
N.F.			
Bismuth subnitrate tablets	$BiPO_4$	0.7665	Bi_2O_3 = 73 to 85 l.a.
Potassium warfarin	K_2SO_4	0.4488	K = 10.9 to 15.1

* Unless otherwise specified, percentages are expressed as *W/W* in all tables.
n.m.t. = not more than; l.a. = labeled amount; (R) = reagent.

molybdate, molybdic acid, and sodium molybdate reagents ($PbMoO_4$); calcium carbonate ($CaCO_3$); and sulfurated potash (CuO).

Sometimes a loss in weight can be used as basis for analysis; e.g., conversion of silicon dioxide, SiO_2, to silicon tetrafluoride, a gas, as in the assay of magnesium trisilicate for SiO_2.

The analytical control of organic medicinal agents by gravimetric methods is indicated, in a few instances, in some of the preceding tables. Frequently the organic constituent of an official preparation is isolated and purified by physical procedures and without any permanent chemical change, e.g., by extraction or by evaporation of solvent.

A list of official substances assayed gravimetrically by separating, purifying, and weighing an organic medicinal agent without any permanent change in composition appears in Table 3. Such procedures are frequently applied to preparations containing organic medicinal agents. Before extraction of the organic medicinal agent, the sample of the powdered tablets is frequently washed with petroleum benzine to remove substances such as binders and lubricants that would be extracted along with the organic medicinal agent by the solvent, chloroform or ether, which is used subsequently. If the organic medicinal agent is basic, e.g., chloroquin in chloroquin phosphate, or acidic, e.g., amobarbital in sodium amobarbital tablets, it is first extracted with an aqueous solution of an

TABLE 3. SUBSTANCES ASSAYED GRAVIMETRICALLY WITHOUT CHEMICAL CHANGE

Substance	Official requirement, percent
U.S.P.	
Amobarbital tablets	94 to 106 l.a.
Aurothioglucose injection	90 to 110 l.a.
Bishydroxycoumarin	98
Caffeine and sodium benzoate	47 to 50 caffeine
Caffeine and sodium benzoate injection	47 to 52 l.a. caffeine
Phenacetin tablets	94 to 106 l.a.
Secobarbital elixir	0.417 to 0.461 *W/V*
Sodium lauryl sulfate	n.m.t. 4 unsulfated alcohols
N.F.	
Amobarbital elixir	0.417 to 0.462 *W/V*
Citrated caffeine	48 to 52 anhydrous caffeine
Citrated caffeine tablets	45 to 55 l.a. anhydrous caffeine
Ephedrine sulfate and phenobarbital capsules	91 to 109 l.a. phenobarbital
Mephobarbital tablets	95 to 110 l.a.
Methylthiouracil tablets	93 to 107 l.a.
Picrotoxin injection	90 to 110 l.a.
Progesterone tablets	90 to 110 l.a.

l.a. = labeled amount; n.m.t = not more than.

acid or base to effect separation from the neutral substances that might be present. The aqueous solution of the salt of the organic medicinal agent is then made alkaline or acidic, as the case may be, and the liberated organic base or acid is extracted with chloroform or ether. Bishydroxycoumarin, precipitated from alkaline solution by acidification, can be filtered, and subsequent extraction is not necessary. If stearic acid or stearates are present, an organic medicinal agent which is acidic (amobarbital) cannot be extracted with sodium hydroxide solution because sodium stearate would be extracted along with the salt of the organic acid. In such a case, saturated solution of barium hydroxide is used, and the insoluble barium stearate can be removed by filtration. The active ingredient of a number of official preparations is a salt of an organic acid. When a solution of a sample of such a preparation is acidified with a strong acid such as hydrochloric acid, the organic acid is liberated and may be extracted with an immiscible solvent such as ether or chloroform. The analysis of official preparations for salts of organic bases may be carried out in an analogous manner; i.e., the organic base can be liberated by sodium hydroxide and extracted with ether or some other solvent. Official preparations assayed for salts of organic acids or bases are summarized in Tables 4 and 5. Specific type methods are also of interest in this connection (see Exercise 16, Chap. 4).

Organic medicinal agents can sometimes be converted quantitatively to derivatives by reactions involving some functional group such as amino, carboxyl, aldehyde or ketone carbonyl, phenolic hydroxyl, etc. Insoluble salts formed by reaction of amino or carboxyl groups may be included in this category.

TABLE 4. SUBSTANCES ASSAYED GRAVIMETRICALLY BY CONVERSION TO FREE ACID

Substance	Weighed as	Gravimetric factor	Official requirement, percent
U.S.P.			
Sodium amobarbital	$C_{11}H_{18}N_2O_3$	1.097	$C_{11}H_{17}N_2NaO_3$ = 98.5
Sodium amobarbital capsules	$C_{11}H_{18}N_2O_3$	1.097	$C_{11}H_{17}N_2NaO_3$ = 90 to 105 l.a.
Sodium amobarbital, sterile	$C_{11}H_{18}N_2O_3$	1.097	$C_{11}H_{17}N_2NaO_3$ = 98.5
Sodium diphenylhydantoin	$C_{15}H_{12}N_2O_2$	1.087	$C_{15}H_{11}N_2NaO_2$ = 98.5
Sodium diphenylhydantoin capsules	$C_{15}H_{12}N_2O_2$	1.087	$C_{15}H_{11}N_2NaO_2$ = 93 to 107 l.a.
Sodium diphenylhydantoin, sterile	$C_{15}H_{12}N_2O_2$	1.087	$C_{15}H_{11}N_2NaO_2$ = 98.5
Sodium fluorescein	$C_{20}H_{12}O_5$	1.132	$C_{20}H_{10}Na_2O_5$ = 98.5
Sodium fluorescein ophthalmic solution	$C_{20}H_{12}O_5$	1.132	$C_{20}H_{10}Na_2O_5$ = 1.86 to 2.1 *W/V*
Sodium pentobarbital	$C_{11}H_{18}N_2O_3$	1.097	$C_{11}H_{17}N_2NaO_3$ = 98.5 to 101
Sodium pentobarbital capsules	$C_{11}H_{18}N_2O_3$	1.097	$C_{11}H_{17}N_2NaO_3$ = 92.5 to 107.5 l.a.
Sodium pentobarbital elixir	$C_{11}H_{18}N_2O_3$	1.097	$C_{11}H_{17}N_2NaO_3$ = 0.375 to 0.425 *W/V*
Sodium pentobarbital injection	$C_{11}H_{18}N_2O_3$	1.097	$C_{11}H_{17}N_2NaO_3$ = 90 to 105 l.a.
Sodium phenobarbital	$C_{12}H_{12}N_2O_3$	1.095	$C_{12}H_{11}N_2NaO_3$ = 98.5 to 101
Sodium phenobarbital injection	$C_{12}H_{12}N_2O_3$	1.095	$C_{12}H_{11}N_2NaO_3$ = 90 to 105 l.a.
Sodium phenobarbital, sterile	$C_{12}H_{12}N_2O_3$	1.095	$C_{12}H_{11}N_2NaO_3$ = 98.5 to 101
Sodium secobarbital	$C_{12}H_{18}N_2O_3$	1.092	$C_{12}H_{17}N_2NaO_3$ = 98.5
Sodium secobarbital capsules	$C_{12}H_{18}N_2O_3$	1.092	$C_{12}H_{17}N_2NaO_3$ = 90 to 105 l.a.
Sodium secobarbital, sterile	$C_{12}H_{18}N_2O_3$	1.092	$C_{12}H_{17}N_2NaO_3$ = 98.5
Tetracaine hydrochloride	$C_{15}H_{24}N_2O_2$	1.138	$C_{15}H_{24}N_2O_2 \cdot HCl$ = 98.5 to 101
N.F.			
Diethylstilbestrol dipropionate	$C_{18}H_{20}O_2$	1.4188	$C_{24}H_{28}O_4$ = 98 to 100.5
Sodium butabarbital	$C_{10}H_{16}N_2O_3$	1.104	$C_{10}H_{15}N_2NaO_3$ = 98.2
Sodium butabarbital capsules	$C_{10}H_{16}N_2O_3$	1.104	$C_{10}H_{15}N_2NaO_3$ = 90 to 105 l.a.
Sodium butabarbital elixir	$C_{10}H_{16}N_2O_3$	1.104	$C_{10}H_{15}N_2NaO_3$ = 0.56 to 0.63 *W/V*
Sodium butabarbital tablets	$C_{10}H_{16}N_2O_3$	1.104	$C_{10}H_{15}N_2NaO_3$ = 94 to 106 l.a.
Sodium pentobarbital tablets	$C_{11}H_{18}N_2O_3$	1.097	$C_{11}H_{17}N_2NaO_3$ = 90 to 105 l.a.
Sodium phenobarbital tablets	$C_{12}H_{12}N_2O_3$	1.095	$C_{12}H_{11}N_2NaO_3$ = 90 to 105 l.a.
Sodium thiamylal for injection	$C_{12}H_{18}N_2O_2S$	1.086	$C_{12}H_{17}N_2NaO_2$ = 93 to 107
Sodium vinbarbital injection	$C_{11}H_{16}N_2O_3$	1.098	$C_{11}H_{15}N_2NaO_3$ = 90 to 105 l.a.

l.a. = labeled amount.

TABLE 5. SUBSTANCES ASSAYED GRAVIMETRICALLY BY CONVERSION TO FREE BASE

Substance	Weighed as	Gravimetric factor	Official requirement, percent
U.S.P.			
Chloroquin phosphate	$C_{18}H_{26}ClN_3$	1.613	$C_{18}H_{26}ClN_3 \cdot 2H_3PO_4$ = 98 to 102
Hydroxychloroquine sulfate	$C_{18}H_{26}ClN_3O$	1.292	$C_{18}H_{26}ClN_3O \cdot H_2SO_4$ = 98 to 102
N.F.			
Metabutethamine HCl	$C_{13}H_{20}N_2O_2$	1.154	$C_{13}H_{20}N_2O_2 \cdot HCl$ = 98.5
Metabutethamine HCl and epinephrine injection	$C_{13}H_{20}N_2O_2$	1.154	$C_{13}H_{20}N_2O_2 \cdot HCl$ = 95 to 105 l.a.
Phenacaine HCl	$C_{18}H_{22}N_2O_2$	1.122	$C_{18}H_{22}N_2O_2 \cdot HCl$ = 98 to 102

l.a. = labeled amount.

Quaternary nitrogen compounds will form insoluble salts with tetraphenylboron (see Chap. 7). These nitrogen-containing compounds, choline chloride and bethanechol chloride, are easily assayed by precipitation with sodium tetraphenylboron.

$$R_4N^+ + B(C_6H_5)_4^- \longrightarrow R_4N{\cdot}B(C_6H_5)_4\downarrow$$

In certain instances it is also possible to obtain uniform substitution products of organic medicinal agents, quantitatively, e.g., phenolphthalein in tablets yields a tetraiodo derivative. The number of organic medicinal agents that can be analyzed in this way is limited because of the reversible nature of the reactions involved and the simultaneous formation of products of side reactions, as well as for other reasons. A list of official substances assayed by such procedures is given in Table 6.

TABLE 6. SUBSTANCES ASSAYED GRAVIMETRICALLY BY CONVERSION TO DERIVATIVES OR SUBSTITUTION PRODUCTS

Substance	Derivative or substitution product	Gravimetric factor	Official requirement, percent
U.S.P.			
Betazole HCl	Phosphotungstate	0.09058	$C_5H_9N_3{\cdot}2HCl$ = 98 to 101
Betazole HCl injection	Phosphotungstate	0.09058	$C_5H_9N_3{\cdot}2HCl$ = 95 to 105 l.a.
Bethanechol chloride injection	Tetraphenylboron	0.4090	$C_7H_{17}ClN_2O_2$ = 95 to 105 l.a.
Bethanechol chloride tablets	Tetraphenylboron	0.4090	$C_7H_{17}ClN_2O_2$ = 90 to 110 l.a.
Choline chloride (R)	Tetraphenylboron	0.3298	$C_5H_{14}ClNO$ = 99.5
Hydroxyamphetamine HBr ophthalmic solution	Diacetyl	0.9866	$C_9H_{13}NO{\cdot}HBr$ = 95 to 105 l.a.
Sodium tetraphenylboron (R)	Potassium salt	0.9550	$(C_6H_5)_4BNa$ = 99.5
Testosterone propionate injection	Semicarbazone	0.8579	$C_{22}H_{32}O_3$ = 88 to 112 l.a.
N.F.			
Camphorated parachlorophenol	Phenylhydrazone	0.4580	$C_{10}H_{16}O$ = 63 to 67
Camphor spirit	Phenylhydrazone	0.4580	$C_{10}H_{16}O$ = 9 to 11 *W/V*
Choline dihydrogen Citrate capsules	Reineckate	0.6989	$C_{11}H_{21}NO_8$ = 93 to 107 l.a.
Choline dihydrogen Citrate tablets	Reineckate	0.6989	$C_{11}H_{21}NO_8$ = 95 to 105 l.a.
Dienestrol	Diacetate	0.7601	$C_{18}H_{18}O_2$ = 98
Diphemanil methylsulfate	Reineckate	0.6527	$C_{20}H_{24}N{\cdot}CH_3SO_4$ = 97
Diphemanil methylsulfate injection	Reineckate	0.6527	$C_{20}H_{24}N{\cdot}CH_3SO_4$ = 92.5 to 107.5 l.a.
Diphemanil methylsulfate tablets	Reineckate	0.6527	$C_{20}H_{24}N{\cdot}CH_3SO_4$ = 92.5 to 107.5 l.a.
Inositol	Hexaacetate	0.4167	$C_6H_{12}O_6$ = 97
Monobenzone	Acetate	0.8265	$C_{13}H_{12}O_2$ = 98 to 102
Monobenzone lotion	Acetate	0.8265	$C_{13}H_{12}O_2$ = 94 to 106 l.a.
Monobenzone ointment	Acetate	0.8265	$C_{13}H_{12}O_2$ = 94 to 106 l.a.
Phenolphthalein tablets	Tetraiodo	0.3873	$C_{20}H_{14}O_4$ = 92.5 to 107.5 l.a.
Phentolamine HCl	Trichloroacetate	0.7146	$C_{17}H_{19}N_3O{\cdot}HCl$ = 98
Phentolamine HCl tablets	Trichloroacetate	0.7146	$C_{17}H_{19}N_3O{\cdot}HCl$ = 93 to 107 l.a.
Progesterone injection	Hydrazone	0.4660	$C_{21}H_{30}O_2$ = 90 to 110 l.a.

l.a. = labeled amount; (R) = reagent.

Questions and Problems

1. Give equations, indicating all reagents and conditions, for the chemical reactions involved in the formation of $Mg_2P_2O_7$ from (*a*) magnesium sulfate; (*b*) sodium phosphate in sodium phosphate solution.

2. Give the composition or chemical formula of each of the following reagents and state the use of each in gravimetric analysis: (*a*) magnesia mixture T.S.; (*b*) dibasic ammonium phosphate; (*c*) 8-hydroxyquinoline T.S.; (*d*) dinitrophenylhydrazine T.S.

3. A 5.00-ml sample of sodium phosphate solution was diluted with water to a volume of 100.00 ml. A 10.00-ml sample of the resulting solution was treated with magnesia mixture T.S. as directed in the official monograph. By ignition of the magnesium ammonium phosphate precipitate, 0.1515 Gm of magnesium pyrophosphate was obtained. Calculate the Na_2HPO_4 content of the sodium phosphate solution in percent *W/V*. Does the sample conform to official specifications? If the specific-gravity were 1.39, what would be the Na_2HPO_4 content in percent *W/W*?

4. Calculate the weight of magnesium hydroxyquinolate that would be obtained in the analysis of magnesium citrate solution containing magnesium citrate equivalent to 1.85 percent *W/V* MgO. Note that in this assay a one-twentieth aliquot part of a 50.00-ml sample is employed.

5. Official bismuth preparations are assayed by one of two types of procedure. Illustrate, and calculate the gravimetric factors involved.

6. In view of the insolubility of calcium oxalate, why are official calcium preparations generally not assayed gravimetrically?

7. In the assay of amobarbital tablets, what change is made in the procedure if stearic acid or stearates are present?

8. Give illustrations of official assay procedures involving (*a*) liberation and extraction of an organic base; (*b*) liberation and extraction of an organic acid.

9. Give the structural formula and the substance precipitated and weighed in each of the following assay procedures and calculate the gravimetric factor: (*a*) phenolphthalein tablets; (*b*) progesterone injection; (*c*) monobenzone; (*d*) inositol.

10. What weight of betazole phosphotungstate might be expected in the assay of an 0.1150-Gm sample of betazole hydrochloride containing 98 percent $C_5H_9N_3 \cdot 2HCl$?

Chapter 3

PRINCIPLES OF TITRIMETRIC (VOLUMETRIC) ANALYSIS

In gravimetric analysis (see Chap. 2), the amount of a reactant, A or B, present in a sample, is determined by *weighing* a chemically equivalent amount of a product, C or D, formed from the reactant in a chemical reaction that goes practically to completion. Using the weight of the product and a gravimetric factor, the weight of the reactant in the sample can be calculated. The amount of one reactant, A, can also be determined by measuring the chemically equivalent amount of the other reactant, B, when the two interact in a chemical reaction that goes practically to completion. This is frequently accomplished by the treatment of a soluble substance A, contained in a suitable vessel, with an appropriate standardized solution of substance B, which is equivalent chemically to the reactant A. The solution of substance B, called a *titrant*, is chosen with respect to its strength such that the volume added is between 30 and 70 percent of the capacity of the buret, i.e., between 15 and 35 ml of a 50-ml buret. This process, known as *titration*, is common to all *titrimetric* or *volumetric* analyses. In titrimetric or volumetric processes, the *volume* of a standard solution containing a chemically equivalent amount of the reactant B is measured. For example, the amount of NaCl in a sample can be calculated from the weight of a chemically equivalent amount of AgCl obtained in gravimetric analysis or from the volume of a standard solution containing a chemically equivalent amount of $AgNO_3$ employed in titrimetric analysis.

Definitions. A *standard solution* is a solution of known normality or molarity. *Normality* is an expression of the concentration of a solution in terms of equivalents per liter or milliequivalents per milliliter. *Molarity* is an expression of the concentration of a solution in terms of moles per liter. The determination of the normality or molarity of a solution is known as *standardization*. This may be accomplished by use of another standard solution known as a *secondary standard*, or by use of a carefully weighed sample of a substance of known purity which is termed a *primary standard*. Thus a solution of hydrochloric acid may be standardized titrimetrically using a standard solution of sodium hydroxide as a secondary standard or using a sample of sodium carbonate of known purity as a primary standard. Titrations may be conducted *directly*, e.g., by addition of a standard solution of $AgNO_3$ in an amount that is chemically equivalent to the NaCl in a sample. They may also be conducted *residually*, e.g., by addition of a carefully measured volume of standard solution of $AgNO_3$ which is in excess of the amount chemically equivalent to the NaCl in a sample followed by the addition of a standard solution of NH_4SCN in an amount that is chemically equivalent to the excess $AgNO_3$. Obviously, in the residual titration, the known amount of $AgNO_3$ is chemically equivalent to the known amount of NH_4SCN plus the unknown amount of NaCl.

In the titration of one reactant with a standard solution of another reactant, the theoretical point at which equivalent amounts of each have reacted is the *stoichiometric point*. The titration is actually stopped at some point, the *end point*, at which there occurs a sudden change in some property of the reaction mixture, a change which is made apparent by use of indicators or potentiometrically. A titration is feasible only when the end point is reasonably close to the theoretical stoichiometric point.

Equivalents of various substances are quantities that are equivalent in a chemical sense, and which may be used to expedite calculations considerably, especially in titrimetric analysis. Whereas 2.016 Gm of hydrogen is required for reaction with 16 Gm of oxygen, or 2 moles of hydrogen with 1 mole of oxygen, 1 equivalent of hydrogen is required for reaction with 1 equivalent of oxygen. An equivalent of a substance may be expressed in weight units, a quantity known as the *equivalent weight*, or mole fractions. Thus an equivalent of hydrogen is 1.008 Gm of hydrogen or 0.5 mole of hydrogen. In general, an equivalent is the quantity of a substance that is chemically equivalent to 1.008 Gm of hydrogen or hydrogen ions. For example, 1 mole or 36.46 Gm of HCl and 0.5 mole or 49.04 Gm of H_2SO_4 contain 1.008 Gm of hydrogen ions; 1 mole or 40.01 Gm of NaOH and 0.5 mole or 85.69 Gm of $Ba(OH)_2$ are capable of neutralizing 1.008 Gm of hydrogen ions; 0.25 mole or 8 Gm of oxygen gains electrons equivalent to those gained by 1.008 Gm of hydrogen ions when they act as oxidizing agent, and 1 mole or 166.01 Gm of KI loses electrons equivalent to those lost by 1.008 Gm of hydrogen when it acts as reducing agent.

A *milliequivalent* (mEq) is one-thousandth of an equivalent and, for HCl, may be variously expressed as 0.001 mole, 0.001 equivalent, 0.03646 Gm, or 1 ml of 1 *N* solution.

Volumetric Apparatus. The apparatus required for volumetric analysis includes the analytical balance described in Chap. 1 and a number of accurately calibrated instruments, as well as beakers, flasks, funnels, etc. Volumetric apparatus is of two types: that made to deliver a definite volume of liquid, such as burets and pipets; and that made to contain a definite volume of liquid, as volumetric flasks and graduated cylinders. The temperature specified in the U.S.P. and N.F. for volumetric measurements is 25°, while the National Bureau of Standards has adopted 20° as the standard temperature for glass volumetric apparatus. It matters little at what temperature the instruments are calibrated, provided the same temperature is used for all, and all measurements are made at the temperature for which the instruments are calibrated. The limit of error in milliliters and percent is given in the U.S.P. and the N.F. for flasks, pipets, and burets of designated volumes.

Units of Capacity. The *milliliter* (ml) represents the one-thousandth part of the liter. The *liter* is defined as the volume occupied by a kilogram of water weighed in a vacuum at 4°. The one-thousandth part of a liter,

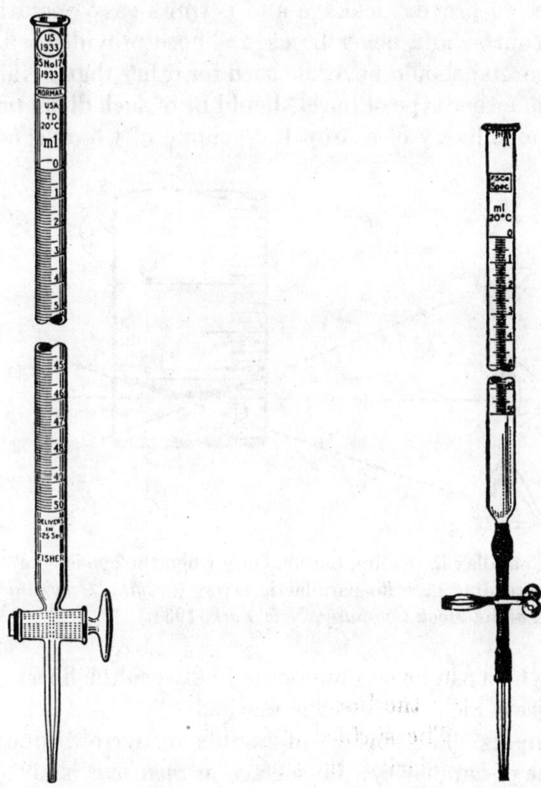

FIG. 9. Buret with glass stopcock.

FIG. 10. Buret with pinchcock. (Mohr buret)

termed a *milliliter*, is used as the unit of capacity. The milliliter has been adopted as the official unit of capacity.

Burets. Graduated glass tubes of uniform bore throughout the whole length, used in the measurement of variable quantities of liquid, are termed *burets* (Figs. 9 and 10). They are graduated into milliliters and tenth-milliliters and are closed at the bottom by a glass or Teflon stopcock or a short, soft rubber tube provided with a glass tip and pinchcock or other suitable device to control the outflow of liquid. The official compendiums formerly gave the following specifications for burets:

All burets should be provided with glass stopcocks and calibrated for use at the standard or other temperature at which they are to be used. Burets provided with rubber tubes, pinchcocks, and glass delivery tubes in place of glass stopcocks may be used for potassium and sodium hydroxide solutions and such other solutions as are best used in this type of buret. The rate of outflow should be regulated so that not less than 2 sec should be consumed for each milliliter delivered.

Burets provided with glass stopcocks should be tested for leakage. A lubricant serves to prevent leakage and permits easy operation, but an excess of lubricant should never be used. Those provided with a rubber outlet (Mohr burets) should never be used for other than alkali solutions. The outlet tip of either type of buret should be of such diameter and taper as to permit the delivery of a drop the volume of which is considerably

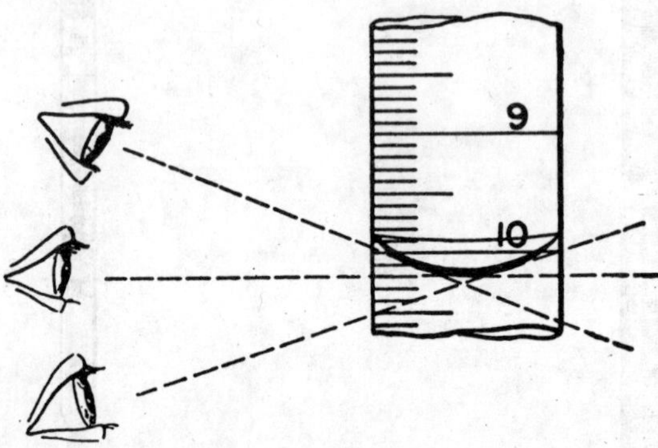

FIG. 11. Effect of parallax in reading burets. Only when the eye is on a level with the meniscus will the reading give no parallactic error. (*Griffin, "Inorganic Quantitative Analysis," McGraw-Hill Book Company, New York,* 1954.)

less than that which can be accommodated between the finest graduations of the scale with which the buret is marked.

Reading Burets. The surface of liquids in narrow tubes is always curved, because of capillarity; the surface or *meniscus* is always concave when the liquid wets the tube and convex when it does not. Readings should always be made at the lowest point of the meniscus, i.e., the graduation mark that coincides with the bottom of the curve, except when measuring highly colored liquids. The eye must be on the same level as the meniscus; otherwise error due to parallax will be introduced (see Fig. 11). Numerous devices have been employed to assist in reading burets correctly. One such device recommended by the National Bureau of Standards is a collar-shaped section of black rubber tubing, cut open at one side and of a size that fits the buret tube so that it can be moved. This black rubber tubing, when placed immediately below the meniscus, renders the meniscus profile dark and clearly observable against a light background.

Pipets. Pipets should be graduated to deliver at standard temperature the volume indicated. Pipets should never be drained by blowing into them unless they are especially graduated for use in this way.

The U.S.P. specifies the following tolerances, accepted by the National Bureau of Standards, for transfer pipets:

Designated volume in ml	1	2	5	10	25	50	100
Limit of error in ml	0.006	0.006	0.01	0.02	0.03	0.05	0.08
Limit of error in percent	0.60	0.30	0.20	0.20	0.12	0.10	0.08

It is essential in using pipets that the same procedure be employed as was used in the calibration.

When transfer or delivery pipets (Fig. 12) are used, they are filled to about 1 cm above the mark. Any liquid adhering to the outer surface is carefully wiped off, and the pipet is allowed to drain down just to the mark. The liquid is drawn up by sucking the upper end of the pipet with the mouth, but this procedure should be avoided and specially designed pipets or some other method of measurement employed when volatile, corrosive, or poisonous liquids are measured.

FIG. 12. *a.* Transfer or delivery pipet. *b.* Measuring pipet.

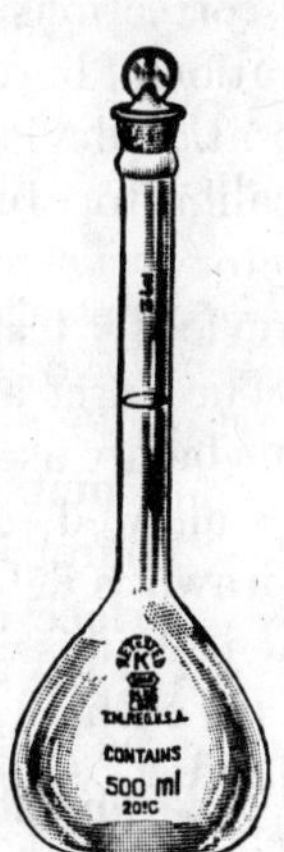

FIG 13. Volumetric flask.

Volumetric Flasks. Standard measuring flasks are calibrated to contain the indicated volume, at 25°, when filled to the mark. The U.S.P. requirements for measuring, or volumetric flasks (Fig. 13) are as follows:

Designated volume in ml	10	25	50	100	250	500	1,000
Limit of error in ml	0.02	0.03	0.05	0.08	0.12	0.15	0.30
Limit of error in percent	0.20	0.12	0.10	0.08	0.05	0.03	0.03

Volumetric flasks are used to make up standard solutions to a given volume. Note that volumetric flasks are graduated to contain, and not to deliver, the calibrated volume.

Graduated Cylinders. Graduated cylinders are used in making approximate measures of volume. They are graduated to contain a given volume of liquid at standard temperature. The inside diameter of the cylinder should not exceed one-fifth of the graduated length.

Cleaning Volumetric Apparatus. New volumetric apparatus should always be cleaned before using, and that in use should be cleaned as often as necessary. The adherence of droplets to the wall of a buret or pipet is positive evidence that the apparatus is dirty. In a clean instrument, the liquid drains down uniformly, wetting the walls so that no droplets are observable. A warm cleaning solution (sodium dichromate in sulfuric acid), solution of sodium hydroxide, or solution of soap powder is the best cleansing agent to use. Hot solutions should be avoided when cleaning accurately calibrated apparatus because of the possible production of a permanent change in volume caused by the heat and known as *thermal aftereffect.*

Calibration of Volumetric Apparatus. The ordinary volumetric apparatus marketed is calibrated at various temperatures and is often unreliable. The analyst should always check the calibration of volumetric apparatus and make corrections if necessary.

Burets. The calibration of burets may be performed in various ways by means of special calibration devices, such as the Ostwald calibrating pipet and the Kiehl buret calibrator. In all practical work, however, the following method is satisfactory:

The clean buret, previously tested for leakage, is filled to the zero mark with distilled water. The buret and water used should be in the room a sufficient length of time before use to acquire room temperature. A definite volume, e.g., 10 ml, is allowed to run into a tared weighing bottle. The weight of water withdrawn is determined, and the buret is filled again to the zero mark, and then 20 ml is withdrawn in exactly the same way. This operation is repeated for each 10 ml until the 50-ml mark is reached; the weight of water withdrawn is determined each time. The results with corrections should be recorded in tabular form showing the relation between the apparent volume and actual weight delivered to each mark on the buret corresponding to an interval of 10 ml. Weighings need not be more accurate than to the second decimal place, since that is the greatest degree of accuracy with which the buret can be read.

A buret calibrated at 20° should deliver 9.9718 Gm of water, or at 25° it should deliver 9.9604 Gm for each 10 ml, since 1 liter of distilled water at 20° weighs 997.177 Gm and at 25°, 996.04 Gm when weighed with brass weights in air of 50 percent humidity at standard atmospheric pressure. The true volume for each 10-ml segment of the buret may be calculated from the weights obtained and recorded on a convenient chart.

Pipets. Clean the pipet thoroughly with soap solution and cleaning mixture and rinse well with distilled water. Fill it with distilled water at 20 or 25°, according to the temperature selected for calibration, and allow it to drain into a tared weighing bottle. Weigh accurately and record the correct volume of the pipet.

Flasks. Clean, dry, and tare the flask, placing the number of gram weights corresponding to its volume on the opposite balance pan, and introduce water at 20 or 25° into the flask in sufficient quantity to counterbalance the added weights. Mark the lowest point of the meniscus if it does not agree with the original calibration.

Sources of Error in the Use of Volumetric Apparatus. Every precaution should be taken to avoid the following common sources of error in the use of volumetric apparatus:

1. Rinse water adhering to walls of apparatus. The apparatus may be dried, but it is more convenient and just as accurate to wash it out with small successive portions of the liquid to be used, discarding the washings.
2. Grease films and dirty apparatus cause irregularity in the delivery of liquids and distort the meniscus.
3. Drainage or afterflow of the liquid adhering to the vessel walls to the liquid below necessitates that the time of drainage be the same as that used in the calibration of the instrument.
4. Parallax must be avoided to secure proper readings of the level of the meniscus.
5. Variations in temperature lead to changes in volume of vessels and liquids. All measurements, therefore, should be made at a temperature closely approximating that at which the apparatus was calibrated.
6. Air bubbles trapped beneath the liquid surface, especially below the stopcock in burets, displace liquid.
7. Heat, such as supplied by hot solutions, causes calibrated volumetric apparatus to suffer a slight permanent change of volume.
8. Most salts when dissolved produce a change in temperature, with concomitant change in volume of the solution. Such solutions should never be measured until they have acquired the temperature at which the instruments used to measure them were calibrated.
9. Failure to use apparatus in a manner approaching as nearly as possible that followed in calibration.

Questions and Problems

1. Define the following terms: (*a*) *standardization;* (*b*) *standard solution;* (*c*) *end point;* (*d*) *residual titration;* (*e*) *primary standard;* (*f*) *equivalent;* (*g*) *equivalent weight.*
2. What should be the relation between end point and stoichiometric point?
3. Burets provided with rubber tubes and pinchcocks in place of glass stopcocks

may be used to measure out sodium or potassium hydroxide solution, but they should never be used in the measurement of solutions of acids or oxidizing agents. Why?

4. A 100-ml volumetric flask calibrated at 25° is used to measure water at 20°. What correction of the volume must be made?

5. A volumetric flask calibrated to contain 100 ml at 25° and 760 mm barometric pressure is used at 18° and 746 mm barometric pressure. Calculate the volume of water contained in the flask at the temperature and pressure at which it is used.

NEUTRALIZATION METHODS—ACIDIMETRY AND ALKALIMETRY

Acidimetry is the measurement of the quantity of base in a given sample by titration with a suitable acid. Conversely, *alkalimetry* is the measurement of the quantity of acid in a given sample by titration with a suitable base. In each case the principle involved is the same, viz., the addition of a chemically equivalent amount of a standard solution of a base or acid to the sample of acid or base, respectively, being assayed. From the volume or the known number of equivalents of standard solution added, the weight of an equal number of equivalents of acid or base present in the sample can be readily calculated. The equivalent weight of the acid or base being assayed can be obtained by examination of the formula of the acid or base or the balanced equation for the reaction involved.

Before undertaking analyses by methods of acidimetry and alkalimetry, the student should become familiar with the theoretical considerations involved.

Titer. Interpretation of the data obtained in titrimetric analysis is facilitated considerably by use of a chemical factor known as the *titer* of a solution, corresponding to the *gravimetric factor* used in gravimetric analysis. *Titer* may be defined as the weight of solute per milliliter of a solution or, more commonly, as the weight of a substance chemically equivalent to 1 ml of a standard solution. Thus 1 ml of 1 N hydrochloric acid contains 0.03646 Gm (0.001 equivalent or 1 mEq) of HCl and is, therefore, equivalent chemically to 0.04000 Gm (0.001 equivalent or 1 mEq) of NaOH.

Titers of standard solutions used in official assay procedures are specified in the various monographs. For example, in the U.S.P. procedure for the assay of calcium hydroxide solution for $Ca(OH)_2$, it is stated: "Each milliliter of 0.1 N hydrochloric acid is equivalent to 3.705 mg. of $Ca(OH)_2$." The $Ca(OH)_2$ titer of 0.1 N hydrochloric acid is, therefore, 3.705 mg per ml, a value that can be calculated as follows: Examination of the formula, $Ca(OH)_2$, or of the equation,

$$\underset{74.09}{Ca(OH)_2} + 2HCl \longrightarrow CaCl_2 + 2H_2O$$

indicates that 1 mole or 74.09 Gm of $Ca(OH)_2$ is 2 equivalents, and the equivalent weight of $Ca(OH)_2$ is 37.05 Gm. Each milliliter of 0.1 N

hydrochloric acid contains 0.0001 equivalent of HCl and is equivalent to 0.0001 equivalent or $0.0001 \times 37.05 = 0.003705$ Gm or 3.705 mg of $Ca(OH)_2$.

Theory. For the purposes of acidimetry and alkalimetry, an *acid* may be defined as an ionic or molecular substance capable of giving up a proton, i.e., a proton donor. A *base* may be defined as an ionic or molecular substance capable of uniting with a proton, i.e., a proton acceptor. For example, H_2O, H_2S, HCl, and H_2SO_4 are molecular acids; NH_3 and H_2O are molecular bases; H_3O^+ and NH_4^+ are cationic acids; $Ag(NH_3)_2^+$ is a cationic base; HSO_4^-, $H_2PO_4^-$, $HPO_4^=$, and HS^- are anionic acids; and OH^-, SH^-, $S^=$, and $SO_4^=$ are anionic bases.

In the water system,

$$2H_2O \rightleftharpoons H_3O^+ + OH^-$$

acid properties, such as taste, effect on indicators, and oxidation of metals high in the electromotive series, depend on the presence of the hydronium ion, H_3O^+. It is generally written as a hydrogen ion, H^+, for convenience. Substances which, on dissolving in water, increase the concentration of the hydronium ion also increase the acid properties of the solution, and are, therefore, acids.

$$HCl + H_2O \longrightarrow H_3O^+ + Cl^-$$
$$NH_4^+ + H_2O \rightleftharpoons H_3O^+ + NH_3$$
$$HSO_4^- + H_2O \rightleftharpoons H_3O^+ + SO_4^=$$

Monoprotic acids are those which are capable of giving up one proton per molecule or ion; diprotic, two protons, etc. The tendency of an acid to give up a proton is the principal determinant of the strength of the acid. Many of the mineral acids appear to have equal strengths in aqueous solution because water is a sufficiently strong base for the following reaction to go practically to completion:

$$HA + H_2O \rightleftharpoons H_3O^+ + A^-$$

In such cases, water is said to have a leveling effect. When weaker acids are dissolved in water, the above reaction is reversible.

Proton acceptors, or bases, when dissolved in water, reduce the concentration of hydronium ions, as illustrated by the following:

$$OH^- + H_3O^+ \rightleftharpoons 2H_2O$$
$$NH_3 + H_3O^+ \rightleftharpoons NH_4^+ + H_2O$$
$$SO_4^= + H_3O^+ \rightleftharpoons HSO_4^- + H_2O$$

Those bases which can accept one proton are monoprotic; those which can accept two protons are diprotic, etc. The stronger a base is, the greater is its tendency to accept protons.

Since a water molecule may donate or accept a proton, it is both an acid and a base. Thus, two water molecules may interact to form ions: $H_2O + H_2O \rightleftharpoons H_3O^+ + OH^-$. When there is simply a transfer of a proton from one molecule to another, the process is called *protolysis*, or is referred to as a *protolytic reaction*. An acid and a base related to each other so that one is formed from the other by the gain or loss of a proton are said to be corresponding or *conjugate to one another*. Thus Cl^- is the *conjugate base* of HCl, CH_3COO^- is the conjugate base of CH_3COOH, and NH_3 is the conjugate base of the acid NH_4^+. Similarly, HCl is the conjugate acid of Cl^-, CH_3COOH is the conjugate acid of CH_3COO^-, and NH_4^+ is the conjugate acid of NH_3. A proton does not tend to leave an acid unless a base is present to accept the proton. The fundamental acid-base relationships may be expressed as follows:

$$\text{Acid} \rightleftharpoons H^+ \text{ (proton)} + \text{base}$$

The transfer of a proton from an acid results in the formation of a new acid and a new base:

$$\text{Acid} + \text{base} \rightleftharpoons \text{acid}' + \text{base}'$$
$$HCl + H_2O \longrightarrow H_3O^+ + Cl^-$$
$$CH_3COOH + OH^- \rightleftharpoons H_2O + CH_3COO^-$$
$$H_2O + NH_3 \rightleftharpoons NH_4^+ + OH^-$$

The weakly basic chloride ion loses its proton to the more strongly basic water molecule. The acetate ion, being a stronger base than the water molecule, retains most of its protons. The ammonia accepts a proton from the water molecule to form the acidic ammonium ion.

For the purpose of brevity, the term *hydrogen ion* and the symbol H^+ are often used in this book for the hydronium ion.

The hydrogen ion concentration influences most reactions employed in analytical methods, whether they are neutralization, precipitation, or oxidation-reduction reactions.

It has been established that in any aqueous solution there are always hydrogen and hydroxide ions present and that, whatever the concentration of the one, the concentration of the other is such that the product of their concentrations is a constant for any given temperature.

$[H^+] \times [OH^-]$ = a constant for any given temperature, and in pure water at about 22°, $[H^+] \times [OH^-] = 1 \times 10^{-14}$. Since this relationship is true, the concentration of hydrogen ions $[H^+]$ expressed in terms of gram-ions per liter in pure water is equal to the square root of $1 \times 10^{-14} = 1 \times 10^{-7}$ (0.0000001), and the hydroxyl ion concentration $[OH^-]$ also is equal to 1×10^{-7}. The degree of acidity or alkalinity of a solution can, therefore, be defined in terms of its concentration of hydrogen ion or hydroxide ion, since when the value of one is known, the value of the other can be calculated; i.e., if $[H^+] = 10^{-5}$, then $[OH^-] = 10^{-9}$, since $[H^+] \times [OH^-]$ must equal 10^{-14}. Solutions which have a hydrogen ion concentration greater than 10^{-7} are acidic, and those which have a hydrogen ion concentration less than 10^{-7} are alkaline. Thus a solution of 0.01 *N* HCl contains 10^{-2} gram-ion of hydrogen per liter, assuming complete dissociation, and is acid in reaction although it contains 10^{-12} gram-ion of OH^-.

To avoid the disadvantage of employing a dual system of definition, it is customary to use only the hydrogen ion concentration to characterize a solution, whether it be acidic, neutral, or alkaline, the hydroxide ion concentration being implied.

pH Value. The reaction (acidity, alkalinity, or neutrality) of a solution may be defined not only in terms of the molecular concentrations of hydrogen ions, numbers frequently so small as to be quite cumbersome, but also in pH values, a system suggested by Sörensen and based on the relationship between the electromotive force and hydrogen ion concentration of certain cells. An important factor in this relationship is the logarithm of the reciprocal of the hydrogen ion concentration (see Chap. 22). This factor has become known as the pH value or simply as pH.

pH is defined as the logarithm of the reciprocal of the hydrogen ion concentration, and is expressed as

$$\text{pH} = \log \frac{1}{[H^+]}$$

Since the logarithm of a quotient equals the difference between the logarithm of the dividend and the logarithm of the divisor,

$$\begin{aligned} \text{pH} &= \log (1) - \log ([H^+]) \\ &= 0 - \log ([H^+]) \\ &= - \log ([H^+]) \end{aligned}$$

Values for pH can be calculated from hydrogen ion concentrations as in the following examples:

1. $[H^+] = 0.0001 = 10^{-4}$ gram-ion per liter

$$\begin{aligned} \text{pH} &= - \log ([H^+]) \\ \text{pH} &= - \log (10^{-4}) \\ &= - (-4) \\ &= 4 \end{aligned}$$

2. $[H^+] = 2.73 \times 10^{-4}$ gram-ion per liter

$$\begin{aligned} \text{pH} &= - \log (2.73 \times 10^{-4}) \\ &= - [\log (2.73) + \log (10^{-4})] \\ &= - \log (2.73) - \log (10^{-4}) \\ &= - 0.4362 + 4 \\ &= 3.5638 \text{ or } 3.56 \end{aligned}$$

Conversely, when the pH value is known, the hydrogen ion concentration can be calculated as in the following examples:

1.
$$pH = 4$$
$$-\log [H^+] = 4$$
$$\log [H^+] = -4$$
$$[H^+] = 10^{-4}$$

2.
$$pH = 3.56$$
$$-\log [H^+] = 3.56$$
$$\log [H^+] = -3.56$$
$$= -4 + 0.44$$
$$[H^+] = 10^{-4} \times 2.75$$

Two practical methods for determining when the hydrogen ion concentration in a solution has acquired a certain value have been developed, viz., the potentiometric method and the colorimetric method (see Chap. 22).

Buffers. In many analytical procedures, a change in the hydrogen ion concentration occurs as a result of chemical reaction. In certain cases, this change in the concentration of hydrogen ion may prevent the reaction from going to completion quantitatively. Since the change in hydrogen ion concentration is caused by the formation of acid or base in the course of the reaction, some provision must be made to eliminate the acid or base formed in such a manner as to maintain a sensibly constant hydrogen ion concentration. This is usually accomplished by means of a buffer. The term *buffer* is applied alike to the salts of weak acids and to the salts of weak bases. Solutions of these salts resist change in hydrogen ion concentration on the addition of small amounts of either acid or base. In the titration of acetic acid with sodium hydroxide, the sodium acetate formed acts as a buffer to prevent a large change in hydrogen ion concentration. Acetic acid in aqueous solution is slightly dissociated, as shown in the following equations:

$$HC_2H_3O_2 \rightleftharpoons H^+ + C_2H_3O_2^- \quad (1)$$
$$Na^+OH^- + H^+ \longrightarrow Na^+ + H_2O \quad (2)$$
$$HC_2H_3O_2 + Na^+OH^- \rightleftharpoons Na^+C_2H_3O_2^- + H_2O \quad (1) + (2)$$

In reaction (1), the slightly dissociated acetic acid forms H^+ and $C_2H_3O_2^-$ which combine with OH^- and Na^+, respectively, to form water and sodium acetate [Eq. (2)]. In the equilibrium mixture, both $HC_2H_3O_2$ and $C_2H_3O_2^-$ are present. If $[H^+]$ is reduced by addition of more NaOH or other base, the $HC_2H_3O_2$ dissociates as in (1) and thus prevents a lasting change in $[H^+]$. On the other hand, if an acid such as HCl is added to the mixture, $[H^+]$ does not increase greatly because of the combination of H^+ with $C_2H_3O_2^-$ [reverse of reaction (1)]. This buffering action, preventing any great change in $[H^+]$, will continue on further addition of acid or base as long as $HC_2H_3O_2$ or $Na^+C_2H_3O_2^-$ is present.

The principle of buffer action is often employed in analytical methods in which acid or base is generated as a result of the analytical reaction. In such cases, the addition of an amount of buffer salt relatively large with respect to the amount of acid or base generated results in a fractional diminution of the buffer ion through its union with the hydrogen ion to form the slightly dissociated acid corresponding to the buffer salt employed. The diminution in the amount of buffer salt will then be so slight that the hydrogen ion concentration will remain practically constant. In iodimetric assays where HI is formed as a product of the reaction, sodium bicarbonate is used as a buffer to maintain an approximately constant hydrogen ion concentration.

Neutralization Reactions. It is important to remember that, in terms of the theory of ionization, a neutral solution contains H^+ and OH^- in the same concentration as in pure water, viz., 0.0000001 gram-ion per liter of solution. Any substance which can furnish excess H^+ when dissolved in water is acidic, and any substance which can furnish excess OH^- when dissolved in water is basic. From the point of view of the theory of ionization, therefore, *neutralization* is the union of H^+ from an acid with OH^- from a base, to form molecules of water. The other ions furnished by the acid or base may or may not combine.

Neutralization reactions must proceed very nearly to completion to be of value in quantitative analysis. Reactions may be caused to go to completion in a number of ways, i.e., by the formation of a slightly dissociated substance as a reaction product; by the removal of one or more of the products of the reaction as a gas; by the removal of one or more of the products of reaction as a precipitate; by the removal of an ionic substance as a complex ion; by changing the charge of an ion; and by adding an excess of one of the reactants.

Solutions of HCl and NaOH in water are dispersions of ions (100 percent ionized), since each of these substances is a strong electrolyte. When two such solutions are mixed, the hydrogen ions of the HCl solution combine with the hydroxide ions of the NaOH solution, forming the very slightly ionized water molecules and leaving an equivalent amount of sodium ions and chloride ions dispersed in the solution as sodium chloride.

$$H^+Cl^- + Na^+OH^- \longrightarrow Na^+Cl^- + H_2O \qquad (3)$$

The neutralization reaction indicated by Eq. (3) goes to completion. In titrating a solution of HCl with a solution of NaOH, an excess of hydrogen ions will be present because of the HCl until an equivalent amount of NaOH has been added. If the stoichiometric point and the end point coincide, hydrogen ions would be present only as a result of dissociation of water molecules and

$$[H^+] = [OH^-] = 1 \times 10^{-7} \qquad \text{or} \qquad pH = 7$$

It is not practical, however, in this titration, to avoid addition of a very slight excess of NaOH, with the result that the pH at the end point is generally considerably higher than 7.

Admixture of solutions of acetic acid, $HC_2H_3O_2$, and NaOH results in an equilibrium mixture at the stoichiometric point in which are present the substances formed by neutralization, viz., $Na^+C_2H_3O_2^-$ and the very slightly ionized H_2O; and those formed by the reverse reaction, hydrolysis, viz., Na^+OH^- and slightly ionized $HC_2H_3O_2$.

$$Na^+OH^- + HC_2H_3O_2 \rightleftharpoons Na^+C_2H_3O_2^- + H_2O$$

This reaction does not go to completion, and there is an excess of OH^- ions at the stoichiometric point when acetic acid is titrated with sodium hydroxide. Such a reversible reaction, however, may be used as the basis for a quantitative analytical procedure, provided that the end point of the titration can be brought close to the stoichiometric point, e.g., by proper choice of indicator.

Indicators. Indicators are complex organic compounds that are used (1) to determine the end point in neutralization processes, (2) to determine hydrogen ion concentrations $[H^+]$ or pH, or (3) to indicate that a desired change in pH has been effected. These organic substances may be acids or bases themselves, capable of existing in two forms of different color that are mutually convertible, one into the other, at given hydrogen ion concentrations. To explain the change in color of indicators, three theories have been proposed which, briefly stated, are as follows:

1. The physicochemical theory attributes the color to certain ions an increase of which causes the appearance of a new color, and the decrease of which causes the disappearance of a color or the appearance of a different color.

2. The organic theory attributes the color of indicators to certain groupings of the elements in a compound, and the change in color to change in molecular structure.

3. The colloidal theory assumes that indicators form colloidal solutions the change in color of which is dependent upon change in size of the colloidal particle.

The point at which an indicator changes color in any given titration is dependent on the hydrogen ion concentration of the solution and may not be indicative of the absolute neutrality or completion of a reaction. Table 7, showing the commonly used indicators, gives the $[H^+]$ expressed in terms of pH between which color changes occur.

This table shows that methyl orange exhibits its acid color, pink, at a pH of 3.2 and its alkaline color, yellow, at a pH of 4.4, while between these pH values the color undergoes transition from one shade to the other. At a pH of 4.4 the solution is slightly acid, since at neutrality pH = 7.

It would appear that an indicator which changes color exactly at the neutral point, pH = 7, would be required in every case, but this is not true. When a strong acid, such as hydrochloric acid, is titrated with a strong base, like sodium hydroxide, the change in concentration of hydrogen ion becomes very rapid as the equivalence point is approached, so that very small amounts of acid or alkali in excess carry the concentration of hydrogen ion far to one side of the neutral point, and the indicator will give a sharp end point. If 25 ml of a solution of 0.1 *N* HCl is titrated with a solution of 0.1 *N* NaOH, the equivalent point would be reached when 25 ml of 0.1 *N* NaOH had been added. Theoretically, the solution at the end of titration should be identical in composition with a solution prepared

TABLE 7. COMMONLY USED pH INDICATORS

Indicator	pH range	Color	
		Acid	Alkaline
Methyl yellow	2.9 to 4.0	Red	Yellow
Bromophenol blue	3.0 to 4.6	Yellow	Blue
Methyl orange	3.2 to 4.4	Pink	Yellow
Bromocresol green	4.0 to 5.4	Yellow	Blue
Methyl red	4.2 to 6.2	Red	Yellow
Bromocresol purple	5.2 to 6.8	Yellow	Purple
Bromothymol blue	6.0 to 7.6	Yellow	Blue
Phenol red	6.8 to 8.2	Yellow	Red
Cresol red	7.2 to 8.8	Yellow	Red
Thymol blue	8.0 to 9.2	Yellow	Blue
Phenolphthalein	8.0 to 10.0	Colorless	Red
Thymolphthalein	8.6 to 10.0	Colorless	Blue

by dissolving a known weight of sodium chloride in the calculated amount of solvent. When methyl orange is used as indicator, the end point as shown by change of color appears when about 24.95 ml of 0.1 *N* NaOH has been added or while the solution is still slightly acidic, but if phenolphthalein is used as indicator, the end point appears when about 25.05 ml of 0.1 *N* NaOH has been added or when the solution is slightly alkaline. The deviation of the end point from the equivalence point is so slight that it is negligible in this case.

When a weak acid is titrated with a strong base, the change in hydrogen ion concentration in the vicinity of the equivalence point is not so abrupt; e.g., if 25 ml of 0.1 *N* acetic acid is titrated with 0.1 *N* NaOH, 25 ml of the alkali should be required to attain the equivalence point. When methyl orange is employed as indicator, the end point is reached when about 17.5 ml of 0.1 *N* NaOH has been added, but if phenolphthalein is used, the end point is reached when practically 25 ml of 0.1 *N* NaOH has been added. Obviously methyl orange would not be a satisfactory indicator in this titration.

If 0.1 ml of 0.1 *N* HCl is added to a liter of 0.1 *N* NaCl, the $[H^+]$ changes from 10^{-7}, the concentration in dilute solutions of neutral salts, to 10^{-4}, the concentration of a 0.0001 *N* HCl solution. If 0.1 ml of 0.1 *N* HCl or NaOH is added to a liter of 0.1 *N* sodium acetate, no appreciable change in $[H^+]$ takes place. Likewise, in the titration of acetic acid with sodium hydroxide, the acetic acid–sodium acetate acts as a buffer solution to prevent large changes in $[H^+]$. The acetate ions from the highly ionized sodium acetate force the reaction $CH_3COOH \rightleftharpoons H^+ + CH_3COO^-$ to the left before an equivalent quantity of NaOH has been added, and the hydrogen ion concentration becomes very small before all the hydrogen of acetic acid has been replaced. Methyl orange, which changes color at a relatively high hydrogen ion concentration, $10^{-4.4}$, will gradually show alkaline color long before the equivalent quantity of alkali has been added, and the end point will not be sharp. Phenolphthalein, however, does not change color until the true neutral point pH = 7 has been passed, and a drop of alkali will then reduce the concentration below $[H^+] = 10^{-9.8}$ or pH = 9.8 where phenolphthalein exhibits a change of color.

A weak base titrated by a strong acid, e.g., ammonium hydroxide titrated with sulfuric or hydrochloric acid, exhibits the reverse of the above condition, as would be expected, and methyl red is a suitable indicator whereas phenolphthalein is not.

Rules for the Use of Indicators. The following rules should be observed in the use of indicators:

1. Use 3 drops of indicator test solution for a titration unless otherwise directed, and add generally as the end point is approached.
2. When a strong acid is titrated with a strong alkali, or a strong alkali with a strong acid, then methyl orange, methyl red, or phenolphthalein may be used.
3. When a weak acid is titrated with a strong alkali, use phenolphthalein as indicator.
4. When a weak alkali is titrated with a strong acid, use methyl red as the indicator.
5. A weak alkali should never be titrated with a weak acid or vice verse, since no indicator will give a sharp end point.
6. The appearance of a color is more easily observable than is the disappearance. Always titrate where possible, therefore, to the appearance of a color.

Mixed Indicators. The criteria of a good indicator for analytical purposes are that there will be a sharp contrast between the two colors which it exhibits in acid and alkaline mediums and that the change in color will take place over a very small range of pH values. Some indicators do not give sharp color changes; in such cases mixed indicators are used to sharpen up the color change. Methylene blue exhibits a color that is complementary to that of methyl red in its red form so that a gray color exists over a narrow range of pH rather than the pink color which is difficult to duplicate. Other combinations of mixed indicators are xylene cyanol–methyl orange, bromocresol green–methyl red, bromocresol green–chlorphenol red, cresol red–thymol blue, and thymol blue–phenolphthalein.

Commonly Used Indicator Solutions

Bromocresol Green T.S. A solution of 50 mg of bromocresol green in 100 ml of alcohol; filter if necessary.

Bromocresol Purple T.S. Dissolve 250 mg of bromocresol purple in 20 ml of 0.05 *N* sodium dydroxide, and dilute to 250 ml.

Bromophenol Blue T.S. A solution of 100 mg of bromophenol blue in 100 ml of diluted alcohol, and filter if necessary.

Bromothymol Blue T.S. Dissolve 100 mg of bromothymol blue in 100 ml of diluted alcohol, and filter if necessary.

Cresol Red T.S. Triturate 100 mg of cresol red in a mortar with 26.2 ml of 0.01 *N* sodium hydroxide until solution is complete, then dilute the solution to 250 ml.

Methyl Orange T.S. A solution of 100 mg of methyl orange in 100 ml of water, and filter if necessary. The color change is sharp unless too much indicator is used. Methyl orange is a good indicator to use in the titration of the salts of weak acids, such as carbonates, borates, sulfides, etc., with a strong acid, since in these cases the acids liberated in the titration reaction are too little ionized to affect the indicator. It does not give the proper end point in the titration of alkaloids or organic acids. This indicator should never be employed in the titration of alcoholic solutions, hot solutions, or very dilute solutions.

Methyl Red T.S. A solution of 100 mg of methyl red, *p*-dimethylaminoazobenzene-*o*-carboxylic acid, in 100 ml of alcohol, filtered if necessary. Two drops of the indicator solution per 100 ml should be used in titrations. Methyl red T.S. is especially useful in titrating solutions of ammonia, other weak bases, and alkaloids, but it is not suitable for the titration of weak organic acids.

Phenol Red T.S. A solution of 100 mg of phenol red, phenolsulfonphthalein, in 100 ml of alcohol, filtered if necessary.

Phenolphthalein T.S. A solution of 1 Gm of phenolphthalein in 100 ml of alcohol. Two to three drops of the indicator solution per 100 ml is used in titrations unless otherwise directed. Phenolphthalein is an excellent indicator to use in titration of weak acids, most alkalies, and alkali salts, but is not satisfactory for titration of ammonia, alkaloids, or cold solutions of carbonates and bicarbonates. It functions well in alcoholic solutions.

Thymol Blue T.S. A solution of 100 mg of thymol blue, thymolsulfonphthalein, in 100 ml of alcohol, filtered if necessary.

Thymolphthalein T.S. A solution of 100 mg of thymolphthalein in 100 ml of alcohol, filtered if necessary.

Sensitivity of Indicator Solutions. Indicator solutions should be tested for sensitivity. This is done by adding 0.15 ml of indicator solution to 25 ml of carbon dioxide–free distilled water and titrating the resulting solution with 0.02 N alkali or acid. Upon the addition of 0.25 ml of the standard alkali or acid solution, the respective characteristic color should develop.

Questions and Problems

1. Define the following terms: acidimetry, titer, milliequivalent, pH, and buffer.
2. Calculate the pH of each of the following solutions: (*a*) 0.025 M hydrochloric acid; (*b*) 0.01 M sodium hydroxide; (*c*) 0.1 M acetic acid, $K_i(HC_2H_3O_2) = 1.8 \times 10^{-5}$; (*d*) 2 percent sodium acetate W/V.
3. Calculate the hydrogen ion concentration of a solution with a pH of 5.69, and the hydroxide ion concentration of a solution with a pH of 8.31.
4. What indicator is best suited for use in the following titrations? Give reasons. (*a*) Alkali carbonate with sulfuric acid; (*b*) acetic acid with sodium hydroxide; (*c*) ammonia solution with hydrochloric acid; (*d*) organic acids with sodium hydroxide.
5. How may the sensitivity of an indicator solution be determined?

Standard Solutions. The concentrations of standard solutions can be expressed in terms of *normality*, or the number of equivalents of solute per liter of solution (milliequivalents per milliliter), or in terms of *molarity*, or the number of moles of solute per liter of solution. In certain types of calculation, the *molality*, or the number of moles of solute per 1,000 Gm of solvent, is used.

A tenth-normal solution of hydrochloric acid, also designated as 0.1 N or $N/10$, contains 0.1 equivalent or 0.1 equivalent weight, 3.646 Gm, of HCl per liter, or 0.1 mEq, 3.646 mg, per ml. Obviously the HCl titer of 0.1 N HCl is 3.646 mg. A number of other titer values are given in the official monographs of bases assayed by use of 0.1 N HCl, e.g., 3.705 mg of calcium hydroxide, etc. Such titer values are of use in the interpretation of analytical data obtained using 0.1 N HCl.

A twentieth-molar (0.05 M) solution of sulfuric acid contains 0.05 mole or 4.904 Gm of H_2SO_4 per liter. Since 1 mole of H_2SO_4 is 2 equivalents, 0.05 M H_2SO_4 contains $2 \times 0.05 = 0.1$ equivalent per liter and is, therefore, 0.1 N. The normality of a solution can be calculated from its molarity by multiplying the latter by the number of equivalents per mole of the solute.

The importance of a correct evaluation of equivalents in terms of moie fractions or grams in all calculations involving standard solutions may be seen in the following examples:

1. The preparation of 1 liter of N H_2SO_4 requires 49.04 Gm of H_2SO_4, since 1 equivalent of H_2SO_4 is needed and since 1 mole or 98.08 Gm of H_2SO_4 is 2 equivalents.
2. Ten percent H_2SO_4 is 1.02 M or 2.04 N; since 1 liter contains 100 Gm, $100/49.04 = 2.04$ equivalents, or $2.04/2 = 1.02$ moles.
3. Each milliliter of 0.1 N H_2SO_4 is equivalent to 1.703 mg of NH_3, since each milliliter of the solution contains 0.0001 equivalent of H_2SO_4,

which is equivalent to $0.0001 \times 17.03 = 0.001703$ Gm or 1.703 mg of NH_3. In this example, typical of the calculation of the titer of a solution, 17.03 Gm was taken as the equivalent weight of NH_3 for the reason indicated by the following equation:

$$\underset{17.03}{NH_3} + H^+ \longrightarrow NH_4^+$$

Calculations in Volumetric Analysis. Interpretation of the data obtained in volumetric analysis is facilitated considerably by the use of milliequivalent factor (mEq wt) corresponding to the gravimetric factor used in gravimetric analysis. It may be defined as the weight of substance chemically equivalent to 1 ml of a 1 *N* solution. In practical analytical work, it is convenient to use a solution which has a normality that may be approximately that called for in the analytical procedure. The exact normality of the solution employed, usually expressed to the fourth decimal place, should, however be known.

For example, a 50.0-ml sample of calcium hydroxide solution required 19.50 ml of a 0.1050 *N* hydrochloric acid solution to neutralize it. Calculate the amount of $Ca(OH)_2$ in 100 ml of the solution (percent *W/V*).

$$Ca(OH)_2 + 2HCl \longrightarrow CaCl_2 + 2H_2O$$

The equation indicates that 1 mole or 74.09 Gm of $Ca(OH)_2$ contains 2 equivalents; the equivalent weight of $Ca(OH)_2$ is 37.05 Gm and the mEq weight is 0.03705 Gm. Then 1 ml of 1 *N* hydrochloric acid is equivalent to 0.03705 Gm of $Ca(OH)_2$. $19.50 \times 0.1050 =$ milliequivalents of HCl required to neutralize a like number of milliequivalents of $Ca(OH)_2$.

$$19.50 \times 0.1050 \times 0.03705 = \text{Gm of } Ca(OH)_2\text{/50-ml sample}$$

$$19.50 \times 0.1050 \times 0.03705 \times \frac{100}{50} = 0.152 \text{ Gm/100 ml}$$

A general formula which can be used for the calculation of the percent purity of a sample determined by direct titration is:

$$\frac{\text{ml} \times N \times \text{mEq wt} \times 100}{\text{Sample weight or volume}} = \text{percent } W/W \text{ or } W/V$$

Titers of the standard solutions used in the official assay procedures in the U.S.P. and the N.F. are specified in the various monographs. For example, in the problem cited above, the U.S.P. assay states: "Each milliliter of 0.1 *N* hydrochloric acid is equivalent to 3.705 mg. of $Ca(OH)_2$." To use this titer value in this problem the normality of the acid solution used must be converted to exactly 0.1000 *N* by use of a correction factor (c.f.), e.g., 19.50×1.050 (c.f.) $= 20.48$ ml of a 0.1000 *N* solution, and $19.50 \times 1.050 \times 3.705 = 75.878$ mg $Ca(OH)_2$/50-ml sample.

The standard acid solutions used in acidimetry and alkalimetry are usually prepared from hydrochloric or sulfuric acid. Either acid may be used in most titrations, but hydrochloric is preferable to sulfuric acid in the titration of compounds that yield a precipitate with the latter, such as barium hydroxide, and sulfuric acid is preferable in hot titrations, since there would be danger of loss of hydrochloric acid due to volatilization.

The standard alkali solutions commonly used are sodium hydroxide, potassium hydroxide, and barium hydroxide. These solutions absorb carbon dioxide from the air, thereby changing rapidly in concentration. Sodium and potassium hydroxides may become contaminated with carbonates that impair their usefulness because of the liberation of carbon dioxide during acid-alkali titration. Barium hydroxide solutions remain free from carbonates, since absorbed carbon dioxide is precipitated as insoluble barium carbonate; this, however, decreases the concentration of barium hydroxide in the solution. Alkali solutions should be prepared carbonate-free, and they should be protected from carbon dioxide by means of soda-lime absorption tubes. All alkali solutions should be restandardized frequently.

Preparation and Standardization of Acid Solutions. Normal hydrochloric acid solution may be standardized titrimetrically against pure sodium carbonate or standard solution of sodium hydroxide which has been recently standardized, or gravimetrically by precipitation of an equivalent amount of silver chloride.

Exercise 9

Object. To Prepare and Standardize 1 *N* Hydrochloric Acid.

Materials Required. 85 ml of hydrochloric acid
1.5 gm of sodium carbonate

Procedure. "Dilute 85 ml. of Hydrochloric Acid to 1000 ml. Standardize as follows: Accurately weigh about 1.5 Gm. of primary standard anhydrous sodium carbonate that has been heated at a temperature of about 270° for 1 hour. Dissolve it in 100 ml. of water, and add 2 drops of methyl red T.S. Add the acid slowly from a buret, with constant stirring, until the solution becomes faintly pink. Heat the solution to boiling, and continue the titration until the faint pink color is no longer affected by continued boiling. Calculate the normality. Each 52.99 mg. of anhydrous sodium carbonate is equivalent to 1 ml. of 1 *N* hydrochloric acid."

Since hydrochloric acid, 1.18, contains about 36.5 percent *W*/*W* if HCl, the solution prepared according to these directions should be approximately 1 *N*. One liter of 1 *N* hydrochloric acid should contain 36.46 Gm or 1 equivalent of HCl. The above solution would contain $85 \times 1.18 \times 0.365 = 36.61$ Gm of HCl. After standardization, a solution may be diluted readily to a desired weaker concentration. Sodium carbonate usually contains varying amounts of moisture and sodium bicarbonate. When a sample is heated to 270°, the moisture is driven off and the $NaHCO_3$ is converted to Na_2CO_3.

$$2NaHCO_3 \longrightarrow Na_2CO_3 + H_2O + CO_2\uparrow$$

The titrated solution is heated to boiling to rid it of dissolved CO_2 which has an adverse effect on the color change of the indicator.

Assuming that a sample of pure sodium carbonate weighing 1.6250 Gm required 30.20 ml of the acid in the titration, the normality of the acid might be calculated as follows: The equivalent weight of sodium carbonate is $^{106}\!/_2 = 53$ Gm, because ½ mole of Na_2CO_3 is chemically equivalent to 1 Gm atom of hydrogen as seen in the following equation:

$$\underset{106}{Na_2CO_3} + 2HCl \longrightarrow 2NaCl + H_2O + CO_2\uparrow$$

Therefore, 1.6250 Gm of Na_2CO_3 is 1.6250/0.053 mEq of Na_2CO_3. Since 30.20 ml of hydrochloric acid solution was found by titration to be chemically equivalent to 1.6250 Gm of Na_2CO_3, there is 1.6250/0.053 mEq of HCl in the 30.20 ml of hydrochloric acid solution. The normality may be calculated as follows:

$$\text{ml} \times N = \text{mEq}$$

$$N = \frac{\text{wt of } Na_2CO_3}{\text{ml} \times 0.053}$$

$$N = \frac{1.6250}{30.20 \times 0.053}$$

$$N = 1.0152$$

If an exactly normal solution is wanted, the dilution required may be calculated as follows: Assuming that 925.00 ml of the 1.0152 N solution remains after standardization, the corresponding volume of 1 N solution would be $925 \times 1.0152 = 939.06$ ml and $939.06 - 925.00 = 14.06$ ml of water must be added to the 925.00 ml of 1.0152 N solution to make a 1 N solution. The dilution may also be carried out as follows: 7.49 ml of water is placed in a dry 500-ml volumetric flask, which is then filled to the mark with the 1.0152 N solution. After dilution, the solution should be thoroughly mixed and restandardized.

Solutions of hydrochloric acid may also be standardized gravimetrically by using the procedure for the gravimetric determination of a soluble chloride. The equivalent weight of AgCl is 143.32 Gm. If a 20.00-ml sample of the hydrochloric acid solution yielded 2.9268 Gm of AgCl, its normality would be

$$\frac{2.9268}{143.32} = \text{equivalents of AgCl formed} = \text{equivalents of HCl in 20.00 ml}$$

$$\frac{2.9268 \times 1{,}000}{143.32 \times 20.00} = \text{equivalents of HCl per liter, or normality}$$

It may sometimes be convenient to standardize a solution of hydrochloric acid using a standard solution of NaOH or KOH as a secondary standard. If 25.20 ml of 0.9505 *N* NaOH is required in titration of 25.00 ml of a solution of hydrochloric acid, the normality would be:

$$\text{ml} \times N = \text{ml} \times N$$
$$25.00 \times 0.9505 = 25.00 \times N$$
$$N = 0.9531$$

In analytical procedures, a standard acid of such concentration usually is chosen so that 35 to 40 ml will be consumed in the titration, because volumes in this range can be measured more accurately with a 50-ml buret than smaller volumes. In the alkalimetric procedures, 1 *N*, 0.5 *N*, 0.05 *N*, 0.02 *N*, or even more dilute solutions of hydrochloric are employed. The more dilute solutions are conveniently prepared by diluting 1 *N* hydrochloric acid and restandardizing the diluted solution by one of the described methods.

Standard solutions of hydrochloric acid preserved in tightly stoppered alkali-free glass bottles retain their strength indefinitely.

Exercise 10

Object. To Prepare and Standardize 1 *N* Sulfuric Acid.

Materials Required. 30 ml of sulfuric acid
1 *N* sodium hydroxide solution

Procedure. Add 30 ml of sulfuric acid in small portions to 1,020 ml of distilled water with constant stirring. Allow the solution to acquire room temperature. Standardize the solution by titration against sodium carbonate as described for 1 *N* hydrochloric acid (Exercise 9) or normal sodium hydroxide solution using phenolphthalein as indicator. Calculate the normality and write it on the label or adjust the solution to exactly 1 *N*.

Since sulfuric acid is a dibasic acid, 1,000 ml of a normal solution should contain 98.08/2 = 49.04 Gm. Sulfuric acid (95 percent) has a specific gravity of about 1.83 at 25°. Consequently, 30 ml of the concentrated acid should contain about $30 \times 1.83 \times 0.95 = 52.15$ Gm of H_2SO_4. The method of calculating the normality of the solution is similar to that described under the preparation of 1 *N* hydrochloric acid (see Exercise 9, earlier in this chapter).

The standardization of sulfuric acid solution can also be effected by titration against pure sodium carbonate, as described under the standardization of hydrochloric acid. It may also be standardized gravimetrically by precipitation and weighing an equivalent amount of barium sulfate, as described under the gravimetric determination of sulfate in a soluble sulfate.

Standard solutions of sulfuric acid preserved in tightly stoppered, alkali-free bottles do not deteriorate.

Preparation and Standardization of Alkali Solution. Solutions of sodium hydroxide or potassium hydroxide may be standardized by use of a standard solution of hydrochloric or sulfuric acid as a secondary standard, or potassium biphthalate as a primary standard.

Exercise 11

Object. To Prepare and Standardize 1 *N* Sodium Hydroxide.

Materials Required. 45 Gm of sodium hydroxide
5 Gm of potassium biphthalate
Phenolphthalein T.S.

Procedure. "Dissolve 45 Gm. of sodium hydroxide in about 950 ml. of water. Add a freshly prepared saturated solution of barium hydroxide until no more precipitate forms. Shake the mixture thoroughly, and allow it to stand overnight in a stoppered bottle. Decant the clear liquid or filter the solution, and standardize it as follows:

"Dry about 5 Gm. of potassium biphthalate at 105° for 3 hours, and weigh accurately. If the potassium biphthalate is in form of large crystals, they should be crushed before drying. Dissolve it in 75 ml. of carbon dioxide-free water, add 2 drops of phenolphthalein T.S., and titrate with the sodium hydroxide solution to the production of a permanent pink color. Calculate the normality. Each 204.2 mg. of potassium biphthalate is equivalent to 1 ml. of 1 *N* sodium hydroxide."

More than 1 equivalent, 40.00 Gm, of sodium hydroxide is weighed out, since it is hygroscopic and if only 1 equivalent were used the resulting solution would be below normal strength. The solution is treated with barium hydroxide to precipitate any carbonate formed through the action of carbon dioxide on the sodium hydroxide. The solution is preserved in a tightly stoppered bottle fitted with a soda-lime tube to protect it from the carbon dioxide of the air. Solutions which contain carbonate are not suitable for titration with phenolphthalein as indicator, but when methyl orange is used the results are the same as if all the sodium present were combined as hydroxide.

Sodium hydroxide or potassium hydroxide solutions can be standardized by use of reagent potassium biphthalate as the primary standard. As indicated in the following equation, 1 mole or 204.23 Gm of the pure dried potassium biphthalate is 1 equivalent; 1 mEq would be 0.20423 Gm.

$$KHC_8H_4O_4 + NaOH \longrightarrow KNaC_8H_4O_4 + H_2O$$

Since it is desirable that approximately 35 ml of the alkali solution be consumed in a titration of a potassium biphthalate sample, the weight of the potassium biphthalate used for standardization of a solution of sodium hydroxide is assumed to be 1 *N*, and the calculation is as follows: 35 ml of 1 *N* NaOH contains 35 mEq of NaOH and is chemically equivalent to 35 mEq of $KHC_8H_4O_4$ or to $35 \times 0.20423 = 7.1477$ Gm of

$KHC_8H_4O_4$. In the standardization procedure about 5 Gm is used. If 24.15 ml of an alkali solution were required in the titration of a 4.9651-Gm sample of dried reagent potassium biphthalate, the normality of the solution would be:

$$\text{ml} \times N = \frac{\text{wt of } KHC_8H_4O_4}{0.20423}$$

$$N = \frac{4.9651}{24.15 \times 0.20423} = 1.0070$$

The solution can be made exactly 1 *N* by dilution in the same manner as that prescribed under the preparation of a 1 *N* hydrochloric acid solution.

Questions and Problems

1. Define the following terms: (*a*) normality; (*b*) molarity; (*c*) milliequivalent; (*d*) equivalent; (*e*) equivalent weight.

2. Calculate the equivalent weight of each of the following substances and, from that, the corresponding titer values of 1 *N* hydrochloric acid. Check the latter by reference to the U.S.P. XIV: (*a*) $CaCO_3$; (*b*) CaO; (*c*) $Na_2B_4O_7$; (*d*) $Na_2B_4O_7 \cdot 10H_2O$; (*e*) $C_2H_8N_2$ (ethylenediamine).

3. What weight of each of the following substances would be required to make 500 ml of 0.5 *N* solution? If the specific gravity is given, calculate the volume also. (*a*) Hydrochloric acid, 36 percent *W/W*, sp gr 1.18; (*b*) $Ba(OH)_2 \cdot 8H_2O$; (*c*) sulfuric acid, 96 percent *W/W*, sp gr 1.84; (*d*) KOH; (*e*) 1 *M* sodium hydroxide.

4. Calculate the normality of each of the following solutions: (*a*) 10 percent *W/V* sulfuric acid; (*b*) the official diluted ammonia solution; (*c*) hydrochloric acid solution with a Na_2CO_3 titer of 26.5 mg; (*d*) sodium hydroxide solution, 36 ml of which is equivalent to 0.7286 Gm of pure potassium biphthalate; (*e*) hydrochloric acid solution, 30 ml of which yields 0.4468 Gm of AgCl.

5. Calculate the molarity of each of the solutions described in problem 4.

6. Calculate the percentage *W/V* concentrations of each of the solutions described in problem 4.

7. What volume of 0.1 *N* solution can be prepared from each of the following? (*a*) 725 ml of 0.4321 *N* hydrochloric acid; (*b*) 30 ml of hydrochloric acid, 36 percent *W/W*, sp gr 1.18; (*c*) 250 ml of 0.1235 *M* sulfuric acid.

8. What volume of a solution of hydrochloric acid believed to be 0.1 *N* would be used in a standardization experiment to produce 0.5 Gm of AgCl?

9. What weight of primary standard sodium carbonate would be used as sample for standardizing 0.01 *N* sulfuric acid, so that approximately 35 ml of the acid would be used in the titration?

10. What weight of 98.5 percent $KHC_8H_4O_4$ would be equivalent to (*a*) 35 ml of 0.01 *N* NaOH; (*b*) 35 ml of 0.1 *M* $Ba(OH)_2$?

11. A 50.00-ml sample of calcium hydroxide solution required 19.50 ml of 0.1050 *N* hydrochloric acid in a titration. Calculate the amount of $Ca(OH)_2$ in the 50.00-ml sample and the $Ca(OH)_2$ concentration in percent *W/V*.

12. Calculate (*a*) the weight; (*b*) the number of moles; (*c*) the volume of 0.1 *N* sulfuric acid required to neutralized 0.5 Gm of CaO.

13. What volume of (*a*) 1 *N*; (*b*) 0.8765 *N*; (*c*) 0.2 *M*; (*d*) 10 percent *W/V* sulfuric acid would be required to neutralize 100 ml of 1.0560 *N* sodium hydroxide?

14. What is the weight in grams of 6.06×10^{23} molecules of HCl? Of 6.06×10^{22} molecules? Of 6.06×10^{10} molecules? Of 7.5×10^{23} molecules?

15. How many molecules of HCl are present in (*a*) 36.46 Gm, (*b*) 0.3646 Gm, (*c*) 1 Gm, and (*d*) 0.25 Gm?

16. What is the weight in grams of 6.06×10^{23} molecules of NaOH? How many gram-molecules are in 40 Gm of NaOH? How many molecules and how many gram-molecules are in 4 Gm of NaOH?

17. How many molecules of HCl will react with 75 molecules of NaOH? With 6.06×10^{23} molecules?

18. What is the calculated pH of a mixture of 500 ml of 5 percent W/V sodium hydroxide and 500 ml of 5 percent W/V hydrochloric acid? What volume of 0.1 N hydrochloric acid or 0.1 N sodium hydroxide would be required to adjust the pH to 7?

19. If successive 5-ml portions of 0.1 N sodium hydroxide are added to 25.00 ml of 0.1 N hydrochloric acid, what would be the pH after each addition?

20. If 25.00 ml of a hydrochloric acid solution were required in titration of a 0.6500-Gm sample of pure sodium carbonate, what weight of each of the following pure reagents would be required to neutralize 25 ml of the hydrochloric acid solution? (*a*) $NaHCO_3$; (*b*) KOH; (*c*) $Ba(OH)_2$; (*d*) CaO.

21. If 25.00 ml of a sodium hydroxide solution required 40.00 ml of a sulfuric acid solution in a titration, and 25.00 ml of the sulfuric acid solution were required in titration of 0.3060 Gm of pure sodium carbonate, what was the normality of the sodium hydroxide solution?

22. If 50.00 ml of a hydrochloric acid solution yielded 0.7167 Gm of AgCl and 30.00 ml of the same solution were required in titration of 25.00 ml of a sodium hydroxide solution, what was the normality of the sodium hydroxide solution?

23. A solution of sulfuric acid was standardized gravimetrically, and 0.3059 Gm of barium sulfate was produced from 25.00 ml of the solution. How much water must be added to 1 liter of the solution to produce a solution exactly 0.1 N?

24. What weight of barium sulfate would be produced from 10.00 ml of a sulfuric acid solution if 50.00 ml of the same solution were found to be equivalent to 0.2650 Gm of pure sodium carbonate?

25. If 25.00 ml of a sulfuric acid solution were found to be equivalent to 20.50 ml of a sodium hydroxide solution, and 35.00 ml of the sodium hydroxide solution were found to be equivalent to 0.7140 Gm of pure potassium biphthalate, what weight of barium sulfate could be produced from 50.00 ml of the sulfuric acid solution?

Chapter 4

ACIDIMETRIC ANALYSIS

Acidimetry, the direct or residual titrimetric analysis of bases using an accurately measured volume of acid, is applied in the analytical control of a considerable number of official substances, both inorganic and organic.

Direct titration is conducted by introducing a standard acid solution gradually from a buret into a solution of the base being assayed until chemically equivalent amounts of each have reacted as shown by some change in properties of the mixture. Such an end point, which must be close to the stoichiometric point, is made evident by a change in the color of some indicator or by potentiometric means.

Residual titration is used whenever the end point of a direct titration deviates appreciably from the stoichiometric point for some reason; e.g., when the basic sample is insoluble in water and the rate of its reaction with the standard acid is relatively slow. In such a procedure, the basic sample is treated with an amount of standard acid known to be in excess of a chemically equivalent amount, and the excess acid is subsequently titrated with a standard base. The amount of the standard acid which is chemically equivalent to the basic sample can be calculated simply by subtracting, from the total amount of standard acid employed, the amount of acid that was neutralized by the standard base in the back titration.

DIRECT TITRATION METHODS

Exercise 12

Object. Assay of Sodium Bicarbonate.

Materials Required. 3 Gm of sodium bicarbonate
1 *N* sulfuric acid

Procedure. "Weigh accurately about 3 Gm. of Sodium Bicarbonate, mix with 25 ml. of water, add methyl orange T.S., and titrate with 1 *N* sulfuric acid. Each milliliter of 1 *N* sulfuric acid is equivalent to 84.01 mg. of $NaHCO_3$."

It is a general rule that when the substance to be assayed is a solid, the accurately weighed quantity to be used should be dissolved in sufficient water to make the solution of about the same equivalent concentration as that of the acid to be used in the titration. Methyl orange is used as indicator because phenolphthalein and most other indicators are affected by the carbonic acid liberated in the reaction, which causes a change of color before the reaction is complete.

$$\underset{2(84.01)}{2NaHCO_3} + H_2SO_4 \longrightarrow Na_2SO_4 + 2H_2O + 2CO_2\uparrow$$

From the equation it is obvious that 2 moles of $NaHCO_3$ is equivalent to 1 mole of H_2SO_4. Therefore 1 mole of $NaHCO_3$ is 1 equivalent and the equivalent weight is equal to the gram-molecular weight, 84.01 Gm. One milliliter of 1 *N* sulfuric acid or 1 mEq is equivalent to 84.01 mg or 1 mEq of $NaHCO_3$.

Calculate the percentage purity of the sample assayed.

$$\frac{\text{ml} \times N \times 0.08401 \times 100}{\text{Sample weight}} = \text{percent NaHCO}_3$$

Questions and Problems

1. Write ionically all equations for all reactions that occur in the above assay.
2. Could 1 *N* acetic acid be used to titrate the bicarbonate in place of sulfuric acid?
3. Look up the pH at which methyl orange exhibits its acid color, and explain why it is a good indicator to use in this titration.
4. If exactly 3.0 Gm of $NaHCO_3$ is dissolved in 25 ml of water in the above assay, what will be the normality of the solution? Does the normality conform to the general rule stated in the discussion?
5. If a 0.2800-Gm sample of sodium bicarbonate (96.5 percent $NaHCO_3$) is titrated with 0.9165 *N* sulfuric acid, what volume of the acid should be required to produce an end point?
6. Calculate the following titer values for 1 *N* sulfuric acid: (*a*) $KHCO_3$; (*b*) K_2CO_3; (*c*) Ca_2CO_3. Check the calculated values with those appearing in the U.S.P. or N.F. assay procedures for these substances.

Determination of Total Alkalinity and Sodium Carbonate Content of Sodium Hydroxide

Exercise 13

Object. Assay of Sodium Hydroxide.

Materials Required. 1.5 Gm of sodium hydroxide
1 *N* sulfuric acid

Procedure. 1. "Dissolve about 1.5 Gm. of Sodium Hydroxide, accurately weighed, in about 40 ml. of recently boiled and cooled carbon dioxide–free water. Cool the solution to 15°, add phenolphthalein T.S., and titrate with 1 *N* sulfuric acid. At the discharge of the pink color of the indicator, record the volume of acid solution required

The base, NaOH, behaves in the same manner when either phenolphthalein or methyl orange is used as the indicator. Consequently, the end point obtained with phenolphthalein as the indicator represents complete neutralization of all the NaOH as represented by the equations:

$$\underset{2(40.00)}{2NaOH} + H_2SO_4 \longrightarrow Na_2SO_4 + 2H_2O \qquad (1)$$

In a cold solution, with phenolphthalein indicator, the end point of titration of sodium carbonate with standard H_2SO_4 is exhibited when the solution is half neutralized or when the Na_2CO_3 is transformed into $NaHCO_3$.

$$\underset{2(106)}{2Na_2CO_3} + H_2SO_4 \longrightarrow Na_2SO_4 + 2NaHCO_3 \qquad (2)$$

2. ". . . add methyl orange T.S., and continue the titration until a persistent pink color is produced. Each milliliter of 1 *N* sulfuric acid is equivalent to 40.00 mg. of total alkali, calculated as NaOH, and each milliliter of acid consumed in the titration with methyl orange is equivalent to 106.0 mg. of Na_2CO_3."

In titration to a phenolphthalein end point, NaOH is neutralized and Na_2CO_3 is converted to $NaHCO_3$. As the Na_2CO_3 is converted to $NaHCO_3$ ($CO_3^{=} + H^+ \rightleftharpoons HCO_3^-$), the [$H^+$] remains low because the $CO_3^{=}$ ion is strongly basic. Consequently, the pH of the mixture remains above the range 8 to 9.8 over which phenolphthalein changes color, until the conversion of Na_2CO_3 to $NaHCO_3$ is complete. The HCO_3^- ion is weakly basic ($HCO_3^- + H^+ \rightleftharpoons H_2CO_3$), and the $NaHCO_3$ formed remains unneutralized even though [H^+] has been increased to the point where phenolphthalein undergoes a change in color. The neutralization of the $NaHCO_3$ is complete only when [H^+] has been increased, by further addition of acid, to the point where methyl orange undergoes a change in color at pH 3.2 to 4.4.

The reaction is indicated by the equation

$$\underset{2(84.01)}{2NaHCO_3} + H_2SO_4 \longrightarrow Na_2SO_4 + 2CO_2\uparrow + 2H_2O \qquad (3)$$

The total volume of 1 *N* sulfuric acid used in the titration was required to neutralize NaOH and Na_2CO_3, converting the latter first to $NaHCO_3$ at the phenolphthalein end point and then to H_2CO_3 at the methyl orange end point. Although part of the standard acid was actually neutralized by Na_2CO_3, the U.S.P. gives specifications for *total alkali* including NaOH and Na_2CO_3, calculated as NaOH. From Eq. (1), it can be seen that the equivalent weight of NaOH is 40.00 Gm. Therefore, each milliliter of the total amount of 1 *N* acid consumed is equivalent to 40.00 mg or 1 mEq of NaOH.

The difference between the total volume of standard acid used and the volume necessary to produce a phenolphthalein end point gives the volume of standard acid necessary to neutralize the $NaHCO_3$ formed from the Na_2CO_3 during titration to the phenolphthalein end point. This same volume was also necessary to convert the Na_2CO_3 to $NaHCO_3$ according to Eq. (2). From this equation, the equivalent weight of Na_2CO_3 is calculated as 106.0 Gm. Consequently, each milliliter of the acid volume calculated by difference is equivalent to 106.0 mg or 1 mEq of Na_2CO_3.

Interpretation of the following data will illustrate the calculations involved. A 965.4-mg sample of sodium hydroxide required 20.80 ml of 1 *N* sulfuric acid in titration to a phenolphthalein end point, and 21.75 ml of 1 *N* sulfuric acid in titration to a methyl orange end point. The 21.75 ml of 1 *N* sulfuric acid consumed is a measure of the total alkalinity, NaOH plus Na_2CO_3. In order to express total alkalinity in terms of NaOH, it is assumed that each milliliter of the 1 *N* acid is equivalent to 40.00 mg of NaOH. The total alkalinity calculated as NaOH is therefore

$$\frac{\text{ml} \times N \times \text{mEq wt} \times 100}{\text{Sample weight}} = \text{percent}$$

$$\frac{21.75 \times 1 \times 0.0400 \times 100}{0.9654} = 90.12 \text{ percent}$$

The volume of 1 *N* sulfuric acid, $21.75 - 20.80 = 0.95$ ml, required to neutralize the $NaHCO_3$ as in Eq. (3) is equal to the volume necessary to produce the $NaHCO_3$ from Na_2CO_3 as in Eq. (2). From Eq. (2), it can be calculated that each milliliter of 1 *N* sulfuric acid is equivalent to 106.0 mg of Na_2CO_3. Therefore the proportion of Na_2CO_3 in the sample is

$$\frac{0.95 \times 1 \times 0.1060 \times 100}{0.9654} = 10.4 \text{ percent}$$

Since 0.95 ml of 1 *N* sulfuric acid was required to convert the Na_2CO_3 to $NaHCO_3$ and an equal volume to neutralize the $NaHCO_3$, then $2 \times 0.95 = 1.90$ ml of 1 *N* sulfuric acid was required for complete neutralization of the Na_2CO_3 according to the following equation, which is a summation of Eqs. (2) and (3):

$$\underset{106.0}{Na_2CO_3} + H_2SO_4 \longrightarrow Na_2SO_4 + H_2O + CO_2$$

The equivalent weight of Na_2CO_3 calculated from this equation is 53.0 Gm. Each milliliter of 1 *N* sulfuric acid required for complete neutralization of Na_2CO_3 is equivalent to 53.0 mg of Na_2CO_3. Since 1.90 ml was required, the proportion of Na_2CO_3 in the sample is

$$\frac{1.90 \times 1 \times 0.053 \times 100}{0.9654} = 10.4 \text{ percent}$$

Questions and Problems

1. In titration of a 1.0900-Gm sample of sodium carbonate (98.5 percent Na_2CO_3) with 0.9050 *N* sulfuric acid, what volumes would be necessary to produce (*a*) phenolphthalein end point; (*b*) methyl orange end point?

2. In titration of a mixture of 0.5000 Gm of pure Na_2CO_3 and 0.4000 Gm of pure $NaHCO_3$ with 1.1050 *N* sulfuric acid, what volume of the acid would be necessary to produce (*a*) phenolphthalein end point; (*b*) methyl orange end point?

3. Calculate the proportion of (*a*) KOH; (*b*) K_2CO_3; (*c*) the total alkalinity in terms of KOH, in a solution of potassium hydroxide, 25.00 ml of which required 26.00 ml of 0.8650 *N* sulfuric acid in titration to a phenolphthalein end point and 31.15 ml of the same acid in titration to a methyl orange end point.

4. Calculate the Na_2CO_3 and $NaHCO_3$ content of a 1.2500-Gm sample requiring 13.80 ml of 0.5250 *N* hydrochloric acid on titration to a phenolphthalein end point and an additional 16.30 ml on further titration to a methyl orange end point.

Determination of Alkali Salt of Water-insoluble Organic Acid using Ether Extraction

Exercise 14

Object. Assay of Sodium Salicylate Tablets.

Materials Required. 20 Sodium Salicylate Tablets
75 ml of ether
0.1 hydrochloric acid

Procedure. "Place not less than 20 Sodium Salicylate Tablets in a 200-ml. volumetric flask, add 100 ml. of water, allow to stand, with frequent agitation, until the tablets disintegrate completely. Dilute to volume, and mix. Filter through a dry filter into a dry flask, rejecting the first 10 ml. of the filtrate. Transfer to a separator an accurately measured volume of the subsequent filtrate, equivalent to about 500 mg. of sodium salicylate, and dilute, if necessary, to make about 25 ml. Add 75 ml. of ether and 10 drops of bromophenol blue T.S. and titrate with 0.1 *N* hydrochloric acid, mixing the water and ether layers by vigorous shaking until a permanent, pale green color is produced in the water layer. Draw off the water layer into a small flask, wash the ether once with 5 ml. of water, and add this to the water layer. Add 20 ml. of ether to the combined water solutions, and mix. Continue the titration with vigorous shaking until a permanent, pale green color is produced in the water layer. Each milliliter of 0.1 *N* hydrochloric acid is equivalent to 16.01 mg. of $C_7H_5NaO_3$."

Upon complete disintegration of the tablets the sodium salicylate is dissolved. In the filtration process the first 10 ml of the filtrate is rejected, since in wetting the filter some of the sodium salicylate will be adsorbed. Sodium salicylate, like sodium carbonate and sodium bicarbonate, is a salt of a strong base and a weak acid, and will neutralize hydrochloric acid by a reaction ($C_7H_5O_3^- + H^+ \rightleftharpoons HC_7H_5O_3$) which is practically complete when $[H^+]$ has been increased to the point where bromophenol blue undergoes a change in color at pH 3.0 to 4.6. The salicylic acid which is formed is not soluble in water, but the ether dissolves it and thus makes possible observation of the color change in the aqueous phase that would otherwise be obscured by the precipitated salicylic acid. Removal of the salicylic acid from the aqueous phase also helps to prevent reversal of the neutralization reaction. For purposes of stoichiometry, the following reaction is used:

$$\underset{160.11}{C_7H_5NaO_3} + HCl \longrightarrow HC_7H_5O_3 + NaCl$$

The separated ethereal layer is washed with water to remove small amounts of sodium salicylate that may remain dissolved therein. Upon washing the combined aqueous solutions and mixing intimately, any remaining salicylic acid partitions itself into the ethereal layer, leaving only the HCl in the aqueous layer to influence the end point.

Calculations. Since 200 ml represents the volume in which the sample of 20 tablets are dissolved, then 10 ml is equivalent to 1 tablet.

$$\frac{\text{ml} \times N \times 0.16011 \times 10}{\text{ml of sample}} = \text{Gm Sodium Salicylate/tablet}$$

The sodium benzoate in Caffeine and Sodium Benzoate and Caffeine and Sodium Benzoate Injection is assayed in the same manner, with the liberated benzoic acid going into the ether layer. The equivalence point for the reaction is approximately the same as that for the sodium salicylate reaction, and methyl orange, which has a transition point about the same as that of bromophenol blue, is the indicator for this assay.

Questions and Problems

1. What are the advantages of bromophenol blue as indicator in direct titrimetric analysis of sodium salicylate?

2. In what ways is the assay procedure described for sodium salicylate not applicable to sodium carbonate or sodium citrate? Explain.

3. The disodium salt of a dibasic organic acid has a molecular weight of 215; what weight of sample would be used for titration to bromophenol blue end point with 0.2500 N acid in order that 35.00 ml of the latter will be required?

4. Twenty sodium salicylate tablets, labeled 5-grain, were dispersed in sufficient water to make 200.0 ml. A 15.0-ml aliquot of the filtrate was titrated to a bromophenol blue end point in the usual way by 29.11 ml of 0.1000 N hydrochloric acid. Calculate the amount of sodium salicylate in each tablet and, from that, the percentage of the labeled amount. Note: 1 grain = 65 mg.

5. Suggest two alternative methods for analyzing a salt of a strong base and weak acid and the advantages or disadvantages of each method.

RESIDUAL TITRATION METHODS

Residual titration or "back titration" is frequently used when a reaction proceeds slowly or when the substance to be assayed does not give a distinct, sharp end point with an indicator by direct titration. Residual titration is carried out by dissolving the substance under examination in an accurately measured quantity of standard solution known to be in excess and titrating the excess of the latter with another standard solution. Numerous examples of this method of titration will be met with in subsequent exercises.

Exercise 15

Object. Assay of Zinc Oxide.

Materials Required. 1.5 Gm of zinc oxide
1 *N* sulfuric acid
1 *N* sodium hydroxide

Procedure. "Dissolve about 1.5 Gm. of freshly ignited Zinc Oxide, accurately weighed, and 2.5 Gm. of ammonium chloride in 50 ml. of 1 *N* sulfuric acid with the aid of gentle heat, if necessary. When solution is complete, add methyl orange T.S., and titrate the excess sulfuric acid with 1 *N* sodium hydroxide. Each milliliter of 1 *N* sulfuric acid is equivalent to 40.68 mg. of ZnO."

The ZnO dissolves in the sulfuric acid, neutralizing an equivalent amount in accordance with the equation

$$\underset{81.37}{ZnO} + H_2SO_4 \longrightarrow ZnSO_4 + H_2O$$

The amount of sulfuric acid neutralized by the ZnO in this way is determined simply by subtracting, from the total amount of sulfuric acid used, the amount that was neutralized by the standard NaOH in the back titration. The equivalent weight of ZnO, as indicated by the equation, is 40.68 Gm; therefore, each milliliter of 1 *N* sulfuric acid, 1 mEq, neutralized by the ZnO is equivalent to 40.68 mg or 1 mEq of ZnO.

Calculate the percentage of zinc and the percentage of zinc oxide in the sample assayed.

$$\frac{(ml_a \times N) - (ml_b \times N) \times mEq \times 100}{\text{Sample weight}} = \text{percent}$$

Questions and Problems

1. Suggest a reason for the use of ammonium chloride in the assay of zinc oxide. In titration of the mixture to a methyl orange end point with standard sodium hydroxide, why is none of the NaOH neutralized by the NH_4Cl?

2. Give a reason explaining why the standard acid used in assay of zinc oxide is 1 *N* while that used in assay of zinc stearate is only 0.1 *N*. Calculate the weight of zinc stearate that would be used as sample if 1 *N* sulfuric acid were employed, so that 35.00 ml of the acid would be neutralized by the zinc stearate.

3. If a 1.2500-Gm sample of zinc oxide, 95.0 percent of ZnO, were treated with 50.00 ml of 1.1230 *N* sulfuric acid in the usual way, what volume of 0.9765 *N* sodium hydroxide would be required in the back titration?

Determination of the Purity of the Alkali Salt of an Organic Acid. In Table 8 are included a number of alkali salts of organic acids which are assayed by a direct acidimetric procedure. A number of others, however, are first ignited to an equivalent amount of carbonate, and this carbonate in the ash formed by ignition is determined residually. Certain

TABLE 8. ACIDIMETRIC ASSAYS. DIRECT TITRATION WITH ACID

Substance	mEq factor, Gm	Indicator	Official requirement percent
U.S.P.			
Aminophylline	0.03005	m.o.	$C_2H_8N_2$ = 14 to 15
Aminophylline injection	0.03005	m.o.	$C_2H_8O_2$ = 20 mg/Gm IV, 60 mg/Gm IM
Ammonia water strong (R)	0.01703	m.r.	NH_3 = 28 to 30
Barium hydroxide (R)	0.15774	phen.	$Ba(OH)_2{\cdot}8H_2O$ = 98
2,3-Butanedione (R)	0.04304	b.p.b.	$CH_3COCOCH_3$ = 97
Caffeine and sodium benzoate	0.14411	m.o.	$C_7H_5NaNO_2$ = 50 to 53
Caffeine and sodium benzoate injection	0.14411	m.o.	$C_7H_5NaNO_2$ = 47.5 to 55.5
Calcium hydroxide solution	0.03705	phen.	$Ca(OH)_2$ = 0.14 *W/V*
Ethylenediamine	0.03005	b.p.b.	$C_2H_8N_2$ = 97
Potassium bicarbonate	0.10012	m.o.	$KHCO_3$ = 99 to 101
Potassium hydroxide	0.05611	phen.	KOH = 85 total alkalinity
	0.06911	m.o.	K_2CO_3 = n.m.t. 3.5
Potassium hydroxide (R)	0.05611	phen.	KOH = 85
	0.06911	m.o.	K_2CO_3 = n.m.t. 2
Sodium bicarbonate tablets	0.08401	m.o.	$NaHCO_3$ = 99
Sodium bicarbonate	0.08401	m.o.	$NaHCO_3$ = 95 to 105 l.a.
Sodium borate	0.19068	m.r.	$Na_2B_4O_7{\cdot}10H_2O$ = 99 to 105
Sodium carbonate, monohydrated	0.05299	m.o.	Na_2CO_3 = 99.5
Sodium hydroxide	0.04000	phen.	NaOH = 95 total alkalinity
	0.05299	m.o.	Na_2CO_3 = n.m.t. 3
Sodium hydroxide (R)	0.04000	phen.	NaOH = 97
	0.05299	m.o.	Na_2CO_3 = n.m.t. 1
Sodium salicylate tablets	0.16011	b.p.b.	$C_7H_5NaO_3$ = 95 to 105 l.a.
Tetramethyl ammonium hydroxide (R)	0.09115	m.r.	$(CH_3)_4NOH$ = 10
Triethanolamine	0.14919	m.r.	$(CH_2CH_2OH)_3N$ = 6.7 to 7.2 ml 1*N* acid/Gm
Tris-(hydroxymethyl)aminomethane (R)	0.12114	pot.	$C_4H_{11}NO_3$ = 99.5 to 100.5
N.F.			
Ammonia solution, strong	0.01703	m.r.	NH_3 = 27 to 30
Aromatic ammonia spirit	0.04804	m.o.	$(NH_4)_2CO_3$ = 3.5 to 4.5 *W/V*
Monoethanolamine	0.06108	m.r.	C_2H_7NO = 97.5
Propylhexadrine	0.15529	m.r.	$C_{10}H_{21}N$ = 98 to 101

n.m.t. = not more than; l.a. = labeled amount; m.o. = methyl orange; m.r. = methyl red; phen. = phenolphthalein; b.p.b. = bromophenol blue; pot. = potentiometer; (R) = reagent.

official acetates, tartrates, and citrates are assayed in this way, and the official compendiums give general directions for the assay of "Alkali Salts of Organic Acids."

Exercise 16

Object. Assay of Potassium Sodium Tartrate.

Materials Required. 3 Gm of potassium sodium tartrate
0.5 *N* sulfuric acid
0.5 *N* sodium hydroxide

Procedure. 1. Dry about 3 Gm of Potassium Sodium Tartrate, accurately weighed, at 150° for 3 hr; it loses not less than 21 percent and not more than 26 percent of its weight.

The $C_4H_4KNaO_6$ content of potassium sodium tartrate is controlled first by specifications pertaining to the water content and then by specifications of the $C_4H_4KNaO_4$ content of the dried material. The latter determination is made according to the general procedure for assay of alkali salts of organic acids using the material dried as described above.

2. "Weigh accurately about 2 Gm. of the organic salt in a tared porcelain crucible, and ignite, gently at first, until the salt is thoroughly carbonized, protecting the carbonized salt from contact with the flame at all times."

The salt is heated slowly at first, since the sample swells and fuses with concomitant decomposition. If strong heat is applied in the initial stages of ignition, there may be loss of a portion of the sample through decrepitation or spattering. After the mass in the crucible is partially charred and white fumes are no longer evolved, it is ignited to dull redness. If heated too strongly or if the flame comes in contact with the carbonized mass, the alkali carbonate may be converted into the oxide.

3. "Cool the crucible, place it in a glass beaker, and break up the carbonized mass with a glass rod. Without removing the glass rod or the crucible, add 50 ml. of water and 50.0 ml. of 0.5 *N* sulfuric acid, cover the beaker, and boil the solution for 30 minutes. Filter, and wash with hot water until the last washing is neutral to litmus. Cool the combined filtrate and washings, add methyl red–methylene blue T.S., and titrate the excess acid with 0.5 *N* sodium hydroxide."

The fused residue is broken up by means of a strong glass rod over glazed, black paper and transferred with the crucible into a 250-ml beaker. About 50 ml of water and 50 ml of sulfuric acid are added along the glass rod, and the mass is completely dissolved.

During the ignition of the salt, an equivalent amount of carbonate is formed,

$$\underset{210.16}{C_4H_4KNaO_6} \longrightarrow KNaCO_3$$

and this neutralizes an equivalent amount of the acid which the ash was treated with.

$$KNaCO_3 + H_2SO_4 \longrightarrow KNaSO_4 + H_2O + CO_2\uparrow$$

It is obvious from the equations that the carbonate from 1 mole of $C_4H_4KNaO_6$ will neutralize 2 equivalents, 1 mole, of sulfuric acid; therefore, the equivalent weight of $C_4H_4KNaO_6$ is 105.08 Gm. Each milliequivalent, then, is 0.10508 Gm, and 1 ml of 0.5 *N* acid is equivalent to ½ mEq or 0.05255 Gm, as stated in the official assay method.

Calculations. The volume of acid consumed is obtained by subtracting the volume of base used to titrate the excess acid.

$$\frac{(\text{ml}_a \times N) - (\text{ml}_b \times N) \times 0.10508 \times 100}{\text{Sample weight, Gm}} = \text{percent } C_4H_4KNaO_6$$

The N.F. requires that potassium sodium tartrate, when rendered anhydrous by drying at 150°, contain not less than 99 percent of $C_4H_4KNaO_6$.

If desired, a 400-mg sample may be used for analysis, in which case 0.1 *N* sulfuric acid and 0.1 *N* sodium hydroxide are used.

Methyl orange indicator has been replaced by methyl red–methylene blue indicator in the N.F. and by phenolphthalein in the U.S.P., since the boiling process rids the solution of any carbon dioxide which otherwise would affect the equivalence point of the titration. Essentially, a strong acid is being titrated with a strong base and any indicator showing a color change between the pH of 4 and 9 is suitable to detect the end point. The mixed indicator of methyl red–methylene blue changes color from red-violet in acid to a dirty blue in the transition state and finally to green on the alkaline side of its color change. The methylene blue serves as a screening dye for the methyl red color change.

Questions and Problems

1. Why cannot the alkali salts of such acids as citric, tartaric, etc., be titrated directly with a strong acid in the same manner as soluble carbonates are titrated?

2. Calculate the weight of each of the following substances which produces sufficient carbonate on ignition to neutralize 1 ml of 0.5 *N* sulfuric acid: (*a*) sodium citrate, $C_6H_5Na_3O_7$; (*b*) potassium acetate, $C_2H_3KO_2$; (*c*) potassium citrate, $C_6H_5K_3O_7$. Compare with the titer values of 0.5 *N* sulfuric acid given in U.S.P. and N.F.

3. What weight of potassium acetate containing 91.5 percent $C_2H_3KO_2$ must be

TABLE 9. ACIDIMETRIC ASSAYS. ALKALI SALTS OF ORGANIC ACIDS DETERMINED BY RESIDUAL TITRATION METHODS

Substance	mEq factor, Gm	Indicator	Official requirement, percent
U.S.P.			
Anticoagulant acid citrate dextrose solution	0.09800	phen.	$C_6H_5Na_3O_7 \cdot 2H_2O$ = 95 to 105 l.a.
N.F.			
Potassic saline injection, lactated	0.08907	m.o.	$C_3H_5O_3$ = 0.437 to 0.501 *W/V*
Potassium acetate	0.09815	m.b./m.r.	$C_2H_3KO_2$ = 99
Potassium citrate	0.10214	m.b./m.r.	$C_6H_5K_3O_7$ = 99
Potassium sodium tartrate	0.10509	m.b./m.r.	$C_4H_4KNaO_6$ = 99
Seidlitz powders	0.14112	m.b./m.r.	$C_4H_4KNaO_6 \cdot 4H_2O$ = 73 to 78
Sodium acetate	0.08204	m.b./m.r.	$C_2H_3NaO_2$ = 99 to 101
Sodium propionate	0.09606	m.b./m.r.	$C_3H_5NaO_2$ = 99
Strontium acetate (R)	0.10736	m.r.	$Sr(CH_3COO)_2 \cdot \frac{1}{2}H_2O$ = 99

l.a. = labeled amount; phen. = phenolphthalein; m.b./m.r. = methylene blue/methyl red; m.o. = methyl orange; (R) = reagent.

used as sample in an assay according to the general method for alkali salts of organic acids if, after treatment of the ash with 50.00 ml of 0.5162 *N* sulfuric acid, it is desirable to consume 30.00 ml of 0.4675 *N* sodium hydroxide in the back titration?

4. A mixture of sodium bicarbonate and potassium sodium tartrate (compound effervescent powders, blue paper) was assayed for $NaHCO_3$ and $KNaC_4H_4O_6 \cdot 4H_2O$. A 2.2440-Gm sample was treated with 20.00 ml of 0.5 *N* sulfuric acid and the mixture titrated to a phenolphthalein end point with 6.17 ml of 0.5 *N* sodium hydroxide. A 1.8760-Gm sample was ignited, the ash taken up in 50.00 ml of 0.5 *N* sulfuric acid, and the mixture titrated to methyl orange end point with 19.84 ml of 0.5 *N* sodium hydroxide. Calculate the $NaHCO_3$ and $KNaC_4H_4O_6 \cdot 4H_2O$ content of the material.

Exercise 17

Object. Assay of Milk of Magnesia.

Materials Required. 5 Gm of Milk of Magnesia
1 *N* sulfuric acid
1 *N* sodium hydroxide

Procedure. "Place about 5 Gm. of Milk of Magnesia, previously well shaken in its original container, in a tared flask, stopper the flask, and weigh accurately. Add 25 ml. of 1 *N* sulfuric acid, and after solution is complete, add methyl red T.S., and titrate the excess acid with 1 *N* sodium hydroxide. Each milliliter of 1 *N* sulfuric acid is equivalent to 29.16 mg. of $Mg(OH)_2$."

The milk of magnesia is dissolved in an accurately measured excess of 1 *N* sulfuric acid solution to ensure complete neutralization of all the magnesium hydroxide with the formation of the soluble magnesium sulfate. The excess acid is then determined by residual titration with 1 *N* sodium hydroxide, using methyl red as indicator.

The reactions which take place when the milk of magnesia is dissolved and upon residual titration are as follows:

$$\underset{58.33}{Mg(OH)_2} + H_2SO_4 \longrightarrow MgSO_4 + 2H_2O$$

$$H_2SO_4 + 2NaOH \longrightarrow Na_2SO_4 + 2H_2O$$

The equivalent weight of $Mg(OH)_2$ is 29.16 Gm; consequently each milliliter of 1 *N* sulfuric acid neutralized by $Mg(OH)_2$ is equivalent to 29.17 mg of $Mg(OH)_2$.

If a sample of magnesia magma weighing 5.2430 Gm when dissolved in 25 ml of 0.9915 *N* H_2SO_4 required 9.85 ml of 1.1402 *N* NaOH to titrate the excess acid, the percent of $Mg(OH)_2$ in the sample could be calculated from the following formula:

$$\frac{[(\text{ml of } H_2SO_4 \times N) - (\text{ml of NaOH} \times N)] \times \text{mEq wt}}{\text{Weight of sample}} \times 100$$

Substituting the figures from the example cited above,

$$\frac{[(25 \times 0.9915) - (9.85 \times 1.1402)] \times 0.02916}{5.2430} \times 100 = 7.54 \text{ percent}$$

The U.S.P. requires that milk of magnesia contain not less than 7 and not more than 8.5 percent of $Mg(OH)_2$. The above sample, therefore, conforms to the official requirements.

Questions and Problems

1. Compare the assay procedures employed for magnesium sulfate, magnesium hydroxide, and milk of magnesia, giving equations for all reactions involved.

2. If 0.5 *N* sulfuric acid were used in assay of milk of magnesia, how would the U.S.P. directions be modified?

3. Calculate the MgO content of milk of magnesia, 12.32 Gm of which was dissolved in 50.00 ml of 1.0340 *N* sulfuric acid, producing a mixture that required 24.60 ml of 1.1255 *N* sodium hydroxide.

Exercise 18

Object. Assay of Methenamine.

Materials Required. 1 Gm of methenamine
1 *N* sulfuric acid
1 *N* sodium hydroxide

Procedure. 1. "Transfer about 1 Gm. of Methenamine, previously dried over phosphorus pentoxide for 4 hours and accurately weighed, to a beaker, add 40 ml. of 1 *N* sulfuric acid, and boil gently, adding water from time to time, if necessary, until the odor of formaldehyde is no longer perceptible."

The weighed sample of methenamine is heated on a water bath with a measured excess of 1 *N* sulfuric acid to decompose it into ammonia and formaldehyde, water being added from time to time to replace the water lost by evaporation. Completion of the decomposition is indicated when the odor of formaldehyde is no longer perceptible. The decomposition reaction may be indicated in two steps as follows:

$$\underset{140.19}{(CH_2)_6N_4} + 10H_2O \longrightarrow 6CH_2O + 4NH_4OH$$

$$4NH_4OH + 2H_2SO_4 \longrightarrow 2(NH_4)_2SO_4 + 4H_2O$$

2. "Cool, add 20 ml. of water, add methyl red T.S., and titrate the excess acid with 1 *N* sodium hydroxide. Each milliliter of 1 *N* sulfuric acid is equivalent to 35.05 mg. of $C_6H_{12}N_4$."

The excess sulfuric acid found by residual titration with 1 *N* sodium hydroxide, subtracted from the amount originally added, gives the amount of acid that combined with the liberated ammonia.

A summation of the above two equations indicates the equivalence of 1 mole of methenamine and 2 moles of sulfuric acid; hence ¼ mole, 35.05 Gm, of methenamine is 1 equivalent and each milliliter of 1 *N* sulfuric acid is equivalent to 35.05 mg or 1 mEq of methenamine.

Calculate the percentage purity of the sample assayed.

TABLE 10. ACIDIMETRIC ASSAYS. RESIDUAL TITRATION OF EXCESS ACID WITH BASE

Substance	mEq factor, Gm	Indicator	Official requirement, percent
U.S.P.			
Ammonium carbonate (R)	0.01703	m.o.	NH_3 = 30
Barium hydroxide (R)	0.09869	m.o.	$BaCO_3$ = n.m.t. 2
Calamine	0.04068	m.o.	ZnO = 98
N-Ethylpiperidine (R)	0.11320	m.r.	$C_5H_{10}NC_2H_5$ = 99
Lidocaine	0.23434	b.c.g./m.r.	$C_{14}H_{22}N_2O$ = 99 to 101
Magnesium oxide	0.02016	m.o.	MgO = 96
Magnesium stearate	0.02016	m.o.	MgO = 6.8 to 8.0
Magnesium trisilicate	0.02016	m.o.	MgO = 20
Magnesium trisilicate	....	m.r.	Acid capacity = 140 to 160 ml 0.1 *N* acid/Gm
Magnesium trisilicate tablets	0.02016	m.o.	MgO = 19 to 21 l.a.
Magnesium trisilicate tablets	....	m.r.	Acid capacity = 130 to 170 ml 0.1 *N* acid/Gm
Milk of magnesia	0.02917	m.r.	$Mg(OH)_2$ = 7.0 to 8.5
Morpholine (R)	0.08712	m.r.	C_4H_9NO = 99 to 100.5
Ringer's injection, lactated	0.11206	m.o.	$C_3H_5NaNO_3$ = 0.29 to 0.33 *W/V*
Sodium morrhuate injection	0.32400	m.o.	93 to 107 l.a.
Thiotepa	0.06037	m.o./phen.	$C_6H_{12}N_3PS$ = 97 to 102
Trichloroacetic acid (R)	0.16339	m.r.	CCl_3COOH = 99
Zinc oxide	0.04069	m.o.	ZnO = 99
N.F.			
Aluminum hydroxide gel, dried	....	b.p.b.	Acid capacity = 250 ml 0.1 *N* acid/Gm
Aluminum phosphate gel	....	t.h.b.	Acid capacity = 6 to 11 ml 0.1 *N* acid/Gm
Ammonia spirit, aromatic	0.01703	m.r.	NH_3 = 1.7 to 2.1 *W/V*
Ammonium carbonate	0.01703	m.o.	NH_3 = 30 to 34
Bromides tablets, three	0.09796	m.r.	NH_4Br = 30.8 to 35.8
Calcium phosphate, tribasic	....	m.o.	Acid capacity = 12.5 to 13.8 ml 1 *N* HCl/Gm
Dihydroxyaluminum aminoacetate	....	b.p.b.	Acid capacity = 140 ml 0.1 *N* acid/Gm
Dihydroxyaluminum aminoacetate magma	....	b.p.b.	Acid capacity = 125 to 175 ml 0.1 *N* acid/Gm
Dihydroxyaluminum aminoacetate tablets	....	b.p.b.	Acid capacity = 140 ml 0.1 *N* acid/Gm
Ichthammol	0.01703	m.r.	NH_3 = 2.5
Magnesium carbonate	0.02026	m.o.	MgO = 40 to 43.5
Magnesium citrate solution	....	phen.	$H_3C_6H_5O_7$ = 23.7 ml 0.5 *N* acid/100 ml
Magnesium phosphate, tribasic	....	m.o.	Acid capacity = 14.8 to 15.4 ml 1 *N* acid/Gm
Methenamine	0.03505	m.r.	$C_6H_{12}N_4$ = 99
Methenamine tablets	0.03505	m.r.	$C_6H_{12}N_4$ = 95 to 105 l.a.
Methenamine and sodium biphosphate tablets	0.03505	m.r.	$C_6H_{12}N_4$ = 92.5 to 107.5 l.a.
Propylhexadrine inhalant	0.15529	m.r.	$C_{10}H_{21}N$ = 90 to 120 l.a.
Seidlitz powders	0.08400	phen.	$NaHCO_3$ = 23 to 27
Silver nitrate solution, ammoniacal	0.01703	m.r.	NH_3 = 9.0 to 9.7 *W/W*
Triethylenemelamine	0.06808	phen.	$C_9H_{12}N_6$ = 98
Tuaminoheptane	0.16426	m.r.	$C_{14}H_{34}N_2{\cdot}H_2SO_4$ = 96.5
Zinc undecylenate	0.21596	m.o.	$C_{22}H_{38}O_4Zn$ = 98 to 102

n.m.t. = not more than; l.a. = labeled amount; (R) = reagent; m.o. = methyl orange; m.r. = methyl red; b.c.g. = bromocresol green; phen. = phenolphthalein; th.b. = thymol blue; b.p.b. = bromophenol blue.

Questions and Problems

1. Give a reason for the addition of water from time to time to replace that lost on heating the mixture of methenamine and standard sulfuric acid.

2. Suggest a possible advantage in the use of standard sulfuric acid rather than hydrochloric acid in the assay of methenamine.

3. If a 0.4340-Gm sample of methenamine were treated in the usual way with 25.00 ml. of 0.6500 *N* sulfuric acid, and 9.45 ml of 0.4500 *N* sodium hydroxide were required in the back titration, what was the percent $C_6H_{12}N_4$ of the sample?

Determination of Distilled Bases in Boric Acid Solution

Exercise 19

Object. Assay of Ammonium Chloride Injection.

Materials Required. 200 mg of ammonium chloride
Sodium hydroxide solution (2 in 5)
Boric acid solution (1 in 25)
0.1 *N* sulfuric acid

Procedure. "Transfer an accurately measured volume of Ammonium Chloride Injection, equivalent to about 200 mg. of ammonium chloride, to a 500-ml. Kjeldahl flask, add 200 ml. of water, mix, and add 50 ml. of sodium hydroxide (2 in 5). Immediately connect the flask by means of a distillation trap to a well-cooled condenser, the delivery tube of which dips into 40 ml. of boric acid solution (1 in 25) contained in a suitable receiver. Heat to boiling, and distill about 200 ml. Cool the liquid in the receiver if necessary, add methyl red T.S., and titrate with 0.1 *N* sulfuric acid. Perform a blank test with the same quantities of the same reagents and in the same manner, and make any necessary correction. Each milliliter of 0.1 *N* sulfuric acid is equivalent to 5.349 mg. of NH_4Cl."

The NaOH reacts with the ammonium chloride, liberating NH_3. The distilling flask should be connected to the condenser at once after adding the NaOH to prevent loss of the volatile ammonia. The NH_3 formed in the reaction

$$NH_4Cl + NaOH \longrightarrow NaCl + NH_3 + H_2O$$

distills with water when the flask is heated. The ammonia is conducted into a receiving flask which contains boric acid solution. The condenser should be fitted tightly with an adapter in such a manner that the distillate will be collected beneath the surface of the boric acid solution. This is done to prevent the loss of NH_3 from the distillate. As soon as the NH_3 in the distillate comes into the boric acid solution it is fixed by the weakly acidic boric acid, as indicated by the equation

$$NH_3 + H_3BO_3 \rightleftharpoons NH_4BO_2 + H_2O$$

Boric acid is such a weak acid that it has no effect on such indicators as bromocresol green or methyl red, and a sharp end point results when the solution is titrated with a quantity of standard sulfuric acid or hydrochloric acid equivalent to the ammonia resulting from complete hydrolysis of ammonium metaborate.

In the assay of a number of official substances such as the neostigmine preparations, ammonia or a volatile amine in organic combination, e.g., as amide or carbamate, is liberated by a quantitative hydrolytic degradation of the organic substance on treatment with sodium hydroxide. Sometimes

TABLE 11. ACIDIMETRIC ASSAYS. TITRATION OF DISTILLED BASE IN BORIC ACID SOLUTION

Substance	mEq factor, Gm	Indicator	Official requirement, percent
U.S.P.			
Ammonium chloride injection	0.05349	m.r.	NH_4Cl = 95 to 105 l.a.
Ammonium chloride tablets	0.05349	m.r.	NH_4Cl = 94 to 106 l.a.
Neostigmine bromide	0.30321	m.r.	$C_{12}H_{19}BrN_2O_2$ = 98 to 102
Neostigmine bromine tablets	0.30321	m.r.	$C_{12}H_{19}BrN_2O_2$ = 93 to 107 l.a.
Neostigmine methylsulfate	0.33439	m.r.	$C_{13}H_{22}N_2O_6S$ = 98 to 102
Neostigmine methylsulfate injection	0.33439	m.r.	$C_{13}H_{22}N_2O_6S$ = 90 to 110 l.a.
Paramethadione	0.15717	m.r.	$C_7H_{11}NO_3$ = 98 *W/W*
Paramethadione capsules	0.15717	m.r.	$C_7H_{11}NO_3$ = 94 to 96 l.a.
Paramethadione solution	0.15717	m.r.	$C_7H_{11}NO_3$ = 94 to 96 l.a.
Pyridostigmine bromide tablets	0.26114	b.c.g./m.r.	$C_9H_{13}BrN_2O_2$ = 95 to 105 l.a.
Trimethadione	0.14314	m.r.	$C_6H_9NO_3$ = 98 to 102
Trimethadione capsules	0.14314	m.r.	$C_6H_9NO_3$ = 94 to 106 l.a.
Trimethadione solution	0.14314	m.r.	$C_6H_9NO_3$ = 94 to 106 l.a.
N.F.			
Nikethamide injection	0.17824	m.r.	$C_{10}H_{14}N_2O$ = 95 to 105 l.a.
Phensuximide oral suspension	0.18922	m.r./m.b.	$C_{11}H_{11}NO_2$ = 5.65 to 6.85 *W/V*
Sodium saccharin tablets	0.24121	m.r.	$C_7H_4NaO_3S \cdot 2H_2O$ = 95 to 110 l.a.
Trimethadione tablets	0.14314	m.r.	$C_6H_9NO_3$ = 94 to 106 l.a.

l.a. = labeled amount; m.r. = methyl red; m.b. = methylene blue; b.c.g. = bromocresol green.

an acid hydrolysis is employed before the ammonia or amine is liberated by addition of sodium hydroxide, as in the assay of nikethamide injection, trimethadione, and sodium saccharin tablets.

Table 11 lists those preparations which have assay procedures in which distillation of ammonia or an amine is involved.

Certain organic bases which are not volatile can be liberated from their salts and extracted with chloroform, ether, or a similar material, in a

TABLE 12. ACIDIMETRIC ASSAYS. SOLVENT-EXTRACTED ORGANIC BASES DETERMINED BY RESIDUAL TITRATION OF EXCESS ACID WITH BASE

Substance	mEq factor, Gm	Indicator	Official requirement, percent
U.S.P.			
Cyclopentolate hydrochloride	0.32785	m.r.	$C_{17}H_{25}NO_3 \cdot HCl$ = 98
Cyclopentolate hydrochloride ophthalmic solution	0.32785	m.r.	$C_{17}H_{25}NO_3 \cdot HCl$ = 95 to 105 l.a.
Dextroamphetamine sulfate	0.18425	m.r.	$(C_9H_{13}N)_2 \cdot H_2SO_4$ = 98 *W/W*
Dimenhydrinate	0.25536	m.r.	$C_{17}H_{21}NO$ = 53.0 to 55.5
Dimenhydrinate syrup	0.46998	m.r.	$C_{17}H_{21}NO \cdot C_7H_7ClN_4O_2$ = 0.295 to 0.330 *W/V*
Lidocaine hydrochloride	0.27082	b.c.g./m.r.	$C_{14}H_{22}N_2O \cdot HCl$ = 99 to 101
Lidocaine hydrochloride injection	0.27082	b.c.g./m.r.	$C_{14}H_{22}N_2O \cdot HCl$ = 95 to 105 l.a.
Lidocaine hydrochloride jelly	0.27082	b.c.g./m.r.	$C_{14}H_{22}N_2O \cdot HCl$ = 95 to 105 l.a.
Lidocaine ointment	0.23434	b.c.g./m.r.	$C_{14}H_{22}N_2O$ = 95 to 105 l.a.
Mecamylamine hydrochloride tablets	0.20376	m.r.	$C_{11}H_{21}N \cdot HCl$ = 90 to 110 l.a.
Methadone hydrochloride injection	0.34592	m.r.	$C_{21}H_{27}NO \cdot HCl$ = 95 to 105 l.a.
Methadone hydrochloride tablets	0.34592	m.r.	$C_{21}H_{27}NO \cdot HCl$ = 93 to 107 l.a.
Methoxamine hydrochloride injection	0.24772	m.r.	$C_{11}H_{17}NO_3 \cdot HCl$ = 93 to 107 l.a.
N.F.			
Cyclopentamine hydrochloride injection	0.17792	m.r.	$C_9H_{19}N \cdot HCl$ = 95 to 105 l.a.
Cyclopentamine hydrochloride solution	0.17792	m.r.	$C_9H_{19}N \cdot HCl$ = 95 to 105 l.a.
Mephentermine sulfate	0.21232	m.r.	$(C_{11}H_{17}N)_2 \cdot H_2SO_4$ = 98
Mephentermine sulfate injection	0.16326	m.r.	$C_{11}H_{17}N$ = 95 to 105 l.a.
Meprylcaine hydrochloride	0.27179	m.r.	$C_{14}H_{21}NO_2 \cdot HCl$ = 98.5 to 101.5 *W/W*
Meprylcaine hydrochloride and epinephrine injection	0.27179	m.r.	$C_{14}H_{21}NO_2 \cdot HCl$ = 90 to 110 l.a.
Metabutoxycaine hydrochloride	0.34489	m.r.	$C_{17}H_{28}N_2O_3 \cdot HCl$ = 98.5 to 101.5 *W/W*
Metabutoxycaine hydrochloride and epinephrine injection	0.34489	m.r.	$C_{17}H_{28}N_2O_3 \cdot HCl$ = 95 to 105 l.a.
Naphazoline hydrochloride	0.24674	m.r.	$C_{14}H_{14}N_2 \cdot HCl$ = 98 *W/W*

l.a. = labeled amount; m.r. = methyl red; b.c.g. = bromocresol green.

manner analogous to the liberation of ammonia or volatile amine from a salt. After concentration of the solution of the organic base in the organic solvent, a measured excess of standard acid is added. The mixture is heated to complete the removal of organic solvent, and the excess acid is determined by back titration with a standard base. Such a procedure is applied in the assay of a number of salts of nonvolatile organic bases listed in Table 12.

Questions and Problems

1. Calculate the pH of a 1 *M* solution of ammonium chloride and of a 1 in 25 solution of boric acid, and explain the significance of the values with respect to the titration of ammonia dissolved in boric acid with standard hydrochloric acid using a methyl red indicator.

2. Give equations for the reactions that take place in the assay of each of the following, and calculate the equivalent weight of the substance determined: (*a*) nikethamide injection; (*b*) saccharin sodium tablets; (*c*) neostigmine bromide; (*d*) trimethadione.

3. The NH_3 from a 0.3256-Gm sample of ammonium valerate was distilled into 25.00 ml of 0.1342 *N* sulfuric acid and the mixture titrated to a methyl red end point with 25.05 ml of 0.0950 *N* sodium hydroxide. Calculate the percent of $C_4H_9COONH_4$ in the sample.

Determination of Nitrogen Content by the Kjeldahl Method. It frequently happens that ammonia is not quantitatively liberated from an

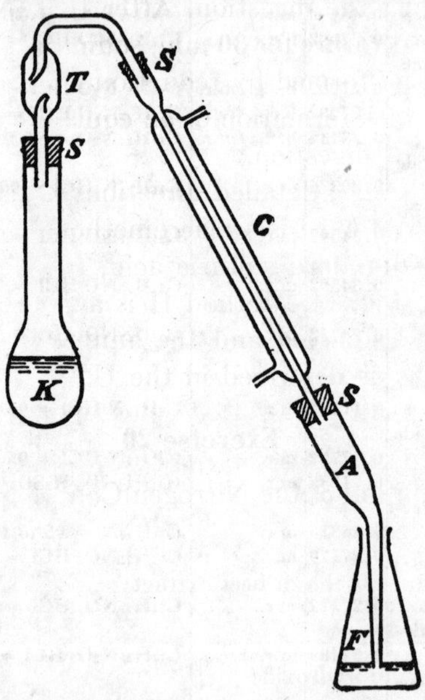

Fig. 14. Apparatus for Kjeldahl distillation. *K*, 500-ml Kjeldahl flask; *S*, tightly fitting rubber stoppers; *T*, spray trap or connecting bulb; *C*, condenser; *A*, adapter; *F*, 500-ml Erlenmeyer flask.

organic nitrogen-containing compound by acid or alkaline hydrolysis as previously described. If the nitrogen of such a compound is in the trivalent state, amino or amido nitrogen, an equivalent amount of ammonia can be obtained as ammonium sulfate by digestion of the sample with concentrated sulfuric acid until decomposition is complete as indicated by the disappearance of carbon which is converted to carbon dioxide during the digestion. When the digestion is complete, the ammonia is liberated by treatment of the residue of ammonium sulfate and excess sulfuric acid with sodium hydroxide, and distilled into excess standard sulfuric acid or boric acid (1 in 25). The distillate in each case is titrated in the usual manner.

The time required for the digestion can be reduced by adding potassium sulfate in a sufficient amount so that the molal ratio of H_2SO_4 to K_2SO_4 is 1.5, thereby raising considerably the temperature at which the digestion is carried out (Gunning modification). The time required for the digestion can also be reduced by addition of catalysts such as copper sulfate or selenium.

If all or part of the nitrogen is present as nitrate, a further modification is necessary. Salicylic acid is added to the digestion mixture; by its reaction with nitric acid liberated by the sulfuric acid, it forms nitrosalicylic acid and prevents loss of the nitric acid along with other volatile substances formed during the digestion. After the mixture has stood with shaking at room temperature for 30 min, sodium thiosulfate is added, and the aminosalicylic acid formed by reduction of nitrosalicylic acid undergoes decomposition with formation of an equivalent amount of ammonia during the subsequent digestion.

In the U.S.P. and N.F., detailed directions for determination of total nitrogen are given. Method I is a macromethod in which the ammonia is distilled into excess standard sulfuric acid, using the apparatus shown diagrammatically in Fig. 14. Method II is a semimicromethod in which smaller samples are employed and the ammonia is distilled into boric acid, using apparatus as described in the U.S.P.

Exercise 20

Object. Determination of the Nitrogen Content of the Alcohol-soluble Substances of Beef Extract.

Materials Required. 10 Gm of beef extract
100 ml of alcohol
50 ml of 0.1 *N* sulfuric acid
100 ml (2 in 5) of sodium hydroxide
10 Gm of powdered potassium sulfate
20 ml. of sulfuric acid

All the chemicals used must be of reagent quality and be free from nitrates and ammonium salts.

Procedure. 1. Weigh about 10 Gm of beef extract accurately, and add sufficient water to make 100 ml.

The beef extract can be weighed conveniently in a small evaporating dish and then washed into a 100-ml volumetric flask with sufficient water to make 100 ml of solution. The solution should be nearly clear and free from sediment to conform to N.F. specifications. This solution is used in the following steps:

2. Distribute 10 ml of the solution over clean, dry sand or asbestos, tared in a porcelain dish, and dry at 105° for 16 hr: the yield of residue is not less than 750 mg, equivalent to 75 percent of total solids in the original extract.

The solution of the extract is distributed over sand or asbestos to expose a large surface area and thereby facilitate evaporation of the water.

3. Transfer 25 ml of the solution to a 100-ml Erlenmeyer flask, add 50 ml of alcohol, and shake the mixture thoroughly. Collect the precipitate upon a counterpoised filter, wash it three times with a mixture of 2 volumes of alcohol and 1 volume of water, and dry at 105° for 2 hr: the weight of this precipitate, representing the alcohol-insoluble solids, does not exceed 10 percent of the total solids in the solution taken. (*Reserve the filtrate and washings for the determination of nitrogen.*)

The alcohol filtrate contains the alcohol-soluble substances of beef extract. This should be made up to a definite volume, so that an accurate aliquot portion may be measured in the next step.

4. Measure a portion of the alcohol filtrate remaining from the test for *Alcohol-insoluble substances*, corresponding to 1 Gm of the alcohol-soluble solids, into a 500-ml Kjeldahl flask.

The aliquot portion corresponding to 1 Gm of the alcohol-soluble solids of beef extract is calculated as follows: If the total solids, step 2, of a 10-ml sample of the solution of the extract weighed 0.8000 Gm, then the total solids of a 25-ml sample used in step 3 must have weighed 2.0000 Gm. If the alcohol-insoluble substances in this 25-ml sample were found to weigh 0.2000 Gm, the 25-ml sample must have contained 2.0000 − 0.2000 or 1.8000 Gm of alcohol-soluble substances. Therefore, $1/1.8 \times 25$ or 13.9 ml of the 25-ml sample of solution must have contained 1 Gm of alcohol-soluble substances. If the filtrate containing the alcohol-soluble substances of the 25-ml sample of solution measured 100 ml, then $13.9/25 \times 100$ or 55.6 ml of the alcohol filtrate should be used, since it contains 1 Gm of alcohol-soluble substances of beef extract.

5. Add about 10 Gm of powdered potassium sulfate or anhydrous sodium sulfate, 500 mg of powdered cupric sulfate, and 20 ml of sulfuric acid. Heat the mixture at a low temperature until frothing has ceased; then raise the temperature and boil the mixture for 30 min until it is clear green or almost colorless.

During this digestion, the organic components are decomposed to CO_2, H_2O, and NH_3. The latter remains in the residue as ammonium sulfate, while the CO_2 and H_2O are for the most part boiled off along with SO_2 resulting from reduction of sulfuric acid.

6. Allow to cool, add 150 ml. of water, mix the contents of the flask, and again cool. Add cautiously 100 ml. of sodium hydroxide solution (2 in 5), in such a manner as to cause the solution to flow down the inner side of the flask to form a layer under the acid solution. Add a few pieces of granulated zinc, and connect the flask, by means of a Kjeldahl connecting bulb, with a condenser, the delivery tube from which dips beneath the surface of 50 ml of boric acid solution (1 in 25) contained in a conical flask or wide-mouthed bottle of about 500-ml capacity. Mix the contents of the Kjeldahl flask by gentle rotation, and distil until about two-thirds of the contents of

TABLE 13. ACIDIMETRIC ASSAYS. NITROGEN DETERMINATIONS

Substance	Method	mEq factor, Gm	Official requirement, percent
U.S.P.			
Aminoacetic acid (R)	II	0.01401	N = 18.4 to 18.8
l-Asparagine (R)	II	0.01401	N = 18.4 to 18.8
Bemegride	II	0.01401	N = 8.9 to 9.12
Calcium caseinate (R)	I	0.01401	N = 12.5 to 14.3
Calcium pantothenate	...	0.01401	N = 5.7 to 6.0
Casein (R)	I	0.01401	N = 15.2 to 16.0
Glucagon	II	0.01401	N = 16.0 to 18.5
Insulin injection	II	0.01401	N = n.m.t. 0.70 mg/100 units* N = n.m.t. 0.85 mg/100 units†
Insulin suspension, isophane	II	0.01401	N = n.m.t. 0.85 mg/100 units
Insulin zinc suspension	II	0.01401	N = n.m.t. 0.70 mg/100 units
Insulin zinc suspension, extended	II	0.01401	N = n.m.t. 0.70 mg/100 units
Insulin zinc suspension, prompt	II	0.01401	N = n.m.t. 0.70/100 units
Insulin zinc suspension, protamine	II	0.01401	N = n.m.t. 1.25 mg/100 units
Pancreatic digest of animal tissue (R)	...	0.01401	N = n.l.t. 10
Pancreatic digest of casein (R)	...	0.01401	N = n.l.t. 10
Peptone, dried (R)	...	0.01401	N = 14.2 to 15.5
Pyridoxal hydrochloride (R)	...	0.01401	N = 6.7 to 7.1
Pyridoxamine dihydrochloride (R)	...	0.01401	N = 11.3 to 11.8
l-Tryptophane (R)	...	0.01401	N = 13.4 to 13.9
Urea	II	0.03003	CH_4N_2O = 98 to 102
Urea, sterile	II	0.03003	CH_4N_2O = 98 to 102
Yeast extract (R)	...	0.01401	N = 7.2 to 9.5
N.F.			
Beef extract (R)	...	0.01401	N = 60 mg/Gm alcohol-soluble solids
Dihydroxyaluminum aminoacetate	II	0.01401	N = 9.9 to 10.6
Dihydroxyaluminum aminoacetate magma	II	0.01401	N = 9.7 to 10.8
Methantheline bromide	II	0.42036	$C_{21}H_{26}BrNO_3$ = 98
Povidone-iodine	...	0.01401	N = 9.5 to 11.5
Tyrosine (R)	...	0.01401	N = 7.5 to 7.9
Urethan tablets	I	0.08909	$C_3H_7NO_2$ = 93 to 107 l.a.
Yeast tablets, dried	I	0.01401	95 l.a.

* Injection made from zinc-insulin crystals.
† Injection made from some other substance.
l.a. = labeled amount; (R) = reagent.

the flask has distilled over. Add about 3 drops of methyl red–methylene blue T.S. to the contents of the receiving vessel, and titrate with 0.5 *N* sulfuric acid. Perform a blank determination, and make necessary corrections. Each milliliter of 0.5 *N* acid consumed is equivalent to 7.003 mg of nitrogen.

Before adding the sodium hydroxide, a strip of litmus paper is dropped into the mixture and all connections, including the outlet tube of the condenser dipping into the boric acid, are made ready. A few pieces of granular zinc or porous plate, placed in the flask prior to distillation, will prevent bumping. The alkali liberates the ammonia from the ammonium sulfate, and upon distillation the ammonia gas is driven over and collected in boric acid solution.

Since 1 gram-atomic weight of nitrogen, 14.01 Gm, is necessary for formation of 1 mole of ammonia, the equivalent weight of nitrogen is 14.01 Gm and each milliliter of 0.5 *N* acid is equivalent to ½ mEq or 0.007003 Gm of nitrogen. Calculate the number of milligrams of nitrogen in 1 Gm of the alcohol-soluble substances of beef extract and compare with the N.F. specifications.

The official acidimetric assays by the macromethod, Method I, and the semimicro method, Method II, are given in Table 13. Where there is no designated method of nitrogen determination, the choice of either method is left up to the analyst.

The Kjeldahl method is applied widely in the analysis of proteins of food, fertilizer, blood, etc. Since the proportion of nitrogen in proteins is fairly constant, multiplication of the percent nitrogen found by a factor, 6.25, provides a rather accurate estimate of the proportion of protein in the sample.

Questions and Problems

1. What weight of each of the following should be used as sample in a semimicro Kjeldahl determination, so that 15 ml of 0.01 *N* sulfuric acid would be required in the titration: (*a*) histidine monohydrochloride; (*b*) carbachol; (*c*) sulfathiazole; (*d*) acetanilid?

2. Explain the purpose of each of the following in a Kjeldahl procedure: (*a*) potassium sulfate; (*b*) copper sulfate; (*c*) salicylate acid; (*d*) benzoic acid.

3. The ammonia from a 6.1500-Gm sample of an organic nitrogen compound was distilled into 50.00 ml of 0.1246 *N* sulfuric acid and the mixture was titrated with 22.42 ml of 0.0962 *N* sodium hydroxide. Calculate the percentage of nitrogen in the sample.

4. A 2.5000-Gm sample of acetanilid (98.5 percent C_8H_4NO) was assayed by a Kjeldahl procedure. If the ammonia from the sample were distilled into 50.00 ml of 1 *N* sulfuric acid, what volume of 1.2450 *N* sodium hydroxide would be required in the titration?

5. Calculate the percent CH_4N_2O in a 0.5000-Gm sample of urea, if 15.35 ml of 0.4967 *N* sodium hydroxide were required in the titration after the ammonia from the sample had been distilled into 50.00 ml of 0.4861 *N* sulfuric acid.

6. If the nitrogen content of the following is determined by a Kjeldahl procedure, what conversion factors would be used in calculation of the corresponding amount of the original compound: (*a*) tyrosine; (*b*) aminoacetic acid?

Chapter 5

ALKALIMETRIC ANALYSIS

Acids are estimated quantitatively by methods analogous to those employed for the quantitative estimation of bases, viz., by directly titrating an exact quantity of the acid, acid salt, or other acidic substance with standard alkali solutions or by adding an excess of the latter and determining the amount in excess by residual titration with standard acid solution. Direct titration is employed whenever practicable, since it requires only one standard solution, the standard base, and fewer measurements of volume.

In assaying acids, the quantity of acid to be taken should be such that about 30 to 40 ml of the standard base will be consumed. As a general principle, it is recommended that the normality of the solution obtained by dissolving the acid sample be approximately the same as that of the titrant. Except when otherwise directed, the liquid to be titrated should be brought to room temperature before titration, as many indicators give different values at different temperatures. For most of the inorganic acids, methyl red or phenolphthalein can be used as indicators, and the alkali should be standardized with the particular indicator used. For organic acids, phenolphthalein is frequently used; however, thymol blue, bromothymol blue, and thymolphthalein are also employed. In addition to the visual methods, potentiometric methods are also used in determining the end point.

DIRECT TITRATION METHODS

Exercise 21

Object. Assay of Hydrochloric Acid.

Materials Required. 3 ml of hydrochloric acid
1 *N* sodium hydroxide

Procedure. "Tare a glass-stoppered flask containing about 20 ml. of water, add about 3 ml. of Hydrochloric Acid, and again weigh. Dilute with about 25 ml. of water. add methyl red T.S., and titrate with 1 *N* sodium hydroxide. Each milliliter of 1 *N* sodium hydroxide is equivalent to 36.46 mg. of HCl."

Concentrated acid solutions are weighed in glass-stoppered flasks to prevent loss of dissolved gases, in some cases, and to prevent the absorption of water, in other cases. Samples are taken by weight rather than by volume, because it is difficult to measure small volumes of concentrated acids accurately, and a small error in the measurement of the sample would cause a large percent error in the result.

$$\underset{36.46}{HCl} + NaOH \longrightarrow NaCl + H_2O$$

The equivalent weight of HCl is 36.46 Gm. Consequently each milliliter of 1 N sodium hydroxide is equivalent to 0.03646 Gm, 1 mEq of HCl, and the proportion of HCl in the sample is

$$\frac{\text{ml base} \times N \times 0.03646 \times 100}{\text{Weight of sample}} = \text{percent, } W/W$$

Questions and Problems

1. On the basis of the assay procedures, why is the concentration of hydrochloric acid, U.S.P., expressed as percent W/W while that of diluted hydrochloric acid N.F. is expressed as percent W/V?

2. Refer to the official monographs for necessary data and calculate the percent W/V concentration of hydrochloric acid, U.S.P., and the percent W/W concentration of diluted hydrochloric acid, N.F.

3. A 4.0520-Gm sample of hydrochloric acid, sp gr 1.18, required 44.15 ml of 0.9035 N sodium hydroxide in a titration. Calculate (*a*) HCl content in percent; (*b*) normality; (*c*) Na_2CO_3 titer of the sample.

4. If 25.00 ml of a hydrochloric acid solution required 26.25 ml of 0.0952 N sodium hydroxide in a titration, what volume of the solution would produce 0.2867 Gm of AgCl?

5. A hydrochloric acid solution has an AgCl titer of 0.1265; what is its Na_2CO_3 titer?

Exercise 22

Object. Assay of Diluted Phosphoric Acid.

Materials Required. 10 ml of Diluted Phosphoric Acid
1 N sodium hydroxide

Procedure. "Pipet 10 ml. of Diluted Phosphoric Acid into a flask, and dilute it to about 50 ml. with water. Add 0.5 ml of thymolphthalein T.S., and titrate with 1 N sodium hydroxide to first appearance of a blue color. Perform a blank determination, and make any necessary correction. Each milliliter of 1 N sodium hydroxide is equivalent to 49.00 mg. of H_3PO_4."

The official diluted acids are assayed, as a rule, by measuring an exact volume of the acid, titrating it after dilution, and calculating the percent of acid on a weight-to-volume basis. This method of sampling is more convenient and rapid than the taking of weighed samples, and since the volume of diluted acid used as sample is relatively large, the relative error in measuring is slight. The phosphoric acid reacts with the sodium hydroxide as indicated:

$$H_3PO_4 + 2NaOH \longrightarrow Na_2HPO_4 + 2H_2O$$

Phosphoric acid is a tribasic acid but behaves as a dibasic acid when titrated to a thymolphthalein end point, because the ionization constant, 5×10^{-13}, for the third proton is very small. The $HPO_4^{=}$ ion is weakly acidic and is not neutralized at the pH at which thymolphthalein undergoes a color change. The pH of a 0.1 *M* solution of dibasic sodium phosphate, Na_2HPO_4, is 9.3; therefore it is expected that the equivalence point of the reaction above will be the same and the indicator of choice, thymolphthalein, which changes color between the pH of 8.6 and 10.0, is proper.

On the basis of the reaction and the choice of indicator, the equivalent weight of H_3PO_4 is 98.00/2 or 49.00 Gm, and that of Sodium Phosphate, Monobasic, NaH_2PO_4, U.S.P. reagent, is 119.98 Gm or 1 mole when it reacts with 1 mole of NaOH to form Na_2HPO_4.

Whenever there is reason to believe that the reagents in the official procedure or the conditions under which the analysis is performed may interfere or compete for the titrant or complement the titrant, then a blank determination is run so that any of the possible interference may be appropriately corrected for. In this analysis, the indicator changes color in a very basic range, and consequently any volume of sodium hydroxide used in the blank to bring about the color change must be accounted for in the actual analysis. The volume of sodium hydroxide used in the analysis of the sample of diluted phosphoric acid must be corrected by subtracting the volume of sodium hydroxide used in the blank determination, and the difference in volumes represents that volume of base that was consumed by the sample alone.

Questions and Problems

1. Hypophosphorous acid, N.F., contains about 31 percent HPH_2O_2, and nitric acid, U.S.P. reagent, contains about 69 percent HNO_3. Both are assayed by titration with 1.0 *N* sodium hydroxide. What weight of each would be used as sample in order that 35 ml of 1.0 *N* sodium hydroxide would be consumed in the titration? Compare answers with the weights suggested in the monographs. Explain the choice of indicators used in these assays.

2. A 10.00-ml sample of sulfuric acid solution required 16.85 ml of a sodium hydroxide solution in a titration. Each milliliter of the sodium hydroxide solution was equivalent to 0.2472 Gm of pure potassium biphthalate. Calculate the H_2SO_4 content of the sulfuric acid solution in percent *W/V*.

Exercise 23

Object. Assay of Boric Acid.

Materials Required. 2 Gm of boric acid
100 ml of glycerin
1 *N* sodium hydroxide

Procedure. "Dissolve about 2 Gm. of Boric Acid, accurately weighed, in 100 ml. of a mixture of equal volumes of glycerin and water, previously neutralized to phenolphthalein T.S. Add phenolphthalein T.S., and titrate with 1 *N* sodium hydroxide. Discharge the pink color by the addition of 50 ml. of glycerin, neutralized to phenolphthalein T.S., and continue the titration until the pink color reappears. Each milliliter of 1 *N* sodium hydroxide is equivalent to 61.83 mg. of H_3BO_3."

Boric acid behaves as a very weak monobasic acid,

$$H_3BO_3 \rightleftharpoons H^+ + BO_2^- + H_2O$$

because the metaborate ion is strongly basic and forces the above reaction to the left. Long before an equivalent amount of sodium hydroxide has been added, the BO_2^- ion concentration is sufficiently high and the H^+ ion concentration consequently sufficiently low that a change in the color of the phenolphthalein indicator occurs. If glycerin is added, however, it forms a slightly dissociated complex with the metaborate ion and thus prevents reversal of the reaction.

$$H_3BO_3 + 2\ \begin{matrix} CH_2OH \\ | \\ HC{-}OH \\ | \\ H_2C{-}OH \end{matrix} \longrightarrow \left[\begin{matrix} CH_2OH & & CH_2OH \\ | & & | \\ HC{-}O & & O{-}CH \\ & \diagdown\ \ \diagup & \\ & B & \\ & \diagup\ \ \diagdown & \\ H_2C{-}O & & O{-}CH_2 \end{matrix} \right]^- + H_3O^+ + 2H_2O$$

As a result, the pH during the titration remains below 8.0 until an equivalent of sodium hydroxide has been added. In other words, in the presence of glycerin, boric acid behaves like a sufficiently strong monobasic acid that its titration with standard alkali is feasible when phenolphthalein is used as indicator.

Since H_3BO_3 behaves as a monobasic acid in this titration, 1 mole is 1 equivalent and the equivalent weight is 61.83 Gm. Each milliliter of 1 *N* sodium hydroxide is equivalent to 61.83 mg or 1 mEq of H_3BO_3.

Questions and Problems

1. What preliminary treatment is necessary before the H_3BO_3 in boric acid ointment can be determined titrimetrically?
2. In the assay of boric acid ointment, about 10 percent H_3BO_3, an aliquot part corresponding to one-fifth of the original weight is titrated in the usual way with 0.1 *N* sodium hydroxide. Explain why an original sample weighing 5 Gm is employed.
3. Why must the glycerin be neutralized? Explain why additional glycerin is added after initially reaching the end point.

Determination of Hydroxy Compounds by Esterification with Acetic Anhydride. Hydroxy compounds can be esterified with acetic anhydride, with the products of the reaction being acetate ester and acetic acid. The acetic acid is determined with base. A blank determination is always run in such determinations. The acetic acid liberated is equivalent to the hydroxy compound being acetylated.

Exercise 24

Object. Assay of Ethohexadiol.

Materials Required. 1 Gm of Ethohexadiol
10 ml of acetic anhydride–pyridine mixture (1 in 3)
0.5 *N* alcoholic potassium hydroxide

Procedure. "Transfer about 1 Gm. of Ethohexadiol, accurately weighed, to an iodine flask, and add 10 ml. of freshly prepared mixture of 1 volume of acetic anhydride and 3 volumes of pyridine. Stopper the flask loosely, and heat on a steam bath for 1 hour. Add 10 ml. of water by means of the well on the top of the flask, and swirl the contents to insure complete contact of the water with the reagent."

During the heating period the two hydroxyl groups of the ethohexadiol are esterified with the liberation of 1 mole of acetic acid per mole of acetic anhydride.

$$\begin{array}{l} C_3H_7\text{—CHOH} \\ \qquad\quad | \\ C_2H_5\text{—CH—}CH_2OH \end{array} + 2 \begin{array}{l} CH_3\text{—C=O} \\ \qquad\quad \diagdown \\ \qquad\qquad O \\ \qquad\quad \diagup \\ CH_3\text{—C=O} \end{array} \longrightarrow \begin{array}{l} C_3H_7\text{—CHOCOCH}_3 \\ \qquad\quad | \\ C_2H_5\text{—CH—}CH_2OCOCH_3 \end{array} + 2CH_3COOH$$

At the end of the heating period the water added will react with the excess acetic anhydride, hydrolizing it to acetic acid. Pyridine acts as a catalyst

$$\begin{array}{l} CH_3C\text{=O} \\ \qquad \diagdown \\ \qquad\quad O \\ \qquad \diagup \\ CH_3C\text{=O} \end{array} + H_2O \longrightarrow 2CH_3COOH$$

in driving the esterification reaction to the right.

2. "After 2 minutes, cool the flask with the stopper partly open to prevent formation of vacuum. Rinse the stopper and the sides of the flask with alcohol, add cresol red–thymol blue T.S., and titrate with 0.5 *N* alcoholic potassium hydroxide. Perform a blank determination."

The blank determination in this analysis is a measure of the volume of alcoholic potassium hydroxide needed to react completely with the acetic anhydride contained in the 10.0-ml mixture. The difference in the volumes of alcoholic potassium hydroxide consumed represents the equivalent amount of acetic anhydride that reacted with the sample in esterifying the two hydroxyl groups.

One mole of the ethohexadiol reacts with 2 moles of acetic anhydride to yield 2 moles of acetic acid; therefore there are 2 equivalents in each mole or 146.23/2 or 73.12 Gm in 1 equivalent weight. Each milliliter of 0.5 *N* alcoholic potassium hydroxide is equivalent to 36.56 mg or 0.03656 Gm (½ mEq). The milliequivalent weight for the compound is 0.07312 Gm.

The mixed indicator gives a much sharper end point change and also accounts for the presence of pyridine in the titration mixture.

This analysis is an example of how an otherwise neutral compound can be analyzed in a conventional manner since in the esterification process an equivalent amount of acid is liberated and determined. The blank determination in this procedure consumes more base than does the sample determination, and the difference in volumes of base used in each determination represents that amount equivalent to the sample.

The calculations for the assay to determine the percent purity are:

$$\frac{(\text{ml}_{\text{blank}} - \text{ml}_{\text{sample}}) \times N \times 0.07312 \times 100}{\text{Sample weight}} = \text{percent } C_8H_{18}O_2$$

Questions and Problems

1. Calculate the weight of a sample of ethohexadiol to be used so that 20.00 ml of 0.5 *N* alcoholic potassium hydroxide will be required to titrate the acetic acid formed on hydrolyzing the excess acetic anhydride.
2. How many milliliters of 0.5 *N* alcoholic potassium hydroxide will 10 ml of the acetic anhydride–pyridine mixture (1 in 3) theoretically combine with in the blank determination performed in the assay?
3. Could an aqueous standard solution of sodium hydroxide be used in place of alcoholic potassium hydroxide? Explain.

Exercise 25

Object. Assay of Tartaric Acid.

Materials Required. 2 Gm of Tartaric Acid
1 *N* sodium hydroxide

Procedure. "Place about 2 Gm. of Tartaric Acid, previously dried over phosphorous pentoxide for 3 hours and accurately weighed, in a conical flask. Dissolve it in 40 ml. of water, add a few drops of phenolphthalein T.S., and titrate with 1 *N* sodium hydroxide. Each milliliter of 1 *N* sodium hydroxide is equivalent to 75.04 mg. of $C_4H_6O_6$."

Phenolphthalein is the most suitable indicator to use in the titration of organic acids. Tartaric acid reacts as a dibasic acid as follows:

$$\underset{150.09}{H_2C_4H_4O_6} + 2NaOH \longrightarrow Na_2C_4H_4O_6 + 2H_2O$$

As indicated by the equation, 2 moles of sodium hydroxide are required to neutralize 1 mole of tartaric acid; hence the equivalent weight of tartaric acid is 75.04 Gm. Each milliliter of 1 *N* sodium hydroxide is equivalent to 0.07504 Gm (1 mEq) of tartaric acid.

In Table 14 are listed a large number of organic acids which are assayed by a direct alkalimetric procedure. Normal or tenth-normal sodium

TABLE 14. ALKALIMETRIC ASSAYS. DIRECT TITRATION WITH BASE

Substance	mEq factor, Gm	Indicator	Official requirement, percent
U.S.P.			
Acetic acid	0.06005	phen.	$C_2H_4O_2$ = 36 to 37 *W/W*
Acetic acid (R)	0.06005	phen.	$C_2H_4O_2$ = 36 to 37 *W/W*
Acetic acid, glacial	0.06005	phen.	$C_2H_4O_2$ = 99.4
Acetic anhydride (R)	....	phen.	$(CH_3CO)_2O$ = 97
Ammonium citrate, dibasic (R)	0.07540	phen.	$(NH_4)_2HC_6H_5O_7$ = 98 to 103
Anticoagulant acid citrate dextrose solution	0.06404	phen.	$C_6H_8O_7$ = 95 to 105 l.a.
Benzaldehyde (R)	0.10613	b.p.b.	C_7H_6O = 98
Benzoic acid	0.12212	phen.	$C_7H_6O_2$ = 99.5
Boric acid	0.06183	phen.	H_3BO_3 = 99.5
Busulfan	0.12315	phen.	$C_6H_{14}O_6S_2$ = 98
Busulfan tablets	0.12315	phen.	$C_6H_{14}O_6S_2$ = 93 to 107 l.a.
2,3 Butanedione (R)	0.08609	b.p./b.	$CH_3COCOCH_3$ = 97
Cellulose, oxidized	0.04502	phen.	COOH = 16 to 24
Chlorambucil	0.30422	phen.	$C_{14}H_{19}Cl_2NO_2$ = 98 to 101
Citric acid	0.06404	phen.	$C_6H_8O_7$ = 99.5
N,N-Dimethyl-*p*-phenylenediamine dihydrochloride (R)	0.10456	pot.	$C_8H_{13}ClN_2$ = 98
Ethohexadiol	0.07311	cr./th.b.	$C_8H_{18}O_2$ = 97
Formic acid (R)	0.04603	phen.	HCOOH = 88
Hexachlorophene	0.40691	pot.	$C_{13}H_6Cl_6O_2$ = 98
Histamine phosphate	0.07678	th.ph.	$C_5H_9N_3{\cdot}2H_3PO_4$ = 98
Hydrochloric acid	0.03646	m.r.	HCl = 35 to 38 *W/W*
Hydrochlorid acid (R)	0.03646	m.o.	HCl = 36.5 to 38
Hydrofluoric acid (R)	0.02001	m.o.	HF = 48 to 51
Hypophosphorous acid (R)	0.0660	m.r.	HPH_2O_2 = 48
Iopanoic acid tablets	0.57094	th.b.	$C_{11}H_{12}I_3NO_2$ = 95 to 105 l.a.
Metaphosphoric acid (R)	0.07998	b.c.g.	HPO_3 = 34 to 36
Methenamine mandelate	0.15214	phen.	$C_8H_8O_3$ = 50
Methenamine mandelate oral suspension	0.29235	phen.	$C_6H_{12}N_4{\cdot}C_8H_8O_3$ = 90 to 110 l.a.
Methimazole	0.11417	b.th.b.	$C_4H_6N_2S$ = 98 to 101
Methimazole tablets	0.11417	b.th.b.	$C_4H_6N_2S$ = 94 to 106 l.a.
Nitric acid (R)	0.06301	m.o.	HNO_3 = 69 to 71
Nitric acid, fuming (R)	0.06301	m.o.	HNO_3 = 90
Perchloric acid (R)	0.10046	phen.	$HClO_4$ = 70 to 72
Phenobarbital	0.23224	pot.	$C_{12}H_{12}N_2O_3$ = 98 to 101
2-Phenoxyethanol (R)	0.13817	phen.	$C_8H_{10}O_2$ = 99
Phosphoric acid (R)	0.0490	th.ph.	H_3PO_4 = 85
Phosphorus pentoxide (R)	0.03549	th.ph.	P_2O_5 = 99
Potassium biphthalate (R)	0.20423	phen.	$KHC_6H_4(COO)_2$ = 99.9 to 100.1
Potassium pyrosulfate	0.04904	m.o.	H_2SO_4 = 35 to 37

TABLE 14. ALKALIMETRIC ASSAYS. DIRECT TITRATION WITH BASE (*Continued*)

Substance	mEq factor, Gm	Indicator	Official requirement, percent
Probenicid	0.28536	phen.	$C_{13}H_{19}NO_4S$ = 98
Propylthiouracil	0.08512	b.th.b.	$C_7H_{10}N_2OS$ = 98
Propylthiouracil tablets	0.08512	b.th.b.	$C_7H_{10}N_2OS$ = 93 to 107 l.a.
Saccharin	0.18319	phen.	$C_7H_5NO_3S$ = 98 to 101
Salicylic acid	0.13812	phen.	$C_7H_6O_3$ = 99.5 to 101
Salicylic acid collodion	0.13812	b.th.b.	$C_7H_6O_3$ = 9.5 to 11.5
Sodium bitartrate (R)	0.19009	phen.	$NaHC_4H_4O_6 \cdot H_2O$ = 99 to 100.5
Sodium hydrosulfite (R)	0.03482	phen.	$Na_2S_2O_4$ = 88
Sodium phosphate, monobasic (R)	0.119	th.ph.	NaH_2PO_4 = 99 to 101
Sulfamic acid (R)	0.09709	phen.	HSO_3NH_2 = 99.5
Sulfuric acid (R)	0.04904	m.o.	H_2SO_4 = 95 to 98
Sulfuric acid, fuming (R)	0.04903	phen.	H_2SO_4
Tolbutamide	0.27035	phen.	$C_{12}H_8N_2O_3S$ = 98 to 101
Trichloroacetic acid	0.16339	phen.	$C_2HCl_3O_2$ = 99
N.F.			
Acetic acid, diluted	0.06005	phen.	$C_2H_4O_2$ = 5.7 to 6.3 *W/V*
Benzaldehyde	0.10613	b.th.b.	C_7H_6O = 98
Bithional	0.35606	pot.	$C_{12}H_6Cl_4O_2S$ = 98
Boric acid solution	0.06183	phen.	H_3BO_3 = 4.25 to *W/V*
Caffeine, citrated	0.06404	phen.	$C_6H_8O_7$ = 48 to 52
Calcium saccharin	0.20223	phen.	$C_4H_8CaN_2O_6S_2$ = 95
Dehydrocholic acid	0.40254	phen.	$C_{24}H_{34}O_5$ = 98.5 to 101
Dehydrocholic acid tablets	0.40254	phen.	$C_{24}H_{34}O_5$ = 94 to 106 l.a.
Ethchlorvynol	0.14460	m.r./m.b.	C_7H_9ClO = 95 to 102
Ethchlorvynol capsules	0.14460	m.r./m.b.	C_7H_9ClO = 90 to 110 l.a.
Ethinamate	0.16721	m.r./m.b.	$C_6H_{13}NO_2$ = 98
Ethinamate tablets	0.16721	m.r./m.b.	$C_6H_{13}NO_2$ = 90 to 110 l.a.
Ethyl biscoumacetate	0.40837	m.r.	$C_{22}H_{16}O_8$ = 97 to 101
Ethyl biscoumacetate tablets	0.40837	m.r.	$C_{22}H_{16}O_8$ = 95 to 105 l.a.
Glutamic acid hydrochloride	0.09180	b.th.b.	$C_5H_9NO_4 \cdot HCl$ = 99 to 101
Glutamic acid hydrochloride capsules	0.09180	b.th.b.	$C_5H_9NO_4 \cdot HCl$ = 93 to 107 l.a.
Hydrochloric acid, diluted	0.03646	m.r.	HCl = 9.5 to 10.5 *W/V*
Hypophosphorous acid	0.06600	phen.	HPH_2O_2 = 30 to 32
Iodoalphionic acid tablets	0.24704	th.b.	$C_{10}H_{15}I_2O_3$ = 94 to 106 l.a.
Maleic acid (R)	0.05804	phen.	$C_4H_4O_4$ = 99
Meprobamate	0.10913	phen.	$C_9H_{18}N_2O_4$ = 97
Meprobamate tablets	0.10913	phen.	$C_9H_{18}N_2O_4$ = 95 to 105 l.a.
Niacin	0.12311	phen.	$C_6H_5NO_2$ = 99.5 to 101
Phenylbutazone	0.30838	phen.	$C_{19}H_{20}N_2O_2$ = 98

TABLE 14. ALKIMETRIC ASSAYS. DIRECT TITRATION WITH BASE *(Continued)*

Substance	mEq factor, Gm	Indicator	Official requirement, percent
Phenylbutazone tablets	0.30838	phen.	$C_{19}H_{20}N_2O_2$ = 93 to 107 l.a.
Phosphoric acid	0.04900	th.ph.	H_3PO_4 = 85 to 88
Phosphoric acid, diluted	0.04900	th.ph.	H_3PO_4 = 9.5 to 10.5 *W/V*
Potassium bitartrate	0.18818	phen.	$C_4H_5KO_6$ = 99 to 101
Potassium phosphate	0.17418	pot.	K_2HPO_4 = 98
Seidlitz powders	0.07505	phen.	$C_4H_6O_6$ = 2 to 2.4 Gm/powder
Sodium biphosphate	0.13799	phen.	$NaH_2PO_4 \cdot H_2O$ = 98 to 103
Sodium dehydrocholate injection	0.42452	th.b.	$C_{24}H_{33}NaO_5$ = 94 to 106 l.a.
Sodium phosphate	0.14196	pot.	Na_2HPO_4 = 98
Sodium phosphate, exsiccated	0.14196	pot.	Na_2HPO_4 = 98
Sodium saccharin	0.20517	phen.	$C_7H_4NNaO_3S$ = 98 to 101
Sulfosalicylic acid (R)	0.12711	phen.	$C_7H_6O_6S \cdot 2H_2O$ = 99 to 101
Tartaric acid	0.07505	phen.	$C_6H_6O_6$ = 99.7
Undecylenic acid ointment, compound	0.18428	phen.	$C_{11}H_{20}O_2$ = 4.5 to 5.5
Vinbarbital	0.22426	th.ph.	$C_{11}H_{16}N_2O_3$ = 98.5
Vinbarbital capsules	0.22426	th.ph.	$C_{11}H_{16}N_2O_3$ = 95 to 105 l.a.

(R) = reagent; l.a. = labeled amount; phen. = phenolphthalein; b.p.b. = bromophenol blue; pot. = potentiometric; c.r. = cresol red; t.h.b. = thymol blue; b.c.g. = bromocresol green; b.t.h.b. = bromothymol blue; m.o. = methyl orange.

hydroxide is generally used as titrant, and, in most titrations, phenolphthalein is used as the indicator.

Table 14 lists those official compounds which are determined by direct titration with base. In addition to the inorganic acids, organic acids, acid anhydrides, and acid salts, there are a number of substances (1) that are acidic in nature, (2) whose otherwise acidic character can be enhanced, and (3) with which a chemical agent will cause the formation of an equivalent amount of acid.

1. Bithionol, Hexachlorophene, and Ethyl Biscoumacetate contain acidic phenolic OH's that are titratable with base. Saccharin, Calcium Saccharin, and Tolbutamide have acidic imide H's which can be neutralized by base.

2. Silver nitrate will react with a number of compounds possessing weakly acidic hydrogens which are displaced by the silver, thus liberating an equivalent amount of HNO_3. Compounds which will undergo this reaction with silver nitrate are acetylenic compounds and certain heterocyclic nitrogen compounds such as Methimazole, and Propylthiouracil.

3. Methenamine will react with acid to yield formaldehyde and an ammonium salt. The reverse of this reaction is used in the assay of

Dibasic Ammonium Citrate and Meprobamate. Formaldehyde will react with the former substance, forming methenamine and citric acid, which is then neutralized with standard acid. Meprobamate is hydrolyzed with acid to ammonium chloride, which is then treated with formaldehyde to form methenamine and hydrochloric acid, which is then neutralized with base.

Benzaldehyde is assayed by reacting it with hydroxylamine hydrochloride, forming the aldoxime and liberating an equivalent amount of HCl, which is then determined.

Blank determinations are necessary in these procedures to correct for the action of the reagent with the titrant.

Questions and Problems

1. Why is bromothymol blue used as indicator in the assay of glutamic acid hydrochloride, while phenolphthalein is used in the assay of tartaric acid? Write the equations for the reactions involved in these assays.

2. Calculate the following titer values for 0.1 *N* sodium hydroxide and compare the calculated values with those given in the official monographs: (*a*) salicylic acid, $C_7H_6O_3$; (*b*) iopanoic acid, $C_{11}H_{12}I_3NO_2$; (*c*) niacin, $C_6H_5NO_2$.

3. Calculate the equivalent weights of three acids assayed by use of 0.1 *N* sodium hydroxid; and of three acids assayed by use of 1 *N* sodium hydroxide. Suggest a reason for the choice of 1 *N* or 0.1 *N* solutions used in acidimetry.

4. What weight of each of the following substances would be used as sample so that 35 ml of 1 *N* sodium hydroxide would be consumed in titration to a phenolphthalein end point? (*a*) phosphoric acid, H_3PO_4; (*b*) sodium biphosphate, $NaH_2PO_4 \cdot H_2O$; (*c*) citric acid, $C_6H_8O_7$, (*d*) tartaric acid, $C_4H_6O_6$. Compare answers with sample weights specified in the official monographs.

RESIDUAL TITRATION METHODS

Residual titration methods are used whenever direct titration methods are not practicable. In general the method is applied to those official compounds which react too slowly with the titrant because of poor solubility; otherwise a heating process is used or a precipitation method is employed to convert the substance for reaction with the standard base. The esters, acid anhydrides, acid chlorides, and aldehydes are some classes of compounds found in this group.

Exercise 26

Object. Assay of Aspirin Capsules.

Materials Required. 20 Aspirin Capsules
0.1 *N* sulfuric acid
0.1 *N* sodium hydroxide

Procedure. 1. "Transfer, as completely as possible, the contents of not less than 20 Aspirin Capsules to a 100-ml volumetric flask. Place the emptied capsules in a small

beaker, add sufficient neutralized alcohol, previously cooled to 15° to 20°, to cover them, and agitate for 10 minutes. Filter into the flask, wash the beaker and filter well with cold neutralized alcohol, receiving the washings in the same flask, add cold neutralized alcohol to make exactly 100 ml., and mix. Transfer an accurately measured volume of the solution, equivalent to about 500 mg. of aspirin, to a conical flask, add cold neutralized alcohol, if necessary, to make about 20 ml., add phenolphthalein T.S., and titrate the solution immediately with 0.1 *N* sodium hydroxide."

The sample is titrated with sodium hydroxide to neutralize any free acid formed by hydrolysis of the acetylsalicylic acid as represented by the equation

$$C_6H_4{\cdot}OCOCH_3{\cdot}COOH + HOH \longrightarrow C_6H_4{\cdot}OH{\cdot}COOH + CH_3COOH$$

and to neutralize the carboxyl group of the acetylsalicylic acid:

$$\underset{180.16}{C_6H_4{\cdot}OCOCH{\cdot}COOH} + NaOH \longrightarrow C_6H_4{\cdot}OCOCH_3{\cdot}COONa + H_2O$$

2. "Add to the titrated solution a volume of 0.1 *N* sodium hydroxide equal to that used in the titration, plus 15 ml. more, and heat the mixture in a bath of boiling water for 15 minutes with frequent agitation. Remove the flask from the water bath, cool it quickly to room temperature, and titrate with 0.1 *N* sulfuric acid. Perform a blank determination with the same quantities of the same reagents and in the same manner. Subtract the number of milliliters of 0.1 *N* sulfuric acid used from the number of milliliters of 0.1 *N* sodium hydroxide added the second time, and multiply the difference by 18.02. The result represents the milligrams of $C_9H_8O_4$ present in the portion of the capsules used for the assay."

After the acetylsalicylic acid and its hydrolysis products, acetic acid and salicylic acid, have been neutralized in the first titration with 0.1 *N* sodium hydroxide, carried out at room temperature or slightly below, the only further reaction of the sample with the standard base added in excess to the titrated mixture involves saponification of the sodium acetylsalicylate as follows:

$$C_6H_4(OCOCH_3)(COONa) + NaOH \longrightarrow C_6H_4(OH)(COONa) + CH_3COONa$$

Consequently each milliliter of 0.1 *N* sodium hydroxide consumed in this reaction is equivalent to 0.1 mEq of sodium acetylsalicylate and, in turn, to 0.1 mEq of acetylsalicylic acid; i.e., $0.1 \times 0.18016 = 0.018016$ Gm of $C_9H_8O_4$.

The aspirin content of the sample employed should be calculated and, from that, the amount in each capsule and the percent of the labeled amount.

A number of official esters and preparations thereof are assayed by a residual alkalimetric procedure, e.g., ethyl acetate, benzyl benzoate, methyl salicylate, methylparaben, propylparaben, etc. Other organic substances which react slowly with sodium hydroxide and which are assayed by residual methods include benzoyl chloride reagent, butyl chloride, chloral hydrate, formaldehyde solution, lactic acid, etc. The chloral hydrate reacts with sodium hydroxide in the following manner:

$$CCl_3CHO \cdot H_2O + NaOH \longrightarrow HCOONa + CHCl_3 + H_2O$$

Formaldehyde is oxidized by hydrogen peroxide to formic acid, which

TABLE 15. ALKALIMETRIC ASSAYS. RESIDUAL TITRATION OF EXCESS BASE WITH ACID

Substance	mEq factor, Gm	Indicator	Official requirement, percent
U.S.P.			
Acetylcholine chloride (R)...	0.04305	phen.	CH_3CO = 23.2 to 24.2
Aluminum acetate solution..	0.06005	phen.	$C_2H_4O_2$ = 4.24 to 5.11 *W/V*
Aluminum subacetate solution	0.06005	phen.	$C_2H_4O_2$ = 5.43 to 6.13 *W/V*
Aspirin....................	0.09008	phen.	$C_9H_8O_4$ = 99.5
Benzoyl chloride (R)........	0.07028	phen.	C_6H_5COCl = 98 to 100.5
Benzyl benzoate...........	0.10613	phen.	$C_{14}H_{12}O_2$ = 99
Benzyl benzoate lotion.....	0.10613	phen.	$C_{14}H_{12}O_2$ = 26 to 30 *W/W*
Chloral hydrate...........	0.16540	phen.	$C_2H_3Cl_3O_2$ = 99.5 to 102.5
Chloral hydrate capsules....	0.16540	phen.	$C_2H_3Cl_3O_2$ = 95 to 110 l.a.
Chloral hydrate syrup......	0.16540	phen.	$C_2H_3Cl_3O_2$ = 95 to 110 l.a.
Formaldehyde solution......	0.03003	b.th.b.	CH_2O = 37 *W/W*
Formaldehyde solution (R)..	0.03003	b.th.b.	CH_2O = 36 to 38
Lactic acid................	0.09008	phen.	$C_3H_6O_3$ = 85 to 90
Methyl salicylate...........	0.15215	phen.	$C_8H_8O_3$ = 98
Methylparaben.............	0.15215	b.th.b.	$C_8H_8O_3$ = 99
Propylparaben.............	0.18021	b.th.b.	$C_{10}H_{12}O_3$ = 99
Succinylcholine chloride injection..................	0.18065	b.th.b.	$C_{14}H_{30}Cl_2N_2O_4$ = 93 to 107 l.a.
N.F.			
Aluminum phosphate gel....	0.00530	phen.	$AlPO_4$ = 4 to 5
Aspirin capsules............	0.18016	phen.	$C_9H_8O_4$ = 93 to 107 l.a.
Calcium phosphate, tribasic..	0.00674	phen.	$Ca_3(PO_4)_2$ = 90
Ethyl acetate..............	0.08811	phen.	$C_4H_8O_2$ = 99
Hydriodic acid, diluted......	0.00287	phen.	HPH_2O_2 = 0.6 to 1.0 *W/W*
Magnesium phosphate, tribasic....................	0.00572	phen.	$Mg_3(PO_4)_2$ = 98 to 101.5
Methacholine chloride.......	0.04305	phen.	CH_3CO = 21.6 to 22.3
Methacholine chloride, sterile	0.04305	phen.	CH_3CO = 21.6 to 22.3
Methenamine and sodium biphosphate tablets..........	0.00600	phen.	$NaH_2PO_4 \cdot H_2O$ = 92.5 to 107.5 l.a.
Paraformaldehyde (R)......	0.03003	b.th.b.	HCHO = 95
Phthalic anhydride (R).....	0.07406	phen.	$C_8H_4O_3$ = 99

l.a. = labeled amount; phen. = phenolphthalein; b.th.b. = bromothymol blue; (R) = reagent.

subsequently reacts with sodium hydroxide in a mole for mole ratio. The acetic acid in aluminum acetate and aluminum subacetate solution is distilled into a measured excess of standard alkali, and the distillate is titrated with standard acid to determine the excess base, a procedure analogous to that used in the assay of preparations containing ammonia, ammonium salts, etc.

The ammonium phosphomolybdate, $(NH_4)_3PO_4 \cdot MoO_3 \cdot H_2O$, produced quantitatively from various phosphates such as aluminum phosphate, calcium phosphate, hypophosphorous acid in diluted hydriodic acid, and magnesium phosphate, is measured by a residual titrimetric procedure according to the following equation:

$$(NH_4)_3PO_4 \cdot 12MoO_3 \cdot H_2O + 23NaOH \longrightarrow 11Na_2MoO_4 + (NH_4)_2MoO_4 + NaNH_4HPO_4 + 12H_2O$$

A list of official substances assayed by residual alkalimetric procedures appears in Table 15.

Questions and Problems

1. The contents of 20 aspirin capsules was dissolved in 100 ml of alcohol. If a 10-ml aliquot was found to contain 0.4880 Gm of aspirin, what was the percent of the labeled amount (5 grains) of aspirin in the capsules?

2. Calculate the weight of each of the following that is equivalent to 1 ml of 0.5 *N* alcoholic potassium hydroxide and check values with those given in the official monographs: (*a*) benzyl benzoate, (*b*) methyl salicylate, (*c*) butyl chloride.

3. Write equations for all reactions involved in the assay of sodium biphosphate in Methenamine and Sodium Biphosphate Tablets, and calculate the equivalent weight of NaH_2PO_4. What is the NaH_2PO_4 titer of 0.5 *N* sodium hydroxide?

Chapter 6

NONAQUEOUS TITRIMETRIC ANALYSIS

In no other field has nonaqueous titrimetry found so wide an application as in the field of quantitative pharmaceutical chemistry. With the introduction of many new complex organic medicinal agents, the problems of quality control increased because these therapeutic agents were in many instances not only poorly soluble but also weakly reactive in water. The amine salts had to be converted to the water-soluble free base, extracted with an organic solvent, and treated with excess standard acid; then the solvent had to be removed and the residual acid determined with standard base. Other organic compounds containing nitrogen were determined by the Kjeldahl method. Those compounds that were official as sodium salts were acidified, the water-insoluble organic acid was extracted with an organic solvent, the solvent was evaporated, and the residue was dried and weighed. These typical quantitative procedures possessed certain drawbacks. With a nonaqueous solvent the disadvantages of poor solubility and weak reactivity were overcome. Not only was it possible to perform many titrations that could not be carried out in an aqueous medium because of weak reactivity, but greater solubility of many substances in these organic solvents made possible a variety of direct titrimetric procedures. Further, by proper choice of solvent and titrant, the physiologically active moiety of a salt, whether the acidic or basic component, could be selectively titrated.

The simplicity, speed, precision, and accuracy of nonaqueous methods are equivalent to those of the classical procedures; and the same laboratory equipment may be employed with the exception that extra precautions should be taken where moisture and carbon dioxide are to be excluded.

Moisture is to be avoided in using nonaqueous procedures, since H_2O, being weakly basic, would compete with the weak nitrogen base for the $HClO_4$

$$H_2O + HClO_4 \rightleftharpoons H_3O^+ + ClO_4^-$$
$$RNH_2 + HClO_4 \rightleftharpoons RNH_3^+ + ClO_4^-$$

and the sharpness of the end point would be lost. Experimentally, it has been found that the moisture content in nonaqueous titrimetry should be held to less than 0.05 percent so as not to have any appreciable effect on the end point.

Greater care in the control of temperature during the standardization and analysis procedures is necessary when utilizing nonaqueous titrimetry because of the higher coefficients of expansion of organic solvents. A correction should be applied for each degree of variation in temperature of operation. In routine practice it will suffice if the temperature at which the standardization is run is kept the same as that at which the analysis is to be performed.

Neutralization. Contrary to the Arrhenius concept that neutralization is a reaction between hydrogen ions and hydroxide ions in aqueous solutions of acids and bases, the titration of a base with a standard solution of an acid, or the titration of an acid with a standard solution of a base when solvents other than water are employed is feasible and frequently offers certain advantages. In aqueous solutions, neutralization may be interpreted as a reaction between proton donors and proton acceptors, i.e., acids and bases, respectively. In nonaqueous solvents, it frequently becomes necessary to regard neutralization as a reaction between weakly protophyllic substances which tend to accept a pair of electrons and highly protophyllic substances which tend to provide a pair of electrons in the formation of covalent bonds by coordination; i.e., a reaction between acids and bases. The titration of an aqueous solution of an amine, R_3N, with an aqueous solution of an acid, HA, may be pictured as proceeding in accordance with the Arrhenius concept, as follows:

$$HA + H_2O \rightleftharpoons H_3O^+ + A^-$$
$$R_3N + H_2O \rightleftharpoons OH^- + R_3NH^+$$
$$H_3O^+ + OH^- \rightleftharpoons 2H_2O$$

A summation of these equations indicates that water is not essential for the reaction and gives the equation depicting the reaction in nonaqueous solvents:

$$HA + R_3N \longrightarrow R_3NH^+A^-$$

and the neutralization of a base with standard acid in nonaqueous titrimetry:

$$BOH + CH_3ONa \longrightarrow BO^-Na^+ + CH_3OH$$

A greater differential in the protophyllic properties of various substances occurs in nonaqueous solutions such as acetonitrile, acetone, and dimethylformamide than in aqueous solutions, because of the "leveling effect" of water in the latter solutions. Thus the least protophyllic substance, i.e., most acidic substance, in aqueous solutions of a variety of acids is the hydronium ion, formed as follows:

$$HA + H_2O \longrightarrow H_3O^+ + A^-$$

Although the following acids are of about equal strength in aqueous solution, in nonaqueous solvents, "acidity" decreases in the following order:

$$HClO_4 > HBr > H_2SO_4 > HCl > HNO_3$$

In dioxane, a neutral solvent, or glacial acetic acid, an acidic solvent, perchloric acid, $HClO_4$, is less protophyllic (more acidic) than HCl, and chlorides (base hydrochlorides) can be titrated with standard perchloric acid even as carbonates can be titrated in aqueous solution with standard HCl.

Bases such as metal alkoxides become stronger bases in those nonaqueous solvents where the alkoxides are least ionized.

Just as the choice of solvent can enhance the basic characteristic of a compound so as to make it more suitable for nonaqueous titration with standard acid, so does the proper choice of solvent enhance the acidic characteristic of a compound to make it more suitable for nonaqueous titration with a standard base. In fact it is possible to titrate mixtures of two or three components selectively with a single titration by the proper selection of solvent for the titration.

End Points. The indicators used in neutralization reactions in water are in themselves weak acids or bases which undergo color changes with changes in pH (hydrogen ion concentration in water). In nonaqueous titrations these same indicators will react with H^+ and produce the expected color changes or give up their H^+ and change color; however the choice of indicator for a particular reaction is for the present empirical. The equivalence point of a given neutralization reaction is ascertained potentiometrically, and then an indicator is chosen which will give the sharpest color change for the least increment of volume of titrant added near the equivalence point. Already certain indicators are known to be effective in the titration of bases and others for the titration of base salts. In the same manner substances of different acid strength require different indicators to detect the end point of titration.

ACIDIMETRY IN NONAQUEOUS SOLVENTS

The titrimetric analysis of a number of weakly basic substances in glacial acetic acid is possible using a standard solution of perchloric acid as the titrant. These compounds are amines, amine salts, heterocyclic nitrogen compounds, alkali salts of organic acids, alkali salts of weak inorganic acids, and amino acids.

Solvents. The solvents used in acidimetric titrimetry are either relatively neutral or acidic in nature; the choice of solvent is determined by the basic character of the substance to be assayed. Neutral solvents such as acetonitrile, alcohols, chloroform, benzene, dioxane, or ethyl acetate are either aprotic or amphiprotic in nature and are used for their solvency action primarily, since they do not enhance dissociation to any great degree. The acidic solvents, such as formic, glacial acetic, and propionic acids, acetic anhydride, and sulfonyl chloride, are used for weak bases and their salts where the protogenic quality of the solvent is utilized. A solvent such as acetic acid reacts with base, B, to yield the conjugate acid of the base, BH^+, and the conjugate base anion, CH_3COO^-,

$$B + CH_3COOH \rightleftharpoons BH^+ + CH_3COO^-$$

The conjugate base anion can then react with perchloric acid in the following manner:

The titrant reacts with the solvent, and the solvent is protonated:

$$HClO_4 + CH_3COOH \rightleftharpoons CH_3COOH_2^+ + ClO_4^-$$

The titration reaction, with the conjugate base anion, is as follows:

$$CH_3COO^- + CH_3COOH_2^+ \rightleftharpoons 2CH_3COOH$$

so that the overall reaction can be represented as

$$B + HClO_4 \longrightarrow BH^+ClO_4^-$$

It is possible to assay a mixture containing two bases of differing strengths, such as ethylenediamine and the much weaker base, theophylline, with the use of two solvents. For example, the titration is initially in acetic acid with perchloric acid, during which only the ethylenediamine is reacted; upon the addition of acetic anhydride, the alkalinity of the theophylline is sufficiently enhanced so that it will now react with the additional perchloric acid. A potentiometer is used to detect both end points. Thus, a single titration procedure would determine selectively the amounts of each constituent in the sample.

Titrant. The most valuable and widely used titrant is perchloric acid in glacial acetic acid or the relatively neutral solvent, dioxane. Perchloric acid is the strongest of the known common acids which gives excellent results with practically all the known weak bases. Hydrogen bromide and certain organic sulfonic acids are also used to a limited extent as titrants; the latter is even stronger than perchloric acid.

Indicators. The indicators which can be used in the titration of weak bases and their salts are crystal violet, methylrosaniline chloride, quinaldine red, α-naphtholbenzein, and malachite green. With relatively stronger bases, indicators such as methyl red, methyl orange, and thymol blue can be used. It has been observed that crystal violet gives a visual end point with less than 0.1 ml of titrant at the potentiometric end point. The choice of indicator is still arbitrary and is dictated only by the proximity of the color transition to the observed potentiometric end point.

Exericse 27

Object. To Prepare and Standardize 0.1 *N* Perchloric Acid.

Materials Required. 8.5 ml of perchloric acid
1,000 ml of glacial acetic acid
30 ml of acetic anhydride
700 mg of potassium biphthalate
Crystal violet T.S.

Procedure. 1. "Mix 8.5 ml. of perchloric acid with 500 ml. of glacial acetic acid and 30 ml. of acetic anhydride. Cool and add glacial acetic acid to make 1000 ml. Allow the prepared solution to stand for 1 day, for the excess acetic anhydride to be combined, and determine the water content by the *Titrimetric Method.* If the water content exceeds 0.05 per cent, add more acetic anhydride. If the solution contains no titratable water, add sufficient water to make the content between 0.02 per cent and 0.05 per cent of water. Allow to stand for 1 day, and again titrate the water content."

Perchloric acid occurs as a 70 to 72 percent mixture with water, having a specific gravity of about 1.6. It undergoes a spontaneous explosive decomposition and, for this reason, is handled in the form of a solution. Further, it is a powerful oxidizing agent, as well as a strong acid, and must be handled with greatest care. $HClO_4$ has a molecular weight of 100.46, and a liter of 0.1 *N* solution would contain one-tenth equivalent weight, or 10.046 Gm. The preparation of the solution calling for 8.5 ml with a specific gravity of about 1.6 and a purity of about 72 percent would calculate out as 9.792 Gm of $HClO_4$ per liter of standard solution.

Dioxane may be used in place of glacial acetic acid and acetic anhydride. It is felt by some chemists that the solutions of perchloric acid in dioxane give sharper end points. Dioxane generally contains reducing substances which must be removed by passing the solution through a column of activated alumina or by some alternate procedure; otherwise the standard solution would be discolored.

The approximately 30 percent water in the perchloric acid solution can be quantitatively converted to acetic acid by the addition of the proper amount of acetic anhydride, in the following manner:

$$H_2O + (CH_3CO)_2O \longrightarrow 2CH_3COOH$$

Thus the amount of water in the standard solution can be carefully regulated.

2. "Standardize the solution as follows: Weigh accurately about 700 mg. of potassium biphthalate, previously dried at 105° for 3 hours, and dissolve it in 50 ml. of glacial acetic acid in a 250-ml. flask. Add two drops of crystal violet T.S., and titrate with the perchloric acid solution until the violet color changes to emerald-green.

Deduct the volume of the perchloric acid consumed by 50 ml. of the glacial acetic acid, and calculate the normality. Each 20.42 mg. of potassium biphthalate is equivalent to 1 ml. of 0.1 *N* perchloric acid."

Potassium biphthalate, $KHC_8H_4O_4$, molecular weight 204.23, is used as a primary standard for bases in aqueous solution because of the weakly protophyllic character of the H^+ produced by ionization:

$$H^+ + OH^- \longrightarrow H_2O$$

or

$$C_6H_4(COOK)COOH + NaOH \longrightarrow H_2O + C_6H_4(COOK)(COONa)$$

In nonaqueous solvents, the same reagent is used as a primary standard for acids because of the highly protophyllic character of the $HC_8H_4O_4^-$ ion:

$$HC_8H_4O_4^- + H^+ \longrightarrow H_2C_8H_4O_4$$

or $$C_6H_4(COOH)COOK + HClO_4 \longrightarrow C_6H_4(COOH)_2 + KClO_4$$

The 700 mg of dried $KHC_8H_4O_4$ is about 3.5 mEq of the base and should consume about 35 ml of the 0.1 *N* perchloric acid. Using the weight of potassium biphthalate sample, the labeled $KHC_8H_4O_4$ content of the potassium biphthalate, and the volume of the perchloric acid solution corrected for the blank titration, the normality of the solution can be calculated. The blank titration corrects for any moisture in the glacial acetic acid that reacts with the perchloric acid.

As can be seen from the standardization procedure, the reaction of the perchloric acid with the primary standard can be applied to almost all alkali salts of organic acids, the only limitation being the solubility of the compound to be assayed.

Exercise 28

Object. Assay of Methamphetamine Hydrochloride.

Materials Required. 400 mg of Methamphetamine Hydrochloride
40 ml of glacial acetic acid
10 ml of mercuric acetate T.S.
Crystal violet T.S.

Procedure. "Dissolve about 400 mg. of Methamphetamine Hydrochloride, accurately weighed, in a mixture of 40 ml. of glacial acetic acid and 10 ml. of mercuric acetate T.S., warming slightly to effect solution. Cool the solution to room temperature, add 5 drops of crystal violet T.S., and titrate with 0.1 *N* perchloric acid. Perform a blank determination, and make any necessary correction. Each milliliter of 0.1 *N* perchloric acid is equivalent to 18.57 mg. of $C_{10}H_{15}N \cdot HCl$."

Perchloric acid is a stronger acid (less protophyllic) than hydrochloric acid in a nonaqueous solvent, and although it could be used to titrate base hydrochlorides, for practical purposes it is necessary to add mercuric acetate T.S. to the titration procedure to prevent the interference of the hydrochloric acid displaced through the formation of the relatively un-ionized $HgCl_2$, thus:

$$2C_{10}H_{15}N \cdot HCl + Hg(OAc)_2 + 2HClO_4 \longrightarrow 2C_{10}H_{15}N \cdot HClO_4 + 2HOAc + 2HgCl_2$$

TABLE 16. ACIDIMETRIC ASSAYS. NONAQUEOUS TITRATIONS WITH PERCHLORIC ACID

Substance	mEq factor Gm	Indicator	Official requirement, percent
U.S.P.			
Bethanechol chloride	0.19668	cr.v.	$C_7H_7ClN_2O_2$ = 98 to 101.5
Caffeine	0.19419	pot.	$C_8H_{10}N_4O_2$ = 98.5 to 101.5
Carbachol	0.18265	cr.v.	$C_6H_{15}ClN_2O_2$ = 99 to 101
Chlorcyclizine hydrochloride	0.16865	pot.	$C_{18}H_{21}ClN_2 \cdot HCl$ = 98
Chloroprocaine hydrochloride	0.30723	pot.	$C_{13}H_{19}ClN_2O_2 \cdot HCl$ = 98 to 102
Chlorpheniramine maleate	0.19544	m.ros.	$C_{16}H_{19}ClN_2 \cdot C_4H_4O_4$ = 98
Chlorpheniramine maleate syrup	0.19544	m.ros.	$C_{16}H_{19}ClN_2 \cdot C_4H_4O_4$ = 0.045 to 0.055 *W/V*
Chlorpromazine hydrochloride	0.35534	pot.	$C_{17}H_{19}ClN_2S \cdot HCl$ = 98 to 101.5
Cyclizine hydrochloride	0.15143	pot.	$C_{18}H_{22}N_2 \cdot HCl$ = 98
Dibucaine hydrochloride	0.18997	cr.v.	$C_{20}H_{29}N_3O_2 \cdot HCl$ = 97
Diethylcarbamazine citrate	0.39142	cr.v.	$C_{10}H_{21}N_3O \cdot C_6H_8O_7$ = 98
Diethylcarbamazine citrate syrup	0.39142	th.b.	$C_{10}H_{21}N_3O \cdot C_6H_8O_7$ = 2.28 to 2.52 *W/V*
Diethylcarbamazine citrate tablets	0.39142	th.b.	$C_{10}H_{21}N_3O \cdot C_6H_8O_7$ = 95 to 105 l.a.
Diphenhydramine hydrochloride	0.29183	cr.v.	$C_{17}H_{21}NO \cdot HCl$ = 98
Edrophonium chloride	0.20170	q.r.	$C_{10}H_{16}ClNO$ = 98
Epinephrine	0.18321	cr.v.	$C_9H_{13}NO_3$ = 98.5
Epinephrine bitartrate	0.33330	cr.v.	$C_9H_{13}NO_3 \cdot C_4H_6O_6$ = 97 to 102
Hydrocortisone sodium succinate	0.02299	cr.v.	Na = 4.6 to 4.89
Hydroxyamphetamine hydrobromide	0.23213	m.ros.	$C_9H_{13}NO \cdot HBr$ = 98 to 101.5
Levarterenol bitartrate	0.31928	cr.v.	$C_8H_{11}NO_3 \cdot C_4H_6O_6$ = 97 to 102
Meclizine hydrochloride	0.23195	q.r.	$C_{25}H_{27}ClN_2 \cdot 2HCl$ = 97
Meclizine hydrochloride tablets	0.23195	q.r.	$C_{25}H_{27}ClN_2 \cdot 2HCl$ = 95 to 105 l.a.
Metaraminol bitartrate	0.31730	cr.v.	$C_9H_{13}NO_2 \cdot C_4H_6O_6$ = 97.5
Methadone hydrochloride	0.34592	cr.v.	$C_{21}H_{27}NO \cdot HCl$ = 98.5
Methamphetamine hydrochloride	0.18570	cr.v.	$C_{10}H_{15}N \cdot HCl$ = 98.5
Methamphetamine hydrochloride tablets	0.18570	m.r.	$C_{10}H_{15}N \cdot HCl$ = 93 to 107 l.a.
Methoxamine hydrochloride	0.24772	cr.v.	$C_{11}H_{17}NO_3 \cdot HCl$ = 98
Niacinamide	0.12213	cr.v.	$C_6H_6N_2O$ = 98.5 to 101
Noscapine	0.41343	cr.v.	$C_{22}H_{23}NO_7$ = 99
Piperazine citrate	0.10711	cr.v.	$(C_4H_{10}N_2)_3 \cdot 2C_3H_8O_7$ = 98
Piperazine citrate syrup	0.10711	cr.v.	$(C_4H_{10}N_2)_3 \cdot 2C_3H_8O_7$ = 10 to 12 *W/V*
Piperazine citrate tablets	0.10711	cr.v.	$(C_4H_{10}N_2)_3 \cdot 2C_3H_8O_7$ = 93 to 107 l.a.
Procainamide hydrochloride	0.13589	cr.v.	$C_{13}H_{21}N_3O \cdot HCl$ = 98
Prochlorperazine edisylate	0.28208	cr.v.	$C_{20}H_{24}ClN_3S \cdot 2C_4H_6O_6S_2$ = 98 to 101.5
Prochlorperazine maleate	0.30305	pot.	$C_{20}H_{24}ClN_3S \cdot 2C_4H_4O_4$ = 98 to 101.5
Promethazine hydrochloride	0.32089	cr.v.	$C_{17}H_{20}N_2S \cdot HCl$ = 97 to 101.5
Propantheline bromide	0.44841	cr.v.	$C_{23}H_{30}BrNO_3$ = 98 to 102
Propantheline bromide tablets	0.44841	cr.v.	$C_{23}H_{30}BrNO_3$ = 95 to 105 l.a.
Propoxyphene hydrochloride	0.37594	cr.v.	$C_{22}H_{29}NO_2 \cdot HCl$ = 98 to 101
Pyridostigmine bromide	0.26113	q.r.	$C_9H_{13}BrN_2O_2$ = 98.5
Pyridoxine hydrochloride	0.20564	cr.v.	$C_8H_{11}NO_3 \cdot HCl$ = 98
Pyrimethamine	0.24872	q.r.	$C_{12}H_{13}ClN_4$ = 98.5
Sodium benzoate	0.14411	cr.v.	$C_7H_5NaO_2$ = 99
Sodium carboxymethylcellulose	0.02299	pot.	Na = 6.98 to 8.50
Sodium citrate	0.08062	pot.	$C_6H_5Na_3O_7$ = 99
Sodium citrate solution, anticoagulant	0.09800	pot.	$C_6H_5Na_3O_7 \cdot 2H_2O$ = 3.8 to 4.2 *W/V*
Sodium lactate injection	0.11206	cr.v.	$C_3H_5NaO_3$ = 95 to 110 l.a.
Sodium salicylate	0.16011	cf.v.	$C_7H_5NaO_3$ = 99.5
Succinylcholine chloride	0.18066	cr.v.	$C_{14}H_{30}Cl_2N_2O_4$ = 98 to 101
Succinylcholine chloride, sterile	0.18066	cr.v.	$C_{14}H_{30}Cl_2N_2O_4$ = 98 to 101
Trihexyphenidyl hydrochloride	0.33794	th.b.	$C_{20}H_{31}NO \cdot HCl$ = 98
Tripelennamine citrate	0.22375	cr.v.	$C_{16}H_{21}N_3 \cdot C_6H_8O_7$ = 98
Tripelennamine hydrochloride	0.14592	cr.v.	$C_{16}H_{21}N_3 \cdot HCl$ = 98

TABLE 16. ACIDIMETRIC ASSAYS. NONAQUEOUS TITRATIONS WITH PERCHLORIC ACID (*Continued*)

Substance	mEq factor, Gm	Indicator	Official requirement, percent
N.F.			
Alphaprodine hydrochloride	0.29783	cr.v.	$C_{16}H_{23}NO_2 \cdot HCl$ = 98
Alphaprodine hydrochloride injection	0.29783	m.r.	$C_{16}H_{23}NO_2 \cdot HCl$ = 93 to 107 l.a.
Aminoacetic acid	0.07507	m.ros.	$C_2H_5NO_2$ = 98.5 to 101.5
Anileridine hydrochloride	0.21270	pot.	$C_{22}H_{28}N_2O_2 \cdot 2HCl$ = 96 to 102
Antazoline phosphate	0.36336	cr.v.	$C_{17}H_{19}N_3 \cdot H_3PO_4$ = 98
Benzocaine	0.16519	cr.v.	$C_9H_{11}NO_2$ = 98 to 101
Benzpyrinium bromide	0.33722	cr.v.	$C_{15}H_{17}BrN_2O_2$ = 98.5
Benztropine mesylate	0.40354	m.r.	$C_{21}H_{25}NO \cdot CH_4O_3S$ = 98
Brompheniramine maleate	0.21767	cr.v.	$C_{16}H_{19}BrN_2 \cdot C_4H_4O_4$ = 98
Brompheniramine maleate elixir	0.21767	cr.v.	$C_{16}H_{19}BrN_2 \cdot C_4H_4O_4$ = 0.038 to 0.042 *W/V*
Brompheniramine maleate tablets	0.21767	cr.v.	$C_{16}H_{19}BrN_2 \cdot C_4H_4O_4$ = 95 to 105 l.a.
Carbetapentane citrate	0.52560	q.r.	$C_{20}H_{31}NO_3 \cdot C_6H_8O_7$ = 98
Carbinoxamine maleate	0.20344	cr.v.	$C_{16}H_{19}ClN_2O \cdot C_4H_4O_4$ = 98 to 102
Carbinoxamine maleate tablets	0.20344	cr.v.	$C_{16}H_{19}ClN_2O \cdot C_4H_4O_4$ = 93 to 107 l.a.
Chlordiazepoxide hydrochloride	0.33622	cr.v.	$C_{16}H_{14}ClN_3O \cdot HCl$ = 99 to 101
Chlorothen citrate	0.24398	pot.	$C_{14}H_{18}ClN_3S \cdot C_6H_8O_7$ = 98
Choline dihydrogen citrate	0.29529	cr.v.	$C_{11}H_{21}NO_8$ = 98
Cyclopentamine hydrochloride	0.17772	cr.v.	$C_9H_{19}N \cdot HCl$ = 98
Cycrimine hydrochloride	0.32391	cr.v.	$C_{19}H_{29}NO \cdot HCl$ = 97
Dexchlorpheniramine maleate	0.19544	cr.v.	$C_{16}H_{19}ClN_2 \cdot C_4H_4O_4$ = 99
Dextromethorphan hydrobromide	0.37034	cr.v.	$C_{18}H_{25}NO \cdot HBr \cdot H_2O$ = 98
Dextromethorphan hydrobromide tablets	0.37034	m.r.	$C_{18}H_{25}NO \cdot HBr \cdot H_2O$ = 90 to 110 l.a.
Dibucaine	0.17174	cr.v.	$C_{20}H_{29}N_3O_2$ = 97
Ethoheptazine citrate	0.45349	cr.v.	$C_{16}H_{23}NO_2 \cdot C_6H_8O_7$ = 97 to 102
Ethoheptazine citrate tablets	0.45349	cr.v.	$C_{16}H_{23}NO_2 \cdot C_6H_8O_7$ = 90 to 110 l.a.
Hydroxyzine hydrochloride	0.22392	q.r.	$C_{21}H_{26}N_2O_2 \cdot 2HCl$ = 98
Hydroxyzine hydrochloride tablets	0.22392	q.r.	$C_{21}H_{26}N_2O_2 \cdot 2HCl$ = 90 to 110 l.a.
Imipramine hydrochloride	0.31688	cr.v.	$C_{19}H_{24}N \cdot HCl$ = 98 to 102
Imipramine hydrochloride injection	0.31688	cr.v.	$C_{19}H_{24}N_2 \cdot HCl$ = 92 to 108 l.a.
Imipramine hydrochloride tablets	0.31688	cr.v.	$C_{19}H_{24}N_2 \cdot HCl$ = 93 to 107 l.a.
Isobucaine and epinephrine injection	0.28582	m.r.	$C_{15}H_{23}NO_2 \cdot HCl$ = 95 to 105 l.a.
Isobucaine hydrochloride	0.28582	cr.v.	$C_{15}H_{23}NO_2 \cdot HCl$ = 98 to 102
Isoproterenol sulfate	0.55663	cr.v.	$C_{22}H_{34}N_2O_6 \cdot H_2SO_4 \cdot 2H_2O$ = 98
Levallorphan tartrate	0.43351	pot.	$C_{16}H_{25}NO \cdot C_4H_6O_6$ = 98.5
Levonordefrin	0.18321	cr.v.	$C_9H_{13}NO_3$ = 98 to 102
Levorphanol tartrate	0.44350	m.r.	$C_{17}H_{23}NO \cdot C_4H_6O_6$ = 99 to 101
Levorphanol tartrate injection	0.44350	m.r.	$C_{17}H_{23}NO \cdot C_4H_6O_6$ = 93 to 107 l.a.
Levorphanol tartrate tablets	0.44350	m.r.	$C_{17}H_{23}NO \cdot C_4H_6O_6$ = 93 to 107 l.a.
Meperidine hydrochloride	0.28380	cr.v.	$C_{15}H_{21}NO_2 \cdot HCl$ = 98 to 101
Meperidine hydrochloride injection	0.28380	cr.v.	$C_{15}H_{21}NO_2 \cdot HCl$ = 95 to 105 l.a.
Meperidine hydrochloride tablets	0.28380	cr.v.	$C_{15}H_{21}NO_2 \cdot HCl$ = 95 to 105 l.a.
Mepivacaine hydrochloride	0.28282	cr.v.	$C_{15}H_{22}N_2O \cdot HCl$ = 98 to 102
Mepivacaine hydrochloride injection	0.28282	m.r.	$C_{15}H_{22}N_2O \cdot HCl$ = 95 to 105 l.a.
Mepivacaine hydrochloride and levonordefrin injection	0.28282	m.r.	$C_{15}H_{22}N_2O \cdot HCl$ = 95 to 105 l.a.
Methacholine bromide	0.24015	cr.v.	$C_8H_{18}BrNO_2$ = 98 to 102
Methacholine bromide tablets	0.24015	cr.v.	$C_8H_{18}BrNO_2$ = 95 to 105 l.a.
Methapyrilene hydrochloride	0.14893	pot.	$C_{14}H_{19}N_3S \cdot HCl$ = 98
Methscopolamine bromide	0.39831	cr.v.	$C_{18}H_{24}BrNO_4$ = 98 to 101
Methylphenidate hydrochloride for injection	0.26977	a.n.b.	$C_{14}H_{10}NO_2 \cdot HCl$ = 95 to 107 l.a.
Methylprednisolone sodium succinate	0.02299	cr.v.	Na = 4.49 to 4.77
Nordefrin hydrochloride	0.21967	cr.v.	$C_9H_{13}NO_3 \cdot HCl$ = 98 to 102
Noscapine hydrochloride	0.44989	m.ros.	$C_{22}H_{23}NO_7 \cdot HCl$ = 99

TABLE 16. ACIDIMETRIC ASSAYS. NONAQUEOUS TITRATIONS WITH PERCHLORIC ACID (*Continued*)

Substance	mEq factor Gm	Indicator	Official requirement, percent
Nylidrin hydrochloride	0.33538	cr.v.	$C_{19}H_{25}NO_2{\cdot}HCl$ = 98 to 102
Nylidrin hydrochloride injection	0.33588	m.r.	$C_{19}H_{25}NO_2{\cdot}HCl$ = 95 to 105 l.a.
Nylidrin hydrochloride tablets	0.33588	m.r.	$C_{19}H_{25}NO_2{\cdot}HCl$ = 95 to 105 l.a.
Penthienate bromide	0.42042	cr.v.	$C_{18}H_{30}BrNO_3$ = 98
Phenindamine tartrate	0.41146	cr.v.	$C_{19}H_{19}N{\cdot}C_4H_6O_6$ = 98 to 101.5
Phenindamine tartrate syrup	0.41146	m.r.	$C_{19}H_{19}N{\cdot}C_4H_6O_6$ = 0.19 to 0.22 *W/V*
Pheniramine maleate	0.17822	cr.v.	$C_{16}H_{20}N_2{\cdot}C_4H_4O_4$ = 99
Pheniramine maleate tablets	0.17822	cr.v.	$C_{16}H_{20}N_2{\cdot}C_4H_4O_4$ = 93 to 107 l.a.
Pipradrol hydrochloride	0.30384	cr.v.	$C_{18}H_{21}NO{\cdot}HCl$ = 98 to 102
Pramoxine hydrochloride	0.32987	cr.v.	$C_{17}H_{27}NO_3{\cdot}HCl$ = 98
Pramoxine hydrochloride cream	0.32987	cr.v.	$C_{17}H_{27}NO_3{\cdot}HCl$ = 90 to 110 l.a.
Pramoxine hydrochloride solution	0.32987	cr.v.	$C_{17}H_{27}NO_3{\cdot}HCl$ = 95 to 105 l.a.
Prochloperazine	0.18698	cr.v.	$C_{20}H_{24}ClN_3S$ = 98 to 101
Pyrilamine maleate	0.20074	cr.v.	$C_{17}H_{23}N_3O{\cdot}C_4H_4O_4$ = 98 to 101
Pyrobutamine phosphate	0.50785	cr.v.	$C_{20}H_{22}ClN{\cdot}2H_3PO_4$ = 98 to 101
Sodium carboxymethylcellulose tablets	0.02299	pot.	Na = 6.8 to 8.6 l.a.
Tetrahydrozoline hydrochloride	0.23675	q.r.	$C_{13}H_{16}N_2{\cdot}HCl$ = 98
Tetrahydrozoline hydrochloride solution	0.23675	m.r.	$C_{13}H_{16}N_2{\cdot}HCl$ = 90 to 110 l.a.
Thenyldiamine hydrochloride	0.14893	pot.	$C_{14}H_{16}N_3S{\cdot}HCl$ = 98 to 102
Thiabendazole	0.20125	cr.v.	$C_{10}H_7N_3S$ = 97 to 101
Thonzylamine hydrochloride	0.16142	pot.	$C_{16}H_{22}N_4O{\cdot}HCl$ = 98
Tridihexethyl chloride	0.35398	th.b.	$C_{21}H_{36}ClNO$ = 98
Trimethobenzamide hydrochloride	0.42493	q.r.	$C_{21}H_{28}N_2O_5{\cdot}HCl$ = 98.5

l.a. = labeled amount: cr.v. = crystal violet: m.r. = methyl red: m.ros. = methylrosaniline chloride: pot. = potentiometer: q.r. = quinaldine red: th.b. = thymol blue: a.n.b. = α-naphtholbenzein.

causing a sufficient shift in the equilibrium so that the titrimetric reaction is quantitative. A blank titration is determined to account for the possible reaction of atmospheric moisture with the titrant, $HClO_4$, and also to account for the titrant used to bring about the blue end point.

One mole of the methamphetamine hydrochloride, $C_{10}H_{15}N{\cdot}HCl$ (mol wt 185.70) reacts with 1 mole of $HClO_4$; therefore 1 mEq of methamphetamine hydrochloride is 1 gram-molecular weight divided by 1,000 or 0.18570 Gm, and 1 ml of a 0.1 *N* solution of perchloric acid is equivalent to 18.57 mg of methamphetamine hydrochloride.

ALKALIMETRY IN NONAQUEOUS SOLVENTS

Many weakly acidic substances can be titrated in an appropriate nonaqueous solvent with a sharp end point. The different types of organic compounds are acid halides, anhydrides, acids, amino acids, enols such as barbiturates and xanthines, imides, phenols, pyrroles, sulfonamides and organic salts of inorganic acids. Boric acid, a weak inorganic acid, can be easily titrated using ethylenediamine as the solvent, and all three H's of H_3BO_3 can be detected using a potentiometer to determine the end points.

Solvents. Just as in acidimetry in nonaqueous media, where the solvent, glacial acetic acid, enhanced the basicity of otherwise weak bases (in water) so that these bases could be titrated with a standard solution of perchloric acid, the proper choice of a base solvent, e.g., ethylenediamine, can enhance the acid strength of otherwise weakly acidic compounds such as phenols so that they can be titrated with a standard solution of lithium or sodium methoxide. Factors which can be considered in choosing a basic solvent are solubility of the substance to be analyzed, the relative base strength of the solvent, the sharpness of the end point change, and the nonreactivity of the solvent. Ethylenediamine, *n*-butylamine, and morpholine are considered strong bases which can be used as solvents for the titration of weak acids such as enols. Somewhat weaker base solvents, such as dimethylformamide and pyridine, are used as solvents for medium-strength acidic substances; strongly acidic compounds may be titrated in any of the above solvents. Sulfonamides which contain an N-alkyl substituent possess low acidity, and consequently ethylenediamine is the solvent of choice, whereas with N-phenyl or N-pyridyl substituted sulfonamides, which are relatively stronger acids, dimethylformamide is the preferred solvent. With the proper selection of basic solvent it is possible to titrate selectively simple mixtures of sulfonamides, e.g., sulfanilamide and sulfathiazole. With the solvent dimethylformamide and thymol blue indicator, only the sulfathiazole will react with the sodium methoxide. A second titration with a sample of the same size in butylamine using azo violet indicator will determine total sulfonamides. The difference in volumes of sodium methoxide consumed in the two determinations is a measure of the amount of sulfanilamide in the mixture.

The assay of Acetic Anhydride U.S.P. reagent illustrates how an alkalimetric analysis of a sample in an aqueous and a nonaqueous solvent by an aqueous and nonaqueous titrant can be used to determine the amount of acetic anhydride alone. The initial titration of the sample with sodium hydroxide in water determines total acidity of the acetic anhydride and acetic acid combined. A second titration of a sample in methyl alcohol with sodium methylate determines only the acetic acid content. The difference in the number of equivalents consumed in each titration is a measure of the amount of acetic anhydride alone.

Titrants. In the titration of acidic compounds the bases that are used officially are sodium methoxide and lithium methoxide. Potassium methoxide, which is a stronger titrant, is not used since it may produce a gelatinous reaction product. In those instances where the sodium methoxide also produces a gelatinous precipitate, lithium methoxide is preferred. Other base titrants are sodium aminomethoxide, which is the strongest base, and sodium triphenylmethane, which is used for weakly acidic compounds such as phenols and pyrroles.

Atmospheric moisture and carbon dioxide will interfere with metal alcoholate titrations. Normal precautions are the use of dried solvents to eliminate the presence of moisture and titration in a closed system under nitrogen to exclude carbon dioxide. A blank determination would account for any possible reaction between the titrant and these substances. The blank should not exceed 0.01 ml of titrant per milliliter of solvent used.

Indicators. The end point of a nonaqueous titration of acidic substances can be determined, as in other titrimetric analyses, with a potentiometer or visually with an indicator. The potentiometric methods are discussed in Chap. 21. The choice of an indicator is at present empirical, as in the nonaqueous acidimetric analyses. A potentiometric titration is currently the best way of determining the equivalence point of a given chemical reaction in nonaqueous titrimetry. It may be located as the midpoint of the inflection part of the curve in which the milliliters of titrant are plotted against the millivolts produced in the reaction. Once this point is determined, the appropriate indicator is chosen by trial and error. Experience has already shown that azo violet is the indicator of choice in the titration of acids of weak or intermediate strength in *n*-butylamine solvent, and that thymol blue is the indicator of choice for the determination of intermediate to strong acids in dimethylformamide solvent. Ethylenediamine is used as the solvent for strong acids. In metal alcoholate titration, azo violet will change color before thymol blue. *o*-Nitroaniline is the indicator used for the titration of very weak acids. A clear blue color is the end point for either azo violet or thymol blue indicators. Other indicators used are thymolphthalein and *p*-hydroxyazobenzene.

Exercise 29

Object. To Prepare and Standardize 0.1 *N* Sodium Methoxide.

Materials Required. 2.5 Gm of sodium
150 ml of methanol
900 ml of benzene
400 mg of benzoic acid
Thymol blue in dimethylformamide (1 in 100)
80 ml of dimethylformamide

Procedure. "Cool in ice-water 150 ml. of methanol contained in a 1000-ml. volumetric flask, and add, in small portions, about 2.5 Gm. of freshly cut sodium metal. When the metal has dissolved, add sufficient benzene to make 1000 ml., and mix. Store preferably in the reservoir of an automatic delivery buret suitably protected from carbon dioxide and moisture. Standardize the solution as follows:

"Dissolve about 400 mg., accurately weighed, of primary standard benzoic acid in 80 ml. of dimethylformamide in a flask. Add 3 drops of a 1 in 100 solution of thymol blue in dimethylformamide, and titrate with sodium methoxide to a blue end point.

Correct for the volume of the sodium methoxide solution consumed by 80 ml. of the dimethylformamide, and calculate the normality. Each 12.21 mg. of benzoic acid is equivalent to 1 ml. of 0.1 *N* sodium methoxide."

The dissolution of sodium metal in methanol is an exothermic reaction; cooling is required in order to keep the reaction under control.

$$Na^{\circ} + CH_3OH \longrightarrow CH_3ONa + H^{\circ}\uparrow$$

Carbon dioxide and moisture are to be excluded because of the following reactions which might ensue with the sodium methoxide:

$$H_2O + CH_3ONa \longrightarrow CH_3OH + NaOH$$
$$H_2CO_3 + 2CH_3ONa \longrightarrow 2CH_3OH + Na_2CO_3$$

giving rise to the turbidity which might form.

The solution of the primary standard, benzoic acid, in dimethylformamide is represented by the equation:

$$C_6H_5COOH + HCON(CH_3)_2 \rightleftharpoons HCON^+H(CH_3)_2 + C_6H_5COO^-$$

The ionization of the sodium methoxide is pictured as

$$CH_3ONa \rightleftharpoons CH_3O^- + Na^+$$

and the reaction between the solvated proton and the methylate ion is shown as

$$HCON^+H(CH_3)_2 + CH_3O^- \longrightarrow HCON(CH_3)_2 + CH_3OH$$

Thus the apparent overall reaction in the standardization may be expressed as follows:

$$C_6H_5COOH + CH_3ONa \longrightarrow C_6H_5COONa + CH_3OH$$

The correction to be applied for the volume of sodium methoxide consumed by the dimethylformamide is in effect a blank determination to account for the interaction between the water in the solvent and the titrant.

One gram-atomic weight of Na (22.99 Gm) is equivalent to 1 gram-molecular weight of CH_3ONa (54.02 Gm). A 0.1 *N* solution contains $\frac{1}{10}$ equivalent per liter; therefore 2.5 Gm of Na would produce about 5.4 Gm or $\frac{1}{10}$ equivalent of sodium methoxide.

Primary standard benzoic acid, C_6H_5COOH, is a monoprotic acid with a molecular weight of 122.12; 1 mEq weight would be 0.12212 Gm, and a 0.400-Gm sample would contain about 3 mEq for the standardization of the sodium methoxide. Standard sodium methoxide solutions, like sodium hydroxide solution, must be restandardized frequently.

Exercise 30

Object. Assay of Diphenylhydantoin.

Materials Required. 500 mg of Diphenylhydantoin
50 ml of dimethylformamide
Azo violet in benzene

Procedure. "Weigh accurately about 500 mg. of Diphenylhydantoin, and transfer to a narrow-mouth, 125-ml. conical flask. Dissolve in 50 ml. of dimethylformamide, add 3 drops of a saturated solution of azo violet in benzene, and titrate with 0.1 *N* sodium methoxide to a blue end point. Perform a blank determination, and make any necessary correction. Each milliliter of 0.1 *N* sodium methoxide is equivalent to 25.23 mg. of $C_{15}H_{12}N_2O_2$."

Diphenylhydantoin is an organic acid whose acidity can be slightly enhanced in dimethylformamide solvent so that it can be titrated readily with sodium methoxide using azo violet to indicate the end point. It behaves as a monoprotic acid and yields monosodium salts with sodium

TABLE 17. ALKALIMETRIC ASSAYS. NONAQUEOUS TITRATIONS WITH LITHIUM OR SODIUM METHOXIDE

Substance	mEq factor, Gm	Indicator	Official requirement, percent
U.S.P.			
Amobarbital	0.22628	th.b.	$C_{11}H_{18}N_2O_3$ = 98.5 to 101
Dichlorphenamide	0.15258	th.ph.	$C_6H_6Cl_2N_2O_4S_2$ = 97
Hydrochlorothiazide	0.14887	az.vi.	$C_7H_8ClN_3O_4S_2$ = 97 to 101.5
Mercaptopurine	0.15219	th.b.	$C_5H_4N_4S$ = 97 to 102
Secobarbital	0.23829	th.b.	$C_{12}H_{18}N_2S$ = 97.5
Sodium pentobarbital	0.22628	th.b.	$C_{11}H_{18}N_2O_3$ = n.m.t. 3.0
Sulfamethoxypyridazine	0.28031	th.b.	$C_{11}H_{12}N_4O_3S$ = 99
Sulfamethoxypyridazine tablets	0.28031	th.b.	$C_{11}H_{12}N_4O_3S$ = 95 to 105 l.a.
Sulfisoxazole	0.26731	th.b.	$C_{11}H_{13}N_3O_3S$ = 99 to 101
Sulfisoxazole tablets	0.26731	th.b.	$C_{11}H_{13}N_3O_3S$ = 95 to 105 l.a.
N.F.			
Aprobarbital	0.21023	th.b.	$C_{10}H_{14}N_2O_3$ = 98.5 to 101
Chlorothiazide	0.29572	th.b.	$C_7H_6ClN_3O_4S_2$ = 97
Diphenylhydantoin	0.25228	az.vi.	$C_{15}H_{12}N_2O_2$ = 98.5
Diphenylhydantoin oral suspension	0.25228	az.vi.	$C_{15}H_{12}N_2O_2$ = 2.25 to 2.75 *W/V*
Mephobarbital	0.24627	th.ph.	$C_{13}H_{14}N_2O_3$ = 98
Methylthiouracil	0.14218	th.b.	$C_5H_6N_2OS$ = 97
Salicylamide	0.13714	th.b.	$C_7H_7NO_2$ = 98 to 102
Salicylamide tablets	0.13714	th.b.	$C_7H_7NO_2$ = 95 to 105 l.a.

l.a. = labeled amount; n.m.t. = not more than; th.b. = thymol blue; th.ph. = thymolphthalein; az.vi. = azo violet.

methoxide. Other hydantoin derivatives would behave similarly. The reaction for the neutralization is as follows:

```
                NH—C=O                                    N═C—O⁻
C6H5          /      |                  C6H5            /    |
     \       /       |                       \         /     |
        C            |   + CH3O⁻ ⇌               C           |
     /       \       |                       /         \     |
C6H5          \      |                  C6H5            \    |
                CO—NH                                    CO—NH
                                                                 + CH3OH
```

A blank determination will account for any water in the solvent, dimethylformamide, which will react with the methoxide ion in the following fashion:

$$H_2O + CH_3O^- \rightleftharpoons OH^- + CH_3OH$$

The calculation for the percent purity of the diphenylhydantoin is based on the mole for mole ratio in the reaction with sodium methoxide.

Chapter 7

PRECIPITATION AND COMPLEXATION METHODS OF ANALYSIS

PRECIPITATION METHODS

In Chaps. 4 and 5 on acidimetry and alkalimetry, a class of reactions was considered that was of value in quantitative analysis because very slightly ionized substances or gases or both were formed. In volumetric precipitimetry a class of reactions is dealt with that requires the formation of relatively insoluble substances or precipitates to cause the reactions to go to sufficient completion to be quantitative in nature. The solubility product principle may be applied to all precipitation reactions, and it should be reviewed before subsequent analyses are undertaken. Complexation methods of analysis will also be considered in this chapter, together with sodium tetraphenylboron titrations.

Determination of the End Point. The end point of a reaction in analysis by precipitation methods may be determined in the following ways:

1. Cessation of precipitation or the appearance of a turbidity
2. Use of internal indicators
3. Instrumental methods, i.e., potentiometric or amperometric

Indicators. The indicators used in the official volumetric precipitation assays are:

1. *Ferric ammonium sulfate T.S.*, prepared by dissolving 8 Gm of reagent ferric ammonium sulfate in sufficient water to make 100 ml. This indicator is used both in direct and residual titrations employing standard ammonium thiocyanate solution. The thiocyanate reacts with the silver or mercuric ions present to form a white precipitate of silver or mercuric thiocyanate, but as soon as all the silver or mercury has been precipitated, the thiocyanate ion reacts with ferric ammonium sulfate to form a red ferric thiocyanate; the first appearance of the red color marks the end point of the reaction.

2. *Potassium chromate T.S.*, prepared by dissolving 10 Gm of reagent potassium chromate in sufficient water to make 100 ml. It forms a red precipitate of silver chromate which is seen against the background of white silver chloride. Using the proper concentration of chromate ion

will ensure its coprecipitation with silver chloride at the point where the chloride ion concentration is reduced to that value calculated as the stoichiometric point.

3. *Starch T.S.*, prepared as directed under Starch Indicator Solution, Chap. 9. Starch is used as an indicator in the assays of various organic iodine compounds. The compounds are converted to iodides by oxidative degradation. The blue color of the starch-iodide-iodine complex persists until the iodide ions have been precipitated by the silver nitrate in the titration. When all the iodide has been converted to silver iodide, the blue color of the complex disappears, which signals the end point..

4. *Adsorption indicators*, such as Dichlorofluorescein T.S., Eosin solution, Eosin Y T.S., and Tetrabromphenolphthalein Ethyl Ester T.S., are used in the analyses of halides by direct titration with silver nitrate solution. They are weak organic acids that vary in strength so that some are more effective in acidic media than others. The end point of the titration is indicated when the color of the silver halide precipitate changes abruptly because of the adsorbed indicator anions. The color changes are best seen in a diffuse light condition.

5. *Potassium iodide solution* (1 in 10) is used as an indicator for cyanide determinations. Silver cyanide, $Ag_2(CN)_2$, will not precipitate until the molar ratio of silver nitrate to the cyanide is greater than 1:2.

$$AgNO_3 + 2KCN \longrightarrow KAg(CN)_2 + KNO_3$$

The potassium argentocyanide is a soluble, colorless complex. Upon addition of the first drop of silver nitrate in excess of the 1:2 ratio, insoluble, canary-yellow silver iodide, AgI, forms in preference to the insoluble silver cyanide, which is kept in solution as an ammonium complex.

Standard Solutions. The standard solutions employed in the official assays by precipitation methods are 0.1 *N* silver nitrate and 0.1 *N* ammonium thiocyanate.

Exercise 31

Object. To Prepare and Standardize 0.1 *N* Silver Nitrate.

Materials Required. 17.5 Gm of silver nitrate
Diluted hydrochloric acid

Procedure. "Dissolve about 17.5 Gm. of silver nitrate in 1000 ml. of water, and standardize the solutions as follows: Pipet, or otherwise measure accurately, about 40 ml. of the silver nitrate solution, and dilute with about 100 ml. of water. Heat the solution, and slowly add, with continuous stirring, diluted hydrochloric acid until precipitation of the silver chloride is complete. Boil the mixture cautiously for about 5 minutes, then allow it to stand in the dark until the precipitate has settled and the supernatant liquid has become clear.

"Transfer the precipitate to a tared filtering crucible, and wash it with small portions of water slightly acidified with nitric acid. Dry the precipitate at 110° to constant weight. From the weight of silver chloride, calculate the normality of the silver nitrate solution. Protect the silver chloride from light as much as possible during the determination."

Since the univalent silver cation, Ag^+, is, in a sense, a chemical equivalent of the hydrogen ion, 1 mole of $AgNO_3$ (169.87) is 1 equivalent. A liter of 0.1 *N* silver nitrate solution should, therefore, contain 16.987 Gm of $AgNO_3$.

$$\underset{169.87}{AgNO_3} + HCl \longrightarrow \underset{143.32}{AgCl}\downarrow + HNO_3$$

Calculation for normality:

$$\frac{\text{Wt of AgCl} \times AgNO_3/AgCl}{40 \text{ ml} \times 0.16987} = \text{mEq/ml or } N$$

The slow addition of the precipitant, hydrochloric acid, with continuous stirring of the heated solution aids in the formation of large particle-size precipitate. The mixture is allowed to stand in the dark to minimize the reducing effect of the light on the silver chloride; free silver is produced. The purplish color which the precipitate acquires on exposure to light is due to the free metal being formed in the precipitate.

Exercise 32

Object. To Prepare and Standardize 0.1 *N* Ammonium Thiocyanate.

Materials Required. 8.0 Gm of reagent ammonium thiocyanate
0.1 *N* silver nitrate
2 ml of ferric ammonium sulfate T.S.

Procedure. Dissolve about 8 Gm of ammonium thiocyanate in 1,000 ml of water. Place 30.00 ml of 0.1 *N* silver nitrate in a 250-ml Erlenmeyer flask. Dilute with 50 ml of water, then add 2 ml of nitric acid and 2 ml of ferric ammonium sulfate T.S., and titrate with the ammonium thiocyanate solution to the first appearance of a red-brown color. Calculate the normality of the ammonium thiocyanate solution, and, if desired, adjust the solution to exactly 0.1 *N*.

Since ammonium thiocyanate is deliquescent, a slight amount in excess of the theoretical quantity required is used in preparation of the solution. If desired, potassium thiocyanate may be used in place of ammonium thiocyanate. As the sample of silver nitrate solution is titrated with the ammonium thiocyanate solution, a white precipitate of AgSCN is formed. As the end point is approached, the precipitate becomes flocculent and settles readily. At the end point, a drop of ammonium thiocyanate solution produces a permanent reddish-brown color which does not disappear on shaking. Equations for the reactions involved follow.

$$AgNO_3 + NH_4SCN \longrightarrow \underset{\text{white}}{AgSCN}\downarrow + NH_4NO_3$$

$$FeNH_4(SO_4)_2 + 3NH_4SCN \longrightarrow Fe(SCN)_3 + 2(NH_4)_2SO_4$$

Oxides of nitrogen give colored salts with ferric alum. Solutions containing oxides of nitrogen and also the nitric acid employed as above should be boiled prior to the addition of the indicator.

Questions and Problems

1. Explain in detail how to standardize a solution of silver nitrate against pure sodium chloride.
2. Why is nitric acid added when NH_4SCN solution is standardized against 0.1 *N* $AgNO_3$ using ferric alum as indicator?
3. Write ionically the reactions involved in the standardization of NH_4SCN against $AgNO_3$ with ferric alum indicator.
4. What must be the normality of a solution of silver nitrate so that each milliliter shall be equivalent to 6 mg of NaCl?
5. How might 0.1 *N* $AgNO_3$ be prepared from pure metallic silver?
6. Why should standard silver nitrate solution be protected from strong sunlight?
7. Why must oxides of nitrogen be absent in titrations with NH_4SCN?
8. How much 0.1 *N* $AgNO_3$ could be prepared from 5 Gm of pure silver?
9. Calculate the weight (milligrams) of each of the following that is equivalent to 1 ml of 0.1 *N* ammonium thiocyanate: (*a*) yellow mercuric oxide; (*b*) silver nitrate; (*c*) nitromersol; (*d*) mild silver protein (Ag). Compare values with those given in the official monographs.

DIRECT TITRATION METHOD

Compounds of silver and mercury that can be readily converted into soluble silver or mercury (Hg^{++}) salts, respectively, may be estimated by direct titration with standard ammonium thiocyanate solution, using ferric ammonium sulfate as indicator. The method is based on the quantitative precipitation of the corresponding thiocyanate, e.g.,

$$AgNO_3 + NH_4SCN \longrightarrow AgSCN\downarrow + NH_4NO_3$$

and

$$Hg(NO_3)_2 + 2NH_4SCN \longrightarrow Hg(SCN)_2 + 2NH_4NO_3$$

When all the metallic ion has been precipitated as thiocyanate, the NH_4SCN reacts with the ferric alum indicator to form red ferric thiocyanate, marking the end point of titration:

$$FeNH_4(SO_4)_2 + 3NH_4SCN \longrightarrow Fe(SCN)_3 + 2(NH_4)_2SO_4$$

The solution must be acidified with nitric acid to prevent the hydrolysis that ferric salts undergo in neutral solution. Chlorides must be absent, since the chlorides of silver and mercury are more soluble than the respective thiocyanates. This method of assay is exemplified in the following exercise.

Exercise 33

Object. Assay of Mild Silver Protein.

Materials Required. 1.5 Gm of Mild Silver Protein
2 ml of ferric ammonium sulfate T.S.
0.1 *N* ammonium thiocyanate

Procedure. 1. "Ignite about 1.5 Gm of Mild Silver Protein, accurately weighed, in a porcelain crucible until all of the carbon is burned off. Cool the crucible, place it in a 400-ml beaker, and fill it with dilute nitric acid (1 in 2). When the reaction has subsided, add water to immerse the crucible, cover the beaker, and heat on a steam bath until all of the silver is dissolved."

Upon ignition, mild silver protein yields a residue of metallic silver and silver oxide. The silver and silver oxide are dissolved in nitric acid:

$$Ag^\circ + 2HNO_3 \longrightarrow AgNO_3 + H_2O + NO_2\uparrow$$
$$Ag_2O + 2HNO_3 \longrightarrow 2AgNO_3 + H_2O$$

2. "Remove and rinse the crucible, and filter the solution into a conical flask. Wash the filter and residue thoroughly with water and cool the filtrate. Add 2 ml. of ferric ammonium sulfate T.S., and titrate with 0.1 *N* ammonium thiocyanate. Each milliliter of 0.1 *N* ammonium thiocyanate is equivalent to 10.79 mg. of Ag."

The mixture is filtered free of the carbonaceous matter, the residue on the filter is washed free of silver nitrate, and the nitrate is titrated with 0.1 *N* ammonium thiocyanate:

$$AgNO_3 + NH_4SCN \longrightarrow AgSCN\downarrow + NH_4SCN$$

The equivalent weight of Ag is 107.87 Gm, since 1 gram-atomic weight of silver produced 1 mole of $AgNO_3$ on ignition of the sample and dissolution in nitric acid. Each milliliter of 0.1 *N* ammonium thiocyanate or 0.1 mEq is equivalent to $0.1 \times 0.10787 = 0.010787$ Gm, or 0.1 mEq of Ag. The N.F. specifies that mild silver protein contain 19 to 23 percent silver. Calculate the percent silver in the sample analyzed, and compare the result with the official requirement.

Mercuric ion does not yield a copious precipitate upon titration with ammonium thiocyanate as does silver ion. This is explained by the fact that mercuric thiocyanate, $Hg(SCN)_2$, complexes with the yet unreacted Hg^{++}. At or near the end point it may or may not precipitate, since it is soluble to the extent of 70 mg percent in water.

The presence of halide ions (Cl, Br, and I) will interfere with the determination of mercuric mercury because they, too, form slightly dissociated compounds, as does thiocyanate, with Hg^{++}. The dissociation constants of the following mercuric compounds decrease in the following order:

$$HgCl_2 > HgBr_2 > Hg(SCN)_2 > HgI_2 > Hg(CN)_2$$

The determination of theophylline in aminophylline suppositories, meralluride, meralluride injection, mercurophylline, mercurophylline injection and tablets is based on the formation of insoluble silver theophyllinate in an ammoniacal solution. The precipitate is then filtered off and collected; the Ag^+ are released upon acidification of the precipitate and determined by a direct titration with ammonium thiocyanate. The silver nitrate reacts with the theophylline molecule on a mole for mole basis, so that 1 mole of theophylline is 1 equivalent weight. An alternative method would be to use a measured excess of standard solution of silver nitrate to react with the sample. Bring the mixture up to a specified volume, filter, and collect an aliquot of the filtrate to determine the amount

TABLE 18. SUBSTANCES ASSAYED BY DIRECT TITRATION WITH AMMONIUM THIOCYANATE

Substance	mEq, Gm	Official requirement, percent*
U.S.P.		
Aminophylline suppositories	0.22823	$C_{16}H_{24}N_{10}O_4 \cdot 2H_2O$ = 90 to 110 l.a.
Meralluride	0.18017	$C_7H_8N_4O_2$ = 94 to 106 l.a.
Meralluride injection	0.19817	$C_7H_8N_4O_2 \cdot H_2O$ = 95 to 105 l.a.
Mercuric acetate (R)	0.15934	$Hg(C_2H_3O_2)_2$ = 98
Mercuric nitrate (R)*	0.16230	$Hg(NO_3)_2$ = 99
Mercuric sulfate (R)*	0.10030	Hg = 67 to 67.5
Mercury, ammoniated	0.12604	$HgNH_2Cl$ = 98
Silver nitrate	0.16987	$AgNO_3$ = 99.8
Silver nitrate ophthalmic solution	0.16987	$AgNO_3$ = 0.95 to 1.05 *W/W*
Silver nitrate, toughened	0.16987	$AgNO_3$ = 94.5
Sodium mercaptomerin	0.30300	$C_{16}H_{25}HgNNa_2O_6S$ = 95 to 105
Sodium mercaptomerin, sterile	0.30300	$C_{16}H_{25}HgNNa_2O_6S$ = 95 to 105
N.F.		
Mercuric oxide, yellow	0.10830	HgO = 99.5
Mercuric oxide ointment, yellow	0.10830	HgO = 0.9 to 1.1
Mercurophylline	0.25497	$C_{14}H_{24}HgNNaO_5$ = 94 to 106
Mercurophylline	0.18017	$C_7H_8N_4O_2$ = 94 to 106
Mercurophylline injection	0.25497	$C_{14}H_{24}HgNNaO_5$ = 94 to 106 l.a.
Mercurophylline injection	0.18017	$C_7H_8N_4O_2$ = 94 to 106 l.a.
Mercurophylline tablets	0.25497	$C_{14}H_{24}HgNNaO_5$ = 90 to 110 l.a.
Mercurophylline tablets	0.18017	$C_7H_8N_4O_2$ = 90 to 110 l.a.
Nitromersol	0.17585	$C_7H_5HgNO_3$ = 98
Nitromersol solution	0.17585	$C_7H_5HgNO_3$ = 0.18 to 0.22 *W/V*
Nitromersol tincture	0.17585	$C_7H_5HgNO_3$ = 0.45 to 0.55 *W/V*
Phenylmercuric nitrate	0.10032	Hg = 62.75 to 63.50
Silver nitrate solution, ammoniacal	0.10788	Ag = 28.5 to 30.5 *W/W*
Silver protein, mild	0.10788	Ag = 19 to 23
Thimerosal	0.20240	$C_9H_9HgNaO_2S$ = 97 to 101
Thimerosal solution	0.20240	$C_9H_9HgNaO_2S$ = 0.095 to 0.105 *W/V*
Thimerosal tincture	0.20240	$C_9H_9HgNaO_2S$ = 0.095 to 0.105 *W/V*

* = ferric nitrate indicator.
l.a. = labeled amount; (R) = reagent.

of unreacted silver nitrate by titration with ammonium thiocyanate, as is done in the assay of aminophylline for the theophylline content. Another alternative would be to titrate the hydrogen ion displaced by the silver ion with standard base. These methods however are being replaced with the newer nonaqueous titrimetric method.

Exercise 34

Object. Assay of Sodium Lauryl Sulfate for Sodium Chloride Content.

Materials Required. 5 Gm of Sodium Lauryl Sulfate
Dilute nitric acid (1 in 20)
2 ml of potassium chromate T.S.
0.1 *N* silver nitrate

Procedure. "Dissolve about 5 Gm., accurately weighed, in about 50 ml. of water. Neutralize the solution with dilute nitric acid (1 in 20), using litmus paper as the indicator, add 2 ml. of potassium chromate T.S., and titrate with 0.1 *N* silver nitrate. Each milliliter of 0.1 *N* silver nitrate is equivalent to 5.844 mg. of NaCl."

In order to employ coprecipitation of a colored compound as an indicator, one must know the chloride ion concentration $[Cl^-]$ at the instant the stoichiometric point is reached.

In this determination;

$$AgNO_3 + NaCl \longrightarrow AgCl\downarrow + NaNO_3$$

the $[Cl^-]$ in solution at the stoichiometric point of this reaction is the same as that in a saturated solution of AgCl at that given temperature.

$$AgCl \rightleftharpoons Ag^+ + Cl^-$$

therefore $$[Ag^+] = [Cl^-]$$

and $$[Ag^+] \times [Cl^-] = K_{sp} = 1.56 \times 10^{-10}$$

Then, $$[Cl^-]^2 = 1.56 \times 10^{-10}$$

or $$[Cl^-] = 1.25 \times 10^{-5} \text{ mole per liter}$$

So a titration of chloride ion, e.g., sodium chloride, by means of silver nitrate should end when the chloride ion concentration is reduced to a value of 1.25×10^{-5} mole per liter.

$$2AgNO_3 + K_2CrO_4 \longrightarrow \underset{\text{Red}}{Ag_2CrO_4\downarrow} + 2KNO_3$$

When silver nitrate is added to a solution containing some chloride ion and chromate ion, only the silver chloride will precipitate initially. Finally, when the chloride ion concentration is reduced to a given point, then silver chromate will coprecipitate along with the silver chloride. Since the K_{sp} of Ag_2CrO_4 is 9×10^{-12} at 25°, it will continue to precipitate as long as the solubility product is exceeded. It is therefore important to adjust the concentration of the chromate ion in the titrating solution so that it will coprecipitate at the instant the chloride ion concentration has been reduced to 1.25×10^{-5} mole per liter.

For coprecipitation:

$$[Ag^+] \times [Cl^-] = 1.56 \times 10^{-10}$$

and $$[Ag^+]^2 \times [CrO_4^{=}] = 9 \times 10^{-12}$$

and in a solution being titrated with silver nitrate the $[Ag^+]$ must be the same for both expressions, so that

$$[Ag^+]^2 \times [Cl^-]^2 = (1.56 \times 10^{-10})^2$$

and

$$[Ag^+]^2 \times [CrO_4^=] = 9 \times 10^{-12}$$

Therefore

$$\frac{[Cl^-]^2}{[CrO_4^=]} = \frac{(1.56 \times 10^{-10})^2}{9 \times 10^{-12}}$$

$$[CrO_4^=] = 3.7 \times 10^8 \times [Cl^-]^2$$

This means that for silver chromate to coprecipitate with silver chloride the chromate ion concentration must be in the order of 3.7×10^8 times that of the chloride ion concentration squared in the same solution.

Since the $[Cl^-]$ at the stoichiometric point is 1.25×10^{-5}, then

$$[CrO_4^=] = 3.7 \times 10^8 \times (1.25 \times 10^{-5})^2$$
$$= 0.0577 \text{ mole per liter}$$

which means that the solution must be so adjusted that at the stoichiometric point of the titration we have a chromate ion concentration of approximately 0.06 *M*. The potassium chromate indicator solution is 10 percent strength, so that 1 ml will furnish about 0.0005 mole of chromate ion and 2 ml will furnish 0.001 mole in approximately 80 ml of titrating solution, resulting in a molar concentration of 12.5×0.001 or 0.0125 *M* or about one-fourth of the amount calculated. If 0.06 *M* concentration of chromate ion were used instead of the 0.0125 *M* concentration called for, the titrating solution would be so yellow as to make the detection of the appearance of red silver chromate difficult.

Questions and Problems

1. Calculate the following titer values for 0.1 *N* ammonium thiocyante with the values given in the official monographs: (*a*) yellow mercuric oxide, HgO; (*b*) meralluride, $C_9H_{16}HgN_2O_6$; (*c*) ammoniated mercury, NH_2HgCl; (*d*) thimerosal solution, $C_9H_9HgNaO_2S$.

2. What preliminary treatment of samples of each of the following is necessary in order to obtain a solution of an equivalent amount of silver or mercury salt prior to titration? (*a*) sodium mercaptomerin; (*b*) nitromersol; (*c*) phenylmercuric nitrate; (*d*) toughened silver nitrate. Explain, by equation if possible, the purpose of the reagents employed.

3. In terms of extra silver nitrate solution, calculate the error introduced by using only 2 ml of the indicator solution, potassium chromate, instead of the 8 ml called for theoretically.

4. In titration of a solution of the ash from a 2.4090-Gm sample of mild silver protein in nitric acid, 18.14 ml of a solution of ammonium thiocyanate was required. Each milliliter of the ammonium thiocyanate solution was equivalent to 9.8410 mg of pure silver. Calculate the Ag content of the sample (percent).

5. What volume of 0.1 *N* ammonium thiocyanate would be used in titrating to a ferric ammonium sulfate end point a solution containing silver or mercury ions equivalent to (*a*) 0.7500 Gm of mercuric acetate; (*b*) 1.0500 Gm of ammoniacal silver nitrate solution; (*c*) 0.1965 Gm of thimerosal?

6. What weight of each of the following would be chosen as sample so that 30.00 ml of 0.1 *N* ammonium thiocyanate would be required in the titration? (*a*) silver nitrate; (*b*) ammoniated mercury; (*c*) yellow mercuric oxide; (*d*) sterile mercaptomerin sodium.

Exercise 35

Object. Assay of Iodoalphionic Acid.

Materials Required. 200 mg of Iodoalphionic Acid
15 ml of sodium hydroxide T.S.
25 ml of potassium permanganate (1 in 15)
10 ml of dilute sulfuric acid (1 in 2)
15 ml of sodium bisulfite (1 in 5)
0.05 *N* silver nitrate

Procedure. 1. "Place about 200 mg. of Iodoalphionic Acid, previously dried over phosphorus pentoxide for 4 hours and accurately weighed, in a 500-ml conical flask, and add 15 ml. of sodium hydroxide T.S. Warm gently until the substance has dissolved, add 25 ml. of a solution of potassium permanganate (1 in 15), then add several glass beads, place a small short-stemmed funnel in the neck of the flask, and boil gently for 10 minutes."

Iodoalphionic Acid is 3-(4-hydroxy-3,5-diiodophenyl)-2-phenylpropionic acid. In the oxidative degradation of the organic compound by heating with alkaline potassium permanganate, the organically bound iodine is converted to an equivalent amount of iodate (IO_3^-).

2. "Allow the mixture in the flask to cool to room temperature, wash the funnel and walls of the flask with 75 ml. of water, and add 10 ml. of dilute sulfuric acid (1 in 2). Add, in one portion, 15 ml. of a solution of sodium bisulfite (1 in 5), and when the solution has become colorless, cool, and then add the solution of potassium permanganate, dropwise, until a yellow color appears. At once add the solution of sodium bisulfite, dropwise, until the yellow color is again discharged. Now add, dropwise, a dilute solution of potassium permanganate, prepared by mixing 1 ml. of the 1 in 15 solution with 49 ml. of water, until a faint yellow color appears."

The mixture is acidified and treated with sodium bisulfite to reduce the iodate to iodide; at the same time the excess potassium permanganate and the manganese dioxide present are reduced to Mn^{++}. The resulting colorless solution is treated alternately with the potassium permanganate and the sodium bisulfite until a very small amount of free iodine is present as indicated by a faint yellow color. Practically all the iodine from the organic iodine compound must be present in this adjusted solution as iodide.

3. "Add 0.5 ml. of starch T.S., and titrate with 0.05 *N* silver nitrate until the blue color is just discharged, leaving a canary-yellow precipitate. Each milliliter of 0.05 *N* silver nitrate is equivalent to 12.35 mg. of $C_{15}H_{12}I_2O_3$."

When the starch is added, a blue color appears, because both iodine and iodide are present. As silver nitrate is run into the mixture, the color gradually fades to green and the precipitated silver iodide coagulates and adsorbs the green color near the end point. At the end point, the precipitate is canary yellow. Such an end point occurs when the silver nitrate has reacted with all the iodide present, including that formed by the reaction

$$3I_2 + 3H_2O \rightleftharpoons HIO_3 + 5HI$$

A small proportion of iodine and iodate is still present at the end point, so that it occurs slightly before the stoichiometric point is reached, and slightly low results are obtained.

Since each molecule of the iodoalphionic acid contains 2 atoms of iodine, which in turn is converted to 2 iodide ions and reacts with 2 moles of silver nitrate, 1 equivalent weight of the acid is equal to ½ mole; 494.07/2 = 247.04 Gm and 1 mEq = 0.24704 Gm.

In Table 19 are listed a number of organic iodine compounds which are assayed by methods like that used for Iodoalphionic Acid.

TABLE 19. SUBSTANCES ASSAYED BY DIRECT TITRATION WITH SILVER NITRATE

Substances	mEq, Gm	Indicator	Official requirement, percent
U.S.P.			
4-Aminoantipyrine hydrochloride (R)	0.02397	d.c.f.	$C_{11}H_{13}N_3O \cdot HCl$ = 100.6 to 108.5
Betazole hydrochloride	0.03545	eo.y.	Cl = 37.8 to 38.9
Hydroxyamphetamine hydrobromide	0.07992	eo.y.	Br = 33.6 to 35.2
Iodochlorhydroxyquin	0.12691	pot.	I = 40 to 42
Iodochlorhydroxyquin	0.03545	pot.	Cl = 11 to 12.2
Iopanoic acid	0.19032	t.b.ph.	$C_{11}H_{12}I_3NO_2$ = 97 to 101
Isoproterenol hydrochloride	0.03545	eo.y.	Cl = 14 to 14.7
Mecamylamine hydrochloride	0.03545	eo.y.	Cl = 17 to 17.8
Mechlorethamine hydrochloride	0.03545	amp.	Cl = 18 to 19.3
Meglumine diatrizoate injection	0.26971	t.b.ph.	$C_{18}H_{26}I_3N_3O_9$ = 90 to 110 l.a.
Meglumine iodipamide injection	0.25504	t.b.ph.	$C_{34}H_{48}I_6N_4O_{16}$ = 90 to 110 l.a.
Meglumine iothalamate injection	0.26971	t.b.ph.	$C_{18}H_{26}I_3N_3O_9$ = 96 to 104 l.a.
Methoxamine hydrochloride	0.03545	eo.y.	Cl = 14.1 to 14.5
Phenylephrine hydrochloride	0.03545	eo.y.	Cl = 17 to 17.7
Potassium cyanide (R)	0.13024	KI	KCN = 96
Propantheline bromide	0.07992	eo.y.	Br = 17.5 to 18.2
Pyridoxine hydrochloride	0.03545	eo.y.	Cl = 16.9 to 17.6
Sodium cyanide (R)	0.09802	KI	NaCN = 95
Sodium diatrizoate injection	0.21197	t.b.ph.	$C_{11}H_8I_3N_2NaO_4$ = 95 to 105 l.a.
Sodium iodipamide injection	0.19728	t.b.ph.	$C_{20}H_{12}I_6N_2NaO_6$ = 90 to 110 l.a.
Sodium iothalamate injection	0.21197	t.b.ph.	$C_{11}H_8I_3N_2NaO_4$ = 97.5 to 102.5 l.a.
Sodium lauryl sulfate	0.05844	K_2CrO_4	NaCl = 8
Succinylcholine chloride	0.03545	eo.y.	Cl = 19.3 to 19.8
N.F.			
Azacyclonol hydrochloride	0.30384	d.c.f.	$C_{18}H_{21}NO \cdot HCl$ = 98 to 102
Iodoalphionic acid	0.24704	st.	$C_{15}H_{12}I_2O_3$ = 98 to 102
Iodopyracet injection	0.12691	st.	I = 61.5 to 63.5
Meperidine hydrochloride	0.03545	eo.s.	Cl = 12.2 to 12.7
Phenmetrazine hydrochloride	0.21371	pot.	$C_{11}H_{15}NO \cdot HCl$ = 98
Sodium cyanide (R)	0.09802	KI	NaCN = 95
Sodium iodomethamate	0.12691	st.	I = 50.5 to 52.5
Sodium iodomethamate injection	0.24646	st.	$C_8H_3I_2NNa_2O_5$ = 95 to 105 l.a.
Sodium methiodal	0.24398	st.	CH_2INaO_3S = 98 to 102
Sodium methiodal injection	0.24398	st.	CH_2INaO_3S = 95 to 105 l.a.
Tetra-*n*-butylammonium iodide (R)	0.36938	pot.	$(C_4H_9)_4NI$ = 99.9

(R) = reagent; d.c.f. = dichlorofluorescein; eo.y. = eosin Y; pot. = potentiometric; amp. = amperometric; t.b.ph. = tetrabromophenolphthalein ethyl ester; st. = starch; l.a. = labeled amount.

The use of dichlorofluorescein in direct titrations with standard silver nitrate is exemplified in the following exercise.

Exercise 36

Object. Assay of Azacyclonol Hydrochloride.

Materials Required. 500 mg of Azacyclonol Hydrochloride
Dichlorofluorescein T.S.
0.1 *N* silver nitrate

Procedure. "Dissolve about 500 mg. of Azacyclonol Hydrochloride, previously dried at 105° for 2 hours and accurately weighed, in a mixture of 40 ml. of alcohol and 20 ml. of water, add dichlorofluorescein T.S., and titrate with 0.1 *N* silver nitrate. Each milliliter of 0.1 *N* silver nitrate is equivalent of 30.38 mg. of $C_{18}H_{21}NO{\cdot}HCl$."

Dichlorofluorescein, like the other adsorptive indicators, is a weak organic acid, and the extent of its ionization is dependent on the pH of the titrated solution. In highly acidic solutions where the pH is lower than 4, the anions present are so few as to make the end point very faint. The pH of a 1 in 200 solution of azacyclonol is between 5 and 7. Alcohol is used in the assay to help keep the precipitated silver chloride in a colloidal state and thus give sharpness to the color change which occurs on the precipitate surface.

$$\begin{array}{ccc} Cl^- \quad Cl^- & & Ag^+ \quad Ag^+ \\ AgCl \quad d.c.f.^- & + Ag^+ \longrightarrow & AgCl \quad d.c.f.^- \\ Cl^- \quad Cl^- & & Ag^+ \quad Ag^+ \\ \text{Little color effect} & & \text{Colored complex} \end{array}$$

A color change does not take place during the titration until the Ag^+ is in excess, at which time the colloidal silver chloride particles bear a positive charge because of a primary adsorption layer of silver cations. A secondary adsorption of the indicator, dichlorofluorescein, anions results in a change in the electron distribution of the anions and a concomitant change in color. Before the end point is reached, halide ions are primarily adsorbed, and the resulting negatively charged particles repel the adsorption of the indicator anions.

Exercise 37

Object. Assay of Sodium Cyanide reagent.

Materials Required. 400 mg of Sodium Cyanide
Potassium iodide solution (1 in 10)
1 ml of stronger ammonia water
0.1 *N* silver nitrate

Procedure. "Weigh accurately about 400 mg., and dissolve in 30 ml. of water. Add 4 drops of potassium iodide solution (1 in 10) and 1 ml. of stronger ammonia water, and titrate with 0.1 *N* silver nitrate to a slight, permanent turbidity: each milliliter of 0.1 *N* silver nitrate is equivalent to 9.802 mg. of NaCN. Not less than 95% is found."

1. $AgNO_3 + 2NaCN \longrightarrow NaAg(CN)_2 + NaNO_3$
2. $AgNO_3 + KI \longrightarrow AgI\downarrow + KNO_3$

Silver cyanide forms immediately upon the addition of silver nitrate but in the presence of excess sodium cyanide in the titration mixture, the soluble complex, sodium argentocyanide, forms and the precipitate dissolves,

$$AgCN\downarrow + NaCN \longrightarrow NaAg(CN)_2$$

The first drop in excess of silver nitrate in reaction (1) then reacts to form insoluble silver iodide, since the ammonia complexes the argentocyanide ion, keeping it in solution. The permanent turbidity is caused by silver iodide, which is a canary-yellow precipitate. 1 mEq of NaCN is 2NaCN/1,000 or 0.09802.

RESIDUAL TITRATION METHOD (VOLHARD METHOD)

This method is based on the complete precipitation of insoluble silver salts from nitric acid solution by the addition of excess standard silver nitrate solution to a soluble salt and the determination of the amount of silver nitrate solution in excess by residual titration with standard ammonium thiocyanate solution, using ferric ammonium sulfate as the indicator.

To Determine the Purity of a Soluble Chloride. The purity of any soluble chloride may be determined by the method given below, provided that other substances which give precipitates with silver nitrate are absent. Substances other than chlorides that produce slightly soluble silver salts are bromides, iodides, cyanides, thiocyanates, sulfides, phosphates, arsenates, carbonates, etc.

Exercise 38

Object. Assay of Sodium Chloride.

Materials Required. 250 mg of sodium chloride
3 ml of nitric acid
2 ml of ferric ammonium sulfate T.S.
50 ml of 0.1 *N* silver nitrate solution
0.1 *N* ammonium thiocyanate solution
5 ml nitrobenzene

Procedure. "Weigh accurately about 250 mg. of Sodium Chloride, dissolve it in 50 ml. of water in a glass-stoppered flask, and add 50 ml. of 0.1 *N* silver nitrate, 3 ml. of nitric acid, 5 ml. of nitrobenzene, and 2 ml. of ferric ammonium sulfate T.S. Shake

well, and titrate the excess silver nitrate with 0.1 *N* ammonium thiocyanate. Each milliliter of 0.1 *N* silver nitrate is equivalent to 5.844 mg. of NaCl."

Nitric acid is added to prevent the precipitation of silver as carbonate, phosphate, etc., and to prevent hydrolysis of the ferric alum.

Silver chloride is sufficiently soluble that the following reaction with ammonium thiocyanate can occur during the back titration:

$$AgCl + NH_4SCN \rightleftharpoons AgSCN\downarrow + NH_4Cl$$

As a result, the ammonium thiocyanate will be consumed not only by the excess silver nitrate present but also, to an appreciable extent, by the silver chloride unless it is removed by filtration prior to the back titration. As a convenient alternative to filtration, nitrobenzene is added to form a film over the precipitated silver chloride particles, thus preventing reaction with the ammonium thiocyanate. Silver bromide and silver iodide are considerably less soluble, and the precipitate does not interfere in the Volhard analysis of soluble bromides and iodides.

The equivalent weight of sodium chloride is 58.44 Gm, since it reacts mole for mole with silver nitrate.

$$\underset{58.44}{NaCl} + AgNO_3 \longrightarrow AgCl + NaNO_3$$

Calculations. The difference between the total milliequivalents of standard silver nitrate (ml × *N*) and milliequivalents of standard ammonium thiocyanate (ml × *N*) is a measure of the number of milliequivalents of NaCl present. The number of milliequivalents of NaCl multiplied by the weight of 1 mEq will give the grams of NaCl in the sample.

$$\frac{(\text{ml}_{ag^+} \times N) - (\text{ml}_{scn^-} \times N) \times 0.05844 \times 100}{\text{Sample weight}} = \text{percent NaCl}$$

Each milliliter of 0.1 *N* silver nitrate, 0.1 mEq, which is consumed by sodium chloride is equivalent to 0.1 × 0.05844 Gm of NaCl.

Questions and Problems

1. Give equations for the reactions which occur in the official methods for analysis of sodium lauryl sulfate (NaCl) and dextrose and sodium chloride injection (NaCl). Compare the methods and explain how the end point is determined in each case.

2. State the important difference in the assay procedures for sodium bromide and for sodium chloride and sodium iodide. Give an explanation for the difference.

3. Using appropriate solubility product constants, calculate the relative solubilities of AgCl, AgBr, AgI, and AgSCN in water.

4. A 0.2182-Gm sample of sodium chloride was assayed by the Volhard method using 50,00 ml of 0.0985 *N* silver nitrate and 11.75 ml of 0.1340 *N* ammonium thiocyanate. Calculate the percent NaCl in the sample.

Exercise 39

Object. Assay of Ammonium Bromide.

Materials Required. 0.4 Gm of ammonium bromide
2 ml of nitric acid
2 ml of ferric ammonium sulfate T.S.
50 ml of 0.1 *N* silver nitrate
0.1 *N* ammonium thiocyanate

Procedure. "Dry about 400 mg. of Ammonium Bromide at 105° for 2 hours and weigh accurately. Dissolve it in about 50 ml. of water, add 50 ml. of 0.1 *N* silver nitrate, 2 ml. of ferric ammonium sulfate T.S., and 2 ml. of nitric acid. Titrate the excess of silver nitrate with 0.1 *N* ammonium thiocyanate. Each milliliter of 0.1 *N* silver nitrate is equivalent to 9.795 mg. of NH_4Br."

The above assay is similar to that for chlorides, except that nitrobenzene need not be added since silver bromide is less soluble than silver thiocyanate.

The reactions and calculations are similar to those of the preceding assay. Write equations for all reactions involved, and calculate the percentage purity of the ammonium bromide.

Questions and Problems

1. If the K_{sp} of AgBr is 5×10^{-13}, what is the solubility of silver bromide in grams per liter in water and in 10^{-3} *M* silver nitrate? Explain the effect of excess silver nitrate on the completeness of the precipitation of AgBr in the assay of ammonium bromide.

2. If ammonium iodide, 95.6 percent NH_4I, is to be assayed by a Volhard procedure using 50.00 ml of 0.0942 *N* silver nitrate, what weight of sample should be employed so that 20.00 ml of 0.1025 *N* ammonium thiocyanate would be required in the back titration?

3. A 250-mg sample of ammonium bromide was found to be equivalent to 25.00 ml of 0.1 *N* silver nitrate in a Volhard determination. What volume of 0.0962 *N* sulfuric acid would be required to neutralize the ammonia produced from a 0.4689-Gm sample of the same material?

Exercise 40

Object. Assay of Hydriodic Acid Syrup.

Materials Required. 25 ml of hydriodic acid syrup
40 ml of 0.1 *N* silver nitrate
5 ml of nitric acid
2 ml of ferric ammonium sulfate T.S.
0.1 *N* ammonium thiocyanate

Procedure. "Place exactly 25 ml. of Hydriodic Acid Syrup in a flask, dilute it with 100 ml. of water, add 40 ml. of 0.1 *N* silver nitrate, agitate the mixture, add 5 ml. of nitric acid, and heat on a water bath until the precipitate has acquired a bright yellow color. Cool, add 2 ml. of ferric ammonium sulfate T.S., and titrate the excess silver nitrate with 0.1 *N* ammonium thiocyanate. Each milliliter of 0.1 *N* silver nitrate is equivalent to 12.79 mg. of HI."

This assay is similar to that for bromides except that a measured volume of liquid is used for the sample.

$$\underset{127.91}{HI} + AgNO_3 \longrightarrow AgI + HNO_3$$

$$AgNO_3 + NH_4SCN \longrightarrow AgSCN + NH_4NO_3$$

Calculate the amount of HI equivalent to 1 ml of 0.1 *N* $AgNO_3$ and the percent of HI in the sample analyzed. The N.F. requires that hydriodic acid syrup contain not less than 1.3 and not more than 1.5 Gm of HI in each 100 ml. Note that this is a *W/V* percentage.

Questions and Problems

1. What preliminary treatment is necessary before standard silver nitrate may be added in the Volhard determination of each of the following: (*a*) tribromoethanol, $C_2H_3Br_3O$; (*b*) iodoform, CHI_3? Give equations where possible.

2. Calculate the titer values for 0.1 *N* silver nitrate for the substances indicated in question 1.

3. A silver nitrate solution had a NaCl titer of 5.569 mg per ml. Each milliliter of the silver nitrate solution would be equivalent to how many milligrams of (*a*) potassium bromide; (*b*) methacholine chloride; (*c*) tribromoethanol; (*d*) calcium chloride?

4. If 25.00 ml of hydriodic acid syrup (1.4 percent *W/V* HI) was treated with 40.00 ml of 0.1012 *N* silver nitrate, what volume of 0.0968 *N* ammonium thiocyanate would be required in the back titration?

5. A 2.00-ml sample of tribromoethanol solution was assayed by the official method. After adding 50.00 ml of 0.1 *N* silver nitrate, 32.39 ml of an ammonium thiocyanate solution was required for titration to a ferric alum end point. In a separate titration, 25.00 ml of the 0.1 *N* silver nitrate solution required 23.79 ml of the ammonium thiocyanate solution. Calculate the percent *W/V* of $C_2H_3Br_3O$ in the tribromoethanol solution.

The determination of theophylline or 8-chlorotheophylline in official preparations such as aminophylline, aminophylline injection and tablets, dimenhydrinate and tablets, theophylline tablets, theophylline sodium acetate and tablets, and theophylline sodium glycinate and tablets is accomplished by the reaction of theophylline with silver nitrate on a mole for mole ratio. An excess of standard silver nitrate is added to an ammoniacal solution of the sample. The silver salt is precipitated and filtered so that an aliquot of the filtrate can be determined for the amount of excess silver nitrate present. From the amount of ammonium thiocyanate needed to titrate all the excess silver nitrate one can readily calculate the amount of silver nitrate that must have reacted with the theophylline in the sample.

In the assay for thiourea and ammonium sulfide solution reagent, silver sulfide is precipitated in an ammoniacal solution. The end point is not sharp enough to adapt these assays to a direct titration method, so the

TABLE 20. SUBSTANCES ASSAYED BY RESIDUAL TITRATION OF SILVER NITRATE WITH AMMONIUM THIOCYANATE

Substances	mEq, Gm	Official requirement, percent
U.S.P.		
Acetylcholine chloride (R)	0.03545	Cl = 19.3 to 19.8
Aminophylline	0.18018	$C_7H_8N_4O_2$ = 84 to 86
Aminophylline injection	0.22823	$C_{16}H_{24}N_{10}O_4{\cdot}2H_2O$ = 93 to 107 l.a.
Aminophylline tablets	0.22823	$C_{16}H_{24}N_{10}O_4{\cdot}2H_2O$ = 93 to 107 l.a.
Ammonium chloride	0.05349	NH_4Cl = 99.5
Ammonium sulfide solution (R)	0.03408	$(NH_4)_2S$ = 16 to 20
Benzene hexachloride, gamma	0.09694	$C_6H_6Cl_6$ = 99
Bethanechol chloride	0.03545	Cl = 17.7 to 18.3
Calcium chloride (R)	0.05549	$CaCl_2$ = 74 to 78
Calcium chloride anhydrous (R)	0.05549	$CaCl_2$ = 96
Chlorobutanol	0.05916	$C_4H_7Cl_3O$ = 98
Chlorophenothane	0.07089	$C_{14}H_9Cl_5$ = 96 to 102
Dextrose and sodium chloride injection	0.05844	NaCl = 95 to 105 l.a.
Dibucaine hydrochloride	0.03545	Cl = 9.1 to 9.5
Dimenhydrinate	0.21462	$C_7H_7ClN_4O_2$ = 44 to 47
Dimenhydrinate tablets	0.46998	$C_{17}H_{21}NO{\cdot}C_7H_7ClN_4O_2$ = 95 to 105 l.a.
Hydriodic acid (R)	0.12791	HI = 47
Methyl iodide (R)	0.14194	CH_3I = 98.5
Potassium chloride	0.07456	KCl = 99
Potassium chloride injection	0.07456	KCl = 95 to 105 l.a.
Potassium chloride tablets	0.07456	KCl = 95 to 105 l.a.
Pyridoxal hydrochloride (R)	0.03545	Cl = 17.2 to 17.7
Pyridoxamine dihydrochloride (R)	0.03545	Cl = 29.1 to 29.6
Ringer's injection	0.05844	NaCl = 0.82 to 0.90 *W/V*
Ringer's injection, lactated	0.05844	NaCl = 0.57 to 0.63 *W/V*
Sodium chloride	0.05844	NaCl = 99.5
Sodium chloride injection	0.05844	NaCl = 0.85 to 0.95 *W/V*
Sodium chloride solution	0.05844	NaCl = 0.85 to 0.95 *W/V*
Sodium chloride tablets	0.05844	NaCl = 95 to 105 l.a.
Sodium sulfobromophthalein	0.07991	Br = 36 to 39
Tetramethylammonium bromide (R)	0.15406	$(CH_3)_4NBr$ = 98
Tetramethylammonium chloride (R)	0.10960	$(CH_3)_4NCl$ = 98
Thiourea (R)	0.03806	$(NH_2)_2CS$ = 99
Trihexyphenidyl hydrochloride	0.03545	Cl = 10 to 10.7
Zinc chloride (R)	0.06814	$ZnCl_2$ = 95
N.F.		
Ammonium bromide	0.09795	NH_4BR = 99 to 101
Bromides elixir, three	0.10590	Total bromides = 23 to 25 *W/V*
Bromides tablets, three	0.07991	Br = 70 to 81 total bromides
Chlorotrianisene	0.38088	$C_{23}H_{21}ClO_3$ = 95 to 105
Fructose and sodium chloride injection	0.05844	NaCl = 95 to 105 l.a.

TABLE 20. SUBSTANCES ASSAYED BY RESIDUAL TITRATION OF SILVER NITRATE WITH AMMONIUM THIOCYANATE (*Continued*)

Substances	mEq, Gm	Official requirement, percent
Gallamine triethiodide	0.29718	$C_{30}H_{60}I_3N_3O_3$ = 98 to 101
Gallamine triethiodide injection	0.29718	$C_{30}H_{60}I_3N_3O_3$ = 95 to 105 l.a.
Hydriodic acid, diluted	0.12791	HI = 9.5 to 10.5 *W/V*
Hydriodic acid syrup	0.12791	HI = 1.3 to 1.5 *W/V*
Iodoform	0.13125	CHI_3 = 99
Methacholine chloride	0.03545	Cl = 17.8 to 18.4
Methacholine chloride, sterile	0.03545	Cl = 17.8 to 18.4
Methantheline bromide	0.07991	Br = 18.6 to 19.2
Methantheline bromide tablets	0.42036	$C_{21}H_{26}BrNO_3$ = 95 to 105 l.a.
Potassic saline injection, lactated	0.03545	Cl = 0.344 to 0.388 *W/V*
Potassium bromide	0.11902	KBr = 99 to 101
Povidone-iodine	0.12690	I = n.m.t. 6.6
Ringer's solution	0.05844	NaCl = 0.82 to 0.90 *W/V*
Sodium bromide	0.10290	NaBr = 99
Sodium bromide elixir	0.10290	NaBr = 16.5 to 18.5 *W/V*
Sodium chloride and dextrose tablets	0.05844	NaCl = 92.5 to 107.5 l.a.
Theophylline sodium acetate	0.18018	$C_7H_8N_4O_2$ = 55 to 65
Theophylline sodium acetate tablets	0.18018	$C_7H_8N_4O_2$ = 53 to 67 l.a.
Theophylline sodium glycinate	0.19818	$C_7H_8N_4O_2 \cdot H_2O$ = 49 to 52
Theophylline sodium glycinate tablets	0.19818	$C_7H_8N_4O_2 \cdot H_2O$ = 47 to 54 l.a.
Theophylline tablets	0.19818	$C_7H_8N_4O_2 \cdot H_2O$ = 94 to 106 l.a.
Tribromoethanol	0.09426	$C_2H_3Br_3O$ = 99 to 101.8
Tribromoethanol solution	0.09426	$C_2H_3Br_3O$ = 95 to 105 *W/W*
Tridihexethyl chloride	0.03545	Cl = n.l.t. 9.81

(R) = reagent; l.a. = labeled amount; n.m.t. = not more than; n.l.t. = not less than.

assay mixture treated with an excess of standard silver nitrate solution is filtered and an aliquot of the filtrate is then determined for the excess silver with ammonium thiocyanate. The reaction for the analysis is as indicated:

$$(H_2N)_2C{=}S + 2AgNO_3 + 2NH_3 \longrightarrow Ag_2S\downarrow + H_2N{-}CN + 2NH_4NO_3$$

Certain organic compounds will give up their halide ion readily if it is present as an ionic halogen, e.g., hydrohalide of a nitrogen base, while other organic halides require intimate contact with silver nitrate for periods up to 10 hr (iodoform, methyl iodide reagent) to complete the reaction with silver nitrate. Others require even more rigorous reaction, i.e., aqueous or alcoholic base and heating, to cleave the carbon-halogen bond before reaction with silver nitrate (γ-benzene hexachloride, chlorobutanol, and tribromoethanol). With chlorophenothane there is need to reflux with metallic sodium in isopropyl alcohol before the halogen reacts with the sodium ion to form sodium chloride and before determination of the chloride ion with standard silver nitrate.

SODIUM TETRAPHENYLBORON TITRATIONS

Sodium tetraphenylboron, $Na(C_6H_5)_4B$, will quantitatively precipitate such organic nitrogen compounds as alkaloids, amines, and quaternary salts, as well as ammonium, potassium, and silver ions.

$$R_4N{\cdot}Cl + Na(C_6H_5)_4B \longrightarrow R_4N{\cdot}(C_6H_5)_4B\downarrow + NaCl$$
$$NH_4^+ + Na(C_6H_5)_4B \longrightarrow NH_4{\cdot}(C_6H_5)_4B\downarrow + Na^+$$

Quaternary compounds will react with certain dyes such as bromophenol blue, an anionic dye, to form a blue, chloroform-soluble complex

$$R_4N{\cdot}Cl + Ind^- \longrightarrow R_4N{\cdot}Ind + Cl^-$$

and this colored complex will react with sodium tetraphenylboron to form an insoluble compound:

$$R_4N{\cdot}Ind + Na(C_6H_5)_4B \longrightarrow R_4N{\cdot}(C_6H_5)_4B\downarrow + Na^+ + Ind^-$$

The titration uses chloroform as an extraction indicator in the same way as it is used in the determination of iodides with potassium iodate where the titration ceases when the chloroform has lost all of the iodine color. In the determination of Benzethonium Chloride or Cetylpyridinium Chloride, the sample is dissolved in water, to which is then added, the indicator solution, chloroform, and the entire mixture is slightly alkalinized before titration with the sodium tetraphenylboron. The end point is reached when the chloroform layer turns colorless as the nitrogen compound is released from the dye complex to precipitate with the tetraphenylboron.

This method of analysis offers some obvious advantages over previous methods using alkaloidal precipitants such as potassium ferricyanide or potassium dichromate and subsequent determination of the excess of these standard solutions by the addition of potassium iodide and titrating the liberated iodine with sodium thiosulfate.

Benzethonium Chloride and Cetylpyridinium Chloride react with the sodium tetraphenylboron on a mole for mole basis so that 1 ml of 0.2 *M* solution is equivalent to 448.09/50,000 or 0.008962 Gm of Benzethonium Chloride and 340.00/50,000 or 0.0068 Gm of Cetylpyridinium Chloride.

The purity rubric for the reagent, sodium tetraphenylboron, is not less than 99.5 percent as determined gravimetrically by precipitation as the potassium salt. A liter of 0.02 *M* solution is made by dissolving 342.23/50 or 6.845 Gm in water. This solution should be freshly prepared before using. A slight haziness does not affect its use.

COMPLEXATION METHODS

Introduction. Historically, the quantitative analysis of inorganic pharmaceutical products containing metal ions such as Al, Bi, Ca, Mg, and Zn was performed using gravimetric methods. These procedures were very time-consuming since they involved precipitation, filtration, washing, and drying or ignition to constant weight. Official calcium preparations were later analyzed volumetrically using an oxalate-permanganate procedure which involved a precipitation technique and titration of a carefully heated solution of oxalate ions. With the introduction of the analytical reagent disodium ethylenediaminetetraacetate, commonly referred to as EDTA, a new volumetric procedure evolved for metal determination employing metal-ion indicators in the same manner that pH indicators are used in acid-base titrations.

Complexation Reaction. When a metal ion combines with a molecule which can donate electrons, the resulting compound is termed a *complex*. If the combining molecule contains two or more groups that donate electrons, this complex is called a *chelate*. EDTA will react with metal ions to form a water-soluble, stable complex or chelate compound. The reaction is rapid and quantitative with such polyvalent metal ions as Al^{3+}, Bi^{3+}, Ca^{++}, Cu^{++}, Hg^{++}, Mg^{++}, and Zn^{++}. Monovalent metal ions yield relatively weak or unstable complexes.

Ionization of ethylenediaminetetraacetic acid.

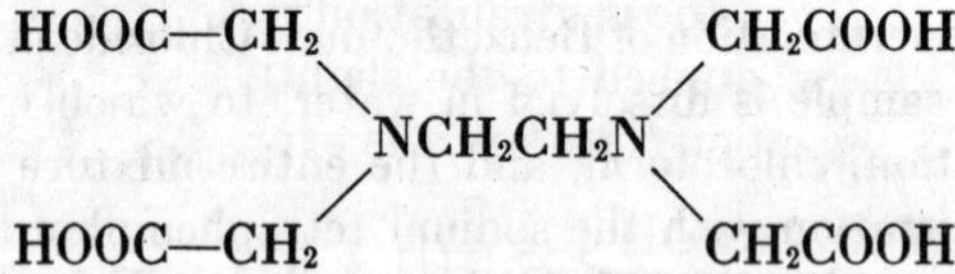

For brevity the symbol H_4Y is used to designate the molecule of ethylenediaminetetraacetic acid; thus Y^{4-} would represent the tetraacetate ion.

$$H_4Y \overset{1}{\rightleftharpoons} H^+ + H_3Y^- \qquad pK_1 = 2.07$$

$$H_3Y^- \overset{2}{\rightleftharpoons} H^+ + H_2Y^= \qquad pK_2 = 2.75$$

$$H_2Y^= \overset{3}{\rightleftharpoons} H^+ + HY^{3-} \qquad pK_3 = 6.24$$

$$HY^{3-} \overset{4}{\rightleftharpoons} H^+ + Y^{4-} \qquad pK_4 = 10.34$$

The predominant form of the dissociated forms of ethylenediaminetetraacetic acid (Y^{4-}, HY^{3-}, $H_2Y^=$, and H_3Y^-) is dependent upon the pH at which a given complexation reaction is being carried out at.

In all EDTA complexation reactions the ratio of the EDTA to the metal ion is 1:1,

$$Y^{4-} + M^{n+} \rightleftharpoons MY^{n-4}$$

With a bivalent metal such as calcium there are six bonds or points of attachment between the metal and the EDTA, of which two are ionic, two are ordinate bonds involving the oxygen atoms, and two are coordinate bonds formed by the unshared pair of electrons of the nitrogen atoms which are donated to the metal. With a trivalent metal ion such as Al^{3+} one of the ordinate bonds would then become ionic in character. The four oxygen and the two nitrogen atoms of the EDTA molecule which are capable of entering a complexation reaction with a metal ion would make it a *hexadentate* molecule. The EDTA molecule which provides groups for attachment to metal ions is called a *ligand.* The stability of any chelate or complex molecule varies with the number of rings which can be formed. The EDTA-metal complexes are very stable, since the five-membered rings which are formed in the complex are strain-free and are the most stable rings involving organic molecules.

In gravimetric methods of analysis the formation of a precipitate was a sufficient driving force to cause a reaction to go far to the right, or to completion. The equilibrium constant of the reaction applied to the insoluble compound formed is called a *solubility product constant*, and the expression used to designate it is in the case of calcium oxalate:

$$K_{sp} = \frac{[Ca^{++}] \times [C_2O_4^{=}]}{[CaC_2O_4]} = [Ca^{++}] \times [C_2O_4^{=}] = 2.6 \times 10^{-9}$$

In acidimetry-alkalimetry, the formation of a very slightly ionized water was the driving force in the neutralization reaction. The equilibrium constant of the reaction applied to the slightly dissociated product, H_2O, formed is called ionization constant; the expression used to designate it is

$$K = \frac{[H^+] \times [OH^-]}{[H_2O]} = [H^+] \times [OH^-] = 1 \times 10^{-14}$$

In a complexometric reaction, the formation of a stable, soluble complex is the driving force in the reaction; the application of the law of mass action to the reaction equilibrium is termed *stability constant* and is expressed:

$$K_s = \frac{[CaY^{=}]}{[Ca^{++}] \times [Y^{4-}]} = 2.5 \times 10^{11}$$

The stability constant is a large number since the complex is very slightly ionized and the concentration of the calcium-EDTA complex is so much greater than its constituent ions. If the reciprocal is designated as the instability constant, then the *K* values would be very small and in the order of magnitude of those constants known for weak acids or insoluble compounds. Since the concentration of Y^{4-} is a function of the pH at which a specific reaction is run, it is obvious that stability constants are valid only for that pH at which they were determined.

The stability constants of those metals which are complexed and determined by EDTA titration in the official assays are (log K values):

$AlY^- = 16.1$ $CaY^= = 10.7$ $HgY^= = 21.8$ $ZnY^= = 16.5$
$BiY^- = 8.6$ $CuY^= = 18.8$ $MgY^= = 8.7$

For successful titrations with EDTA, the stability constants of the complex formed must be greater than 8. Li and Na form very weak complexes and cannot be titrated. Ag and Ba stability constants are just below 8. Anions which can form insoluble precipitates with certain metal ions ($MgNH_4PO_4$, $BaSO_4$, $AgAsO_4$, etc.) can be assayed indirectly by determining the amount of metal liberated from the precipitate or by determining the amount of metal by back titration of the excess metal ion after treatment of an anion with a fixed amount. Metal ions such as Fe^{3+} and Cu^{++} that are colored and yield highly colored complexes cannot easily be titrated visually. The factors influencing the EDTA reaction are the activity of the metal ion, the pH at which the titration is run, and the presence of interfering ions such as cyanide, citrate, tartrate, fluoride, and other complex forming agents. The higher the stability constant, the more acid the pH at which the complexation titration can be run. Organic solvents also increase the stability of the complex. Neutral salts such as NaCl decrease the stability.

Reagent. The disodium salt is preferred over the free acid in preparing the standard solution of EDTA. It is much more water-soluble, nonhygroscopic, and very stable. Disodium Ethylenediaminetetraacetate, $C_{10}H_{14}N_2Na_2O_8 \cdot 2H_2O$, has a molecular weight of 372.24; consequently a $\frac{1}{20}$ *M* solution would contain 372.24/20 or 18.612 Gm dissolved in 1 liter of solution. The water should be metal-free; therefore glass-distilled water is preferred. Solutions are best stored in polyethylene containers or in glass containers previously prepared by boiling the container in an alkaline 2 percent solution of EDTA to remove any metal ions. The standardization and the analysis of official calcium compounds are performed in the same manner.

EDTA Titration. Just as in an acid-base titration where the equivalence point is indicated by an abrupt change in pH, EDTA titration behaves similarly at its stoichiometric point. Plotting pM versus milliliters of titrant similar to pH versus milliliters of acid or base, a sharp change in pM occurs at the equivalence point, corresponding to an inflection in the titration curve. The equivalence point can be detected visually with an indicator or instrumentally, by electrometric or photometric techniques. The higher the value for a stability constant of a complex, the more abrupt the plot of the titration curve. Similarly, complexes with low stability constants would exhibit plots of rather flat titration curves.

In certain instances an indirect or residual titration method is more practical. An excess of EDTA is added to complex the metal ion to be determined, and the residual EDTA is determined by titration with a suitable metal-ion standard solution. Here the indicator is complexed with the first drop of excess metal ion to signal the end point with a sharp color change.

Indicators. EDTA titration indicators are organic compounds which form colored complex ions with the metal ion in high dilution. The qualities of a good indicator are sharpness of color change at the end point, specificity of the indicator for the metal ion under the conditions of the analysis, and a stability constant smaller than that of the metal-EDTA complex, i.e., the indicator must give up the metal ion to the titrant EDTA for complexing and not compete with it.

$$M^{2+} + HIn^{=} \rightleftharpoons MIn^{-} + H^{+}$$

$$\underset{\text{Red}}{MIn^{-}} + H_2Y^{=} \rightleftharpoons MY^{=} + \underset{\text{Blue}}{HIn^{=}} + H^{+}$$

The end point may be sharpened by the addition of a nonchelating screening dye that is complementary to the color observed either just before or just after the end point is reached.

Where the metal-indicator complex color is similar to the free indicator color, the color change occurs when one or more hydrogen ions are replaced by the metal ion, e.g., azo dyes, phthaleins, triphenylmethane dyes, etc.

The colored forms for eriochrome black T are:

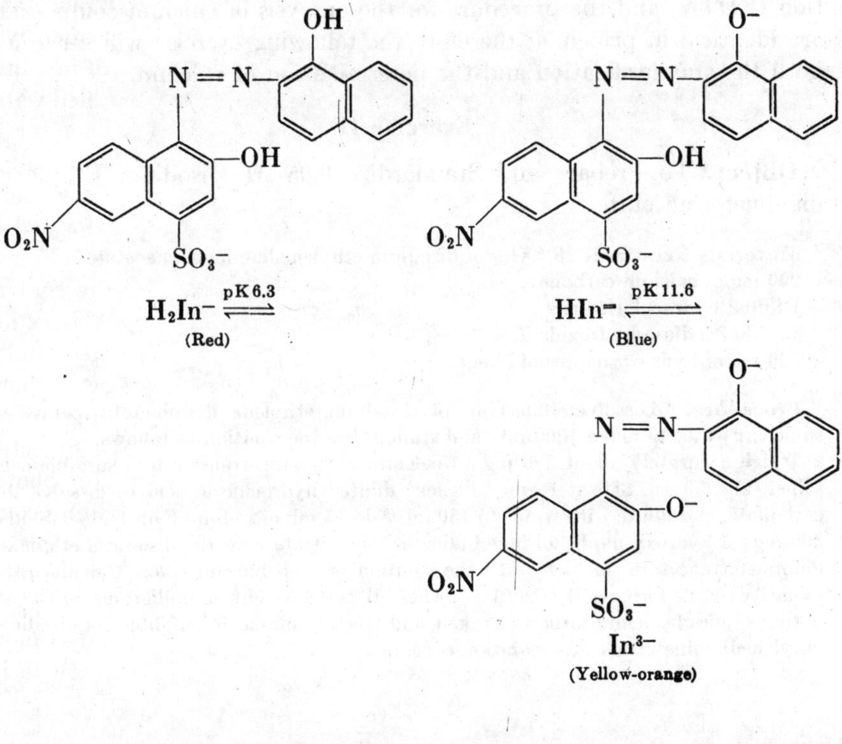

OH, N=N, OH, O_2N, SO_3^-

H_2In^- $\overset{pK\ 6.3}{\rightleftharpoons}$
(Red)

O^-, N=N, OH, O_2N, SO_3^-

$HIn^{=}$ $\overset{pK\ 11.6}{\rightleftharpoons}$
(Blue)

O^-, N=N, O^-, O_2N, SO_3^-

In^{3-}
(Yellow-orange)

As can be seen, this dye is sensitive to changes in metal concentration as well as to changes in pH. Red complexes are formed with Mg^{++} and Zn^{++} and many other metals, such as Al, Ca, Pb, and Hg.

DIRECT TITRATION METHODS

The metal ions which are determined by direct titration with EDTA are Ca^{++}, Mg^{++}, and Zn^{++}. The mercuric ion–EDTA complex reaction is used in the analysis of Calcium Disodium Edetate and its official preparation. A standard solution of mercuric nitrate is used as the titrant. The 0.1 *M* mercuric nitrate solution is standardized against 0.1 *N* ammonium thiocyanate solution using ferric ammonium sulfate T.S. as the indicator. The Hg^{++} displaces the Ca^{++} from the EDTA complex quantitatively at pH 2. This strong acid weakens the Ca-EDTA complex, and the very stable Hg-EDTA complex forms readily.

$$CaY^{=} + Hg^{++} \longrightarrow HgY^{=} + Ca^{++}$$

A comparison of the respective stability constants of the Ca and Hg complexes indicates that the mercuric complex forms readily over the calcium complex. The end point is a purplish complex formed between the Hg^{++} and the diphenylcarbazone indicator.

The standardization of the disodium ethylenediaminetetraacetate solution (EDTA) and the procedure for the analysis of calcium compounds are identical in principle; therefore the following exercise will serve for both the standardization and the determination of calcium.

Exercise 41

Object. To Prepare and Standardize 0.05 *M* Disodium Ethylenediaminetetraacetate.

Materials Required. 18.6 Gm of disodium ethylenediaminetetraacetate
200 mg of calcium carbonate
Diluted hydrochloric acid
15 ml of sodium hydroxide T.S.
300 mg of hydroxy naphthol blue

Procedure. "Dissolve 18.6 Gm. of disodium ethylene diaminetetraacetate in sufficient water to make 1000 ml., and standardize the solution as follows:

Weigh accurately about 200 mg. of calcium carbonate, transfer to a suitable container, add 50 ml. of water and sufficient diluted hydrochloric acid to dissolve the carbonate, and dilute with water to 150 ml. Add 15 ml. of sodium hydroxide T.S. and 300 mg. of hydroxy naphthol blue indicator, and titrate with the disodium ethylenediaminetetraacetate solution until the solution is deep blue in color. Calculate the molarity by the formula $W/(100.1\ V)$, where W is the weight, in milligrams, of $CaCO_3$ in the sample of calcium carbonate taken, and V is the volume, in milliliters, of disodium ethylenediaminetetraacetate solution consumed."

The hydrochloric acid solubilizes the calcium carbonate by converting it to calcium chloride. The sodium hydroxide alkalinizes the solution to a pH of about 13 so that the Ca-EDTA complex would be stable and any magnesium which might be present as a contaminant would not react. After all the Ca^{++} has been complexed with the EDTA, the colored calcium-indicator complex gives up the calcium to the EDTA, liberating the free, uncomplexed indicator, which is blue.

Calculations. The weight of $CaCO_3$ in milligrams divided by 100.1 (mol wt of $CaCO_3$) is the number of millimoles of $CaCO_3$ which is equivalent to a like number of millimoles of EDTA. This number divided by the milliliters of titrant equals millimoles per milliliter, or molarity.

Magnesium ions form a relatively weak complex with EDTA (log K 8.7), but at pH 10 the reaction is rapid and quantitative. In order to maintain this pH, an ammonia–ammonium chloride buffer T.S. is used. This solution acts as a buffer toward the H^+ being displaced by the Mg^{++} as it complexes with the EDTA. The solution further prevents the precipitation of magnesium hydroxide, and, further, the ammonia serves as a ligand, complexing agent, with its unshared pair of electrons.

Zinc metal ions in zinc sulfate readily form a very stable EDTA complex (log K 16.5) under the same conditions as Mg determination. The indicator for each metal determination is eriochrome black T.S. The assay for Zinc Sulfate Ophthalmic Solution is run at pH 5 to 5.5. Copper, a colored ion, forms a highly colored complex with pyridylazonaphthol indicator that is stable in acid media. It is added to the assay as an EDTA complex, and the reaction is as follows:

$$Zn^{++} + CuY^{=} + \underset{\text{Yellow}}{HIn^{=}} \rightleftharpoons ZnY^{=} + \underset{\text{Violet}}{CuHIn} \qquad (\text{pH 5 to 5.5})$$

The zinc ions would normally produce a red color with the indicator, but the copper ions displace the zinc ions from the color complex to form the more highly colored violet complex.

Masking is the term used to indicate the determination of a metal in the presence of another metal. The masking agent itself is a compound capable of entering a complexation reaction. The masking can be accomplished by adjusting the pH of the titration media so that it will be favorable for complexation of the metal being determined and not of the other metal, Bi will complex at pH 2 in the presence of most other metals and without interference from them. Zn will not interfere with Ca titration at a pH of 13. Auxiliary complexing agents also serve as masking agents. Triethanolamine will impart enough alkalinity to suppress an Al-EDTA complexation in the presence of Mg-EDTA reaction, as well as serving as a ligand with its unshared pair of electrons. Thioglycols will inactivate metals such as Hg and Cu which react with the sulfhydryl groups, thus allowing the titration of Zn at pH 6. Potassium cyanide can be used to

inactivate or mask or screen such metals as Co, Ni, Cu, and Zn. Ammonium fluoride will mask Ca, Mg, and Al to allow titration of Zn. Ascorbic acid, citrates, and tartrates are also used as masking agents.

The determination of Al in Aluminum Paste is an example of an assay of one metal in the presence of another, Zn.

TABLE 21. SUBSTANCES ASSAYED BY DIRECT TITRATION WITH EDTA

Substance	*M* of titrant	Equivalent factor, mg	Indicator	Official requirement, percent
U.S.P.				
Calcium carbonate, precipitated	0.05	5.004	h.n.b.	$CaCO_3$ = 98
Calcium chloride	0.05	7.351	h.n.b.	$CaCl_2{\cdot}2H_2O$ = 99 to 107
Calcium gluconate	0.05	21.52	h.n.b.	$C_{12}H_{22}CaO_{14}$ = 98 to 102
Calcium gluconate tablets	0.05	21.52	h.n.b.	$C_{12}H_{22}CaO_{14}$ = 95 to 105 l.a.
Calcium hydroxide	0.05	3.705	h.n.b.	$Ca(OH)_2$ = 95
Calcium lactate (R)	0.05	10.91	h.n.b.	$C_6H_{10}CaO_6$ = 98
Calcium pantothenate	0.05	2.004	h.n.b.	Ca = 8.2 to 8.6
Calcium pantothenate tablets	0.05	2.004	h.n.b.	Ca = 7.9 to 9.6
Calcium pantothenate, racemic	0.05	2.004	h.n.b.	Ca = 8.2 to 9.6
Magnesium sulfate	0.05	6.019	e.b.	$MgSO_4$ = 99
Magnesium sulfate injection	0.05	12.32	e.b.	$MgSO_4{\cdot}7H_2O$ = 93 to 107 l.a.
Ringer's injection	0.005	0.7351	h.n.b.	$CaCl_2{\cdot}2H_2O$ = 0.03 to 0.036 *W/V*
Ringer's injection, lactated	0.005	0.7351	h.n.b.	$CaCl_2{\cdot}2H_2O$ = 0.018 to 0.022 *W/V*
Zinc chloride	0.05	6.814	e.b.	$ZnCl_2$ = 97
Zinc oxide ointment	0.05	4.068	e.b.	ZnO = 18.5 to 21.5
Zinc oxide paste	0.05	4.068	e.b.	ZnO = 24 to 26
Zinc stearate	0.05	4.068	e.b.	ZnO = 12.5 to 14.0
Zinc sulfate	0.05	8.072	e.b.	$ZnSO_4$ = 55.6 to 61.0
Zinc sulfate ophthalmic solution	0.01	2.875	p.z.n.	$ZnSO_4{\cdot}7H_2O$ = 95 to 105 l.a.
N.F.				
Calcium carbonate tablets	0.05	5.004	h.n.b.	$CaCO_3$ = 92.5 to 107.5 l.a.
Calcium chloride injection	0.05	7.351	h.n.b.	$CaCl_2{\cdot}2H_2O$ = 95 to 105 l.a.
Calcium citrate (R)	0.05	8.307	h.n.b.	$Ca_3(C_6H_5O_7)_2{\cdot}4H_2O$ = 97.5 to 101
Calcium gluconate injection	0.05	21.52	h.n.b.	$C_{12}H_{22}CaO_{14}$ = 92 to 102 l.a.
Calcium lactate	0.05	10.91	h.n.b.	$C_6H_{10}CaO_6$ = 98 to 101
Calcium lactate tablets	0.05	15.42	h.n.b.	$C_6H_{10}CaO_6{\cdot}5H_2O$ = 94 to 106 l.a.
Lime	0.05	2.804	h.n.b.	CaO = 95
Ringer's solution	0.005	0.7351	h.n.b.	$CaCl_2{\cdot}2H_2O$ = 0.03 to 0.036 *W/*

(R) = reagent; h.n.b. = hydroxynaphthol blue: l.a. = labeled amount: e.b. = eriochrome black; p.z.n. = 1-(2-pyridylazo)-2-naphthol.

RESIDUAL TITRATIONS

The indirect or residual titration method is applied to the analysis of aluminum and bismuth compounds, since conditions for direct titration may introduce errors because of the precipitation of the metal as hydroxides in alkaline media. In such cases an excess of the standard solution of EDTA is added to the sample; then the pH is adjusted for the back titration with a metal-ion solution, zinc sulfate, using an indicator that is sensitive to the titrant. The metal being determined remains complexed with the EDTA and does not interfere with the Zn-EDTA complexation. Bismuth forms a highly stable complex and can be titrated at a pH as low as 1 or 2. At a pH of about 4 the bismuth reacts quantitatively with EDTA, forming a stable complex, and the indicator, dithizone, which has a transition state of color at pH 4.6, detects the end point. The bismuth ion forms an inner complex with the sulfur and azo nitrogen atoms of dithizone, but the bismuth is readily liberated to form the more stable complex with EDTA, and then the indicator takes on its uncomplexed color. Interference from other heavy metals can be avoided by masking with ascorbic acid. Antimony can be masked with tartaric acid. Metal complexes of the indicator dithizone are largely insoluble in water; therefore alcohol must be added to ensure the solubility of the color complex. An organic solvent could serve as an extraction indicator in this instance.

Exercise 42

Object. Assay of Aluminum Paste.

Materials Required. 2.5 Gm of Aluminum Paste
35 ml of dilute hydrochloric acid (3 in 7)
50 ml of dilute hydrochloric acid (1 in 5)
10 ml of stronger ammonia water
0.05 *M* disodium ethylenediaminetetraacetate (EDTA)
5 ml of triethanolamine solution (3 in 10)
20 ml of ammonium acetate buffer T.S.
50 ml of alcohol
0.05 *M* zinc sulfate
5 ml of ammonia–ammonium chloride buffer T.S.
Dithizone T.S.
Eriochrome black T.S.

Procedure. 1. *Assay Preparation.* "Transfer about 2.5 Gm., accurately weighed, of Aluminum Paste to a 250-ml. beaker, and add 35 ml. of dilute hydrochloric acid (3 in 7). Heat gently until the sample melts, then boil gently until the oils separate. Cool to solidify the oils, and decant the aqueous layer through a filter into a 250-ml. volumetric flask. Add 25 ml. of dilute hydrochloric acid (1 in 5) to the residue in the beaker, and repeat the boiling, cooling, and filtering steps, using the same filter and collecting the filtrate in the same volumetric flask. Repeat the washing with a second 25-ml. portion of dilute hydrochloric acid (1 in 5), boiling, cooling, and filtering as before. Finally wash the beaker and the filter with several portions of cold water, collecting the washings with the filtrates in the flask. Dilute to volume, and mix."

In this portion of the procedure the aluminum and zinc oxide are converted to soluble $AlCl_3$ and $ZnCl_2$, respectively, and separated from the oils, which are solidified by chilling. The 25-ml aliquots used subsequently will represent one-tenth of the original weighed sample.

2. "*Determination of aluminum and zinc*—Pipet 25 ml. of *Assay Preparation* into a 250-ml. beaker, and add stronger ammonia water dropwise until a faint turbidity appears. Clear the solution by the dropwise addition of hydrochloric acid, and add a few drops in excess. Add in the order named and with continuous stirring, 50.0 ml. of

0.05 *M* disodium ethylenediaminetetraacetate, 20 ml. of ammonium acetate buffer T.S., 50 ml. of alcohol, and 2 ml. of dithizone T.S. Titrate the solution with 0.05 *M* zinc sulfate until the color changes from green-violet to rose-pink. Record the volume of 0.05 *M* disodium ethylenediaminetetraacetate consumed."

Aluminum and zinc have about the same stability constants and also behave similarly at about pH 4. The faint turbidity produced by the ammonia water is insoluble aluminum hydroxide. It is much more insoluble than zinc hydroxide. The acid redissolves it; then the pH of the titration mixture is controlled by the ammonium acetate buffer. Aluminum hydroxide would precipitate at a pH more alkaline than 4.5. Alcohol is used to solubilize the metal-indicator complex, which is insoluble in water. Alcohol will also increase the stability of the metal-EDTA complex. The pH for the transition stage of the dithizone indicator is 4.6.

3. "*Determination of zinc*—Pipet 25 ml. of *Assay Preparation* into a 250-ml. beaker and add 10 ml. of stronger ammonia water, 5 ml. of triethanolamine solution (3 in 10), and 5 ml. of ammonia–ammonium chloride buffer T.S. Cool the mixture to below 5° in a cooling bath, add 0.1 ml. eriochrome black T.S., and titrate with 0.05 *M* disodium ethylenediaminetetraacetate to a blue color, maintaining the temperature of the solution below 5° during the titration. Record the volume of 0.05 *M* disodium ethylenediaminetetraacetate consumed by the zinc present."

The titration of the zinc alone in the presence of the aluminum is accomplished using the masking technique. When the sample is treated with the stronger ammonia water, triethanolamine, and the ammonia-ammonium chloride buffer, the resulting pH is about 9 to 11, and the temperature at below 5° highly favors the formation of triethanolamine-aluminum complex which will not compete with the zinc for the EDTA nor for the eriochrome indicator. EDTA reacts with all metals in a 1:1 ratio. The volume of titrant used in the zinc determination subtracted from the volume of EDTA consumed by the aluminum and zinc titration is the volume equivalent to just the aluminum present in the 25-ml aliquot or one-tenth of the original sample weight. The atomic weight of Al is 26.9815, and 1 ml of a $\frac{1}{20}$ *M* solution would be equal to $26.9815/20 \times 1{,}000$ or 1.349 mg of Al.

Questions and Problems

1. Define complexation and chelation with simple schematic equations.
2. What volume of 0.05 *M* EDTA solution would be equivalent to 100 ml of a 0.025 *M* solution of titratable divalent metal cations? Trivalent? Tetravalent?
3. How could EDTA be used to determine quantitatively such anions as sulfate and arsenate?
4. How many grams of disodium ethylenediaminetetraacetate ($C_{10}H_{14}N_2Na_2O_8 \cdot 2HO$) is required to prepare 250 ml of a 0.025 *M* solution? What would the titer of this solution be in terms of $CaCl_2$; $2H_2O$; $MgSO_4$; and $ZnCl_2$?
5. A 200-mg sample of pure calcium carbonate was acidified and dissolved in 500 ml of solution. A 50-ml sample required 25.0 ml of an EDTA solution for titration. Find the molarity of the EDTA solution.

TABLE 22. SUBSTANCES ASSAYED BY RESIDUAL TITRATION OF EDTA WITH ZINC SULFATE

Substance	Equivalent factor, mg	Molarity of titrant	Indicator	Official requirement, percent
U.S.P.				
Aluminum	1.349	0.05	d.th.	Al = 95
Aluminum acetate solution	2.549	0.05	d.th.	Al_2O_3 = 1.2 to 1.45 *W/V*
Aluminum hydroxide gel	2.549	0.05	d.th.	Al_2O_3 = 3.6 to 4.4
Aluminum paste	1.349	0.05	d.th.	Al = 9.5 to 10.5
Aluminum subacetate solution	2.549	0.05	d.th.	Al_2O_3 = 2.3 to 3.6 *W/V*
Aluminum sulfate	16.66	0.05	d.th.	$Al_2(SO_4)_3 \cdot 18H_2O$ = 99.5 to 112
N.F.				
Alum	22.67	0.05	d.th.	$AlNH_4(SO_4)_2 \cdot 12H_2O$ = 99.5
Alum	23.72	0.05	d.th.	$AlK(SO_4)_2 \cdot 12H_2O$ = 99.5 to 105
Alum, exsiccated	11.86	0.05	d.th.	$AlNH_4(SO_4)_2$ = 96.5
Alum, exsiccated	12.91	0.05	d.th.	$AlK(SO_4)_2$ = 96.5 to 105
Aluminum chloride	12.07	0.05	d.th.	$AlCl_3 \cdot 6H_2O$ = 95 to 102
Aluminum hydroxide gel, dried	2.549	0.05	d.th.	Al_2O_3 = 50
Aluminum hydroxide gel tablets, dried	3.900	0.05	d.th.	$Al(OH)_3$ = 95 to 105 l.a.
Glycobiarsol	10.45	0.05	d.th.	Bi = 38 to 42.5
Glycobiarsol tablets	10.45	0.05	d.th.	Bi = 38 to 42.5
Substances Assayed by Direct Titration with Mercuric Nitrate				
U.S.P.				
Calcium disodium edetate	37.43	0.1	d.p.c.	$C_{10}H_{12}CaN_2Na_2O_8$ = 97 to 102
Calcium disodium edetate injection	37.43	0.1	d.p.c.	$C_{10}H_{12}CaN_2Na_2$ = O_8 90 to 110 l.a.
Disodium Edetate versus Known Weight of Calcium Carbonate				
U.S.P.				
Disodium edetate	......	....	h.n.b.	$C_{10}H_{14}N_2Na_2O_8 \cdot 2H_2O$ = 99
Disodium edetate injection	......	....	h.n.b.	$C_{10}H_{14}N_2Na_2O_8$ = 90 to 110 l.a.

l.a. = labeled amount; d.p.c. = diphenylcarbazone; d.th. = dithizone; h.n.b. = hydroxynaphthol blue.

FERRIC CHLORIDE TITRATION

Sodium Fluoride determination is based on the reaction of sodium fluoride with ferric chloride to form a reasonably stable complex of the type FeF_6^{3-}.

The sample of sodium fluoride is dissolved in water with the aid of a little warming and is then neutralized with acid or base (whichever is needed), phenolphthalein being used as the indicator. After the solution has been cooled to room temperature, it is treated with 20 Gm of sodium chloride to repress the ionization of NaF. The hydrolysis of the sodium fluoride may produce an alkaline condition which would interfere with the titration by precipitating the ferric chloride as ferric hydroxide. Potassium thiocyanate is used as the indicator. An alcohol-ether mixture serves as an extraction solvent. The first drop of ferric chloride added in excess reacts with the thiocyanate ion, forming a ferric-thiocyanate complex that is soluble and is concentrated in the organic solvent mixture, imparting a red color for the end point detection.

$$FeCl_3 + 6NaF \longrightarrow Na_3FeF_6 + 3NaCl$$

The assay of sodium fluoride calls for the preparation of a fresh solution of ferric chloride (10 Gm per 1,000 ml of solution). This solution, which will be approximately 0.037 *N*, is standardized against sodium thiosulfate using potassium iodide. One equivalent weight of $FeCl_3$ is 1 mole; thus a 1 *N* and a 1 *M* solution are of the same concentration. One milliliter of a 1 *N* solution is equivalent to 6 × 41.99/1,000 or 0.25194 Gm of NaF. To calculate the quantity of NaF in the sample,

$$\text{ml} \times N \times \text{mEq} = \text{Gm NaF}$$
$$\text{ml} \times 0.25194 = \text{Gm NaF}$$

The U.S.P. directs that formula 252*FV* be used, where *F* = the normality of the ferric chloride and *V* = the volume consumed to calculate the milligrams in the sample. The figure 252 is the factor used to express the results in milligrams.

Chapter 8

OXIDATION-REDUCTION METHODS

The chemical reactions that occur in neutralization and precipitation methods of analysis take place without change in valence. Oxidation-reduction methods of analysis involve a change in valence of the reacting substances. A study of the principles underlying the latter type of reaction is an essential preliminary if the methods based on them are to be understood.

Theory. The simplest type of oxidation-reduction reaction is the direct combination of elements. Thus, when oxygen gas unites with hydrogen gas, forming water, $O_2 + 2H_2 \longrightarrow 2H_2O$, the oxygen is reduced and the hydrogen oxidized. When carbon burns in the presence of oxygen to form carbon dioxide, $C + O_2 \longrightarrow CO_2$, it is oxidized and the oxygen is reduced, and when carbon unites with sulfur to form carbon disulfide, $C + 2S \longrightarrow CS_2$, the carbon is oxidized and the sulfur reduced.

From the above simple illustrations, it is obvious that reduction need not imply a reaction of hydrogen, since oxygen is reduced by carbon, and that oxidation need not imply a reaction of oxygen, since carbon is oxidized by sulfur. When one substance is oxidized, some other substance must be correspondingly reduced, and, conversely, when one substance is reduced, some other substance must be correspondingly oxidized.

The oxidation-reduction reactions that take place in the official assay processes are, for the most part, between electrolytes in aqueous solution. Their quantitative value is based upon the fact that metals, nonmetals, and their ions, under suitable conditions, can be made to undergo a change in the quantity of electric charge associated with them and that in the change there exists a simple relationship between the quantity of electricity lost or gained and the weight of the reacting substances. This relationship may be expressed according to Faraday's law as follows: *A change of charge of one is equivalent to the gain or loss of* 96,500 *coulombs of electricity for each formula weight of element or group of elements involved.* Since in every oxidation-reduction reaction the charge lost or gained by one substance must necessarily be gained or lost by another, it follows that there is always a transfer of electrons in oxidation-reduction reactions.

The reactant which loses electrons in an oxidation-reduction reaction is the *reducing agent* and can be identified in the equation for the reaction as the reactant containing a constituent atom or atoms converted to a higher state of oxidation, e.g.,

$$Fe^{++} \longrightarrow Fe^{3+} + e$$
$$2I^{-} \longrightarrow I_2 + 2e$$

The electrons lost by the reducing agent are gained by the *oxidizing agent*, the reactant containing a constituent atom or atoms which are converted to a lower state of oxidation, e.g.,

$$Ce^{4+} + e \longrightarrow Ce^{3+}$$
$$Fe^{3+} + e \longrightarrow Fe^{++}$$

A combination of oxidizing and reducing agents results in a reaction which may be used as basis for the quantitative measurement of one of the reactants. For example, ferrous sulfate may be assayed by its reaction with ceric sulfate:

$$Fe^{++} + Ce^{4+} \longrightarrow Fe^{3+} + Ce^{3+}$$

This equation may be divided into two half-equations showing the loss of electrons by the ferrous ion and the gain of electrons by the ceric ion:

$$Fe^{++} \rightleftharpoons Fe^{3+} + e \quad \text{and} \quad Ce^{4+} + e \rightleftharpoons Ce^{3+}$$

Ferric chloride may be assayed by its reduction with potassium iodide:

$$2I^{-} \longrightarrow I_2 + 2e \quad \text{and} \quad 2Fe^{3+} + 2e \longrightarrow 2Fe^{++}$$

The combined equation may be represented:

$$2I^{-} + 2Fe^{3+} \longrightarrow I_2 + 2Fe^{++}$$

The amount of the reducing agent or oxidizing agent in the sample can be determined by measuring a chemically equivalent amount of the oxidizing agent or reducing agent, respectively.

In the equation for the oxidation of ferrous sulfate with ceric sulfate, each molecule of ferrous sulfate, on oxidation, loses one electron. One mole of ferrous sulfate, therefore, loses 6.02×10^{23} electrons equivalent to 1 faraday or 96,500 coulombs. This quantity of electricity is identical with that necessary to reduce 1 mole of ceric sulfate. Consequently 1 mole of ferrous sulfate and 1 mole of ceric sulfate are chemical equivalents One gram of hydrogen, acting as a reducing agent, loses electrons equivalent to 96,500 coulombs. Therefore, the *equivalent weight of a reducing agent* is that weight which loses electrons equivalent to 96,500 coulombs and it can be calculated by dividing the gram-molecular weight by the number of electrons lost by each molecule; e.g., the equivalent weight of ferrous sulfate oxidized to ferric sulfate, $Fe^{++} \longrightarrow Fe^{3+} + e$, is 151.91 Gm or 1 gram-molecular weight.

An *equivalent weight of an oxidizing agent* is that weight which gains electrons equivalent to 1 faraday, or to the electrons gained by 1 Gm of hydrogen ions, $2H^+ + 2e \longrightarrow H_2$. It can be calculated by dividing the gram-molecular weight by the number of electrons gained by each molecule; e.g., the equivalent weight of ceric sulfate, $Ce^{4+} + e \longrightarrow Ce^{3+}$, is 1 gram-molecular weight, 332.24 Gm, and that of potassium permanganate, $MnO_4^- + 5e \longrightarrow Mn^{++}$, is ⅕ gram-molecular weight or 31.61 Gm.

Standard Solutions. The standard solutions used in the official assays by oxidation-reduction methods include ferric ammonium sulfate, potassium permanganate, potassium dichromate, potassium bromate, potassium iodate, potassium ferricyanide, ceric sulfate, iodine and bromine as oxidizing agents, and ferrous ammonium sulfate, oxalic acid, potassium arsenite, titanium chloride, and sodium thiosulfate as reducing agents.

The number of electrons gained by a given oxidizing agent sometimes depends on the conditions, e.g., pH, under which the reaction is carried out. Certain oxidizing agents, nitric acid, for example, do not undergo a single uniform reaction even though conditions are carefully controlled. Such substances cannot be used as standard oxidizing agents or assayed by oxidation-reduction methods. The equivalent weights of oxidizing agents which determine the concentrations of standard solutions can be calculated as indicated in the five following equations:

1. Permanganate ion⟶manganous ion
   $MnO_4^- + 5e \longrightarrow Mn^{2+}$

Since each molecule of $KMnO_4$ gains 5 electrons when permanganate is reduced in acid mediums, ⅕ mole, 158.04/5 = 31.61 Gm, is 1 equivalent. A liter of 0.1 *N* solution should contain, therefore, 3.161 Gm of $KMnO_4$.

2. Dichromate ion⟶chromous ion
   $Cr_2O_7^= + 6e \longrightarrow 2Cr^{3+}$

One-sixth mole, 294.19/6 = 49.032 Gm, is 1 equivalent of potassium dichromate, $K_2Cr_2O_7$.

3. Bromate ion⟶bromide ion
   $BrO_3^- + 6e \longrightarrow Br^-$ (in presence of H_2SO_4)

One-sixth mole, 167.01/6 = 27.84 Gm, is 1 equivalent of potassium bromate, $KBrO_3$.

4. Ceric ion⟶cerous ion
   $Ce^{4+} + e \longrightarrow Ce^{3+}$

One mole, 332.24 Gm, is 1 equivalent of ceric sulfate, $Ce(SO_4)_2$.

5. Iodine⟶iodide ion
   $I_2 + 2e \longrightarrow 2I^-$

One-half mole or 1 gram-atomic weight, 126.90 Gm, is 1 equivalent of iodine.

The reduction of the iodine in potassium iodate to lower oxidation states varies considerably, depending on conditions specified for the various assays in which it is employed. Therefore, the concentrations of standard solutions of potassium iodate are usually expressed in terms of molarity rather than normality.

Reducing agents used for the preparation of standard solutions are oxidized:

1. Oxalate ion⟶carbon dioxide

$$C_2O_4^{=} \longrightarrow 2CO_2 + 2e$$

One-half mole, 126.07/2 = 63.035 Gm, is 1 equivalent of oxalic acid, $H_2C_2O_4 \cdot 2H_2O$.

2. Thiosulfate ion⟶tetrathionate ion

$$2S_2O_3^{=} \longrightarrow S_4O_6^{=} + 2e$$

The net oxidation number (valence number) of the four sulfur atoms in two thiosulfate ions is 8^+, while that of the four sulfur atoms in the tetrathionate ion is 10^+; hence two thiosulfate ions must lose 2 electrons in forming one tetrathionate ion, or each thiosulfate ion loses 1 electron. One mole, 248.18 Gm, is 1 equivalent of sodium thiosulfate, $Na_2S_2O_3 \cdot 5H_2O$.

3. Arsenite ion⟶arsenate ion

$$AsO_2^- + H_2O \longrightarrow AsO_3^- + 2H^+ + 2e$$

Arsenic trioxide (As_2O_3) is solubilized with sodium or potassium hydroxide and is represented by the arsenite ion. AsO_2^-, which shows arsenic in the trivalent form. In a neutral or alkaline medium the arsenite ion is a reducing agent that gives up 2 electrons per atom of arsenic. In an acid medium the reaction is reversed, with the pentavalent arsenic in the arsenate ion serving as an oxidizing agent.

One equivalent of arsenic trioxide, As_2O_3, is 197.84/4 or 49.46 Gm.

4. Titanous ion⟶titanic ion

$$Ti^{3+} \longrightarrow Ti^{4+} + 1e$$

One equivalent of titanium trichloride, $TiCl_3$, is 154.26 Gm.

5. Ferrous ion⟶ferric ion

$$Fe^{++} \longrightarrow Fe^{3+} + 1e$$

One equivalent of ferrous ammonium sulfate, $Fe(NH_4)_2(SO_4)_2 \cdot 6H_2O$ is 1 gram-molecular weight or 392.14 Gm.

PERMANGANATE METHODS

Potassium permanganate solution can be standardized easily, and it retains its concentration over long periods of time when proper precautions are observed in its preparation and preservation. The reactions of permanganate in solution are rapid. It also serves as an indicator in titrations where it is used, since a very slight excess of permanganate imparts to solutions a distinct pink color.

Exercise 43

Object. To Prepare and Standardize 0.1 *N* Potassium Permanganate.

Materials Required. About 3.3 Gm of potassium permanganate
About 1 Gm of reagent sodium oxalate

Procedure. 1. "Dissolve about 3.3 Gm. of potassium permanganate in 1000 ml. of water in a flask, and boil the solution for about 15 minutes. Stopper the flask, allow it to stand for at least 2 days, and filter through asbestos."

Any organic matter that may be present in distilled water is decomposed by potassium permanganate. Consequently, after solution of the permanganate is effected, the liquid is allowed to stand for 2 days to ensure the completion of the decomposition reaction. The solution is then filtered through asbestos to remove all traces of manganese dioxide, which if present acts as a catalyst to promote the formation of more manganese dioxide at the expense of permanganate ion. The formation of MnO_2 as a reduction product in neutral solution may be represented as

$$4KMnO_4 + 2H_2O \longrightarrow 4KOH + 4MnO_2 + 3O_2$$

2. "Standardize the solution as follows:

"Weigh accurately about 200 mg. of sodium oxalate, previously dried at 110° to constant weight, and dissolve it in 250 ml. of water. Add 7 ml. of sulfuric acid, heat to about 70°, and then slowly add the permanganate solution from a buret, with constant stirring, until a pale pink color which persists for 15 seconds is produced. The temperature at the conclusion of the titration should not be less than 60°. Calculate the normality."

Titrate the mixture by adding the permanganate solution from a glass-stoppered buret, with continuous stirring of the titration mixture until a drop of the permanganate causes a permanent pink color. Read the top of the meniscus in the buret instead of the lowest part. Using the first sample as a preliminary run, titrate two other portions of oxalate in the same manner and base the calculations of normality upon them.

Sodium oxalate is the best standard to use in the standardization of potassium permanganate, since it can be obtained in very pure condition. Sufficient sulfuric acid must be added to keep the hydrogen ion concentration reasonably constant throughout the titration, thereby preventing the formation of manganese dioxide and supplying the hydrogen ions used up in the reduction of the permanganate ion. Thus, for every mole of permanganate ion reduced, 8 moles of hydrogen ion is required, as shown in the equation below. The top of the meniscus is read, since the color of the permanganate solution renders impossible accurate readings of the lowest point.

The half-equations are

$$MnO_4^- + 8H^+ + 5e \longrightarrow Mn^{++} + 4H_2O$$

and

$$C_2O_4^{=} \longrightarrow 2CO_2 + 2e$$

In order to balance and cancel out electrons the half-equation for reduction of permanganate is multiplied by 2 and that for the oxidation of oxalate is multiplied by 5. The combined equation may be written

$$2MnO_4^- + 16H^+ + 5C_2O_4^{=} \longrightarrow 2Mn^{++} + 8H_2O + 10CO_2$$

or

$$5Na_2C_2O_4 + 2KMnO_4 + 8H_2SO_4 \longrightarrow 2MnSO_4 + 5Na_2SO_4 + K_2SO_4 + 10CO_2 + 8H_2O$$

The balanced equation shows that 5 molecules of sodium oxalate lose 10 electrons on oxidation with potassium permanganate; therefore, the equivalent weight of sodium oxalate is one-half the gram-molecular weight, or 134.00/2 = 67.00 Gm. If a 0.2500-Gm sample of sodium oxalate required 34.20 ml of a permanganate solution in a titration, there must have been 0.2500/67.00 equivalents of $KMnO_4$ in the 34.20 ml of the solution, or

$$\frac{1{,}000 \times 0.2500}{34.20 \times 67.00} = 0.1091 \text{ equivalent of } KMnO_4 \text{ per liter}$$

Hence the permanganate solution is 0.1091 *N*.

When permanganate solutions are diluted to an exact normality, water freshly distilled from permanganate solution should be used to avoid the introduction of organic matter.

Potassium permanganate solution prepared as directed will retain its value for a long time if it is protected from sunlight and from the entry of organic matter. It is best, however, to restandardize the solution from time to time. Solution withdrawn from the stock bottle should never be returned to it. Solutions of oxidizing agents should never be titrated from Mohr burets, since the oxidizing agent attacks the rubber and the $KMnO_4$ titer decreases. The buret used to measure the permanganate solution should be washed out immediately after use to prevent the formation of a deposit of manganese dioxide on the buret wall.

Questions and Problems

1. Define the following terms: (*a*) oxidation; (*b*) reducing agent; (*c*) equivalent weight of an oxidizing agent.

2. Refer to the U.S.P. and give the formula and equivalent weight of the primary standards used in standardizing (*a*) 0.1 *N* iodine; (*b*) 0.1 *N* potassium permanganate; (*c*) 0.1 *N* ceric sulfate.

3. A 0.2295-Gm sample of sodium oxalate (97.8 percent $Na_2C_2O_4$) required 35.12 ml of a potassium permanganate solution in a titration. What was the normality of the permanganate solution?

4. What volume of 0.1225 *N* sodium hydroxide would be required in titration of a 0.1065-Gm sample of oxalic acid if a 0.1126-Gm sample of the oxalic acid required 24.65 ml of 0.0965 *N* potassium permanganate in a titration?

DIRECT TITRATION METHODS

To Determine the Purity of Ferrous Sulfate. Potassium permanganate is especially well suited to the estimation of iron and ferrous salts. When the titration is carried out in solution acidified with sulfuric acid, the ferrous ion is oxidized quantitatively to ferric ion.

Exercise 44

Object. Assay of Ferrous Sulfate.

Materials Required. About 1 Gm of ferrous sulfate
25 ml of diluted sulfuric acid
0.1 potassium permanganate

Procedure. "Dissolve about 1 Gm. of Ferrous Sulfate, accurately weighed, in 25 ml. of diluted sulfuric acid and 25 ml. of water, and titrate with 0.1 *N* potassium permanganate until a permanent pink color is produced. Each milliliter of 0.1 *N* potassium permanganate is equivalent to 15.19 mg. of $FeSO_4$ or to 27.80 mg. of $FeSO_4 \cdot 7H_2O$."

After dissolving the sample in 25 ml of diluted sulfuric acid, it is best to wash down the sides and neck of the flask with about 25 ml of cold water. The titration must be conducted in acid solution to maintain a reasonably constant hydrogen ion concentration, as explained in Exercise 43. The pink color of permanganate is distinct when a drop in excess of that required to oxidize all the ferrous salt to the ferric state has been added.

The equation for the reactions that occur may be balanced by writing the reactions as two half-equations, one showing the gain of electrons by the oxidizing agent and the other showing the loss of electrons by the reducing agent, e.g.:

$$5Fe^{++} \longrightarrow 5Fe^{3+} + 5e$$

and

$$MnO_4^- + 8H^+ + 5e \longrightarrow Mn^{++} + 4H_2O$$

Adding the two half-equations gives

$$MnO_4^- + 8H^+ + 5Fe^{++} \longrightarrow Mn^{++} + 4H_2O + 5Fe^{3+}$$

If the complete molecular equation is required, the combined half-equations must be multiplied by 2:

$$2KMnO_4 + 8H_2SO_4 + 10FeSO_4 \longrightarrow 2MnSO_4 + 8H_2O + 5Fe_2(SO_4)_3 + K_2SO_4$$

The equation reveals that 10 molecules of ferrous sulfate lose 10 electrons; hence the equivalent weight of $FeSO_4$ is 1 mole, 151.91 Gm, and that of $FeSO_4 \cdot 7H_2O$ is 278.02 Gm. Each milliliter of 0.1 *N* potassium permanganate, 0.1 mEq, is equivalent to $0.1 \times 0.15191 = 0.015191$ Gm or 0.1 mEq of $FeSO_4$, or to $0.1 \times 0.27802 = 0.027802$ Gm of $FeSO_4 \cdot 7H_2O$. If the amount of iron present in the sample in the Fe^{++} state is desired, its equivalent weight would be 55.85 Gm because 1 gram-atomic weight of iron is equivalent to 1 mole of ferrous sulfate.

Calculate the percent of $FeSO_4 \cdot 7H_2O$ in the sample assayed and compare the value with the official specifications.

Questions and Problems

1. What weight of exsiccated ferrous sulfate, 80 percent $FeSO_4$, would be used as sample so that 35.00 ml of 0.1 *N* potassium permanganate would be required in the titration? Compare the calculated value with the amount specified in the U.S.P. assay procedure.

2. Calculate the percent of iron present as Fe^{++} in ferrous sulfate, 99.8 percent $FeSO_4 \cdot 7H_2O$, and in dried ferrous sulfate, 86 percent $FeSO_4$.

3. Consult available texts on volumetric analysis, and suggest an indicator which may be used when titrating ferrous sulfate with very dilute standard permanganate solutions.

4. A 1.1021-Gm sample of iron wire was dissolved in hydrochloric acid and the solution treated with zinc in such a way that all the iron was present as Fe^{++}. The resulting solution required 37.63 ml of 0.5220 *N* potassium permanganate in a titration. Calculate the Fe content of the sample.

Exercise 45

Object. Assay of Hydrogen Peroxide Solution.

Materials Required. 2 ml of hydrogen peroxide solution
20 ml of diluted sulfuric acid
0.1 *N* potassium permanganate

Procedure. "Measure accurately 2 ml. of Hydrogen Peroxide Solution, and transfer it to a suitable flask containing 20 ml. of water. Add 20 ml. diluted sulfuric acid, and titrate with 0.1 *N* potassium permanganate. Each milliliter of 0.1 *N* potassium permanganate is equivalent to 1.701 mg. of H_2O_2."

The reasons for the various steps in the above assay procedure ha already been given under the assay of ferrous sulfate. Although t hydrogen peroxide generally acts as an oxidizing agent and is reduced to H_2O, in this assay it reduces the potassium permanganate while it is being oxidized to oxygen, O_2.

$$\underset{34.01}{5H_2O_2} + 2KMnO_4 + 3H_2SO_4 \longrightarrow 5O_2\uparrow + 2MnSO_4 + K_2SO_4 + 8H_2O$$

The equation shows that 1 molecule of H_2O_2 loses 2 electrons; hence the equivalent weight of H_2O_2 is ½ gram-molecular weight or 17.01 Gm. Consequently each milliliter of 0.1 *N* potassium permanganate is equivalent to $0.1 \times 0.01701 = 0.001701$ Gm or 0.1 mEq of H_2O_2.

To comply with U.S.P. definition, solution of hydrogen peroxide should contain not less than 2.5 and not more than 3.5 percent *W/V* of H_2O_2. Calculate the percentage of H_2O_2 in the sample assayed and compare the strength of the solution with the U.S.P. requirement.

Questions and Problems

1. A 2.00-ml sample of hydrogen peroxide solution required 8.5 ml of a permanganate solution in titration. If each milliliter of the permanganate solution is equivalent to 0.007294 Gm of Fe, what percent *W/V* of H_2O_2 was in the sample?

2. When sodium perborate, $NaBO_3 \cdot 4H_2O$, is assayed by titration with standard permanganate, each mole is equivalent to 1 mole of H_2O_2: $NaBO_3 \cdot 4H_2O \longrightarrow H_2O_2 + NaBO_2 + 3H_2O$. (*a*) How many grams of "available" oxygen are in 10 Gm of sodium perborate? (*b*) If a sodium perborate sample were found on analysis to contain 9.0 percent "available" oxygen, what was the $NaBO_3 \cdot 4H_2O$ content of the sample?

3. If 31.00 ml of a potassium permanganate solution were required in titration of 0.4617 Gm of 85 percent $NaBO_3 \cdot 4H_2O$, what volume of the permanganate solution would be required in titration of 2.00 ml of 3 percent H_2O_2, *W/V*?

INDIRECT TITRATION METHODS

The indirect method of permanganate oxidation is employed with those compounds that can be converted through chemical reactions to an equivalent amount of oxalate, which in turn can then be quantitatively oxidized by permanganate. In the assay for the malic acid content of Cherry Juice the malic acid is converted to an equivalent amount of calcium salt, which is converted to insoluble calcium oxalate; the oxalate is then liberated and oxidized with permanganate.

Exercise 46

Object. Assay of Cherry Juice for Malic Acid.

Materials Required. 10 ml of Cherry Juice
1 Gm of calcium carbonate
1 ml of ammonia T.S.
5 ml of ammonium oxalate T.S.
25 ml of diluted ammonia T.S. (1 in 49)
30 ml of dilute sulfuric acid (1 in 3)
0.1 *N* potassium permanganate

Procedure. 1. "Place 10 ml. of Cherry Juice, accurately measured, in a 125-ml. flask, and add 1 Gm. of calcium carbonate. Heat on steam bath for 15 minutes, mixing occasionally, and filter."

The malic acid reacts with the calcium carbonate, forming the soluble calcium malate, which is in the filtrate, while the excess insoluble calcium carbonate is filtered off.

$$CaCO_3 + C_4H_6O_5 \longrightarrow CaC_4H_4O_5 + CO_2\uparrow + H_2O$$

2. "Wash the filter with five 5 ml. portions of water, and to the combined filtrate and washings add 1 ml. of ammonia T.S. and 15 ml. of ammonium oxalate T.S. Heat on a steam bath for 15 minutes, filter through hardened filter paper, and wash the filter with five 5 ml. portions of a solution prepared by mixing 1 ml. of ammonia T.S. with 49 ml. of water."

By treatment of the calcium malate with the ammonium oxalate, an equivalent amount of reasonably pure calcium oxalate, $CaC_2O_4 \cdot H_2O$, is precipitated.

$$CaC_4H_4O_5 + (NH_4)_2C_2O_4 \longrightarrow CaC_2O_4 \cdot H_2O\downarrow + (NH_4)_2C_4H_4O_5$$

Ammonia T.S. is used to alkalinize the solution and keep the calcium oxalate insoluble. The calcium oxalate crystals form slowly and grow sufficiently large so that filtration is readily accomplished after the brief heating period.

3. "Puncture the filter paper, and wash the precipitate into the same flask by means of hot water, followed by 30 ml. of dilute sulfuric acid (1 in 3). Heat the solution to 80°, and titrate with 0.1 *N* potassium permanganate. Each milliliter of 0.1 *N* potassium permanganate is equivalent to 6.704 mg. of $C_4H_6O_5$."

The calcium oxalate, equivalent to the calcium malate, which in turn is equivalent to the malic acid in the sample, could be measured gravimetrically after drying the crystals carefully at 100° or after ignition to calcium carbonate at 450 to 500° or to calcium oxide at even higher temperatures. However, because of the difficulties in obtaining uniform substances in this way, the titrimetric procedure, which entails no loss of accuracy, is preferred. Calcium oxalate is dissolved in sulfuric acid, and the solution reduces the permanganate at a sufficiently rapid rate for successful titration.

$$CaC_2O_4 + H_2SO_4 \longrightarrow CaSO_4 + H_2C_2O_4$$
$$5H_2C_2O_4 + 2KMnO_4 + 3H_2SO_4 \longrightarrow 10CO_2\uparrow + 2MnSO_4 + K_2SO_4 + 8H_2O$$

Since each mole of malic acid produces 1 mole of calcium malate, which in turn produces 1 mole or 2 equivalents of calcium oxalate, the equivalent weight of malic acid is ½ gram-molecular weight or 67.04 Gm; 1 mEq is 0.06704 Gm, and 1 ml of 0.1 *N* potassium permanganate is equivalent to 6.704 mg.

This is an indirect method of determination, since neither the calcium ion nor the malate ion is measured directly: an equivalent amount of oxalate ion reduces the standard permanganate in the titration. Calculate the $C_4H_6O_5$ content of the sample and compare the results with the U.S.P. specifications.

The same oxidation-reduction reactions occur in the determination of Dibasic Calcium Phosphate. The pH during this analysis must be carefully controlled to minimize the coprecipitation effects of sodium, potassium, and magnesium oxalates.

Questions and Problems

1. Write the equations for all reactions in the assay of Dibasic Calcium Phosphate. Calculate the titer for 1 ml of a 0.1 N solution of potassium permanganate.

2. In the titration of the solution of CaC_2O_4 obtained from a 0.2600-Gm sample of Calcium chloride, 39.40 ml of 0.0950 N potassium permanganate was required. Calculate the $CaCl_2$ and the $CaCl_2 \cdot 2H_2O$ content of the sample.

3. What are the advantages of a chelometric determination of calcium salts as against a gravimetric or permanganate-oxalate method of analysis?

Exercise 47

Object. Assay of Manganese Dioxide, Precipitated.

Materials Required. 100 mg of Manganese Dioxide, Precipitated
3 ml of sulfuric acid
20.0 ml of diluted hydrogen peroxide (1 in 40)
0.1 N potassium permanganate

Procedure. "Weigh accurately about 100 mg., transfer to a 300-ml flask with the aid of 20 ml. of water, and add 3 ml. of sulfuric acid previously mixed with 10 ml. of water. In a similar flask place a mixture of 3 ml. of sulfuric acid and 30 ml. of water. To each flask add 20.0 ml of diluted hydrogen peroxide (made by diluting 30 percent hydrogen peroxide with 40 volumes of water), allow to stand in the dark with occasional agitation until no more black particles are visible, and titrate the contents of each flask with 0.1 N potassium permanganate: the difference in the volumes of the potassium permanganate consumed. multiplied by 4.347, represents the quantity, in milligrams, of MnO_2 in the sample weight and corresponds to not less than 90%."

$$MnO_2 + H_2O_2 + H_2SO_4 \longrightarrow MnSO_4 + 2H_2O + O_2\uparrow$$

The hydrogen peroxide is added in measured excess and the residual peroxide is determined by titration with standard potassium permanganate. A blank is run under the same conditions to correct for any reactions of the peroxide such as decomposition.

Calculations

$$\frac{(ml_{blank} - ml_{assay}) \times N \times (86.94/2{,}000) \times 100}{\text{Sample weight}} = \%\ MnO_2$$

This is an example of how an oxidizing substance (MnO_2) can be determined indirectly with another oxidizing agent, potassium permanganate,.

by measuring the amount of unreacted hydrogen peroxide in the sample as compared with that found in the blank. The difference, as measured by the volumes of potassium permanganate, represents the equivalent amount of MnO_2 in the sample.

Exercise 48

Object. Assay of Titanium Dioxide.

Materials Required. 500 mg of Titanium Dioxide
20 ml of sulfuric acid
8 Gm of ammonium sulfate
10 ml of stronger ammonia water
Jones reductor tube
20 to 30-mesh zinc
Mercuric chloride solution (1 in 50)
50 ml of ferric ammonium sulfate T.S.
100 ml of diluted sulfuric acid
0.1 *N* potassium permanganate

Procedure. 1. "Accurately weigh about 500 mg. of Titanium Dioxide, transfer to a 250-ml. beaker, and add 20 ml. of sulfuric acid and 7 to 8 Gm. of ammonium sulfate. Mix, and heat on a hot plate until fumes of sulfur trioxide appear, and continue heating over a strong flame until solution is complete or it is apparent that the undissolved residue is siliceous matter. Cool, cautiously dilute with 100 ml. of water, stir, heat carefully to boiling while stirring, and allow the insoluble matter to settle. Filter, transfer the entire residue to the filter, and wash thoroughly with cold, diluted sulfuric acid. Dilute the filtrate to 200 ml., and cautiously add about 10 ml. of stronger ammonia water to reduce the acid concentration to about 5 percent, by volume, of sulfuric acid."

Titanium dioxide is soluble in hot sulfuric acid; the silicic acid is insoluble and is filtered off.

2. "Prepare a zinc amalgam column in a 25-cm. Jones reductor tube, placing a pledget of glass wool in the bottom of the tube, and filling the constricted portion of the tube with zinc amalgam prepared as follows: Add 20 to 30-mesh zinc to mercuric chloride solution (1 in 50), using about 100 ml. of the solution for each 100 Gm. of zinc, and after about 10 minutes, decant the solution from the zinc, then wash the zinc by decantation. Wash the zinc amalgam with 100-ml. portions of diluted sulfuric acid until 100 ml. of the washing does not decolorize 1 drop of 0.1 *N* potassium permanganate."

When the color of the potassium permanganate persists it means that there are no reducing substances left in the column.

3. "Place 50 ml. of ferric ammonium sulfate T.S. in a 500-ml. flask, and add 0.1 *N* potassium permanganate until a faint pink color persists for 5 minutes. Attach the Jones reductor tube to the neck of the flask, and pass 50 ml. of diluted sulfuric acid through the reductor at a rate of about 30 ml. per minute. Pass the prepared titanium solution through the reductor at the same rate, and follow with 100 ml. each of diluted sulfuric acid and of water. During these operations, keep the reductor filled with solu-

tion or water above the upper level of the amalgam. Gradually release the suction, wash down the outlet tube of the reductor and the sides of the receiver, and titrate immediately with 0.1 N potassium permanganate. Perform a blank determination, substituting 200 ml. of dilute sulfuric acid (1 in 20) for the sample solution, and make any necessary corrections. Each milliliter of 0.1 N potassium permanganate is equivalent to 7.990 mg. of TiO_2."

$$TiO_2 + 2H_2SO_4 \longrightarrow Ti(SO_4)_2 + 2H_2O$$
$$Ti^{4+} + \text{zinc-amalgam} \longrightarrow Ti^{3+}$$
$$Ti^{3+} + Fe^{3+} \longrightarrow Ti^{4+} + Fe^{2+}$$

The reduction of an equivalent amount of Fe^{3+} to Fe^{2+} by the Ti^{3+} serves as a means by which titanium dioxide can be determined. The volume of permanganate found to be equivalent to the ferrous iron in the assay is in turn equivalent to the titanium dioxide in the sample.

The calculations for the assay are:

$$\frac{\text{ml} \times N \times \text{mEq} \times 100}{\text{Sample weight}} = \%\ TiO_2$$

The solution containing an equivalent amount of Ti^{3+} must be collected in an inert atmosphere because of the relative ease with which Ti^{3+} is oxidized in air. Consequently, the receiving tube must be immersed in the flask containing the ferric ammonium sulfate T.S. When this is done, the Ti^{3+} reduces an equivalent amount of Fe^{3+} to Fe^{2+}, which is reasonably stable in air and can be measured by titration with standard potassium permanganate.

In the assay of Hydroxylamine Hydrochloride reagent, an equivalent amount of ferric ammonium sulfate is reduced to the ferrous state and the amount of ferrous iron is determined with a standard solution of potassium permanganate, as was done in the titanium dioxide analysis.

$$2NH_2OH + 4Fe(NH_4)(SO_4)_2 \longrightarrow 4FeSO_4 + 4(NH_4)HSO_4 + H_2O + N_2O\uparrow$$

Ammonium Persulfate reagent is reacted with excess, freshly standardized ferrous sulfate T.S., and the residual ferrous sulfate is determined with standard potassium permanganate.

$$(NH_4)_2S_2O_8 + 2FeSO_4 \longrightarrow Fe_2(SO_4)_3 + (NH_4)_2SO_4$$

The pentavalent vanadium atom in Ammonium Vanadate reagent is first reduced to quadrivalent vanadium with sulfur dioxide, then oxidized quantitatively with potassium permanganate, as is Vanadyl Sulfate reagent.

$$VO_4^{3-} + 6H^+ + 1e \rightleftharpoons VO^{++} + 3H_2O$$
$$10(VO)SO_4 + 2KMnO_4 + 22H_2O \longrightarrow 10H_3VO_4 + 2MnSO_4 + K_2SO_4 + 7H_2SO_4$$

RESIDUAL TITRATION METHODS

The residual titrations in the official assays with potassium permanganate solution are of two types, viz.: (1) titration in which an excess of standard potassium permanganate solution is employed to oxidize a substance, and the amount in excess is determined by reduction with (*a*) excess standard oxalic acid or (*b*) excess ferrous ammonium sulfate and back titration with more standard potassium permanganate; (2) titration in which an excess of standard oxalic acid solution is added to the substance and the excess oxalic acid is titrated with standard potassium permanganate.

In the assays of potassium nitrite and sodium nitrite reagents in the U.S.P., an excess of potassium permanganate is used to oxidize the nitrite to nitrate. The residual permanganate is in turn reduced with an excess of ferrous ammonium sulfate, and the residual excess of ferrous ammonium sulfate is oxidized by titration with additional potassium permanganate.

The assay of sodium nitrite (see Exercise 50) is an example of a residual titration method in which an excess of potassium permanganate solution is employed to oxidize a substance, and the amount in excess is determined by reduction with excess standard oxalic acid and back titration with more of the standard potassium permanganate. The advantage of ferrous ammonium sulfate over oxalic acid is that no heating is required to initiate reaction of the reducing reagent with the permanganate. In the assay of Potassium Permanganate U.S.P., a measured excess of standard oxalic acid is used to reduce the potassium permanganate sample; then the excess is measured with standard potassium permanganate. Since the permanganate-oxalate reaction is run at elevated temperatures, it is not practical to titrate permanganate directly with a standard oxalate solution.

Exercise 49

Object. To Prepare and Standardize 0.1 *N* Oxalic Acid.

Materials Required. 6.45 Gm of oxalic acid
7 ml of sulfuric acid
0.1 *N* potassium permanganate

Procedure. "Dissolve 6.45 Gm. of oxalic acid in sufficient water to make 1000 ml. Standardize by titration against freshly standardized 0.1 *N* potassium permanganate as directed under 0.1 *N* potassium permanganate."

The purest oxalic acid obtainable varies in composition because of differences in its moisture content. If the acid is heated sufficiently to drive off absorbed moisture, it tends to lose water of crystallization. Consequently, a solution of approximate concentration is prepared, and its normality is determined by titration against a standard solution, in this case, 0.1 *N* $KMnO_4$. The titration is conducted as follows:

Measure accurately from a buret about 30 ml of the oxalic acid solution into a conical flask and dilute with about 200 ml of water. Add 7 ml of sulfuric acid, heat to about 70°, and then slowly add the standard permanganate solution from a glass-stoppered buret with constant stirring, until a pale pink color is produced which persists for 15 sec. The temperature at the conclusion of the titration should be not less than 60°. Calculate the normality of the solution and, if desired, adjust the solution to exactly 0.1 *N*. The calculations and reactions are similar to those described in the preparation of 0.1 *N* $KMnO_4$.

The error is minimized by standardizing the oxalic acid solution against the same permanganate solution with which it is to be used in analyses. The normality of an oxalic acid solution may be determined by titration with standard alkali as follows:

Dilute 25 ml of 0.1 *N* NaOH accurately measured from a buret with 25 ml of distilled water, add 2 to 4 drops of phenolphthalein T.S. as indicator, and heat the solution to boiling. Titrate this solution with 0.1 *N* oxalic acid until the pink color disappears. The solution should be brought to a gentle boil at the end to expel all CO_2, which, if present, would affect the phenolphthalein indicator.

The equivalent weight of oxalic acid is the same whether it neutralizes sodium hydroxide or reduces potassium permanganate; hence the same normality would be calculated from data obtained in either of the above methods of standardization.

Oxalic acid solution deteriorates on standing and must be restandarded from time to time.

Questions and Problems

1. Why is hot titration employed?
2. What is the equivalent weight of oxalic acid as an acid; as a reducing agent?
3. Write the reaction involved in the standardization of oxalic acid against sodium hydroxide solution.
4. How much oxalic acid is required to prepare 1,000 ml of solutions of the following respective normalities: (*a*) 0.5000; (*b*) 0.2500; (*c*) 0.200?

Exercise 50

Object. Assay of Sodium Nitrite.

Materials Required. 1 Gm of sodium nitrite
50 ml of 0.1 *N* potassium permanganate
5 ml of sulfuric acid
0.1 *N* oxalic acid

Procedure. 1. "Dissolve about 1 Gm. of Sodium Nitrite, accurately weighe water to make 100.0 ml. Pipet 10 ml. of this solution into a mixture of 50 ml. c potassium permanganate, 100 ml. of water, and 5 ml. of sulfuric acid. When adding the sodium nitrite solution, immerse the tip of the pipet beneath the surface of the permanganate mixture."

Sodium nitrite is deliquescent and is therefore directed to be dried in a desiccator over sulfuric acid previous to weighing. Because of this property, the sample should be weighed in a stoppered weighing bottle. Nitrous acid is oxidized by the permanganate to nitric acid. If the solution of sodium nitrite were acidified with sulfuric acid, nitrous acid, which is volatile, would be lost. The sodium nitrite is therefore introduced beneath the surface of the acidified permanganate solution.

2. "Warm the liquid to 40°, allow it to stand for 5 minutes, and add 25.0 ml. of 0.1 *N* oxalic acid. Heat the mixture to about 80°, and titrate with 0.1 *N* potassium permanganate. Each milliliter of 0.1 *N* potassium permanganate is equivalent to 3.450 mg. of $NaNO_2$.

The oxidation of nitrous to nitric acid proceeds slowly at ordinary temperature, but it is rapid at 40°. The warm mixture is allowed to stand for 5 min to ensure completion of the oxidation. Instead of titrating the excess permanganate with standard oxalic acid to disappearance of the permanganate color, the excess permanganate is reduced by adding an excess of standard oxalic acid and titrating the excess oxalic acid with more of the standard permanganate to appearance of permanganate color. The volume of standard permanganate reduced by the sodium nitrite is then calculated by subtracting it from the sum of the two volumes of permanganate employed, which gives the volume reduced by the standard oxalic acid.

$$\underset{69.00}{10NaNO_2} + 4KMnO_4 + 11H_2SO_4 \longrightarrow 10HNO_3 + 4MnSO_4 + 2K_2SO_4 + 5Na_2SO_4 + 6H_2O$$

Each molecule of sodium nitrite loses 2 electrons, and the equivalent weight of sodium nitrite is 34.50 Gm or ½ gram-molecular weight. Therefore, each milliliter of 0.1 *N* potassium permanganate is equivalent to $0.1 \times 0.0345 = 0.00345$ Gm or 0.1 mEq of $NaNO_2$.

Questions and Problems

1. Why is sodium nitrite not assayed by direct titration with standard potassium permanganate in the usual way?

2. A 1.1000-Gm sample of sodium nitrite was dissolved in sufficient water to make 100 ml. A 10-ml sample of the solution was added to 50.00 ml of 0.1 *N* potassium permanganate in the presence of sulfuric acid. The mixture was treated with 25.00 ml of 0.0975 *N* oxalic acid and titrated with 4.58 ml of 0.1 *N* potassium permanganate. Calculate the percent $NaNO_2$ in the sample.

3. If it were desired to measure the total excess of oxalic acid rather than an aliquot part, how would the above procedure be modified?

4. Outline a procedure for analysis of potassium permanganate for $KMnO_4$ which involves use of standard oxalic acid solution, and calculate the $KMnO_4$ titer of 0.1 *N* oxalic acid. Compare with the U.S.P. assay procedure for potassium permanganate.

TABLE 23. SUBSTANCES ASSAYED BY DIRECT TITRATION WITH POTASSIUM PERMANGANATE

Substance	mEq, Gm	Official requirement, percent
U.S.P.		
Calcium phosphate, dibasic	0.06803	$CaHPO_4$ = 98
Cherry juice	0.06704	$C_4H_6O_5$ = 1
Ferrous sulfate	0.27802	$FeSO_4 \cdot 7H_2O$ = 99.5 to 104.5
Ferrous sulfate, dried	0.15190	$FeSO_4$ = 80
Hydrogen peroxide (R)	0.01701	H_2O_2 = 29 to 32
Hydrogen peroxide solution	0.01701	H_2O_2 = 2.5 to 3.5 *W/V*
Hydroxylamine hydrochloride (R)	0.03475	$NH_2OH \cdot HCl$ = 96
Manganese dioxide, precipitated (R)	0.04347	MnO_2 = 90
Sodium peroxide (R)	0.03899	Na_2O_2 = 93
Titanium dioxide	0.07990	TiO_2 = 99
Vanadyl sulfate (R)	0.16302	$VOSO_4$ = 97
N.F.		
Ammonium vanadate (R)	0.11698	$NH_4V_4O_3$ = 98
Lithium oxalate (R)	0.05095	LiC_2O_4 = 99
Sodium perborate	0.07693	$NaBO_3 \cdot 4H_2O$ = 86.5
Sodium perborate	0.0080	0 = 9
Substances Assayed by Residual Titration of Ferrous Sulfate, Ferrous Ammonium Sulfate, or Oxalic Acid with Potassium Permanganate		
U.S.P.		
Ammonium persulfate (R)	0.11410	$(NH_4)_2S_2O_8$ = 98
Potassium nitrite (R)	0.04255	KNO_2 = 94
Potassium permanganate	0.03161	$KMnO_4$ = 99
Potassium permanganate tablets	0.03161	$KMnO_4$ = 95 to 105 l.a.*
Sodium nitrite	0.03450	$NaNO_2$ = 97 to 101
Sodium nitrite (R)	0.03450	$NaNO_2$ = 97

* = 93 to 107 percent if less than 0.3 mg tablet.

l.a. = labeled amount; (R) = reagent.

5. If potassium permanganate is to be assayed by treatment of a sample with 50.00 ml of 0.0952 *N* oxalic acid, what should be the weight of the permanganate sample (assumed to be 100 percent $KMnO_4$) so that 10.00 ml of 0.0915 *N* potassium permanganate would be required in the back titration?

6. Write half-equations and the combined equation showing the electron transfer that occurs in the assay of lime.

CERIC SULFATE TITRATION METHODS

A solution of ceric sulfate in diluted sulfuric acid is a strong oxidizing agent and considerably more stable than standard permanganate solutions, provided sufficient sulfuric acid is present to prevent hydrolysis and precipitation of basic salts. Although permanganate can be reduced to any of several oxidation states, Ce^{3+} always results on reduction of Ce^{4+}.

Solutions of ceric sulfate combine many of the advantages of permanganate and dichromate, viz: (1) the solutions are stable even on boiling; (2) they react quantitatively with oxalate or arsenite ion, and either sodium oxalate or arsenic trioxide may be used as a primary standard; (3) the cerous ion is colorless and does not obscure the indicator end point; (4) no intermediate products are formed in the reduction of ceric cerium; (5) rather high concentrations of chloride ion are not oxidized by ceric salts, so that ferrous iron can be determined in the presence of chlorides; (6) the ferrous phenanthroline ion (ferroin) is a very satisfactory indicator in titrations with ceric salts.

***o*-Phenanthroline Test Solution.** "Dissolve 150 mg. of orthophenanthroline in 10 ml. of a solution of ferrous sulfate prepared by dissolving 1.48 Gm. of clear crystals of ferrous sulfate in 100 ml. of water. The ferrous sulfate solution must be prepared immediately before dissolving the orthophenanthroline. Store the solution in well-closed containers."

The base *o*-phenanthroline dissolves readily in aqueous solutions of ferrous salts, 3 molecules combining with 1 ferrous ion to form complex ions, known as ferroin, which have an intensely red color. Strong oxidizing agents convert the ferrous to a ferric complex, which has a slight blue color, the intensity of which is much less than that of the ferrous complex, so that it functions practically as a one-color indicator.

$$\underset{\substack{o\text{-Phenanthroline}\\ \text{(colorless)}}}{C_{12}H_8N_2} + Fe^{++} \longrightarrow \underset{\substack{\text{Ferrous complex}\\ \text{(red)}}}{Fe(C_{12}H_8N_2)_3^{++}} \rightleftharpoons \underset{\substack{\text{Ferric complex}\\ \text{(blue)}}}{Fe(C_{12}H_8N_2)_3^{3+}} + e$$

The color change, red to blue, is reversed by reducing agents. The blue oxidized form is very stable to further change by strong oxidizing agents.

One drop of this indicator solution is equivalent to less than 0.01 ml of 0.1 *N* oxidizing agent, so that the indicator blank is negligible. Stock solutions of the indicator are stable.

Exercise 51

Object. To Prepare and Standardize 0.1 *N* Ceric Sulfate.

Materials Required. 59 Gm of ceric ammonium nitrate
31 ml of sulfuric acid
200 mg of arsenic trioxide
25 ml of sodium hydroxide (2 in 25)
10 ml of dilute sulfuric acid (1 in 3)
o-Phenanthroline T.S.
Osmium tetroxide in 0.1 *N* sulfuric acid (1 in 400)

Procedure. "Transfer 59 Gm. of ceric ammonium nitrate to a beaker, add 31 ml. of sulfuric acid, mix, and cautiously add water, in 20-ml portions, until solution is complete. Cover the beaker, allow to stand overnight, filter through a fine-porosity, sintered-glass crucible, dilute to 1000 ml., and mix.

Standardization—Weigh accurately 200 mg. of arsenic trioxide, previously dried at 100° for 1 hour, and transfer to a 500-ml. conical flask. Wash down the inner walls of the flask with 25 ml. of sodium hydroxide solution (2 in 25), swirl to dissolve the sample, and when solution is complete, add 100 ml. of water, and mix. Add 10 ml. of dilute sulfuric acid (1 in 3), then add 2 drops each of orthophenanthroline T.S. and a 1 in 400 solution of osmium tetroxide in 0.1 *N* sulfuric acid, and slowly titrate with the ceric sulfate solution until the pink color is changed to a very pale blue. Each 4.946 mg. of arsenic trioxide is equivalent to 1 ml. of 0.1 *N* ceric sulfate."

The trace of osmium tetroxide is added to catalyze the otherwise slow reaction between ceric ion and the arsenite ion.

One equivalent weight of ceric ammonium nitrate, $Ce(NO_3)_4 \cdot 2NH_4NO_3 \cdot 2H_2O$, is 1 gram-molecular weight or 584.26 Gm, and a liter of 0.1 *N* solution would require $\frac{1}{10}$ equivalent or approximately 59 Gm. One mole of ceric sulfate, $Ce(SO_4)_2$, or 332.24 Gm is equivalent to 1 mole of ceric ammonium nitrate, and $\frac{1}{10}$ equivalent of ceric sulfate in 1,000 ml would contain about 33 Gm in solution.

$$2Ce(SO_4)_2 + Na_3AsO_3 + H_2O \longrightarrow Ce_2(SO_4)_3 + Na_3AsO_4 + H_2SO_4$$

$$As_2O_3 + 6NaOH \longrightarrow 2Na_3AsO_3 + 3H_2O$$

As can be seen from the equations above, 4 equivalents of ceric sulfate is needed to oxidize 1 mole of arsenic trioxide; therefore 1 equivalent weight of arsenic trioxide is $\frac{1}{4}$ mole or 197.84/4 or 49.46 Gm, and 1 mEq would contain 49.46 mg or 0.04946 Gm.

Calculation for the normality of ceric sulfate solution is as follows:

$$N = \frac{\text{wt of arsenic trioxide}}{\text{ml} \times 0.04946}$$

Exercise 52

Object. Assay of Ferrous Sulfate Tablets.

Materials Required. 20 Ferrous Sulfate Tablets
40 ml of diluted sulfuric acid
o-Phenanthroline T.S.

Procedure. "Weigh and finely powder not less than 20 Ferrous Sulfate Tablets. Dissolve a portion of the powder, equivalent to about 500 mg. of ferrous sulfate and accurately weighed, in a beaker containing a mixture of 20 ml. of diluted sulfuric acid and 80 ml. of freshly boiled and cooled water. Filter the solution rapidly as soon as all soluble ingredients in the tablets are dissolved, and wash the container and the filter with small portions of a mixture of 20 ml. of diluted sulfuric acid and 80 ml. of freshly boiled and cooled water. Add orthophenanthroline T.S., and immediately titrate the combined filtrate and washings with 0.1 *N* ceric sulfate. Each milliliter of 0.1 *N* ceric sulfate is equivalent to 27.80 mg. of $FeSO_4 \cdot 7H_2O$."

The following oxidation-reduction reaction occurs between the ceric sulfate and the ferrous sulfate:

$$2FeSO_4 + Ce(SO_4)_2 \longrightarrow Fe_2(SO_4)_3 + Ce_2(SO_4)_3$$

The equivalent weight of the ferrous sulfate is 1 gram-molecular weight or 278.02 Gm; therefore each milliliter of 0.1 *N* ceric sulfate is equivalent to 0.1 × 0.27802 or 0.027802 Gm or 27.80 mg.

Since the estimation of the iron salts is based on the oxidation of the ferrous ion to the ferric ion, the variations in the procedures for ferrous gluconate and ferrous fumarate are necessary to ensure that the iron is in the ferrous state before oxidation with ceric sulfate.

The analysis of Ceric Ammonium Nitrate reagent is based on the same chemical reaction. The ceric salt is reacted with a standard solution of ferrous sulfate, and its purity is determined.

The assay of Menadione U.S.P. is based on, first, the reduction of the quinone, in solution in glacial acetic acid, to the corresponding hydroquinone by means of zinc and hydrochloric acid, and then oxidation of the hydroquinone back to a quinone with ceric sulfate. Menadiol Sodium Diphosphate U.S.P., the diphosphate of the hydroquinone from menadione, is titrated potentiometrically with standard ceric sulfate, since it is a highly colored compound and a visual color end point would not be so sharp as that determined by a potentiometer.

OPO_3Na / CH_3 / OPO_3Na $\xrightarrow{HOH}$ OH / CH_3 / OH $\underset{[red]}{\overset{[O]}{\rightleftharpoons}}$ O / CH_3 / O $+ 2H^+$

Menadiol sodium diphosphate — Menadiol (hydroquinone) — Menadione (quinone)

In the analysis of tocopherols using diphenylamine as the indicator, the assays are based on the oxidation of the free alcohol or hydroquinone to the quinone structure.

The quantitative oxidation of the nitrite ion to nitrate has been adapted for the assay of the KCl content of Ringer's injection and solution in the N.F. The stoichiometry in this procedure can be understood from the following equations:

$$2KCl + Na_3Co(NO_2)_6 \longrightarrow K_2NaCo(NO_2)_6\downarrow + 2NaCl$$

$$K_2NaCo(NO_2)_6 + 12Ce(SO_4)_2 + 6H_2O \longrightarrow Co(NO_3)_3 + 2KNO_3 + NaNO_3 + 6Ce_2(SO_4)_3 + 6H_2SO_4$$

$$2Ce(SO_4)_2 + 2Fe(NH_4)_2(SO_4)_2 \longrightarrow Ce_2(SO_4)_3 + 2FeNH_4(SO_4)_2 + (NH_4)_2SO_4$$

The potassium ion is quantitatively precipitated, the precipitate is collected and treated with a measured excess of ceric sulfate, which subsequently oxidizes the nitrite to nitrate, and the residual ceric sulfate is determined with ferrous ammonium sulfate standard solution. From the equations it can be seen that 2KCl is equivalent to $12Ce(SO_4)_2$; therefore, 1 equivalent weight of KCl is $\frac{1}{6}$ mole or 74.56/6 or 12.426 Gm, and 1 ml of 0.02 N ceric sulfate is equivalent to 0.2485 mg of KCl.

Assay of Ascorbic Acid. A solution of dichlorophenol-indophenol, which has been standardized using U.S.P. ascorbic acid reference stand-

TABLE 24. SUBSTANCES ASSAYED BY DIRECT TITRATION WITH CERIC SULFATE

Substance	mEq, Gm	Indi-cator	Official requirement, percent
U.S.P.			
Ferrous fumarate	0.16991	o.ph.	$C_4H_2FeO_4$ = 96.5 to 101
Ferrous fumarate tablets	0.16991	o.ph.	$C_4H_2FeO_4$ = 95 to 110 l.a.
Ferrous sulfate tablets	0.27802	o.ph.	$FeSO_4 \cdot 7H_2O$ = 95 to 110 l.a.
Hydroquinone (R)	0.05506	d.ph.	$C_6H_4(OH)_2$ = 99
Menadiol sodium diphosphate	0.21105	pot.	$C_{11}H_8Na_4O_8P_2$ = 97.5 to 102
Menadiol sodium diphosphate injection	0.26509	pot.	$C_{11}H_8Na_4O_8P_2 \cdot 6H_2O$ = 95 to 110 l.a.
Menadione	0.08609	o.ph.	$C_{11}H_8O_2$ = 98.5 to 101
Menadione tablets	0.08609	o.ph.	$C_{11}H_8O_2$ = 95 to 110 l.a.
N.F.			
Ferrous gluconate	0.44615	o.ph.	$C_{12}H_{22}FeO_{14}$ = 95
Ferrous gluconate tablets	0.48218	o.ph.	$C_{12}H_{22}FeO_{14} \cdot 2H_2O$ = 93 to 107 l.a.
Iron, reduced (R)	0.05585	o.ph.	Fe, 93
Menadiol sodium diphosphate tablets	0.26509	pot.	$C_{11}H_8Na_4O_8P_2 \cdot 6H_2O$ = 95 to 110 l.a.
Menadione sodium bisulfite	0.13813	o.ph.	$C_{11}H_8O_2 \cdot NaHSO_3$ = 94
Menadione sodium bisulfite injection	0.16514	o.ph.	$C_{11}H_8O_2 \cdot NaHSO_3 \cdot 3H_2O$ = 90 to 110 l.a.
Tocopherol, *dl-α*	0.21536	d.ph.	$C_{29}H_{50}O_2$ = 96
Tocopheryl acetate, *d-α*	0.23638	d.ph.	$C_{31}H_{52}O_5$ = 96
Tocopheryl acetate, *dl-α*	0.23638	d.ph.	$C_{31}H_{52}O_3$ = 96
Tocopheryl acid succinate, *d-α*	0.26540	d.ph.	$C_{33}H_{54}O_5$ = 96

(R) = reagent; l.a. = labeled amount; o.ph. = *o*-phenanthroline; pot. = potentiometric; d.ph. = diphenylamine.

ard, is used as a standard oxidizing agent in determination of the ascorbic acid content of ascorbic acid tablets, capsules, and injection, as well as of the ascorbic acid content in Hexavitamin and Decavitamin capsules and tablets. No indicator is necessary, since the oxidized form of the dichlorophenol-indophenol is colored blue in alkali and red in acid, while the reduced form is colorless. A solution of ascorbic acid from the sample is titrated in the presence of metaphosphoric acid and acetic acid with the dichlorophenol-indophenol standard solution until a distinct rose-pink color persisting for at least 5 sec results.

$$C_6H_8O_6 + O{=}C_6H_4{=}N{-}C_6H_2Cl_2{-}OH \longrightarrow$$

(Pink)

$$C_6H_6O_6 + HO{-}C_6H_4{-}\underset{H}{\underset{|}{N}}{-}C_6H_2Cl_2{-}OH$$

(Colorless)

A change in color is also produced when a number of other organic substances are reduced. Thus amaranth, amaranth solution, sodium indigotindisulfonate, and sodium indigotindisulfonate injection are assayed by titration with a reducing agent, 0.1 *N* titanium trichloride. In the assay of gentian violet and gentian violet solution an excess of titanium trichloride in an inert atmosphere is used, and the excess is back-titrated with standard ferric ammonium sulfate solution. The end point is the production of a red color because of the formation of a ferric thiocyanate complex.

Standard potassium ferricyanide solution is used to assay methyprylon, methyprylon elixir, and methyprylon tablets. Methyprylon is oxidized in the following manner by the ferricyanide reagent:

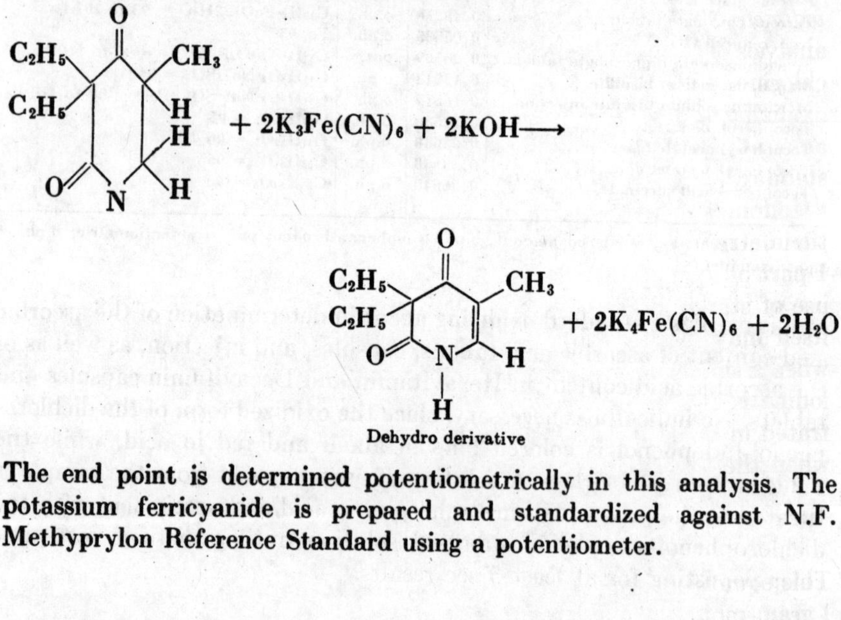

O; C_2H_5; C_2H_5; CH_3; H; H; O; N; H

$$+ 2K_3Fe(CN)_6 + 2KOH \longrightarrow$$

O; C_2H_5; C_2H_5; CH_3; O; N; H; H

$$+ 2K_4Fe(CN)_6 + 2H_2O$$

Dehydro derivative

The end point is determined potentiometrically in this analysis. The potassium ferricyanide is prepared and standardized against N.F. Methyprylon Reference Standard using a potentiometer.

Chapter 9

OXIDATION-REDUCTION: IODIMETRIC AND IODOMETRIC METHODS

The reversible reaction $2I^- \rightleftharpoons I_2 + 2e$ can be applied in analysis of reducing agents such as thiosulfates and arsenites by the use of standard solutions of iodine. Such a direct procedure is sometimes designated as *iodimetry*. The same reversible reaction can be applied indirectly in the analysis of oxidizing agents such as ferric and cupric salts, and "available" chlorine. In the indirect procedure, known as *iodometry*, the sample of oxidizing agent is reduced with excess potassium iodide and an equivalent amount of iodine is produced. The iodine thus formed is titrated with a standard solution of sodium thiosulfate.

Iodometric methods include some of the most accurate procedures in titrimetric analysis, because, under proper conditions, the presence of 1 part of iodine in several million parts of solution is readily detected by use of starch indicator solution. In some procedures, the color of iodine itself may be used in observing the end point of a titration, especially when a small volume of carbon tetrachloride or carbon disulfide, in which iodine is soluble, is added to the mixture. The iodine which is concentrated in the immiscible solvent layer imparts a distinct color to it, even when the total amount of iodine in the mixture is very small.

$$I_2 + \underset{\text{Sodium thiosulfate}}{2Na_2S_2O_3} \longrightarrow 2NaI + \underset{\text{Sodium tetrathionate}}{Na_2S_4O_6}$$

This equation shows that 1 gram-atomic weight of iodine, 126.9 Gm, and 1 gram-molecular weight of sodium thiosulfate, $Na_2S_2O_3 \cdot 5H_2O$, 248.18 Gm, are equivalent.

Iodine is not very soluble in water but dissolves readily in the presence of potassium iodide because of the formation of the triiodide ion $I_2 + I^- \longrightarrow I_3^-$. For this reason standard iodine solutions used in iodimetry are solutions of iodine in potassium iodide solution. Moreover, excess potassium iodide is used as a reducing agent in iodometry so that the iodine formed will remain in solution through interaction with potassium iodide.

In alkaline solutions, reaction of iodine with OH^- produces first hypoiodite, and finally iodate, ions.

$$I_2 + OH^- \longrightarrow HI + IO^-$$

$$3IO^- \longrightarrow IO_3^- + 2I^-$$

These ions oxidize thiosulfate, at least partially, to a higher oxidation state, e.g., sulfate, and the stoichiometry indicated above no longer holds true. Consequently, iodometric procedures are never carried out in strongly basic solutions.

The uniformity of the reaction of iodine with sodium thiosulfate accounts for the use of standard solution of iodine in the analysis of sodium thiosulfate and the use of standard solution of sodium thiosulfate in the titrimetric determination of iodine in various iodimetric and iodometric procedures.

Iodine quantitatively oxidizes certain other reducing agents, and standard solutions of iodine may be used in their direct titrimetric analysis, e.g., arsenious acid:

$$H_3AsO_3 + H_2O + I_2 \rightleftharpoons H_3AsO_4 + 2H^+ + 2I^-$$

If a residual method of iodimetry is employed, the excess iodine may be titrated with standard sodium thiosulfate, e.g., the assay of Calomel:

$$2HgCl + I_2 + 6KI \longrightarrow 2K_2HgI_4 + 2KCl$$
$$I_2 + 2Na_2S_2O_3 \longrightarrow 2NaI + Na_2S_4O_6$$

In the iodometric analysis of oxidizing agents, an equivalent amount of iodine is produced when the sample of the oxidizing agent oxidizes potassium iodide in the presence of an acid, e.g., assay of cupric sulfate:

$$2CuSO_4 + 4KI \longrightarrow 2CuI\downarrow\ I_2 + 2K_2SO_4$$

The equivalent amount of iodine formed in this way can be measured by titration with standard sodium thiosulfate. When potassium iodide is oxidized in strongly acid solution, care must be taken to avoid simultaneous oxidation of the iodide by atmospheric oxygen; otherwise values which are too high will be obtained.

Starch Indicator Solution. This solution is prepared as follows: Triturate 1 Gm of arrowroot starch with 10 ml of cold water in a mortar to form a thin paste. Heat about 200 ml of water to boiling and add the starch paste to it with constant stirring. Boil the mixture gently for about 30 min or until it forms a thin translucent liquid. Since a solution of starch deteriorates rapidly, it should be prepared freshly each day.

In hot water, starch granules burst and form a colloidal dispersion of β-amylose, known as soluble starch, and should be separated by decantation from α-amylose, amylopectin, which is insoluble.

Interaction of iodine and iodide ion with the colloidal β-amylose results in the production of an intensely blue-colored sol, and this color change is reversible, the color being discharged when the iodine is reduced with sodium thiosulfate or other reducing agent. The sensitivity of the indicator is greater in slightly acid medium and is markedly decreased by temperatures above 25°, strong solutions of electrolytes, alcohol, and other organic solvents. The reversibility of color formation is decreased when the iodine concentration is high. For this reason the indicator should not be added in residual iodimetric procedures or in iodometric procedures until most of the iodine has been reduced, as shown by the change from a brown to a straw-colored solution.

Standard Solutions. The standard solutions employed in the official assays by iodometric methods are 0.1 *N* iodine and 0.1 *N* sodium thiosulfate, and 0.1 *N* potassium arsenite.

DIRECT TITRATION WITH IODINE, SODIUM THIOSULFATE, AND POTASSIUM ARSENITE

Exercise 53

Object. To Prepare and Standardize 0.1 *N* Iodine Solution.

Materials Required. 14 Gm of iodine
36 Gm of potassium iodide
150 mg of arsenic trioxide
20 ml of 1.0 *N* sodium hydroxide
Diluted hydrochloric acid
2 Gm of sodium bicarbonate
3 ml of starch T.S.

Procedure. 1. "Dissolve about 14 Gm. of iodine in a solution of 36 Gm. of potassium iodide in 100 ml. of water, add 3 drops of hydrochloric acid, dilute to 1000 ml., and standardize as follows:"

One gram of iodine is soluble in 2,950 ml of water. In the presence of sufficient potassium iodide, however, its solubility is greatly increased because of the formation of the complex I_3^- anion:

$$KI + I_2 \rightleftharpoons KI_3$$

In the presence of reducing agents, the reaction proceeds quantitatively to the left, and the solution reacts like a solution of iodine alone.

2. "Weigh accurately about 150 mg. of arsenic trioxide, and dissolve in 20 ml. of 1.0 *N* sodium hydroxide by warming if necessary. Dilute with 40 ml. of water, add 2 drops of methyl orange T.S., and follow with diluted hydrochloric acid until yellow color is changed to pink."

Arsenic trioxide is slowly soluble in cold water, more rapidly soluble in boiling water, and readily soluble in sodium hydroxide solution with the formation of sodium arsenite.

$$As_2O_3 + 6NaOH \longrightarrow 2Na_3AsO_3 + 3H_2O$$

If iodine is added to this alkaline solution, some of it tends to form sodium hypoiodite, NaIO, or similar compounds, which do not react readily with arsenious ion.

$$2NaOH + I_2 \longrightarrow NaIO + NaI + H_2O$$

The excess sodium hydroxide is therefore neutralized with hydrochloric acid using methyl orange as the indicator.

3. "Then add 2 Gm. of sodium bicarbonate, dilute with 50 ml. of water, add 3 ml. of starch T.S. Slowly add the iodine solution from a buret until a permanent blue color is produced. Calculate the normality. Each 4.946 mg. of arsenic trioxide is equivalent to 1 ml. of 0.1 *N* iodine."

Sodium bicarbonate is added to neutralize the hydriodic acid, HI, formed in the reversible reaction:

$$H_3AsO_3 + H_2O + I_2 \rightleftharpoons H_3AsO_4 + 2HI$$

The $NaHCO_3$ removes the HI as rapidly as it is formed, causing the reaction to go to completion toward the right:

$$Na_3AsO_3 + I_2 + 2NaHCO_3 \longrightarrow Na_3AsO_4 + NaI + 2CO_2\uparrow + H_2O$$

Under conditions of the assay, sodium bicarbonate does not react with iodine, but if an excess of iodine is added, it may react slightly. The reaction between the arsenious ion and iodine proceeds quantitatively to completion best in a neutral solution. Since HI is formed in the reaction, a buffer must be added to keep the solution neutral. The sodium bicarbonate in this case is the buffer.

The calculation for determining the normality of the iodine solution is as follows:

$$N = \frac{\text{wt of } As_2O_3 \text{ sample}}{\text{ml of Iodine} \times 0.04946}$$

Each arsenic atom loses 2 electrons when it is oxidized; therefore each mole of As_2O_3 contains 4 equivalents, and 1 mEq is equal to 197.84/4,000 or 0.04946 Gm or 4.946 mg.

Questions and Problems

1. Why may iodine containing chlorine or bromine as impurity be used to prepare standard solutions of iodine?
2. What is the normality of a solution containing 0.0136 Gm of iodine per milliliter?
3. What weight of arsenic trioxide (98.73 percent As_2O_3) would be used as sample so that 35.00 ml of 0.1 *N* iodine would be needed to titrate it?
4. A solution of iodine has an iodine titer of 11.43 mg. What is its As_2O_3 titer? What is the normality of the iodine solution?
5. Why is starch T.S. not added during the titration until the solution is but slightly yellow?

Exercise 54

Object. Assay of Antimony Potassium Tartrate.

Materials Required. 500 mg of Antimony Potassium Tartrate
25 ml of saturated solution of sodium bicarbonate
0.1 *N* iodine
Starch T.S.

Procedure. "Dissolve about 500 mg. of Antimony Potassium Tartrate, accurately weighed, in 30 ml. of water, and 25 ml. of a cold, saturated solution of sodium bicarbonate and a few drops of starch T.S., and immediately titrate with 0.1 *N* iodine to the production of persistent blue color. Each milliliter of 0.1 *N* iodine is equivalent to 16.70 mg. of $C_4H_4KO_7Sb \cdot \frac{1}{2}H_2O$."

The sodium bicarbonate is added to neutralize the HI formed in the reaction in a manner analogous to the reaction of arsenious ion with iodine so that the reaction may proceed quantitatively to completion:

$$KOOC \cdot CHOH \cdot CHOH \cdot COO(SbO) + I_2 + H_2O \rightleftharpoons KOOC \cdot CHOH \cdot CHOH \cdot COO(SbO_2) + 2HI$$

$$NaHCO_3 + HI \longrightarrow NaI + CO_2\uparrow + H_2O$$

Calculations. Since 1 mole of antimony potassium tartrate reacts with 2 equivalents of iodine, 1 equivalent weight of antimony potassium tartrate is ½ mole or 333.93/2 or 167.0 Gm and 1 mEq = 0.1670 Gm.

$$\frac{ml \times N \times (333.93/2{,}000) \times 100}{\text{Sample weight}} = \text{percent } C_4H_4KO_7Sb \cdot 1/2H_2O$$

Questions and Problems

1. Summarize the equations for the reactions involved in the analysis of arsenic trioxide in a single equation which shows the chemical equivalence of 1 mole of As_2O_3 and 2 moles or 4 equivalents of iodine.

2. Calculate the equivalent weights and 0.1 *N* iodine titer values for each of the following, and check the calculated values with those given in the U.S.P. and N.F.: (*a*) antimony and potassium tartrate, $C_4H_4KO_7Sb \cdot \frac{1}{2}H_2O$; (*b*) ascorbic acid, $C_6H_8O_6$; (*c*) stibophen, $C_{12}H_4Na_5O_{16}S_4Sb$.

3. Outline the steps in the assay of Carbarsone, indicating the chemical changes involved in each step. How could this procedure be adapted for analysis of an inorganic pentavalent arsenical such as sodium arsenate?

4. What sample weight of each of the following would be used so that 35 ml of 0.1 *N* iodine would be consumed in the titration? Compare with the sample weights specified in the U.S.P. and N.F. assay procedures: (*a*) sodium thiosulfate; (*b*) stibophen; (*c*) sodium ascorbate.

Ascorbic acid is a fairly strong reducing agent and is determined by simple titration with iodine solution, which oxidizes the ascorbic acid to dehydroascorbic acid; the iodine is reduced to iodide.

$$I_2 + C_6H_8O_6 \longrightarrow 2HI + C_6H_6O_6$$

Organic arsenicals can be determined as trivalent or pentavalent arsenic. The determination of arsenic in its pentavalent form will be taken up later. The organic arsenical is titrated directly in determining the amount of trivalent arsenic in Oxophenarsine Hydrochloride. In Carbarsone the arsenic is in its trivalent form when it is digested with sulfuric acid and potassium sulfate in the manner of a Kjeldahl digestion. As the organic material is oxidized the arsenic is reduced to trivalent arsenic. With Tryparsamide and Glycobiarsol, nitric acid is used in addition to sulfuric acid for the digestion. Following the digestion the excess oxidizing agent and the arsenic are reduced with starch or potassium iodide, the iodine formed from the potassium iodide is reduced with sodium thiosulfate, and then the pH is adjusted, sodium bicarbonate is added, and the trivalent arsenic is titrated with the iodine solution. Potassium sodium tartrate is used in the Glycobiarsol assay for arsenic to complex the bismuth, which might otherwise interfere with the analysis.

Sulfur dioxide is converted to sodium sulfite by treatment with a sodium hydroxide solution, and the sulfite ion is oxidized to sulfate with iodine.

$$SO_3^{=} + I_2 \rightleftharpoons SO_4^{=} + 2HI$$

TABLE 25. SUBSTANCES ASSAYED BY DIRECT TITRATION WITH IODINE

Substance	mEq, Gm	Official requirement, percent
U.S.P.		
Antimony and potassium tartrate	0.16697	$C_4H_4KO_7Sb \cdot \frac{1}{2}H_2O$ = 99 to 103
Ascorbic acid	0.08807	$C_6H_8O_6$ = 99
Carbarsone	0.13004	$C_7H_9AsN_2O_4$ = 97.5 to 101
Carbarsone tablets	0.13004	$C_7H_9AsN_2O_4$ = 93 to 107 l.a.
Dimercaprol	0.06211	$C_3H_8OS_2$ = 97
Dimercaprol	0.04676	$C_3H_8S_3$ = n.m.t. 1.5
Dimercaprol injection	0.06211	$C_3H_8OS_2$ = 9 to 11 *W/V*
Oxophenarsine hydrochloride	0.03746	As = 29.5 to 32
Oxophenarsine hydrochloride for injection	0.11775	$C_6H_6AsNO_2 \cdot HCl$ = 92.5 to 107.5 l.a.
Sodium ascorbate	0.09905	$C_6H_7NaO_6$ = 99 to 101
Sodium thioglycollate (R)	0.11410	$HSCH_2COONa$ = 75
Sodium thiosulfate	0.15811	$Na_2S_2O_3$ = 99
Sodium thiosulfate injection	0.24818	$Na_2S_2O_3 \cdot 5H_2O$ = 95 to 105 l.a.
Stibophen	0.38457	$C_{12}H_4Na_5O_{16}S_4Sb$ = 98.5 to 102
Stibophen injection	0.44761	$C_{12}H_4Na_5O_{16}S_4Sb \cdot 7H_2O$ = 94 to 106 l.a.
Sulfur dioxide	0.03203	SO_2 = 97 *V/V*
Tryparsamide	0.14904	$C_8H_{10}AsN_2NaO_4$ = 99 to 101
Tryparsamide, sterile	0.14904	$C_8H_{10}AsN_2NaO_4$ = 99 to 101
N.F.		
Carbarsone capsules	0.13004	$C_7H_9AsN_2O_4$ = 93 to 107 l.a.
Glycobiarsol	0.03746	As = 14 to 16

(R) = reagent; n.m.t. = not more than; l.a. = labeled amount.

The sulfur in Dimercaprol and reagent sodium thioglycollate is oxidized with iodine. The sulfhydryl group is oxidized to disulfide in a reaction such as

$$2RSH + I_2 \longrightarrow R{-}S{-}S{-}R + 2HI$$

Exercise 55

Object. To Prepare and Standardize 0.1 *N* Sodium Thiosulfate.

Materials Required. 26 Gm of sodium thiosulfate ($Na_2S_2O_3{\cdot}5H_2O$)
200 mg of sodium carbonate
210 mg of potassium dichromate
3 Gm of potassium iodide
2 Gm of sodium bicarbonate
5 ml of hydrochloric acid
Starch T.S.

Procedure. 1. "Dissolve about 26 Gm. of sodium thiosulfate and 200 mg. of sodium carbonate in 1000 ml. of recently boiled and cooled water."

The sodium thiosulfate is dissolved in water which has been recently boiled, primarily for purposes of sterilization, since deterioration of the solution results mainly from the action of bacteria. Boiling also expels carbon dioxide, which, if present as carbonic acid, may act like any other acid in causing hydrolysis and decomposition of the sodium thiosulfate.

$$Na_2S_2O_3 + 2H_2CO_3 \longrightarrow 2NaHCO_3 + H_2S_2O_3$$
$$H_2S_2O_3 \longrightarrow H_2SO_3 + S\downarrow$$

The sodium carbonate is added to the solution as a preservative to prevent acid-catalyzed hydrolysis.

2. "Weigh accurately about 210 mg. of primary standard potassium dichromate, previously pulverized and dried at 120° for four hours, and dissolve in 100 ml. of water in a glass-stoppered, 500-ml. flask. Swirl to dissolve the sample, remove the stopper, and quickly add 3 Gm. of potassium iodide, 2 Gm. of sodium bicarbonate, and 5 ml. of hydrochloric acid. Stopper the flask, swirl to mix, and allow to stand in dark for 10 minutes. Rinse the stopper and the inner walls of the flask with water, and titrate the liberated iodine with the sodium thiosulfate until the solution is only faintly yellow in color. Add starch T.S., and continue the titration to the discharge of the blue color. Calculate the normality."

The potassium dichromate oxidizes the iodide ion in the presence of acid to an equivalent amount of iodine. However the high acidity may introduce an error caused by the atmospheric oxidation of the hydriodic acid to iodine; the addition of sodium bicarbonate minimizes this side reaction by filling the atmosphere with inert carbon dioxide. The chromic acid is reduced to chromous chloride, which is green.

The half-equations

$$Cr_2O_7^{=} + 14H^+ + 6e \rightleftharpoons 2Cr^{3+} + 7H_2O$$

and

$$6I^- \rightleftharpoons 3I_2 + 6e$$

are the basis for the molecular equation

$$K_2Cr_2O_7 + 6KI + 14HCl \longrightarrow 3I_2 + 2CrCl_3 + 8KCl + 7H_2O$$

The iodine formed in this reaction will oxidize a chemically equivalent amount of sodium thiosulfate. In other words, the volume of sodium thiosulfate used in titration of the iodine is chemically equivalent to the sample weight of potassium dichromate. Therefore, the normality of the sodium thiosulfate may be calculated:

$$\text{ml} \times N = \frac{\text{wt of } K_2Cr_2O_7}{0.04903 \text{ (1 mEq wt)}}$$

or

$$N = \frac{\text{wt of sample}}{\text{ml} \times 0.04903}$$

One mole of $K_2Cr_2O_7$ contains 6 equivalents, so that 1 mEq would then be 1/6,000 of a gram-molecular weight, or 294.19/6,000 or 0.04903 Gm.

Potassium iodide in excess of the amount necessary for reduction of the $K_2Cr_2O_7$ is employed so that the liberated iodine will be held in solution as KI_3. The solution is diluted before titration to make the end point more readily observed, since the mixture does not become colorless when all the iodine is reduced but rather changes from the deep blue color of starch-iodide-iodine complex to that of the light green Cr^{3+}.

The change in the titer of sodium thiosulfate solutions on standing results primarily from bacterial action. The pH of the solution prepared as above directed using sodium carbonate is unfavorable to the growth of the bacteria in question; but, as a further precaution, specially cleaned glassware and freshly boiled water should be used in preparation of the solution, and subsequent gross bacterial contamination of the solution should be avoided.

Sodium thiosulfate solution may also be standardized against pure iodine dissolved in potassium iodide solution or against iodine set free from acidified solution of potassium iodide by standard solutions of potassium permanganate or various primary standard oxidizing agents such as potassium iodate, etc.

Questions and Problems

1. Calculate the titer values of each of the following substances for 0.1 *N* sodium thiosulfate and the size of the sample of each that would liberate sufficient iodine to consume 35 ml of 0.1 *N* sodium thiosulfate in a titration: (*a*) $K_2Cr_2O_7$; (*b*) KIO_3; (*c*) 0.1 *N* $KMnO_4$.

2. What is the iodine titer of 0.1 *N* sodium thiosulfate? If all the $Na_2S_2O_3$ of the 0.1 *N* solution were converted to Na_2SO_3 in accordance with the equations given, what would be the iodine titer of the resulting solution?

3. In the procedure for standardization of the sodium thiosulfate solution, 30 ml of 0.1 *N* potassium dichromate is employed. Calculate the theoretical amount of KI necessary for reduction of the $K_2Cr_2O_7$, and compare this value with the 2 Gm of potassium iodide actually used. Account for the discrepancy.

4. Why do sodium thiosulfate solutions undergo a change in titer, and what measures are taken to prevent this?

Iodine and preparations containing iodine may be assayed by direct titration with 0.1 *N* sodium thiosulfate or 0.1 *N* potassium arsenite. Iodine, U.S.P., or Providone-Iodine, N.F., is assayed by a procedure which is quite similar to that used for standardization of 0.1 *N* iodine solution.

The use of standard potassium arsenite solution for the titrimetric analysis of the iodine content is illustrated by the procedure for Strong Iodine Solution U.S.P.

Exercise 56

Object. To Prepare 0.1 *N* Potassium Arsenite Solution.

Materials Required. 5 Gm of arsenic trioxide
75 ml of 1 *N* potassium hydroxide solution
40 Gm of potassium bicarbonate

Procedure. "Dissolve 4.9455 Gm. of arsenic trioxide, which has been pulverized and dried at 100° to constant weight, in 75 ml. of 1 *N* potassium hydroxide. Add 40 Gm. of potassium bicarbonate, dissolved in about 200 ml. of water, and dilute with water to exactly 1000 ml."

The arsenic trioxide is neutralized by potassium hydroxide and forms potassium arsenite.

$$As_2O_3 + 2KOH \longrightarrow 2KAsO_2 + H_2O$$

From the reaction it is evident that there are 2 equivalents per mole of arsenic trioxide and in 1 liter of standard solution or 5 Gm of As_2O_3 there would be about 50 mEq, which in turn will readily react with the 75 mEq of potassium hydroxide used to dissolve it. The potassium bicarbonate is used to provide the pH of about 9 which is necessary for accurate results and a rapid reaction with iodine.

When used as a reducing agent in the titration with iodine, the meta-arsenite is oxidized to arsenate as follows:

$$HAsO_2 + 2H_2O \longrightarrow H_3AsO_4 + 2H^+ + 2e$$

The molecular equation for the titration may be written

$$KAsO_2 + I_2 + 2H_2O \longrightarrow KH_2AsO_4 + 2HI$$

and the potassium bicarbonate reacts with the HI formed, thus shifting the equilibrium of the reaction far to the right so as to make it a quantitative reaction.

$$KHCO_3 + HI \longrightarrow KI + H_2O + CO_2\uparrow$$

A summation of the equations shows the chemical equivalence of 1 mole of As_2O_3, 2 moles of $KAsO_2$, and 2 moles or 4 equivalents of iodine. Consequently the equivalent weight of As_2O_3 is 49.460 Gm, and in 1,000 ml of 0.1 *N* potassium arsenite, 0.1 × 49.460 = 4.9460 Gm, the amount of arsenic trioxide required.

Exercise 57

Object. Assay of Strong Iodine Solution.

Materials Required. 5 ml of Strong Iodine Solution
0.1 *N* potassium arsenite
Starch T.S.

Procedure. "Pipet 5 ml. of Strong Iodine Solution into a glass-stoppered, 500-ml. flask, add 10 ml. of water, and titrate with 0.1 *N* potassium arsenite, adding starch T.S. as the end point is neared. Each milliliter of 0.1 *N* potassium arsenite is equivalent to 12.69 mg. of I."

The explanation for the procedure and for the reactions is similar to that given for the preparation of potassium arsenite solution.

Reagent Lead Dioxide N.F. is an example of the determination using an excess of potassium arsenite and back-titrating the excess with standard potassium bromate.

$$PbO_2 + KAsO_2 + 2HCl \longrightarrow PbCl_2 + KH_2AsO_4$$
$$KBrO_3 + 3KAsO_2 + 3H_2O \longrightarrow KBr + 3KH_2AsO_4$$

The first drop of excess bromate reacts with bromide to form bromine, which in turn reacts with the dye, destroying its color for the end point.

TABLE 26. SUBSTANCES ASSAYED BY DIRECT TITRATION WITH POTASSIUM ARSENITE

Substance	mEq factor	Indicator	Official requirement, percent
U.S.P.			
Iodine solution, strong	0.12690	st.	I = 4.5 to 5.5 *W/V*
Iodine tincture	0.12690	st.	I = 1.8 to 2.2 *W/V*
N.F.			
Iodine ampuls	0.12690	st.	I = 1.8 to 2.2 *W/V*
Iodine solution	0.12690	st.	I = 1.8 to 2.2 *W/V*

st. = starch.

RESIDUAL TITRATION OF EXCESS IODINE WITH SODIUM THIOSULFATE

Exercise 58

Object. Assay of Calomel.

Materials Required. 700 mg of Calomel
50 ml of 0.1 *N* iodine

4 Gm of potassium iodide
1 ml of diluted hydrochloric acid
0.1 *N* sodium thiosulfate
Starch T.S.

Procedure. 1. "Dry about 700 mg. of Calomel over phosphorous pentoxide for 5 hours, weigh accurately, mix well with 10 ml. of water in a glass-stoppered flask, and add 50.0 ml of 0.1 *N* iodine and 4 Gm. of potassium iodide dissolved in 10 ml. of water."

Since mercurous chloride, Calomel, is practically insoluble in water, the purity of the salt cannot be determined by the precipitation method used in determination of the chloride ion in a soluble chloride. When mercurous chloride is treated with a solution of iodine in potassium iodide, it dissolves quite readily, with the formation of a double salt of mercuric and potassium iodides:

$$\underset{2(236.04)}{2HgCl} + 6KI + I_2 \longrightarrow 2HgK_2I_4 + 2KCl$$

2. "Stopper the flask, shake vigorously until solution is complete, add 1 ml. diluted hydrochloric acid, and then titrate the excess iodine with 0.1 *N* sodium thiosulfate, using starch T.S. as the indicator. Each milliliter of 0.1 *N* iodine is equivalent to 23.60 mg. of HgCl."

In the assay process, the mercurous chloride mixture is allowed to stand with occasional agitation until the reaction has gone to completion as indicated by its complete solution. The excess 0.1 *N* iodine added is then determined by residual titration. The formula of the double salt of mercuric potassium iodide may be written $HgI_2 \cdot 2KI$, and the ionic reaction may be represented as

$$2Hg^+ + I_2 \longrightarrow 2Hg^{++} + 2I^-$$

Since each molecule of mercurous chloride loses 1 electron on oxidation, its equivalent weight is 236.04 Gm, 1 gram-molecular weight. Each milliliter of 0.1 *N* iodine reduced by the mercurous chloride in the sample is equivalent to $0.1 \times 0.23604 = 0.02360$ Gm or 0.1 mEq of HgCl.

Questions and Problems

1. Why is it impracticable to assay the following substances by direct titration with standard iodine solution: (*a*) mild mercurous chloride; (*b*) sodium bisulfite?

2. Give equations for the reaction of iodine with $NaHSO_3$ and Na_2SO_3, calculate the weight of each of these substances that is equivalent to 1 ml of 0.1 *N* iodine, and compare the values with those given in the official assay procedures.

3. A 0.6500-Gm sample of mild mercurous chloride was treated with 50.00 ml of 0.0145 *N* iodine and excess potassium iodide. The resulting solution was titrated with 26.24 ml of 0.0985 *N* sodium thiosulfate. Calculate the percent HgCl in the sample.

4. Acetone may be assayed by treatment with excess standard iodine solution in presence of a base and titration of the excess iodine with standard sodium thiosulfate. Give equations, calculate the equivalent weight of acetone, and prepare an outline of a procedure specifying exact quantities. Compare the latter with the procedure given in N.F. VIII.

The chemical reactions for the assays of **Hydrazine Sulfate** reagent and **Isoniazide** and its preparations are similar if one considers the isonicotinic acid hydrazide as a substituted hydrazine.

$$\text{(pyridine ring, N)}\text{—}C(=O)\text{—}NH\text{—}NH_2 + H_2O \longrightarrow \text{(pyridine ring, N)}\text{—}C(=O)\text{—}OH + H_2N\text{—}NH_2$$

$$H_2N\text{—}NH_2 + 2I_2 + 4NaHCO_3 \longrightarrow N_2\uparrow + 4NaI + 4H_2O + 4CO_2\uparrow$$

The assay for **Methenamine Mandelate** involves the hydrolysis of the methenamine with hydrochloric acid, producing formaldehyde, which in turn is allowed to reduce an equivalent amount of mercuric ion to mercury. The mercury so produced is then oxidized with standard iodine solution, and the excess is determined by sodium thiosulfate. 1 mole compound = 6HCHO = 6Hg = 12I°; therefore 1 mole of methenamine mandelate contains 12 equivalents.

Potassium Xanthogenate is oxidized by iodine in the following manner:

$$2C_2H_5OC(S)SK + I_2 \longrightarrow 2KI + [C_2H_5OC(S)S]_2$$

TABLE 27. SUBSTANCES ASSAYED BY RESIDUAL TITRATION OF EXCESS IODINE WITH SODIUM THIOSULFATE

Substance	mEq, Gm	Official requirement, percent
U.S.P.		
Hydrazine sulfate (R)	0.03253	$(NH_4)_2{\cdot}H_2SO_4$ = 99
Isoniazid	0.03429	$C_6H_7N_3O$ = 98 to 101
Isoniazid injection	0.03429	$C_6H_7N_3O$ = 95 to 110 l.a.
Isoniazid tablets	0.03429	$C_6H_7N_3O$ = 93 to 107 l.a.
Methenamine mandelate	0.02436	$C_6H_{12}N_4{\cdot}C_8H_8O_3$ = 96
Methenamine mandelate tablets	0.02436	$C_6H_{12}N_4{\cdot}C_8H_8O_3$ = 95 to 105 l.a.
Potassium xanthogenate (R)	0.16030	C_2H_5OSSK = 87
Sodium bisulfite	0.03203	SO_2 = 58.5 to 67.4
Sodium metabisulfite (R)	0.04753	$Na_2S_2O_5$ = 95
Sodium sulfite (R)	0.06302	Na_2SO_3 = 98
Sulfurous acid (R)	0.03203	SO_2 = 6
N.F.		
Calomel	0.23604	HgCl = 99.6
Calomel tablets	0.23604	HgCl = 92.5 to 107.5 l.a.
Methionine	0.07461	$C_5H_{11}NO_2S$ = 99
Methionine capsules	0.07461	$C_5H_{11}NO_2S$ = 90 to 110 l.a.
Methionine tablets	0.07461	$C_5H_{11}NO_2S$ = 90 to 110 l.a.

(R) = reagent; l.a. = labeled amount.

Sodium Metabisulfite may be considered as being $Na_2O \cdot 2SO_2$, and its oxidation with iodine proceeds as in the case of sodium bisulfite, sodium sulfite, and sulfurous acid in which the sulfur in each compound gives up 2 electrons and is oxidized from tetravalent to hexavalent sulfur.

LIBERATION OF IODINE FROM POTASSIUM IODIDE AND TITRATION WITH SODIUM THIOSULFATE

Substances that are oxidizing agents which cannot be determined by direct titration with a standard solution of a reducing agent are assayed indirectly. In this indirect method an equivalent amount of iodine produced by the oxidation of the iodide ion in an acid medium by the substance is determined by a standard solution of a reducing agent such as sodium thiosulfate.

Exercise 59

Object. Assay of Selenium Sulfide.

Materials Required. 100 mg of Selenium Sulfide
25 ml of fuming nitric acid
10 Gm of urea
10 ml of potassium iodide solution (1 in 10)
0.05 *N* sodium thiosulfate
Starch T.S.

Procedure. "Place about 100 mg. of Selenium Sulfide, accurately weighed, in a suitable container, add 25 ml. of fuming nitric acid, and digest over gentle heat until no further solution occurs. Cool, transfer the solution to a 250-ml. volumetric flask containing 100 ml. of water, cool again, and dilute to volume. Pipet 50 ml. of the solution into a small flask, add 25 ml. of water and 10 Gm. urea and heat to boiling. Cool, add 10 ml. of potassium iodide solution (1 in 10), and immediately titrate with 0.05 *N* sodium thiosulfate, adding starch T.S. as the end point is neared. Perform a blank titration, and make any necessary correction. Each milliliter of 0.05 *N* sodium thiosulfate is equivalent to 0.9870 mg. of Se."

Selenium sulfide is practically insoluble in water and organic solvents but is soluble in fuming nitric acid, which oxidizes the compound to selenious acid and sulfuric acid, as shown in the reaction:

$$SeS_2 + 16HNO_3 \longrightarrow H_2SeO_3 + 2H_2SO_4 + 16NO_2\uparrow + 5H_2O$$

After cooling and diluting to 250 ml, a 50-ml aliquot or one-fifth of the original weighed sample is treated with excess urea to form the salt, urea nitrate, with the nitric acid and to destroy the nitrous acid as shown:

$$CO(NH_2)_2 + HNO_3 \longrightarrow CO(NH_2)_2 \cdot HNO_3$$
$$CO(NH_2)_2 + 2HNO_2 \longrightarrow CO_2\uparrow + 2N_2\uparrow + 3H_2O$$

The selenious acid formed is an oxidizing agent which will liberate iodine from potassium iodide, and the amount of sodium thiosulfate consumed by the iodine in the titration is equivalent to the selenium in the sample.

$$H_2SeO_3 + 4KI + 4HNO_3 \longrightarrow 2I_2 + Se\downarrow + 3H_2O + 4KNO_3$$
$$I_2 + 2Na_2S_2O_3 \longrightarrow 2NaI + Na_2S_4O_6$$

Blank determination is run to account for any iodine liberated from the potassium iodide by the reagents used in the analysis, which in turn would consume some sodium thiosulfate.

Calculations. One mole of selenious acid liberates 2 moles or 4 equivalents of iodine, so that 1 mole of selenious acid or selenium contains 4 equivalents. Therefore, 1 equivalent weight is Se/4, 1 mEq of selenium is 78.96/4,000 or 0.01974 Gm, and 1 ml of 0.05 *N* sodium thiosulfate would be equivalent to $\frac{1}{20}$ of 1 equivalent or 0.00987 Gm. The titer value given in the U.S.P. takes into account the fact that only a 50-ml aliquot of the 250 ml of solution (one-tenth) is actually taken for analysis.

$$\frac{\text{ml} \times N \times (78.96/4{,}000) \times 100}{\text{Sample weight}/5} = \text{percent Se}$$

Questions and Problems

1. Calculate the theoretical selenium content of selenium sulfide. How does this value compare to the limits set forth in the U.S.P.?

2. In the oxidation of the selenium sulfide with the fuming nitric acid, how many electrons are given up by the selenium sulfide? Why must the acidity in this analysis be controlled? Explain.

Exercise 60

Object. Assay of Sodium Hypochlorite Solution.

Materials Required. 3 ml of Sodium Hypochlorite Solution
2 Gm of potassium iodide
10 ml of acetic acid
0.1 *N* sodium thiosulfate
Starch T.S.

Procedure. "Weigh accurately, in a glass-stoppered flask, about 3 ml. of Sodium Hypochlorite Solution, and dilute it with 50 ml. of water. Add 2 Gm. of potassium iodide, 10 ml. of acetic acid, and titrate the liberated iodine with 0.1 *N* sodium thiosulfate, using starch T.S. as the indicator. Each milliliter of 0.1 *N* sodium thiosulfate is equivalent to 3.722 mg. NaClO."

Hypochlorous acid which is formed from sodium hypochlorite reacts with the potassium iodide in the presence of acetic acid to liberate iodine.

$$HOCl + 2KI + HOAc \longrightarrow I_2 + KCl + KOAc + H_2O$$

In the reaction, Cl^+ is reduced to Cl^- as it takes up 2 electrons. Since 1 gram-atomic weight of chlorine is equivalent to 1 mole of hypochlorous acid and to 1 mole or 2 equivalents of iodine, the equivalent weight of NaClO is one-half its molecular weight.

Calculation

$$\frac{\text{ml} \times N \times (\text{NaClO}/2{,}000) \times 100}{\text{Sample weight}} = \text{percent NaClO}$$

In the stoichiometry involved in the assay of Halazone, each Cl^+ is converted to Cl^-; as a result each mole of Halazone produces 2 moles or 4 equivalents of iodine, and the equivalent weight of Halazone is 67.52 Gm or ¼ gram-molecular weight.

$$HOOC-C_6H_4-SO_2-NCl_2 + 2H^+ + 4e \longrightarrow HOOC-C_6H_4-SO_2NH_2 + 2Cl^-$$

Questions and Problems

1. By means of half-equations show the electronic transfer in the assay of chlorinated lime.

2. Calculate the equivalent weight, 0.1 N sodium thiosulfate titer, and the weight of substance that must be in the sample used in the assay procedure, for each of the following, and compare with the corresponding values appearing in the official procedures: (1) Halazone tablets ($C_7H_5Cl_2NO_4S$); (2) sodium hypochlorite (NaClO); (3) chlorinated lime ($CaOCl_2$).

3. A 4.0050-Gm sample of chlorinated lime was mixed with enough water to make 1,000 ml. A 100-ml sample of the mixture was treated with potassium iodide and acetic acid, and titrated with 22.61 ml of sodium thiosulfate solution. A 20-ml sample of the sodium thiosulfate solution was found to be equivalent to 0.3680 Gm of pure iodine. Calculate the "available" chlorine content of the chlorinated lime.

Exercise 61

Object. Assay of Cupric Sulfate.

Materials Required. 1 Gm of cupric sulfate
4 ml of acetic acid
3 Gm of potassium iodide
0.1 N sodium thiosulfate
Starch T.S.

Procedure. "Weigh accurately about 1 Gm. of Cupric Sulfate, dissolve it in 50 ml. of water, add 4 ml. of acetic acid and 3 Gm. of potassium iodide, and titrate the liberated iodine with 0.1 N sodium thiosulfate, using starch T.S. as the indicator. Perform a blank determination, and make any necessary correction. Each milliliter of 0.1 N sodium thiosulfate is equivalent to 24.97 mg. of $CuSO_4 \cdot 5H_2O$."

This assay is based on the reaction between cupric sulfate and potassium iodide in which the copper is precipitated as cream-colored cuprous iodide and one atom of iodine is liberated for each cupric ion present:

$$2Cu^{++} + 4I^- \longrightarrow 2Cu^+ + 2I^- + I_2$$

or

$$\underset{2(249.68)}{2CuSO_4 \cdot 5H_2O} + 4KI \longrightarrow 2CuI\downarrow + I_2 + 2K_2SO_4 + 10H_2O$$

The equation for the oxidation of potassium iodide with cupric sulfate in the presence of acetic acid shows that the equivalent weight of $CuSO_4 \cdot 5H_2O$ is 249.68 Gm, 1 gram-molecular weight.

Each milliliter of 0.1 *N* sodium thiosulfate is equivalent to $0.1 \times 0.24968 = 0.02497$ Gm or 0.1 mEq of $CuSO_4 \cdot 5H_2O$.

The official cupric sulfate contains $CuSO_4$ corresponding to not less than 98.5 percent and not more than 104.5 percent of $CuSO_4 \cdot 5H_2O$. Calculate the $CuSO_4 \cdot 5H_2O$ content of the sample assayed.

Questions and Problems

1. Suggest a method for standardization of 0.1 *N* sodium thiosulfate using pure copper, and specify the weight of sample employed so that approximately 35 ml of the thiosulfate solution will be used in titration.
2. What volume of 0.1056 *N* sodium thiosulfate would be required in titration of the iodine equivalent to 0.2 Gm of $CuSO_4$?

Drocarbil N.F., containing acetarsone, an organoarsenical, can be converted quantitatively by oxidative degradation, using potassium permanganate and sulfuric acid, to a water-soluble inorganic arsenate. The excess permanganate is reduced with hydrogen peroxide, and the mixture is heated to evaporate any of the excess remaining after all the permanganate has been reduced. By careful treatment of the mixture at this stage, first with 0.1 *N* permanganate and then with 0.1 *N* oxalic acid, a solution containing all the arsenic of the sample in the pentavalent form, H_3AsO_4, is obtained. The arsenic is then measured iodometrically.

$$H_3AsO_4 + 2KI + 2H^+ \longrightarrow H_3AsO_3 + I_2 + H_2O + 2K^+$$

Since 1 mole, 2 equivalents, of arsenic acid is formed for each As atom in the organoarsenical molecule, the equivalent weight of acetarsone, $C_8H_{10}AsNO_5$ is 137.54 Gm, ½ gram-molecular weight, and the equivalent weight of organically bound arsenic is 37.46 Gm, ½ gram-atomic weight.

A residual type of iodometric procedure may be used for substances which react quantitatively with such oxidizing agents as potassium ferricyanide, potassium dichromate, or potassium periodate. The sample is treated with a carefully measured excess of a standard solution of the oxidizing agent, and the excess oxidizing agent is subsequently determined iodometrically. For example, **Lead Monoxide,** U.S.P. reagent, is converted to lead acetate which reacts with potassium dichromate in the presence of acetic acid as follows:

$$PbO + 2HC_2H_3O_2 \longrightarrow Pb(C_2H_3O_2)_2 + H_2O$$
$$2Pb(C_2H_3O_2)_2 + K_2Cr_2O_7 + H_2O \longrightarrow 2PbCrO_4\downarrow + 2KC_2H_3O_2 + 2HC_2H_3O_2$$

It should be noted that lead acetate reacts with potassium dichromate not by reducing it but by a reaction which goes to completion because of formation of the insoluble lead chromate. The excess potassium dichromate is measured iodometrically. Each milliliter of 0.1 *N* potassium dichromate contains 0.1 mEq or 0.1/6 = 0.0166 mM of $K_2Cr_2O_7$ and is equivalent to 2 × 0.0166 = 0.0333 mM or 0.0333 × 0.22319 = 0.00744 Gm of PbO.

Quinacrine Hydrochloride assay also involves a precipitation with potassium dichromate:

$$\text{Quinacrine}\cdot 2HCl + K_2Cr_2O_7 \longrightarrow \text{quinacrine}\cdot H_2Cr_2O_7\downarrow + 2KCl$$

The reaction is at a mole for mole basis, and an aliquot of the excess of potassium dichromate is determined with potassium iodide and sodium thiosulfate. Care must be exercised in this analysis in that the equipment must be dry; the first run of filtrate is discarded so as to make certain that the subsequent 50 ml is of uniform concentration. The precipitation is best run in a buffered acid medium to ensure a quantitative reaction with the potassium dichromate. Each milliliter of 0.1 *N* potassium dichromate contains 0.1 mEq or 0.1/6 = 0.0166 mM of $K_2Cr_2O_7$ and is equivalent to 0.0166 mM or 0.0166 × 508.92 or 0.008482 Gm.

Sugar alcohols such as mannitol selectively reduce an acidified solution of potassium periodate. **Mannitol** and mannitol injection are assayed by treatment with excess potassium periodate reagent, followed by iodometric measurement of the excess oxidizing agent.

$$C_6H_{14}O_6 + 5HIO_4 \longrightarrow 2HCHO + 4HCOOH + 5HIO_3 + H_2O$$

When potassium iodide is added, the excess iodate and periodate are determined by measuring the amount of liberated iodine. The blank determination will liberate more iodine and consequently consume more sodium thiosulfate. The difference in volumes of sodium thiosulfate consumed is a measure of the mannitol in the sample. The chemical reaction in the blank determination proceeds as follows:

$$HIO_4 + 7HI \longrightarrow 8I_2 + 8H_2O$$

Exercise 62

Object. Assay of Thyroid.

Materials Required. 1 Gm of thyroid
17 Gm of anhydrous potassium carbonate

7 ml of freshly prepared bromine T.S. (prepared by agitating 2 or 3 ml of bromine with 100 ml of cold water in a glass-stoppered flask)
5 ml of a solution of Phenol (1 in 20)
50 ml of diluted phosphoric acid (1 in 2)
5 ml of potassium iodide T.S.
0.01 *N* sodium thiosulfate
Starch T.S.

Procedure. 1. "Weigh accurately about 1 Gm. of Thyroid in a porcelain crucible, add 7 Gm. of anhydrous potassium carbonate, mix carefully, and gently tap the crucible several times to compact the mixture. Overlay with an additional 10 Gm. of potassium carbonate, and again compact the material thoroughly by tapping. Ignite the mixture for 25 minutes at 675° to 700° in a muffle furnace preheated to that temperature."

The active principles of dried thyroid gland are iodine-substituted organic compounds such as thyroxine and diiodotyrosine. When thyroid is subjected to pyrolysis in the presence of potassium carbonate, potassium iodide equivalent to the iodine present in the thyroid sample results. The potassium carbonate employed must be anhydrous; otherwise decrepitation and loss of material would result. The temperature of the muffle furnace and the duration of heating must be carefully regulated, since sodium iodide is appreciably volatile at high temperatures and some may be lost on prolonged heating.

2. "Cool, add 20 ml. of water, or more if necessary, heat gently to boiling, and decant through a filter into a conical flask of suitable size. Repeat the extraction by boiling with 20 ml. of water, then wash the crucible and the char on the filter with hot water until the filtrate measures approximately 200 ml. Add 7 ml. of freshly prepared bromine T.S., then slowly add 40 ml. of dilute phosphoric acid (1 in 2), and boil until starch iodide paper is no longer colored blue by the vapors. During the boiling add water from time to time, as necessary, to maintain a volume of at least 200 ml. Wash down the walls of the flask with water, and continue the boiling for 5 minutes. Cool, add 5 ml. of a solution of phenol (1 in 20), again rinse the walls of the flask and allow to stand for 5 minutes."

After carbonization is complete and the crucible has cooled, the residue is treated with water, which dissolves the KI, carbonates, and other soluble compounds. The filtered solution is treated with bromine T.S. in the presence of phosphoric acid to oxidize the iodide to iodate:

$$I^- + 3Br_2 + 3H_2O \longrightarrow IO_3^- + 6HBr$$

Most of the excess bromine is removed by heating the solution until the vapors give a negative test with starch iodide paper. The solution must be acidic at this stage; otherwise bromine will be fixed in the form of potassium hypobromite, KBrO. The last traces of bromine are reduced by treatment of the solution with phenol, forming bromophenols such as tribromophenol.

$$C_6H_5OH + 3Br_2 \longrightarrow C_6H_2Br_3OH + 3HBr$$

3. "Add 2 ml. of dilute phosphoric acid (1 in 2) and 5 ml. of potassium iodide T.S., and titrate immediately with 0.01 *N* sodium thiosulfate, adding 3 ml. of starch T.S. as the end point is neared. Perform a blank determination with the same quantities of the same reagents and in the same manner, and make any necessary correction. Each milliliter of 0.01 *N* sodium thiosulfate is equivalent to 0.2115 mg. of iodine (I)."

In the presence of a weak acid IO_3^- oxidizes potassium iodide to an equivalent amount of iodine.

$$IO_3^- + 5I^- + 6H^+ \longrightarrow 3I_2 + 3H_2O$$

Care must be taken to remove the excess bromine in the previous step; otherwise iodide would be oxidized by the bromine and high assay values would result. The iodine formed in the above reaction is titrated with 0.01 *N* sodium thiosulfate in the usual way.

Each gram-atomic weight of iodine in thyroid is converted to 1 mole of iodate and subsequently to 3 moles, 6 equivalents, of iodine. Consequently the equivalent weight of the iodine in the thyroid is 21.15 Gm, ⅙ gram-atomic weight, and each milliliter of 0.01 *N* sodium thiosulfate is equivalent to 0.01 × 0.02115 or 0.0002115 Gm of iodine.

The above procedure is also applied with some modifications in the assay of Thyroid Tablets, Iodized Oil, Diiodohydroxyquin, Diiodohydroxyquin Tablets, Iophendylate Injection, Sodium Levothyroxine, Sodium Levothyroxine Tablets, Sodium Lyothyronine, and Sodium Lyothyronine Tablets.

Questions and Problems

1. Why is phosphoric acid used in the oxidation of iodide to iodate and the subsequent oxidation of iodide by the iodate? What reaction would have taken place if a stronger acid, such as hydrochloric, were used?

2. What purpose is served by starch iodide paper and phenol solution in the assay of thyroid?

3. Calculate the normality of a sodium thiosulfate solution, 48.00 ml of which was required for titration of the iodine formed from KI and 0.3000 Gm of pure potassium iodate.

4. Consult the N.F. concerning the procedure used for assay of Iodized Oil. Outline the steps in the procedure, giving the purpose for each step and each reagent employed.

A number of chemical reagents and colorimetric solutions (C.S.) are assayed with the following reactions taking place:

Ferric Chloride C.S., ferric citrate, and ferric sulfate:

$$Fe^{3+} + I^- \rightleftharpoons Fe^{++} + I^0$$

Cobaltous Chloride C.S. is first oxidized with hydrogen peroxide to the cobaltic state and then determined iodometrically with potassium iodide:

$$Co^{3+} + I^- \rightleftharpoons Co^{++} + I^0$$

TABLE 28. SUBSTANCES ASSAYED BY TITRATING THE LIBERATED IODINE FROM POTASSIUM IODIDE WITH SODIUM THIOSULFATE

Substance	mEq, Gm	Official requirement, percent
U.S.P.		
Arsanilic acid (R)	0.10850	$C_6H_8AsNO_3$ = 98.5 to 100.8
Diiodohydroxyquin	0.03308	$C_9H_5I_2NO$ = 96
Diiodohydroxyquin tablets	0.03308	$C_9H_5I_2NO$ = 95 to 105 l.a.
Ferric citrate (R)	0.05585	Fe = 16.5 to 118.5
Iodine pentoxide (R)	0.02782	I_2O_5 = 98.5
Iophendylate injection	0.06939	$C_{19}H_{29}IO_2$ = 98
Potassium bromate	0.02783	$KBrO_3$ = 99.8
Potassium periodate (R)	0.02875	KIO_4 = 99.8 to 100.3
Propyliodone	0.03725	$C_{10}H_{11}I_2NO_3$ = 99 to 101
Propyliodone oil suspension, sterile	0.03725	$C_{10}H_{11}I_2NO_3$ = 95 to 105 l.a.
Selenium sulfide	0.01974	Se = 52 to 55.5
Selenium sulfide detergent suspension	0.03577	SeS_2 = 2.25 to 2.75
Sodium chromate (R)	0.07801	$Na_2CrO_4{\cdot}4H_2O$ = 99
Sodium hypochlorite T.S. (R)	0.03723	NaOCl = 4
Sodium levothyroxine	0.03329	$C_{15}H_{10}I_4NNaO_4$ = 97 to 103
Sodium levothyroxine tablets	0.03329	$C_{15}H_{10}I_4NNaO_4$ = 90 to 115 l.a.
Sodium liothyronine	0.03739	$C_{15}H_{11}I_3NNaO_4$ = 95 to 101
Sodium liothyronine tablets	0.03739	$C_{15}H_{11}I_3NNaO_4$ = 90 to 110 l.a.
Thyroid	0.02115	I = 0.17 to 0.23
Thyroid tablets	0.02115	I = 0.17 to 0.23 l.a.
N.F.		
Chlorinated lime (R)	0.03545	Cl = n.l.t. 30 available Cl
Cupric sulfate	0.24968	$CuSO_4{\cdot}5H_2O$ = 98.5 to 104.5
Drocarbil	0.13755	$C_8H_{10}AsNO_5$ = 64 to 67
Ferric sulfate (R)	0.05585	Fe = 21 to 23
Halazone	0.06752	$C_7H_5Cl_2NO_4S$ = 91.5
Halazone tablets	0.06752	$C_7H_5Cl_2NO_4S$ = 90 to 135 l.a.
Iodized oil	0.02115	I = 38 to 42
Mannitol	0.01822	$C_6H_{14}O_6$ = 98 to 102
Mannitol injection	0.01822	$C_6H_{14}O_6$ = 95 to 105 l.a.
Povidone-iodine	0.12690	I = 9 to 12
Povidone-iodine solution	0.12690	I = 85 to 120 l.a.
Sodium hypochlorite solution	0.03722	NaClO = 4 to 6
Sodium hypochlorite solution, diluted	0.03722	NaClO = 0.45 to 0.50 *W/V*
Sodium hypochlorite T.S.	0.03723	NaClO = 4
p-Sodium periodate (R)	0.03674	$Na_3H_2IO_6$ = 98.5

n.l.t. = not less than; l.a. = labeled amount; (R) = reagent.

Cupric Sulfate C.S.:

$$Cu^{++} + I^- \rightleftharpoons Cu^+ + I^0$$

Iodine Pentoxide:

$$I_2O_5 + H_2O \longrightarrow 2HIO_3$$

$$IO_3^- + 5I^- + 6H^+ \rightleftharpoons 3I_2 + 3H_2O$$

Potassium Periodate:

$$KIO_4 + 8KI + 8HCl \longrightarrow 4I_2 + KI + 8KCl + 4H_2O$$

Sodium Chromate:

$$CrO_4^{=} + 3I^- + 8H^+ \longrightarrow Cr^{3+} + 3I^0 + 4H_2O$$

Sodium Periodate:

$$Na_3H_2IO_6 + 8KI + 10HCl \longrightarrow 4I_2 + NaI + 8KCl + 2NaCl + 6H_2O$$

OXIDATION-REDUCTION METHODS WITH 0.1 *N* BROMINE

In the assay of compounds such as aniline, phenol, and resorcinol, bromine is employed as an oxidizing agent in place of iodine, since it is reduced by the readily oxidized organic substances in a reaction that proceeds quantitatively with the production of water-insoluble bromine substitution products such as tribromoaniline, tribromophenol, etc. The standard solution employed does not contain bromine as such but rather an equivalent amount of potassium bromate and excess potassium bromide. Bromine is liberated when the solution is acidified.

Exercise 63

Object. To Prepare and Standardize 0.1 *N* Bromine, Koppeschaar's Solution.

Materials Required. 3 Gm of potassium bromate
15 Gm of potassium bromide
5 ml of hydrochloric acid
5 ml of potassium iodide T.S. (16.5 Gm of KI in 100 ml)
0.1 *N* sodium thiosulfate
Starch T.S.

Procedure. Dissolve 3 Gm of potassium bromate and 15 Gm of potassium bromide in sufficient water to make 1,000 ml and standardize the solution as follows:

Pipet or otherwise measure accurately about 25 ml of the solution into a 500-ml iodine flask, and dilute with 120 ml of water. Add 5 ml of hydrochloric acid, stopper the flask, and shake it gently. Then add 5 ml of potassium iodide T.S., restopper, shake the mixture, allow it to stand for 5 min, and titrate the liberated iodine with 0.1 *N* sodium thiosulfate, adding starch T.S. as the indicator. Calculate the normality.

When the solution of potassium bromide and potassium bromate is acidified with hydrochloric acid, bromine is set free:

$$5KBr + KBrO_3 + 6HCl \longrightarrow 6KCl + 3Br_2 + 3H_2O$$

and the liberated bromine oxidizes iodide to an equivalent amount of iodine:

$$2KI + Br_2 \longrightarrow 2KBr + I_2$$

The flask should be tightly stoppered after the addition of the acid to prevent the escape of bromine vapors.

The free iodine is titrated with 0.1 *N* sodium thiosulfate to standardize the solution:

$$I_2 + 2Na_2S_2O_3 \longrightarrow 2NaI + Na_2S_4O_6$$

From the volume of sodium thiosulfate solution consumed and its normality the normality of the bromine solution may be calculated in the usual way.

The equations show that ½ mole of bromine, I gram-atomic weight, is 1 equivalent, and ⅙ mole or 27.835 Gm of potassium bromate, when treated with potassium bromide in the presence of hydrochloric acid, is required to produce 1 equivalent of bromine. To make 1,000 ml of 0.1 *N* bromine, $0.1 \times 27.835 = 2.7835$ Gm, 0.1 equivalent of $KBrO_3$ is required. However, 3 Gm of potassium bromate is actually used, and the resulting solution should be slightly more concentrated than 0.1 *N*. After standardization, the solution may be diluted as necessary, to produce a 0.1 *N* solution. Restandardization after diluting is required.

Exercise 64

Object. Assay of Phenol.

Materials Required. 2 Gm of Phenol
30 ml of 0.1 *N* bromine
5 ml of hydrochloric acid
5 ml of potassium iodide solution (1 in 5)
1 ml of chloroform
0.1 *N* sodium thiosulfate
Starch T.S.

Procedure. 1. "Place about 2 Gm. of Phenol, accurately weighed, in a 1000-ml. volumetric flask, and dilute to volume. Pipet 20 ml. of the solution into an iodine flask, add 30.0 ml. of 0.1 *N* bromine, then 5 ml. of hydrochloric acid, and immediately insert the stopper. Shake the flask repeatedly during 30 minutes, allow it to stand for 15 minutes, add quickly 5 ml. of potassium iodide solution (1 in 5), taking precautions against the escape of bromine vapor, and at once stopper the flask."

When the HCl is added bromine is set free:

$$5KBr + KBrO_3 + 6HCl \longrightarrow 6KCl + 3Br_2 + 3H_2O$$

The bromine reacts with phenol to form a white crystalline precipitate of tribromophenol and hydrobromic acid:

$$\underset{94.11}{C_6H_5OH} + 3Br_2 \longrightarrow C_6H_2Br_3OH + 3HBr$$

The flask should be tightly stoppered to prevent the escape of the volatile bromine and shaken gently during 30 min to mix the contents thoroughly and allow the oxidation of phenol to tribromophenol to go to completion.

2. "Shake thoroughly, remove the stopper, and rinse it and the neck of the flask with a little water, so that the washings flow into the flask. Add 1 ml. of chloroform, shake the mixture well, and titrate the liberated iodine with 0.1 *N* sodium thiosulfate, adding starch T.S. as the end point is neared. Perform a blank determination with the same quantities of the same reagents and in the same manner. Each milliliter of 0.1 *N* bromine is equivalent to 1.569 mg. of C_6H_6O."

After the addition of potassium iodide, 1 ml of chloroform is added to dissolve the precipitated tribromophenol which would otherwise interfere with the clear observation of the end point, particularly in the assay of old, colored phenol. The excess bromine liberates iodine in accordance with the following equation:

$$2KI + Br_2 \longrightarrow 2KBr + I_2$$

The free iodine reacts with sodium thiosulfate:

$$I_2 + 2Na_2S_2O_3 \longrightarrow 2NaI + Na_2S_4O_6$$

Since 1 mole of phenol reacts with 3 moles, 6 equivalents, of bromine under the conditions of this assay, the equivalent weight of phenol is 15.685 Gm, ⅙ gram-molecular weight. Therefore, each milliliter of 0.1 *N* bromine consumed in the reaction with phenol is equivalent to $0.1 \times 0.015685 = 0.001569$ Gm or 0.1 mEq of phenol (C_6H_6O).

A number of official phenolic substances are assayed in this manner, e.g., hexylresorcinol, phenol, resorcinol, parachlorophenol, and preparations containing these phenols as well as salicylic acid (Table 29).

The reaction of salicylic acid with acidified 0.1 *N* bromine is somewhat unusual, since decarboxylation and formation of tribromophenol occur.

$$C_7H_6O_3 + 3Br_2 \longrightarrow C_6H_3Br_3O + 3HBr + CO_2$$

The equivalent weight of salicylic acid is ⅙ gram-molecular weight.

Questions and Problems

1. In order to make 1,000 ml of 0.1 *N* bromine, what weight of potassium bromide is theoretically required? How much is actually used?

2. What substance determines the normality of a standard bromine solution?

3. When assayed using 0.1 *N* bromine, the products obtained from each of the following substances are given in parentheses. Give equations for the reactions involved, and calculate the 0.1 *N* bromine titer value for each: (*a*) hexylresorcinol ($C_{12}H_{16}Br_2O_2$); (*b*) salicylic acid plaster ($C_7H_6O_3$); (*c*) *p*-chlorophenol ($C_6H_3Br_2ClO$).

4. How much $KBrO_3$ must be present in a solution which, when treated with potassium bromide and hydrochloric acid, will liberate bromine equivalent to 45 ml of 0.0915 *N* sodium thiosulfate? If the volume of the $KBrO_3$ solution were 35 ml, what would be its normality?

5. A 1.4500-Gm sample of liquefied phenol was dissolved in enough water to make 1,000 ml. A 30-ml sample of the solution was treated with 30 ml of Koppeschaar's solution and hydrochloric acid. The mixture was treated with potassium iodide and titrated with 7.38 ml of 0.1 *N* sodium thiosulfate. It was also found that 21 ml of 0.1 *N* sodium thiosulfate was required in titration of the iodine liberated when 20 ml of the Koppeschaar's solution was treated with potassium iodide and hydrochloric acid. Calculate the percent of phenol in the original sample.

Exercise 65

Object. Assay of Calcium Cyclobarbital.

Materials Required. 500 mg of Calcium Cyclobarbital
5 ml of glacial acetic acid
50 ml of 0.1 *N* bromine
10 ml of potassium iodide solution (1 in 10)
0.1 *N* sodium thiosulfate
Starch T.S.

Procedure. 1. "Transfer about 500 mg. of Calcium Cyclobarbital, accurately weighed, to a 250-ml. glass-stoppered flask, add 5 ml. of water and 5 ml. of glacial acetic acid, and swirl until dissolved, warming if necessary. Cool to room temperature, and add 50.0 ml. of 0.1 *N* bromine, followed by 10 ml. of hydrochloric acid. Stopper immediately, mix, and cool in an ice bath for 15 minutes, shaking occasionally."

Upon addition of the hydrochloric acid, active bromine is liberated from the standard solution of bromine, which in turn attacks the double bond in the cyclohexane ring, forming the dibromo derivative. The flask must be stoppered immediately to prevent any bromine vapors from escaping into the atmosphere, since all excess bromine must be accounted for in the residual titration. The mixture is cooled in an ice bath to minimize further any loss of bromine vapors. The following reaction occurs in this analysis:

$$KBrO_3 + 5KBr + 6HCl \longrightarrow 6Br^0 + 6KCl + 3H_2O$$

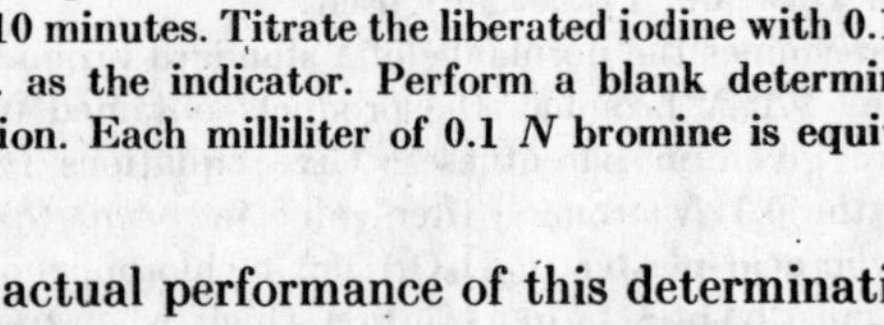

$$2\,(\text{cyclohexene ring with R and H on the double bond}) + 2Br_2 \longrightarrow 2\,(\text{cyclohexane ring with R, Br and Br, H})$$

2. "Add 10 ml. of potassium iodide solution (1 in 10), mix, and allow to stand in ice for an additional 10 minutes. Titrate the liberated iodine with 0.1 *N* sodium thiosulfate, using starch T.S. as the indicator. Perform a blank determination, and make any necessary correction. Each milliliter of 0.1 *N* bromine is equivalent to 12.77 mg. of $C_{24}H_{30}CaN_4O_6$."

Since in the actual performance of this determination there is a possibility of some loss of bromine vapors as well as some loss of iodine vapors because of the action of excess bromine on potassium iodide, it is necessary to perform a blank determination so that these effects can be accounted for in the assay of the sample.

If these losses then do occur in the analysis and in the blank determination to the same extent, the net effect is one of canceling out this condition.

Calculations. The 50.0 ml of bromine used in the blank determination will consume a certain volume of sodium thiosulfate solution; with the sample determination, a certain amount of the bromine will react with the sample, thus leaving less bromine (excess) to react in the residual titration with the sodium thiosulfate. This difference in volumes of sodium thiosulfate consumed in the analysis is equivalent to the bromine that reacted with the cyclobarbital. Each mole of calcium cyclobarbital reacts with 4 atoms of bromine; therefore 1 mEq of calcium cyclobarbital is 510.61/4,000 or 0.12765 Gm.

$$\frac{\text{ml}_{\text{difference}} \times N \times (510.61/4{,}000) \times 100}{\text{Sample weight}} = \text{percent } C_{24}H_{30}CaN_4O_6$$

Sodium Hexobarbital will behave in a similar manner with standard bromine solution in that the bromine will add across the double bond of the cyclohexenyl group. Since there is only one cyclohexenyl group in the hexobarbital versus the two groups in the calcium cyclobarbital, it will react with only 1 mole of bromine, and consequently 1 mole of the sodium hexobarbital contains 2 equivalents.

TABLE 29. SUBSTANCES ASSAYED BY USING BROMINE SOLUTION AND DETERMINING THE EXCESS USING POTASSIUM IODIDE AND SODIUM THIOSULFATE

Substance	mEq, Gm	Official requirement, percent
U.S.P.		
Phenol	0.01569	C_6H_6O = 99
Phenol, liquefied	0.01569	C_6H_6O = 89
Phenylephrine hydrochloride	0.03395	$C_9H_{13}NO_2 \cdot HCl$ = 97.5 to 102.5
Phenylephrine hydrochloride injection	0.03395	$C_9H_{13}NO_2 \cdot HCl$ = 95 to 105 l.a.
Phenylephrine hydrochloride solution	0.03395	$C_9H_{13}NO_2 \cdot HCl$ = 95 to 105 l.a.
Resorcinol	0.01835	$C_6H_6O_2$ = 99
Salicylic acid plaster	0.02302	$C_7H_6O_3$ = 90 to 110 l.a.
N.F.		
Calcium cyclobarbital	0.12765	$C_{24}H_{30}CaN_4O_6$ = 97.5 to 102.5
Calcium cyclobarbital	0.12765	$C_{24}H_{30}CaN_4O_6$ = 94 to 106 l.a.
Hexylresorcinol	0.04857	$C_{12}H_{18}O_2$ = 98
Hexylresorcinol pills	0.04857	$C_{12}H_{18}O_2$ = 90 to 110 l.a.
Parachlorophenol	0.03214	C_6H_5ClO = 99
Sodium hexobarbital	0.12913	$C_{12}H_{15}N_2NaO_3$ = 98.5 to 101
Sodium hexobarbital, sterile	0.12913	$C_{12}H_{15}N_2NaO_3$ = 98.5 to 101

l.a. = labeled amount.

OXIDATION-REDUCTION METHODS WITH POTASSIUM IODATE

Potassium iodate may be used as the oxidizing agent in the assay of a number of substances, such as iodides, arsenites, and other reducing agents. The method depends upon the formation of iodine monochloride in strong hydrochloric acid solution.

Exercise 66

Object. To Prepare 200 ml of 0.05 *M* Potassium Iodate.

Materials Required. About 2.2 Gm of potassium iodate

Procedure. Dry 2.2 Gm of potassium iodate to constant weight at 110°. Dissolve 2.140 Gm of the dried salt in sufficient water to make exactly 200 ml at 25°. The resulting solution should be exactly 0.05 *M*.

The potassium iodate is dried to constant weight at 110° to ensure its freedom from moisture. Since potassium iodate is a very stable salt and can be obtained in a very pure condition, standard solutions of this reagent can be prepared by dissolving the calculated weight of the salt in water and diluting to the proper volume.

Standard iodate solutions of known molarity rather than those of known normality are usually employed, because the normality varies depending on the nature of the reaction.

The uniform reduction of potassium iodate to iodide, unlike the reduction of potassium bromate to bromide, is generally not feasible in a direct titrimetric procedure and has no application in official assay methods.

$$IO_3^- + 6e + 6H^+ \longrightarrow I^- + 3H_2O \qquad (1)$$

In such a reaction, 1 mole of KIO_3 is 6 equivalents and a 0.05 *M* solution would be 0.3 *N*.

If excess potassium iodate is used, any iodide formed [Eq. (1)] is oxidized to iodine, and the excess iodate and the iodine are determined iodometrically. The reduction of the iodate to iodine may be represented as follows:

$$2IO_3^- + 10e + 12H^+ \longrightarrow I_2 + 6H_2O \qquad (2)$$

In this reaction, 1 mole of iodate is 5 equivalents and a 0.05 *M* solution would be 0.25 *N*. This reaction of iodate is not employed in official assay procedures.

Under certain conditions, the iodide and iodine formed as reduction products of iodate [Eqs. (1) and (2)] are quantitatively converted to I^+. In official procedures, this conversion is accomplished by carrying out the titration in the presence of a rather high concentration of hydrochloric acid. The iodine initially formed by reduction of iodate [Eq. (2)][1] undergoes solvolysis in a polar solvent:

$$I_2 \rightleftharpoons I^+ + I^-$$

The iodine cation forms iodine monochloride when the HCl concentration

[1] When iodides are assayed using KIO_3, iodine is also formed by oxidation of the iodide.

is sufficiently great,[2] and this is further stabilized by complex ion formation:

$$I^+ + HCl \rightleftharpoons ICl + H^+$$
$$ICl + HCl \rightleftharpoons ICl_2^- + H^+$$

or, by summation of these two equations,

$$I^+ + 2HCl \rightleftharpoons ICl_2^- + 2H^+$$

Chloroform or carbon tetrachloride is added to make the end-point observation possible. In the early stages of the titration, iodine is produced and the chloroform layer becomes colored. When all the reducing agent being assayed has been oxidized, the iodate completes the oxidation of iodine and iodide to I^+ and color disappears from the chloroform layer.

If HCN formed from an equivalent amount of KCN in acid solution is used instead of HCl, iodine cyanide, which is much less dissociated than iodine chloride, results. The reaction may be carried out in the presence of not less than 1 *N* concentrations of HCl, and starch may be used as indicator. Other substances function in the same way as HCl or HCN in controlling the reduction of IO_3^- to I^+, e.g., Br^- and acetone.

The iodine monochloride procedure is applied in the analysis of a number of official substances. In these procedures, the reduction of KIO_3 may be represented as follows:

$$IO_3^- + 4e + 6H^+ \longrightarrow I^+ + 3H_2O$$

In this reaction, 1 mole of potassium iodate is 4 equivalents and a 0.05 *M* solution would be 0.2 *N*.

Exercise 67

Object. Assay of Potassium Iodide.

Materials Required. 0.5 Gm potassium iodide
35 ml of hydrochloric acid
5 ml of chloroform
0.05 *M* potassium iodate

Procedure. "Dissolve about 500 mg. of Potassium Iodide, accurately weighed, in about 10 ml. of water, add 35 ml. of hydrochloric acid and 5 ml. of chloroform. Titrate with 0.05 *M* potassium iodate until the purple color of iodine disappears from the chloroform. Add the last portions of the iodate solution dropwise, agitating vigorously and continuously. After the chloroform has been decolorized, allow the mixture to stand for 5 minutes. If the chloroform develops a purple color, titrate further with the iodate solution. Each milliliter of 0.05 *M* potassium iodate is equivalent to 16.60 mg. of KI."

A number of official iodides are assayed by use of standard potassium iodate, employing the iodine monochloride procedure discussed and illus-

[2] If the solution is saturated with potassium chloride, lower concentrations of hydrochloric acid may be used.

trated by the assay of potassium iodide. As the end point in the titrations is approached, vigorous shaking is necessary because of the heterogeneous nature of the mixture with iodine and iodate in different phases. The oxidation of potassium iodide by iodate may be represented as follows:

$$\underset{2(166.01)}{2KI} + KIO_3 + 6HCl \longrightarrow 3ICl + 3KCl + 3H_2O$$

As pointed out previously, 1 mole of KIO_3 in this reaction is 4 equivalents; hence a 0.05 *M* solution is 0.2 *N*. The equivalent weight of potassium iodide is 83.01 Gm, ½ gram-molecular weight. Each milliliter of 0.05 *M*, 0.2 *N*, potassium iodate is equivalent to $0.2 \times 0.08301 = 0.016602$ Gm or 0.2 mEq of KI.

Official substances assayed by this procedure (Table 30) include water-soluble inorganic iodides and organic iodine compounds.

In the assay of **Benzalkonium Chloride** using potassium iodate as the titrant, the sample is treated with an exact volume of a potassium iodide solution. The chloroform-soluble benzalkonium iodide complex is removed and the excess potassium iodide determined with the potassium iodate. A blank determination is run with water substituted for the sample. The difference in the two volumes in titration represents the amount of KIO_3 equivalent to the benzalkonium chloride in the sample. Each mole of KIO_3 is equivalent to 2 moles of KI, which in turn is equivalent to 2 moles of benzalkonium chloride. Therefore 1 ml of a $\frac{1}{20}$ *M* solution of KIO_3 is equivalent to $2 \times \frac{360}{20} \times 1{,}000$ or 0.036 Gm of benzalkonium chloride.

Hydralazine Hydrochloride is a substituted hydrazine and as such is capable of being oxidized by potassium iodate in the manner of hydrazine:

$$NH_2 - NH_2 + KIO_3 + 2HCl \longrightarrow ICl + N_2\uparrow + KCl + 3H_2O$$

Based on the mole per mole ratio of reaction, 1 ml of $\frac{1}{50}$ *M* solution of KIO_3 is equivalent to $196.64/50 \times 1{,}000$ or 0.003933 Gm of Hydralazine Hydrochloride.

The sulfite content expressed as percent sulfur in **Sodium Glucosulfone Injection, U.S.P.,** is determined by the reaction of the sulfite with iodine monochloride. The iodide so formed in the reaction is in turn determined quantitatively by titration with KIO_3, with the following reactions:

$$SO_3^{=} + ICl + H_2O \longrightarrow SO_4^{=} + HI + HCl$$
$$KIO_3 + 2HI + 4HCl \longrightarrow 3ICl + KCl + 3H_2O$$

Each mole of KIO_3 is equivalent to 2 moles of HI and 2 moles of $SO_3^{=}$. One mole of $SO_3^{=}$ is equivalent to 1 mole of S; therefore 1 ml of a 0.01667 *M* or $\frac{1}{60}$ *M* of KIO_3 is equivalent to $2 \times \frac{32}{60} \times 1{,}000$ or 0.001069 Gm of S.

TABLE 30. SUBSTANCES ASSAYED BY DIRECT TITRATION WITH POTASSIUM IODATE

Substance	*M* of titration	1 ml equivalent, mg	Official requirement, percent
U.S.P.			
Benzalkonium chloride	0.05	36.0	$[C_6H_5CH_2N(CH_3)_3R]Cl$ = 97 to 103
Benzalkonium chloride solution	0.05	36.0	$[C_6H_5CH_2N(CH_3)_3R]Cl$ = 95 to 105 l.a.*
Echothiophate iodide	0.01	22.99	$C_9H_{23}INO_3PS$ = 88
Echothiophate iodide for ophthalmic solution	0.002	4.60	$C_9H_{23}INO_3PS$ = 95 to 115 l.a.
Iodine solution, strong	0.05	16.60	KI = 9.5 to 10.5 *W/V*
Iodine tincture	0.05	14.99	NaI = 2.1 to 2.6 *W/V*
Mercuric iodide, red (R)	0.05	22.72	HgI_2 = 99
Potassium iodide	0.05	16.60	KI = 99 to 101.5
Sodium iodide	0.05	14.99	NaI = 99 to 101.5
N.F.			
Hydralazine hydrochloride	0.02	3.933	$C_8H_8N_4{\cdot}HCl$ = 98
Hydralazine hydrochloride injection	0.02	3.933	$C_8H_8N_4{\cdot}HCl$ = 95 to 105 l.a.
Hydralazine hydrochloride tablets	0.02	3.933	$C_8H_8N_4{\cdot}HCl$ = 95 to 105 l.a.
Iodine solution	0.05	14.99	NaI = 2.1 to 2.6 *W/V*
Iodine ampuls	0.05	14.99	NaI = 2.1 to 2.6 *W/V*
Potassium iodide solution	0.05	16.60	KI = 97 to 103 *W/V*

* If more than 1 percent; 93 to 107 percent of labeled amount if less than 1 percent.
(R) = reagent; l.a. = labeled amount.

OXIDATION-REDUCTION METHODS WITH POTASSIUM BROMATE

Potassium bromate, $KBrO_3$, is employed as a standard oxidizing agent in the assay of official phenolic compounds (see Table 29), in the form of standard "bromine" solution. Standard solutions of potassium bromate, without the potassium bromide present, are used for the analysis of official organoarsenicals which can be converted quantitatively into water-soluble trivalent inorganic arsenic compounds. The arsenite obtained from the organoarsenical reduces potassium bromate uniformly to bromide:

$$3AsO_3^{3-} + BrO_3^- \longrightarrow 3AsO_4^{3-} + Br^-$$

At the end point when all the stronger reducing agent (AsO_3^{3-}) has been oxidized, a slight excess of the potassium bromate solution oxidizes Br^- to Br_2.

$$5Br^- + BrO_3^- + 6H^+ \longrightarrow 3Br_2 + 3H_2O$$

Only a small amount of bromine need be formed in this way to discharge the color of the methyl orange or methyl red added as indicator. The azo dye is oxidized by the bromine to colorless products by a reaction which is somewhat slow; consequently the standard bromate solution must be added very slowly as the end point is approached.

When a sample of Carbarsone suppositories is digested in a Kjeldahl flask with H_2SO_4 and K_2SO_4, the Carbarsone, $C_7H_9AsN_2O_4$, is converted to arsenite, AsO_3^{3-}. The foregoing equations show that a mole of arsenite, hence a mole of Carbarsone, is 2 equivalents; and the equivalent weight of Carbarsone is 130.04 Gm. Each milliliter of 0.01 *N* $KBrO_3$, 0.01 mEq, is equivalent to $0.01 \times 0.1300 = 0.0013$ Gm of $C_7H_9AsN_2O_4$.

Oxophenarsine Hydrochloride is assayed for total arsenic by oxidation of the sample with 30 percent hydrogen peroxide, sulfuric acid, and heat. The excess oxidizing agent is destroyed by reaction with the reducing agent hydrazine sulfate, and the excess reducing agent is in turn destroyed by boiling, leaving the arsenic in the trivalent state, much like the inorganic state of arsenic in carbarsone suppositories after treatment. Each milliliter of 0.1 *N* $KBrO_3$ is equivalent to $As/2 \times 1{,}000$ or $74.916/2{,}000$ or 0.03746 Gm of As.

Questions and Problems

1. Give equations for the reduction of potassium iodate in the presence of phosphoric acid (see Thyroid assay, Exercise 62), and for its reduction in the presence of high concentrations of hydrochloric acid.

2. What purpose is served by the hydrochloric acid in the assay of potassium iodide? What other substances may be used in its place?

3. State the differences in composition and application between standard bromine solutions and standard potassium bromate solutions.

4. Forty tablets of red mercuric iodide labeled "1/15-grain tablets of mercuric iodide" weighed 12.4220 Gm. A 10.0000-Gm sample of the powdered tablets required 14.45 ml of 0.02 *M* KIO_3. Calculate the amount of HgI_2 per tablet. Calculate the percent of deviation from the labeled amount.

5. The following substances are assayed using 0.05 *M* KIO_3 as above described for potassium iodide. Calculate the equivalent weight and the 0.05 *M* KIO_3 titer value for each: (*a*) hydralazine hydrochloride; (*b*) red mercuric iodide; (*c*) benzalkonium chloride.

Diazotization Assays with Nitrite. The assays of the sulfa drugs and other compounds containing an arylamino group by diazotization with sodium nitrite are not strictly iodometric methods. They may be included at this point because the end point is determined by the liberation of iodine from iodide.

Exercise 68

Object. To Prepare and Standardize 0.1 *M* Sodium Nitrite Solution.

Materials Required. 7.5 Gm of sodium nitrite
500 mg of U.S.P. Sulfanilamide Reference Standard

5 ml of hydrochloric acid
25 Gm of crushed ice
Starch iodide paper

Procedure. "Dissolve 7.5 Gm. of sodium nitrite in sufficient water to make 1000 ml., and standardize as follows:

Weigh accurately about 500 mg. of U.S.P. Sulfanilamide Reference Standard, previously dried at 105° for three hours, and transfer to a beaker or a casserole. Add 50 ml. of water and 5 ml. of hydrochloric, and stir well until dissolved. Cool to 15°, add about 25 Gm. of crushed ice, and titrate slowly with sodium nitrite solution, stirring vigorously, until a glass rod dipped into the titrated solution produces an immediate blue ring when touched to starch iodide paper. When the titration is complete, the end point should be reproducible after the mixture has been allowed to stand for 1 minute. Calculate the molarity. Each 17.22 mg. of sulfanilamide is equivalent to 1 ml. of 0.1 *M* sodium nitrite."

Since the standardization reaction and procedure are identical with the assay method, the possibility of an end point error is minimized. It is usually desirable to conduct an exploratory titration before the actual titration. This may be done on a weighed sample, using the outside indicator after the addition of each milliliter of sodium nitrite. In the final standardization titration, sodium nitrite may be added to within 1 or 2 ml of the end point found in the exploratory titration, and as the end point is approached, in amounts of 0.1 ml, 0.05 ml, and finally dropwise, testing with the outside indicator after each addition.

Nitrous acid, formed when the sodium nitrite solution is introduced into the acid reaction mixture, reacts with the primary amino group of sulfanilamide, forming an unstable nitrite which decomposes with formation of a diazonium salt. This product is also unstable, and if the reaction mixture is not kept cold, it will decompose, forming phenol products which are capable of reacting further with nitrous acid. The formation of the diazonium salt may be represented as follows:

$$NaNO_2 + HCl \longrightarrow HNO_2 + NaCl$$

$$\underset{172.2}{H_2NSO_2-C_6H_4-NH_2} + HNO_2 + HCl \longrightarrow H_2NSO_2-C_6H_4-N_2Cl + 2H_2O$$

Since 1 mole of sulfanilamide is equivalent to 1 mole of sodium nitrite under the conditions of diazotization, each milliliter of 0.1 *M* sodium nitrite, 0.1 millimole, is equivalent to 0.1 × 0.1722 or 0.01722 Gm or 0.1 millimole of $C_6H_8N_2O_2S$.

At the end point a slight excess of nitrous acid, persisting for at least 1 min, is present. This can be detected by use of starch iodide paste T.S.

$$2I^- + 2HNO_2 + 2H^+ \longrightarrow I_2 + 2NO\uparrow + 2H_2O$$

A number of official "sulfa" drugs are assayed by this method, including phthalylsulfathiazole and succinylsulfathiazole, which can be diazotized only after a preliminary hydrolysis. The method is applicable also in the assay of many other compounds having a primary arylamino grouping.

TABLE 31. SUBSTANCES ASSAYED BY DIRECT TITRATION WITH SODIUM NITRITE

Substance	1 ml equivalent, mg	Official requirement, percent
U.S.P.		
Aminohippuric acid	19.42	$C_9H_{10}N_2O_3$ = 98
Aminosalicylic acid	15.31	$C_7H_7NO_3$ = 98.5
Aminosalicylic acid tablets	15.31	$C_7H_7NO_3$ = 95 to 105 l.a.
Dapsone	12.42	$C_{12}H_{12}N_2O_2S$ = 99
Dapsone tablets	12.42	$C_{12}H_{12}N_2O_2S$ = 92.5 to 107.5 l.a.
p-Aminobenzoic acid (R)	13.71	$C_7H_7NO_2$ = 98.5
Phthalylsulfathiazole	40.34	$C_{17}H_{13}N_3O_5S_2$ = 98
Phthalylsulfathiazole tablets	40.34	$C_{17}H_{13}N_3O_5S_2$ = 94 to 106 l.a.
Primaquine phosphate	45.53	$C_{15}H_{21}N_3O{\cdot}2H_3PO_4$ = 98 to 102
Primaquine phosphate tablets	45.53	$C_{15}H_{21}N_3O{\cdot}2H_3PO_4$ = 93 to 107 l.a.
Procainamide hydrochloride capsules	27.18	$C_{13}H_{21}N_3O{\cdot}HCl$ = 95 to 105 l.a.
Procainamide hydrochloride injection	27.18	$C_{13}H_{21}N_3O{\cdot}HCl$ = 95 to 105 l.a.
Procaine hydrochloride	27.28	$C_{15}H_{20}N_2O_2{\cdot}HCl$ = 99 to 101
Sodium aminohippurate injection	21.62	$C_9H_9N_2NaO_3$ = 95 to 105 l.a.
Sodium aminosalicylate	17.51	$C_7H_6NNaO_3$ = 98 to 101
Sodium aminosalicylate tablets	21.12	$C_7H_6NNaO_3{\cdot}2H_2O$ = 95 to 105 l.a.
Sodium glucosulfone injection	39.03	$C_{24}H_{34}N_2Na_2O_{18}S_3$ = 38.4 to 41.6 *W/V*
Sodium sulfacetamide	23.62	$C_8H_9N_2NaO_3S$ = 99
Sodium sulfacetamide ophthalmic ointment	25.42	$C_8H_9N_2NaO_3S{\cdot}H_2O$ = 9.5 to 10.5
Sodium sulfacetamide ophthalmic solution	25.42	$C_8H_9N_2NaO_3S{\cdot}H_2O$ = 28.5 to 31.5
Sodium sulfadiazine	27.23	$C_{10}H_9N_4NaO_2S$ = 99
Sodium sulfadiazine injection	27.23	$C_{10}H_9N_4NaO_2S$ = 95 to 105 l.a.
Sodium suramin	23.82	$C_{51}H_{34}N_6Na_6O_{23}S_6$ = 97.5
Sodium suramin, sterile	23.82	$C_{51}H_{34}N_6Na_6O_{23}S_6$ = 97.5
Succinylsulfathiazole	35.54	$C_{13}H_{13}N_3O_5S_2$ = 99
Succinylsulfathiazole tablets	37.34	$C_{13}H_{13}N_3O_5S_2{\cdot}H_2O$ = 95 to 105 l.a.
Sulfadiazine	25.03	$C_{10}H_{10}N_4O_2S$ = 99
Sulfadiazine tablets	25.03	$C_{10}H_{10}N_4O_2S$ = 95 to 105 l.a.
Sulfamerazine	26.43	$C_{11}H_{12}N_4O_2S$ = 99
Sulfamethazine	27.83	$C_{12}H_{14}N_4O_2S$ = 99
Sulfapyridine	24.93	$C_{11}H_{11}N_3O_2S$ = 99
Sulfapyridine tablets	24.93	$C_{11}H_{11}N_3O_2S$ = 95 to 105 l.a.
Tetracaine	26.44	$C_{15}H_{24}N_2O_2$ = 98 to 101
Trisulfapyrimidines oral suspension	26.43	Total sulfas = 9.3 to 10.7 *W/V*
Trisulfapyrimidines tablets	26.43	Total sulfas = 95 to 105 l.a.
N.F.		
Acetyl sulfisoxazole	30.93	$C_{13}H_{15}N_3O_4S$ = 98
Acetyl sulfisoxazole oral suspension	26.73	$C_{11}H_{13}N_3O_3S$ = 9.3 to 10.7 *W/V*

TABLE 31. SUBSTANCES ASSAYED BY DIRECT TITRATION WITH SODIUM NITRITE (*Continued*)

Substance	1 ml equivalent, mg	Official requirement, percent
N.F.		
Aminobenzoic acid	13.71	$C_7H_7NO_2$ = 98.5
Aminobenzoic acid tablets	13.71	$C_7H_7NO_2$ = 94 to 106 l.a.
Calcium aminosalicylate	17.22	$C_{14}H_{12}CaN_2O_6$ = 98
Calcium aminosalicylate tablets	19.92	$C_{14}H_{12}CaN_2O_6 \cdot 3H_2O$ = 95 to 105 l.a.
Calcium cyclamate	19.83	$C_{12}H_{24}CaN_2O_6S_2$ = 98 to 101
Calcium cyclamate solution, compound	21.63	$C_{12}H_{24}CaN_2O_6S_2 \cdot 2H_2O$ = 5.4 to 6.6 *W/V*
Calcium cyclamate tablets, compound	21.63	$C_{12}H_{24}CaN_2O_6S_2 \cdot 2H_2O$ = 90 to 110 l.a.
Glycobiarsol	49.91	$C_8H_9AsBiNO_6$ = 97 to 103
Glycobiarsol tablets	49.91	$C_8H_9AsBiNO_6$ = 95 to 105 l.a.
Phthalylsulfacetamide	36.24	$C_{16}H_{14}N_2O_6S$ = 99
Phthalylsulfacetamide tablets	36.24	$C_{16}H_{14}N_2O_6S$ = 95 to 105 l.a.
Procaine, *tetracaine*, and nordefrin hydrochlorides injection	30.08	$C_{15}H_{24}N_2O_2 \cdot HCl$ = 95 to 105 l.a.
Procaine, tetracaine, and nordefrin hydrochlorides injection	27.28	$C_{13}H_{20}N_2O_2 \cdot HCl$ = 95 to 105 l.a.
Propoxycaine hydrochloride	33.09	$C_{16}H_{26}N_2O_3 \cdot HCl$ = 98 to 102
Sodium cyclamate	20.12	$C_6H_{12}NNaO_3S$ = 98 to 101
Sodium cyclamate solution, compound	20.12	$C_6H_{12}NNaO_3S$ = 5.4 to 6.6 *W/V*
Sodium cyclamate tablets, compound	20.12	$C_6H_{12}NNaO_3S$ = 90 to 110 l.a.
Sodium sulfabromomethazine	39.72	$C_{12}H_{12}BrN_4NaO_2S \cdot H_2O$ = 97 to 102
Sodium sulfamerazine injection	28.63	$C_{11}H_{11}N_4NaO_2S$ = 95 to 105 l.a.
Sulfacetamide	21.42	$C_8H_{10}N_2O_3S$ = 99
Sulfacetamide tablets	21.42	$C_8H_{10}N_2O_3S$ = 95 to 105 l.a.
Sulfacetamide, sulfadiazine, and sulfamerazine suspension	21.42	$C_8H_{10}N_2O_3S$ = 90 to 110 l.a.
Sulfacetamide, sulfadiazine, and sulfamerazine tablets	21.42	$C_8H_{10}N_2O_3S$ = 90 to 110 l.a.
Sulfadimethoxine	31.03	$C_{12}H_{14}N_4O_4S$ = 99
Sulfadimethoxine oral suspension	31.03	$C_{12}H_{14}N_4O_4S$ = 93 to 107 l.a.
Sulfadimethoxine tablets	31.03	$C_{12}H_{14}N_4O_4S$ = 95 to 105 l.a.
Sulfamerazine tablets	26.43	$C_{11}H_{12}N_4O_2S$ = 95 to 105 l.a.
Sulfamethizole	27.03	$C_9H_{10}N_4O_2S_2$ = 99
Sulfamethizole suspension	27.03	$C_9H_{10}N_4O_2S_2$ = 90 to 110 l.a.
Sulfamethizole tablets	27.03	$C_9H_{10}N_4O_2S_2$ = 95 to 105 l.a.

(R) = reagent; l.a. = labeled amount.

Chapter 10

GASOMETRIC ANALYSIS

The assay processes that involve the volumetric measurement of gases are of two types, viz.:

1. The purity of a gas is determined by absorbing one of the components of a gaseous mixture by means of a suitable absorbing agent and measuring the quantity of gas absorbed by the change in volume.

2. The purity of a given substance is determined by measuring the quantity of gas that it may be made to evolve in a given chemical reaction.

Theory. In assay procedures in which determination of the quantity of a gas is involved, the volume of the gas can be more conveniently and accurately measured than the weight. A simple relationship exists between the volume of a sample of a gas and its weight or mole fraction, and this relationship may be applied in interpretation of experimental data provided corrections are made for temperature and pressure, factors which also determine the volume of a sample of a gas. The relationships between weight or number of molecules, pressure, temperature, and volume of a sample of gas are understandable in the light of the *kinetic-molecular theory.*

The pressure exerted by a gas depends on the number of molecules which strike the walls of the container in a given period of time, on the weight of the molecules, and on their average velocity. The weight and number of molecules in a given sample of gas are, of course, constant, and the velocity of the molecules depends on temperature. If the temperature is constant, the pressure exerted by a sample of gas depends on the volume, and vice versa. This is true because the number of molecules striking the walls of a small container in a given period of time is greater than it would be if the container were larger, since the molecules need to travel shorter distances in a small container before striking a wall. Therefore, a given weight of gas in which the number of molecules is constant will exert a pressure which is inversely proportional to its volume, provided the temperature is constant. Boyle's law defines the relationship between the volume of a given weight of gas and the pressure exerted by the gas. It may be stated as follows: *the volume of a given weight of gas is inversely proportional to the pressure, provided temperature remains constant.*

$$P_1V_1 = P_2V_2 \quad \text{and} \quad V_2 = \frac{P_1V_1}{P_2} \tag{1}$$

A given weight of gas will expand or increase in volume when heated at constant pressure, because the average velocity of the molecules and consequently the number of molecules striking the walls of a container are greater at higher temperatures. If the container's volume is fixed, then the gas will exert a greater pressure at the higher temperature. However, if the volume can change, as it can in a gas buret, then the gas expands until it exerts the same pressure in a larger volume which it exerted in a smaller volume before heating. Charles's law, deduced from the kinetic-molecular theory, defines the relationship between the volume of a given weight of gas and the temperature, viz., *the volume of a given weight of gas is directly proportional to the absolute temperature, provided pressure is constant.*

$$\frac{V_1}{T_1} = \frac{V_2}{T_2} \quad \text{and} \quad V_2 = \frac{T_2 V_1}{T_1} \tag{2}$$

Since the volume of a given weight of gas varies inversely as the pressure (Boyle's law) and directly with the temperature (Charles's law),

$$\frac{P_1 V_1}{T_1} = \frac{P_2 V_2}{T_2} = K \tag{3}$$

The constant value, K, of the ratio of the product of pressure and volume to absolute temperature of an ideal gas depends on the number of molecules, their mass, and their velocity. Few gases behave as an ideal gas, and, in very precise measurements, corrections must be applied for the volume of the molecules and intermolecular attraction.

The value of K is such that 6.02×10^{23} molecules, 1 mole, of an ideal gas, at 0°C or 273° absolute and a pressure of 760 mm of mercury (known as *standard conditions*), occupy a volume of 22,400 ml. It follows that 22,400 ml of any ideal gas, at 0° and 760 mm, contains 6.02×10^{23} molecules or 1 mole, or *equal volumes of gases measured under the same conditions of temperature and pressure contain the same number of molecules.* This is Avogadro's hypothesis. In order to make use of the 22,400-ml volume of 1 mole of gas as a conversion factor in calculating mole fractions or weights of a sample of gas from the volume measured in the course of an assay, it is frequently necessary to calculate the volume of the sample at standard conditions from the volume measured under the conditions of the experiment. For example, 40 ml of NO, measured at 20° and 740 mm pressure, was obtained in the assay of a 0.2000-Gm sample of amyl nitrite, $C_5H_{11}NO_2$.

1. What would be the volume of the same sample of NO under standard conditions?

$$V_0 = \frac{P_1 \times V_1 \times T_0}{P_0 \times T_1} \qquad \text{[from Eq. (3)]}$$

where V_0 = volume of NO at standard conditions

$P_0 = 760$
$T_0 = 273$
$P_1 = 741$
$T_1 = 293$

Note that degrees C + 273 = degrees absolute.

$$V_0 = \frac{40 \times 740 \times 273}{760 \times 293} = 36.3 \text{ ml}$$

2. What fraction of a mole and what weight of NO were obtained from the amyl nitrite sample?

$$\frac{36.3}{22{,}400} = 0.00162 \text{ mole} \qquad \frac{36.3}{22{,}400} \times 30.01 = 0.0486 \text{ Gm}$$

3. What was the $C_5H_{11}NO_2$ content of the sample?

The following equation shows the stoichiometric relationship between 1 mole of NO and 1 mole of $C_5H_{11}NO_2$.

$$\underset{2(117.15\ \text{Gm})}{2C_5H_{11}NO_2} + 2KI + H_2SO_4 \longrightarrow 2C_5H_{11}OH + \underset{2(22{,}400\ \text{ml})}{2NO} + I_2 + K_2SO_4$$

Since 36.3 ml of NO was obtained from a 0.2000-Gm sample of amyl nitrite, the $C_5H_{11}NO_2$ content of the material may be calculated as follows:

$$\frac{36.3 \times 117.15 \times 100}{22{,}400 \times 0.2} = 94.89 \text{ percent}$$

GAS ABSORPTION METHOD

A number of official gases are assayed by treating them with a suitable absorption agent and measuring the difference in volume of gas after treatment.

Apparatus. The volumetric measurement of gases in the official assay processes is made by the use of a buret of the type illustrated in Fig. 15. The buret should be of 100-ml capacity and graduated in 0.2-ml divisions. The buret may be enclosed in a water jacket, as illustrated, to permit the control of temperature. The lower end of the gas buret is attached to an open leveling or equilibrium tube by means of a heavy-walled rubber tube.

The gases are absorbed in a gas pipet similar to that shown in Fig. 16. The absorption bulb of the pipet is provided with an opening at the bottom, which can be closed by means of a rubber stopper and through which solid absorbing agents may be introduced.

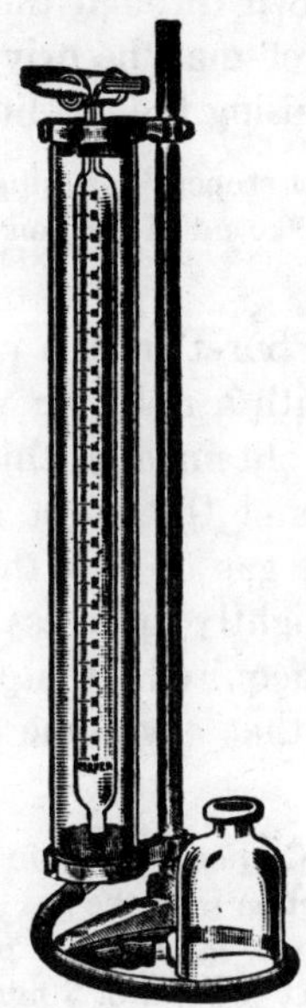

FIG. 15. Gas buret with leveling flask.

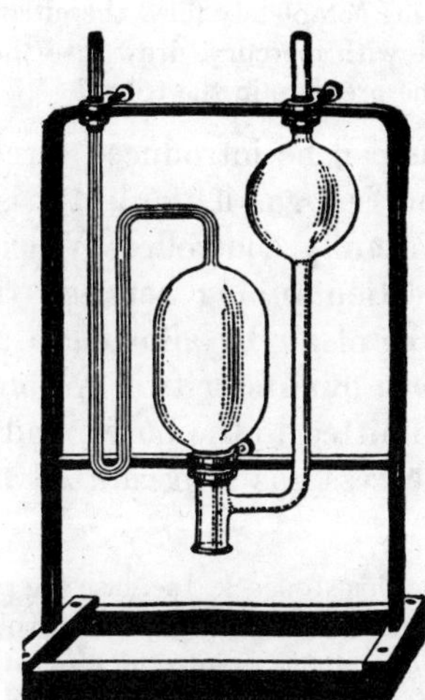

FIG. 16. Simple gas pipet.

Exercise 69

Object. Assay of Carbon Dioxide.

Materials Required. Carbon dioxide compressed in a metallic cylinder, which is provided with a reducing valve
A gas buret with leveling tube
A gas pipet
About 2 lb of mercury
150 ml of 50 percent of KOH solution

Procedure. 1. "Place a sufficient quantity of mercury in a 100-ml. nitrometer or gas buret provided with a two-way stopcock and a two-way outlet, and properly connected with a balancing tube. Connect one of the outlet tubes of the nitrometer with a gas pipet of suitable capacity. Place in the pipet about 125 ml. of potassium hydroxide solution (1 in 2). Draw the liquid (free from air bubbles) through the capillary opening, connection, and stopcock opening in the nitrometer by reducing the pressure in the nitrometer tube and opening the stopcock controlling the connection with the gas pipet. Then close the stopcock."

The gas buret is filled with mercury by raising the leveling flask containing the mercury until the stopcock opening is filled with the mercury. The stopcock is then closed. The gas pipet is filled with 50 percent KOH so that the absorption bulb and capillary tube are completely full and the expansion bulb is about one-tenth full and is connected to one of the outlet tubes of the gas buret. Upon opening the stopcock connecting the pipet and the buret and lowering the leveling flask, the KOH solution and any trapped air in the connection are drawn through the stopcock opening. Any air bubbles drawn into the buret may be driven out by opening the stopcock to the other outlet and raising the leveling flask.

2. "Having completely filled the nitrometer, the other stopcock opening, and other intake tube with mercury, draw into the nitrometer 100.0 ml of Carbon Dioxide by reducing the pressure in the tube."

The gas can be introduced directly into the buret from a cylinder of the compressed gas if the latter is provided with a reducing valve that can be carefully controlled. When the gas is obtained in this manner, the connection tubing between the cylinder and the buret should be flushed free of air by allowing a stream of the gas to flow through the tubing for a minute or two. A volume of gas slightly in excess of 100 ml is then admitted to the buret, and after it has acquired a temperature of 25°, the excess gas is permitted to escape so that a volume of 100 ml remains.

3. "Close this stopcock. Increase the pressure on the Carbon Dioxide in the nitrometer tube, and open the stopcock controlling the connection with the gas pipet. Force the entire volume of gas into the pipet. Close the stopcock, and rock the pipet gently, providing frequent contact of the liquid and the gas. At the end of 5 minutes, when most of the Carbon Dioxide will have been absorbed by the liquid, facilitate the absorption of the remainder by drawing some of the liquid into the nitrometer tube, and forcing the residual gas back upon the surface of the liquid in the gas pipet. Again rock the pipet until no further diminution in the volume of gas occurs. Draw the residual gas, if any, into the nitrometer tube and measure its volume: not more than 1 ml. of gas remains."

The CO_2 is absorbed by the KOH forming K_2CO_3, leaving other gases which are not absorbed by KOH as residual gas. If the initial volume of the sample was exactly 100 ml, the decrease in volume is numerically equal to the percent of CO_2 in the sample.

TABLE 32. SUBSTANCES ASSAYED BY GASOMETRIC METHODS

Substance	Method of analysis	Official requirement, percent
U.S.P.		
n-Butyl nitrite (R)	Nitric oxide production	$C_4H_9NO_2$ = 90
Carbon dioxide	Absorption, 50% KOH	CO_2 = 99 *V/V*
Cyclopropane	Absorption, H_2SO_4	C_3H_6 = 99 *V/V*
Nitrogen	Washing, $NH_4Cl—NH_4OH$ T.S., Cu coil	N_2 = 99 *V/V*
Nitrous oxide	Condensation	N_2O = 95 *V/V*
Oxygen	Absorption, $NH_4Cl—NH_4OH$ T.S., Cu coil	O_2 = 99 *V/V*
N.F.		
Amyl nitrite	Nitric oxide production	$C_5H_{11}NO_2$ = 95 to 105
Ethylene	Absorption, bromine T.S.	C_2H_4 = 99 *V/V*

(R) = reagent.

Official gases which may be assayed by procedures comparable with that used for the assay of carbon dioxide are listed in Table 32. The methods employed for the assay of cyclopropane, ethylene, and oxygen are quite similar to that used for carbon dioxide, differing primarily in regard to the reagent used for absorption. Nitrogen is washed free of other gases, especially oxygen, by use of ammonium chloride—ammonium hydroxide T.S. in the presence of a copper wire oil. Nitrous oxide is condensed by use of a specially constructed gas buret and a condensation bulb which is immersed in liquid air (see U.S.P.).

GAS EVOLUTION METHOD

In the assay of Amyl Nitrite and *n*-Butyl Nitrite reagent, the quantity of gas evolved in a chemical reaction is measured.

Apparatus. The volumetric measurement of evolved gases may be made by the use of an instrument known as a *nitrometer*. The nitrometer of the type devised by Lunge (Fig. 17) consists of a gas buret *A* of at least 50 ml capacity, graduated into 0.1 ml, with a special glass stopcock *B* at the top, to which is sealed a glass cup *D*. The gas buret is connected with an open leveling or equilibrium tube *E* by means of a heavy-walled rubber tube *F*.

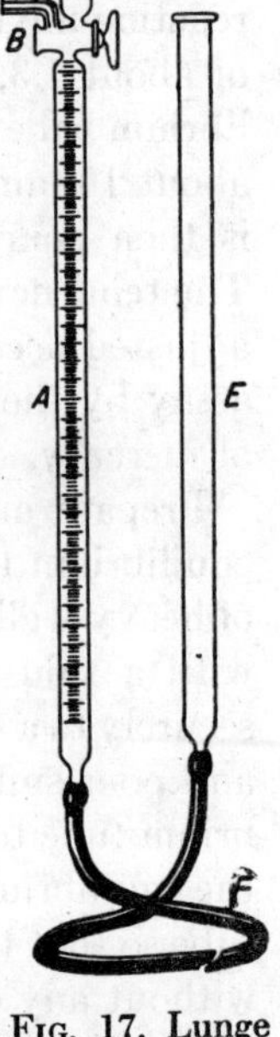

FIG. 17. Lunge nitrometer.

In the use of the simple Lunge apparatus described above, both tubes of the instrument are partly filled with some liquid in which the gas to be measured is insoluble; generally mercury is employed, but in some cases a saturated salt solution is used. The latter may be used in the measurement of the nitric oxide liberated from *n*-Butyl Nitrite or Amyl Nitrite.

When mercury is used, enough of it is poured into the equilibrium tube to fill the buret and leave a slight excess. The stopcock is turned to connect the gas buret with the glass cup, and the equilibrium tube is raised until the mercury level reaches the bottom of the cup to expel all the air from the buret and stopcock bore, and the stopcock is closed. The sample to be assayed, a liquid or a solid in solution, is weighed and poured into the cup from a weighing bottle or weight pipet. Upon opening the stopcock and lowering the equilibrium tube slightly, the sample is drawn into the buret. Care must be exercised to prevent the entrance of air into the passage of the stopcock. The cup is rinsed two or three times with 3-ml portions of the reagents used to liberate the gas, each being drawn into the buret as described above. With the stopcock closed, the equilibrium tube is lowered sufficiently to cause a slight diminution of the pressure in the buret, and the latter is shaken vigorously for about 5 min to bring the reagents and sample into intimate contact. When no more gas is evolved on further shaking, the buret is allowed to stand about 30 min to permit its contents to acquire room temperature. The volume is then measured at atmospheric pressure. Since there is a column of impure reaction mixture above the mercury in the buret, some procedure must be employed to adjust the level of the equilibrium tube so that the weight of the column of mercury in it will just equal the weight of the column of mercury and reaction mixture in the buret. If the reagent mixture has a specific gravity of about 1.3, as it does in the assay for amyl or butyl nitrite, the equilibrium tube is set on the assumption that 1 mm of mercury will balance about 10 mm of the reaction mixture. About 5 ml of dilute sulfuric acid is then poured into the cup, and the stopcock is carefully opened to it. The tendency to draw in or expel gas is noted, and the equilibrium tube is adjusted accordingly. This tedious leveling procedure is obviated in the assay by the use of a saturated solution of sodium chloride in the place of mercury.

Prepare and test a nitrometer as follows: Clean the gas buret and equilibrium tube of the nitrometer thoroughly, connect them by a piece of heavy-walled rubber tubing about 30 in. long, grease the buret stopcock with a thin film of petrolatum, and clamp the assembled instrument securely to a support in a vertical position. Open the stopcock of the buret, and pour sufficient saturated solution of sodium chloride into the equilibrium tube to fill the buret and the bore of the stopcock completely when the equilibrium tube is raised. Carefully adjust the level of the equilibrium tube so that the stopcock passage is completely filled with the salt solution without any of it being present in the glass cup, and close the stopcock. Lower the equilibrium tube so that the solution level in it corresponds approximately with the 50-ml graduation on the buret, note the solution level in the buret, and, after the instrument has stood for 20 min, take another reading. If the two readings do not correspond, the stopcock is leaky. If the leakage cannot be corrected by the use of stopcock grease, another instrument should be employed.

Exercise 70

Object. Assay of Amyl Nitrite.

Materials Required. Lunge nitrometer
10-ml delivery pipet
100-ml volumetric flask
100 ml of alcohol
10 ml of potassium iodide T.S.
5 ml of diluted sulfuric acid

Procedure. 1. "Shake about 3 ml. of Amyl Nitrite with 500 mg. of anhydrous potassium carbonate, decant carefully, transfer to a 100-ml. volumetric flask tared with about 20 ml. of alcohol, and weigh accurately. Add alcohol to volume, and mix."

In the presence of moisture, amyl nitrite is partially hydrolyzed, with the formation of amyl alcohol and nitrous acid:

$$C_5H_{11}NO_2 + H_2O \longrightarrow C_5H_{11}OH + HNO_2$$

The nitrous acid if present would liberate carbon dioxide from the potassium carbonate:

$$2HNO_2 + K_2CO_3 \longrightarrow 2KNO_2 + H_2O + CO_2\uparrow$$

The potassium nitrite formed is absorped in the moisture onto the potassium carbonate and eliminated. The sample is diluted with alcohol to 100 ml after weighing, so that an aliquot portion of the solution representing a definite weight of the sample may be easily measured into the nitrometer.

2. "Fill a nitrometer with a saturated solution of sodium chloride, pipet into the funnel top 10 ml. of the alcohol solution, and, after drawing this into the measuring tube of the nitrometer without the admission of air, follow it successively with 5 ml. of alcohol as a rinse, then with 10 ml. of potassium iodide T.S., and afterwards with 5 ml. of diluted sulfuric acid, introducing each reagent separately into the measuring tube. When the volume of gas has become constant in about (30 to 60 minutes), note the amount collected, the temperature at the nitrometer, and the barometric pressure."

Adjust the nitrometer prepared for use in the preceding exercise so that the level of the salt solution in the equilibrium tube is about even with the 15-ml graduation mark on the buret, the latter as well as the stopcock passage being filled with the salt solution. Fill a 10-ml pipet with the diluted sample exactly to the graduation mark and transfer the liquid into the nitrometer cup. Open the stopcock at once to allow the solution to be drawn into the buret, using care to prevent the entrance of air at the same time. Rinse the cup at once with two 5-ml portions of potassium iodide solution (1.65 Gm of KI in 10 ml of H_2O), drawing each portion into the buret by lowering the equilibrium tube, if necessary, and then place about 6 ml of diluted sulfuric acid in the cup and draw about 5 ml of it into the buret in the same manner. The excess acid remaining in the cup serves to prevent the escape of the nitric oxide gas, which begins to form as soon as the first of the acid is introduced, and it prevents the entrance of air. After the stopcock has been tightly closed and the equilibrium tube well clamped to the support, remove the buret from the clamp, incline it at an angle of about 45 degrees, and shake it vigorously for about 5 min to mix the reagents thoroughly with the amyl nitrite.

Potassium iodide, in the presence of sulfuric acid, reduces amyl nitrite ($C_5H_{11}ONO$) to an equivalent amount of nitric oxide (NO). The nitric oxide gas which is formed, being insoluble in saturated salt solution, rises to the top of the buret, and after it has acquired room temperature, as shown by the constancy of its volume when two successive readings taken at 10-min intervals agree, the volume of gas liberated is recorded.

When the final reading of the volume of gas is taken, the liquids in the buret and equilibrium tube should be on the same level.

$$\underset{117.15}{2C_5H_{11}ONO} + 2KI + H_2SO_4 \longrightarrow 2C_5H_{11}OH + \underset{2(22{,}400\ ml)}{2NO} + I_2 + K_2SO_4$$

3. "Multiply the volume of gas in milliliters by 4.8, and divide the product by the weight of Amyl Nitrite taken. At 25° and 760 mm. pressure, the quotient represents the percentage of $C_5H_{11}NO_2$ in the liquid. The temperature correction is $\frac{1}{298}$ of the total percentage just found for each degree, added if the temperature is below 25°, and subtracted if it is above 25°. The barometric correction is $\frac{1}{760}$ of the total percentage, for each millimeter, added if it is above 760 mm., and subtracted if it is below 760 mm."

If the nitric oxide, collected at 25° and 760 mm pressure, has a volume V_{25}, the volume at 0° would be

$$V_0 = V_{25} \times \frac{273}{298}$$

Since 117.15 Gm or 1 mole of $C_5H_{11}ONO$ would have produced 22,400 ml of NO at 0° and 760 mm pressure, the amount of $C_5H_{11}ONO$ which produced the above corrected volume of NO must have been $[V_{25} \times (273/298)]/(22{,}400) \times 117.15$ or $V_{25} \times 273/298 \times 117.15/22{,}400$ or $V_{25} \times 0.0048$ (Gm of $C_5H_{11}ONO$). In other words, each milliliter of nitric oxide, collected at 25° and 760 mm, is equivalent to 0.0048 Gm of $C_5H_{11}ONO$. This amount of $C_5H_{11}ONO$ was in $\frac{1}{10}$ aliquot part of the sample. Therefore, in 100 Gm of the material, there must have been the following percent of $C_5H_{11}ONO$:

$$\frac{V_{25} \times 0.0048}{\text{Sample}/10} \times 100 = \frac{V_{25} \times 0.48}{\text{Sample}/10} = \text{percent } C_5H_{11}ONO$$

The N.F. directs that the volume observed at 25° and 760 mm be multiplied by 4.8; the product divided by the weight of Amyl Nitrite taken results in the percentage of $C_5H_{11}ONO$ in the liquid. As indicated above, the result is the number of grams of $C_5H_{11}ONO$ in 100 Gm of substance, i.e., percent *W/W*.

The corrections for the deviation of temperature from 25° and the pressure from 760 mm are made in accordance with the gas laws, as directed in the procedure. For example, a 3.0000-Gm sample of amyl nitrite yielded 60 ml of NO, measured at 21° and 740 mm. If the NO had been measured at normal conditions, the $C_5H_{11}ONO$ content would have been

$$\frac{60 \times 4.8}{3.0} = 96 \text{ percent}$$

After correcting for the deviation of temperature,

$$96.0 + \left(\frac{4}{273} \times 96.0\right) = 97.41$$

and for the deviation in pressure,

$$97.41 - \left(\frac{20}{760} \times 96.0\right) = 94.88$$

the $C_5H_{11}ONO$ content of the sample is calculated as 94.88 percent.

Comment. When a gas is collected over a liquid, the pressure of the gaseous mixture is equal to the pressure of the gas plus the pressure of the vapors of the liquid in accordance with Dalton's law of partial pressures. In the above experiment, no correction was made for vapor pressure of the liquid over which the nitric oxide was collected. The liquid consisted of a saturated solution of sodium chloride to which had been added, in the course of the experiment, a small amount of water contained in potassium iodide T.S. and diluted sulfuric acid and a small amount of alcohol. Since dissolved salts lower the vapor pressure of the solvent, and since the liquid used in this experiment was practically saturated with sodium chloride, any error arising from the vapor pressure of the liquid is very small and probably compensated by the slight solubility of nitric oxide in the liquid.

Questions and Problems

1. A mixture of the following three samples of gas was prepared: 50 ml of carbon dioxide, measured at 21°, 750 mm; 25 ml of oxygen, measured at 28°, 740 mm; and 30 ml of nitrogen, measured at 26°, 765 mm. (*a*) What is the volume of the mixture at normal conditions? (*b*) What is the volume at standard conditions after shaking the mixture with 50 percent potassium hydroxide?

2. If 2.5 Gm of oxygen is collected over water at 25° and 760 mm under conditions such that the gas is saturated with water vapor, what is the volume of the mixture of oxygen and water vapor? What would be the volume of the oxygen if no water vapor were present?

3. By what factor must the volume of nitric oxide produced by a sample of butyl nitrite, $C_4H_9NO_2$, and collected at 25° and 760 mm, be multiplied so that the ratio of this product to the weight of the sample of butyl nitrite employed will be the percentage concentration of $C_4H_9NO_2$ in the sample? Compare the calculated value with that given in the N.F. assay procedure.

4. What chemical reactions take place in the assay of cyclopropane and ethylene?

PART 2

SPECIAL METHODS USED IN OFFICIAL PHARMACEUTICAL ANALYSES

Quantitative analyses of crude drugs and of the products derived from them are made to establish purity or to determine the amount of therapeutically active constituents present for the purpose of standardization. The special methods employed in the analyses of this type may be classified as follows:

Chemical methods, such as those employed in the determination of ash, moisture, crude fiber, extractive obtained with different solvents, the estimation of alkaloidal content, etc.

Biological methods, or those in which the effects of drugs upon microbes, animals, or animal tissues are measured and which are employed in the absence of satisfactory chemical methods for standardization. Official standards for drugs or preparations include biological assays or tests for such medicinal agents as adrenal cortex, corticotropin, digitalis, heparin sodium, insulin, oxytocin, parathyroid, and vasopressin. Microbiological assays are required for the antibiotics, liver, vitamin B_{12}, and other related drugs. Tests are required for the toxicity and potency of the serological preparations such as the vaccines, serums, toxoids, and antitoxins. Biological methods require special apparatus for their performance and a technique that is not chemical in nature; hence a discussion of them does not fall within the scope of this text.

Many tests of antibiotics which are based on chemical reactions are no longer included in the official compendiums, because they are now available in the Federal Register. Under Public Law 87-781, all antibiotics intended for use in man are subject to production and testing controls under Federal regulation. The Food and Drug Administration publishes these regulations in the Federal Register, and all the antibiotics recognized in the official compendiums are subject to the provisions of these regulations. Therefore, analytical methods for the determination of purity, moisture, and other constants are obtained from this publication.

Chapter 11

ASH AND WATER DETERMINATIONS

Ash Content. The ash content of a crude drug is generally taken to be the residue remaining after incineration. It usually represents the inorganic salts naturally occurring in the drug and adhering to it, but it may also include inorganic matter added for the purposes of adulteration. There is a considerable difference in the ash content of different drugs, but the difference varies within narrow limits in the case of the same individual; hence, an ash determination furnishes a basis of judging the identity and cleanliness of a drug and gives information relative to its adulteration with inorganic matter. Ash standards have been established for a number of the official drugs; usually these standards set a maximum limit on the total ash or on the acid-insoluble ash permitted. The *total ash* is the residue remaining after incineration, the *acid-insoluble ash* is that part of the total ash which is insoluble in diluted hydrochloric acid. The Pharmacopeia specifies that all vegetable drugs shall be separated by mechanical means insofar as possible from lumps of dirt or other foreign inorganic matter before they are ground or powdered.

The ash or residue yielded by an organic chemical compound is, as a rule, a measure of the amount of inorganic matter present as impurity. In most cases, the inorganic matter is present in small amounts, which are difficult to remove in the purification process and which are not objectionable if only traces are present.

The careful control of temperature is the most important analytical factor to regulate in making ash determinations. All determinations should be made in such a manner as to duplicate insofar as possible the conditions under which standards are established. When an electric furnace is used for ignitions, the following temperature equivalents, which are only approximate, will be found of value: Very dull red heat = 500 to 550°; dull red heat = 550 to 700°; bright red heat = 800 to 1000°; yellow red heat = 1000 to 1200°; and white heat = 1200 to 1600°. Practically all these temperatures are within the limits of the ordinary Bunsen burner, depending on the construction of the burner and the nature of the gas burned. When it is considered that calcium carbonate is converted to calcium oxide at about 825°, that calcium phosphate is converted to calcium pyrophosphate at about 1550°, that lithium carbonate forms lithium oxide at about 600°, and that potassium chloride sublimes at about 1500°, it becomes apparent that ignitions should be carried out at low temperatures if possible. The ash limits of some of the official vegetable drugs are given in Table 33.

TABLE 33. SUBSTANCES WITH ASH LIMITS

Substance	Total ash, percent	Acid-insoluble ash, percent
U.S.P.		
Acacia	4.0	0.5
Agar	6.5	0.5
Aloe	4.0	
Belladonna leaf		3.0
Benzoin (Sumatra)	.....	1.0
Benzoin (Siam)	.....	0.5
Digitalis		5.0
Powdered digitalis		5.0
Glycyrrhiza		2.5
Methylcellulose	1.0	
Podophyllum	.....	2.0
N.F.		
Areca	2.5	
Caramel	8.0	
Caraway	.....	1.5
Cardamom seed		4.0
Eriodictyon	.....	2.0
Glycyrrhiza extract	8.0	
Honey	0.3	
Plantago seed	4.0	1.0
Rauwolfia serpentina		2.0
Powdered Rauwolfia serpentina		2.0
Senna		3.0
Sodium alginate	18–24	
Dried yeast	8.0	

Exercise 71

Object. To Determine the Total and Acid-insoluble Ash Content of Digitalis.

Materials Required. 2 to 4 Gm of the ground drug [see Sampling, Chap. 1]
2 ashless filter papers

Procedure. 1. "Accurately weigh a quantity of the ground drug, representing from 2 to 4 Gm. of the air-dried material, in a tared crucible and incinerate at a low temperature, not to exceed very dull redness, until free from carbon, and determine the weight of the ash."

The crucible in which the drug is to be ignited should be heated to dull redness until of constant weight, cooled in a desiccator, and weighed. This procedure drives off the moisture and the absorbed gases that are present on the surface of the crucible. The drug is incinerated at a temperature not to exceed a dull redness. Very often, the ash consists largely of calcium carbonate formed from the calcium oxalate contained in the drug. When the incineration is carried out at a dull red heat, the calcium oxalate is converted into carbonate, but when the incineration is carried out at a bright red heat, varying amounts of the carbonate are converted into the oxide with consequent variable results. If the ash contains alkali chlorides, which it frequently does, there may be a loss of some chloride by volatilization when a high ignition temperature is used.

2. "If a carbon-free ash cannot be obtained in this way, exhaust the charred mass with hot water, collect the insoluble residue on an ashless filter paper, incinerate the residue and filter paper until the ash is white or nearly so, then add the filtrate, evaporate it to dryness and heat the whole to a low redness. If a carbon-free ash cannot be obtained in this way, cool the crucible, add 15 ml. of alcohol, break up the ash with a glass rod, burn off the alcohol, and again heat the whole to a low redness. Cool, weigh the ash, and calculate the percentage of total ash from the weight of the drug taken."

When the ignition residue is not heated above dull redness, it is frequently very difficult to burn off all the carbon, some of which may become surrounded by fused inorganic matter, thereby keeping it from coming in contact with the oxygen of the air. In such cases, the residue is treated with hot distilled water to dissolve most of the inorganic matter, and the insoluble portion is collected on an ashless filter and incinerated. The filtrate is then added to this residue, evaporated, and the whole is ignited to burn off any colloidal carbon that may have passed into the filtrate. Occasionally, it may be necessary to treat the residual mass with alcohol to disintegrate the fused lump, especially in the ignition of resinous drugs.

The weight of the crucible and ash minus the weight of the crucible gives the weight of total ash obtained from the sample of drug taken. This total ash, which usually contains carbonates, phosphates, sulfates, chlorides, oxides, etc., of calcium, magnesium, potassium, sodium, aluminum, iron, and other metallic elements, does not necessarily represent all the inorganic constituents of the drug, since ammonium salts, some alkali iodides and nitrates, etc., are volatilized or converted into carbonates, etc.

3. "Boil the ash obtained with 25 ml. of diluted hydrochloric acid for five minutes, collect the insoluble matter on a tared filter crucible or on an ashless filter, wash with hot water, ignite, and weigh. Determine the percent of acid-insoluble ash calculated from the weight of drug taken."

The diluted hydrochloric acid dissolves the calcium carbonate, alkali chlorides, etc., leaving an acid-insoluble residue that consists almost entirely of silica derived from the soil, adhering to the drug.

Questions and Problems

1. From what different sources may the ash of a drug be derived?
2. When no specific ash limit is set by the official standards, what percent of acid-insoluble ash is the drug permitted to contain?

3. Why should the incineration residue not be heated above a dull redness?
4. Name the important inorganic plant constituents that will not be found in the ash.
5. What does the quantity of acid-insoluble ash in a drug indicate?

Residue on Ignition. The ash content of chemicals is determined by ignition to dull redness in the same manner as in the determination of the ash content of a crude drug. It is economically impracticable to make ash determinations on large samples of expensive chemicals. However, it is

TABLE 34. SUBSTANCES WITH NONVOLATILE RESIDUE LIMITS

Substance	Amount used, Gm or ml	Residue dried at °C	Official requirement, mg
U.S.P.			
Acetic acid	20	105	1.0
Camphor	2.0	120	1.0
Cherry juice	5.0	S.G., V.	n.l.t. 500
Glacial acetic acid	20	105	1.0
Hydrogen peroxide solution	20	105	30
Iodine	.....	105	0.05%
Iron dextran injection 50	1.0	105	28–32%
Iron dextran injection 75	1.0	105	35–40%
Iron dextran injection 100	1.0	105	37–43%
Menthol	2	105	1
Phenol	5	105	0.05%
Raspberry juice	5	S.G., V.	250
Sulfur dioxide	300	Evaporate	7.5
Tetrachloroethylene	50	105	1
Trichloroethylene	50	105	1
N.F.			
Acetic acid, diluted	20	105	1
Acetone	50	105	2
Alcohol, rubbing	25	105	n.l.t. 89
Ammonia solution, strong	10	105	5
Amylene hydrate	10	105	2
Butyl chloride	10	105	1
Chloroform	50	105	1
Ethyl acetate	.....	105	0.02%
Ethyl oxide	50	105	1
Isopropyl alcohol	50	105	2.5
Isopropyl rubbing alcohol	50	105	50
Myristica oil, East Indian	3	105	60
Myristica oil, West Indian	3	105	50
Orange flower water	100	105	15
p-Chlorophenol	1	105	0.01%
Thymol	2	100	10.05%
Vinyl ether	10	105	2

n.l.t. = not less than; V. = vacuum; S.G. = silica gel.

necessary to control the amount of inorganic matter that may be contained in such substances as alkaloidal salts, especially when the substance is intended for hypodermic administration. For this reason, the official standards usually require that small amounts of expensive substances be used and that the ash yielded be negligible. The term *negligible* is defined as a quantity not exceeding 0.5 mg. A residue on ignition limit is given for almost 400 drugs official in the U.S.P. XVII and N.F. XII. According to their respective monographs, acetazolamide may have an upper residue limit of 0.1 percent, whereas aspirin may have an upper residue limit of 0.05 percent. For this determination 1- to 2-Gm samples of the official substances are usually taken. In the case of Cocaine HCl, however, a 0.5-Gm sample is taken and the residue which remains must not exceed 0.5 mg. In other words, for this drug, only a negligible amount of residue may remain. Nonvolatile residue limits for a number of solvents and pharmaceutical preparations which are obtained in a similar manner are listed in Table 34.

Loss on Ignition. A number of official substances which have definite limitations on the amounts of volatile matter which they will lose when heated are U.S.P. calamine (2.0), calcium phosphate, dibasic (24.5 to 26.5), magnesium oxide (10.0), magnesium sulfate (40.0 to 52.0), talc (5.0), titanium dioxide (0.5), zinc oxide (1.0); N.F. calcium phosphate, tribasic (8.0), kaolin (15.0), lime (10.0), magnesium hydroxide (30.0 to 33.0), magnesium phosphate, tribasic (20.0 to 27.0). The permitted loss in percent weight is given in parentheses.

Water Content. Many drugs official in the U.S.P. and N.F. contain varying quantities of water, either as water of crystallization (hydrates) or as water in the adsorbed form. In order to assure uniformity in the official drugs, it is necessary that the U.S.P. and N.F. specify certain water-content limits in the drug monographs.

Water content of drugs may also be important for other reasons. When crude drugs, for example, are sold with a guaranteed assay for active constituent, the percent active constituent must be calculated on the basis of the moisture-free drug. In scientific work, also, moisture or water content of crude drugs and other substances must usually be accurately determined, since variable quantities of moisture would adversely affect the calculated results.

Water content is generally determined by one of four official methods: (1) gravimetric, for drugs containing no constituents, other than water, volatile at 105°, (2) gravimetric, for drugs containing ether-soluble constituents volatile at 105°, (3) azeotropic (toluene distillation), and (4) titrimetric (Karl Fischer).

For those drugs which contain no constituents volatile at 105°, other than water, a specified quantity of drug is accurately weighed and then dried in an oven at a specified temperature and time. The loss in weight of the drug represents the moisture present in the sample. Since, however, most vegetable drugs contain variable amounts of volatile substances other than water, such as essential oils, ethers, ester, and amines, this method is not generally applicable. When the drug contains matter, other than water, which is volatile at 105° the N.F. directions require that the volatile ether-soluble extractive shall be determined and the weight of the latter subtracted from the weight lost by the drug upon drying, the difference to be taken as the moisture content of the drug. Even this procedure is not free from error. Since water is slightly miscible with ether, some of the moisture contained in the drug is extracted by the anhydrous ether and determined as volatile ether-soluble extractive. A better method, known as the *xylene method* and developed by the U.S. Forestry Service, has been modified and is one of the official methods for moisture determination. It is referred to as the *azeotropic method* in the U.S.P. and the *moisture method by toluene distillation* in the N.F. Figure 18 shows a drawing of the apparatus used for moisture determination by this technique. This method is specified for the determination of moisture content of many vegetable drugs containing 2 or more percent of moisture, but it has the disadvantage that a comparatively large amount of drug, in some cases 50 to 100 Gm, must be used for the determination in order to secure a volume of water that can be measured conveniently without considerable error.

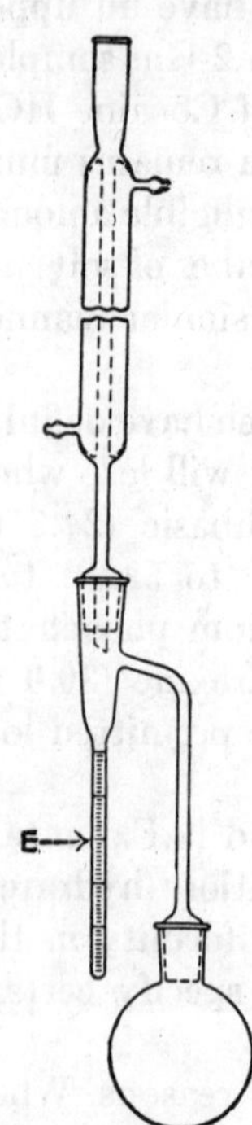

FIG. 18. Toluene moisture tube *E*.

Most of the official chemical compounds are relatively stable in respect to their moisture content at room temperatures; i.e., they neither gain nor lose more than slight traces of water. Some of them are hygroscopic and absorb water readily, e.g., aluminum chloride. Others lose a part or all of their water of crystallization unless properly preserved, e.g., copper sulfate. Control of the amount of water contained in chemicals is important in making solutions of definite concentration and in the determination of the dosage of those substances used as medicaments. Therefore many crystalline compounds that contain water of hydration or that contain adsorbed water are subject to a limit test under the heading "Water" which specifies a determination of the water content by titration with Karl Fischer reagent. Table 35 lists some of the official compounds with water limits. Where it is uncertain that the loss is due to water, or when it is known to represent other volatile substances, the test is headed "Loss on Drying." Nearly 400 official drugs have loss-on-drying limits. Table 36 lists some of these official compounds.

TABLE 35. SUBSTANCES WITH WATER LIMITS

Substance	Temperature, °C, and other conditions	Water limit, percent
U.S.P.		
Acacia	105 (5)	15.0
Acetazolamide	T.M.	0.5
Agar	105 (5)	20.0
Aloe	105 (5)	12.0
Aminophylline, anhydrous	T.M.	0.5
Aminophylline, hydrous	T.M.	7.9
Benzalkonium chloride	T.M.	15.0
Caffeine, anhydrous	80 (4)	0.5
Caffeine, hydrous	80 (4)	8.5
Caffeine and sodium benzoate	80 (4)	3.0
Calcium disodium edetate	T.M.	13.0
Chlorambucil	T.M.	0.5
Chlorobutanol, hydrous	T.M.	6.0
Citric acid, anhydrous	T.M.	0.5
Citric acid, hydrous	T.M.	8.8
Dextrose, anhydrous	105 (16)	0.5
Dextrose, hydrous	105 (16)	9.5
Digitalis	105 or A.M. (5)	6.0
Lanolin	S.B.	25.0 to 30.0
Magnesium trisilicate	C.W.	17.0 to 34.0
N.F.		
Acetaminophen	T.M.	0.5
Aluminum aspirin	T.M.	8.0
Amodiaquine hydrochloride	T.M.	7.0 to 9.0
Amylene hydrate	$CuSO_4$	No blue color
Calcium aminosalicylate	T.M.	14.5
Calcium cyclamate	T.M. or 140 (2)	6.0 to 9.0
Calcium saccharin	T.M.	15.0
Carbetapentane citrate	T.M.	1.0
Cetylpyridinium chloride	T.M.	4.5 to 5.5
Choline dihydrogen citrate	T.M. or P_2O_5, V. (4)	0.25
Sodium saccharin	T.M.	15.0
Testosterone enanthate	T.M.	0.05
Thiabendazole	T.M.	0.5
Triethylenemelamine	T.M.	0.3
Urethan	T.M.	1.0
Yeast, dried	100 (5)	7.0

T.M. = titrimetric method; A.M. = azeotropic method; S.B. = steam bath; C.W. = dried to constant weight; V. = vacuum. Time of drying, in hours, is indicated by number in parentheses.

TABLE 36. SUBSTANCES WITH LOSS-ON-DRYING LIMITS

Substance	Temperature, °C, and other conditions	Limit percent
U.S.P.		
Aluminum monostearate	80 (16)	2.0
Amaranth	120 C.W.	10.0
Aminohippuric acid	105 (2)	0.25
Ammonium chloride	S.G. (4)	0.5
Amobarbital	105 (4)	1.0
Aspirin	S.G. (5)	0.5
Aurothioglucose	P_2O_5 (24)	1.0
Bentonite	105 (2)	5.0 to 8.0
γ-benzene hexachloride	60, P_2O_5 10 mm (4)	0.5
Betazole hydrochloride	105 (3)	1.0
Bethanechol chloride	105 (2)	1.0
Bishydroxycoumarin	105 (3)	0.5
Bismuth subcarbonate	105 (3)	2.0
Boric acid	S.G. (5)	0.5
Busulfan	60, V. C.W.	2.0
Calcium carbonate, precipitated	200 (4)	2.0
Calcium gluconate	105 (16)	3.0
Calcium pantothenate	105 (3)	5.0
Carbachol	105 (2)	2.0
Carbasone	80 (6)	1.5
Cellulose, oxidized	P_2O_5, V. (18)	15.0
N.F.		
Syrosingopine	60, V. (3)	0.5
Tannic acid	105 (2)	12.0
Tartaric acid	P_2O_5 (3)	0.5
Testosterone	P_2O_5, V. (4)	1.0
Tetrahydrozoline hydrochloride	105 (2)	1.0
Thenyldiamine hydrochloride	105 (3)	1.0
Theobromine	80 (2)	0.5
Theophylline	105 (4)	0.5
Theophylline sodium glycinate	105 (4)	2.0
Thimerosal	P_2O_5, V. C.W.	0.5
Thonzylamine hydrochloride	105 (2)	0.5
Tribromoethanol	S.G. (4)	1.0
Tridihexethyl chloride	105 (2)	0.5
Trimethobenzamide hydrochloride	105 (4)	0.5
Trypsin crystallized	60, V. (4)	5.0
Tuaminoheptane sulfate	105 C.W.	1.0
Vinbarbital	105 (2)	1.5
Zinc undecylenate	105 (2)	1.25

S.G. = silica gel; V. = vacuum; C.W. = dried to constant weight; Time of drying, in hours, is indicated by number in parentheses.

For determinations of moisture by the titrimetric method, see Karl Fischer electrometric titration, Chap. 23.

Exercise 72

Object. To Determine the Water Content of Acacia.

Materials Required. 10 Gm of Acacia

Procedure. 1. Powder the acacia in a mortar.

In the case of other unground or unpowdered drugs, prepare about 10 Gm of the official sample by cutting, or shredding, so that the parts are about 3 mm in thickness. Seeds or fruits smaller than 3 mm should be cracked. High-speed mills, since they generate considerable heat, should not be used for preparing the sample, and care should be taken that no appreciable amount of moisture is lost during the preparation and that the portion taken is representative of the official sample.

The sample should be ground in order to expose as large a surface of the drug as possible so that the moisture may be driven off readily.

2. Accurately weigh about 10 Gm of the drug, as prepared, in a tared evaporating dish. Dry at 105° for 5 hr, and weigh. Continue the drying and weighing at 1-hr intervals until the loss is not more than 0.25 percent in 1 hr of drying.

The moisture and any other volatile constituents of acacia are driven off.

Calculate the percent of moisture contained in the sample of acacia used.

Questions and Problems

1. Why is the loss in weight of a drug dried at 100° not always a true measure of its moisture content?

2. If an air-dried drug containing 10 percent of moisture yielded 2.5 percent ash calculated on the basis of the air-dried drug, what percent of ash would the moisture-free drug contain?

Exercise 73

Object. Determination of the Moisture Content of Digitalis by the Toluene Distillation Method.

Materials Required. 50 Gm of digitalis leaf

200 ml of toluene previously shaken with water, separated from any excess water, and distilled

Toluene Moisture Apparatus. Use a 500-ml flask preferably of Pyrex glass, a straight-tube Liebig condenser of about 500-mm length, and a moisture-tube receiver calibrated to 0.1 ml of the type illustrated (Fig. 18). Clean the condenser and moisture tube with cleaning mixture, rinse with distilled water and then with alcohol, and dry them in an oven at 100°.

Procedure. 1. "Place in the dry flask a quantity of the substance, weighed accurately to the nearest centigram, which it is estimated will yield from 2 to 4 ml. of water. If the substance is of a pasty character, it is best weighed in a boat of metal foil of a size that will just pass through the neck of the flask. If the substance is likely to cause bumping, add enough dry, washed sand to cover the bottom of the flask, or a number of capillary melting-point tubes, about 100 mm. in length, sealed at the upper end. Place about 200 ml. of toluene in the flask, connect the apparatus as illustrated, and fill the receiving tube *E* with toluene poured through the top of the condenser. Heat the flask gently for 15 minutes, and when the toluene begins to boil, distil at the rate of about 2 drops per second until most of the water has passed over; then increase the rate of distillation to about 4 drops per second."

Sufficient drug must be taken to yield from 2 to 4 ml of water because smaller amounts would be difficult to determine without a relatively large percent of error. Thus, an error of 0.05 ml in the calibration or reading of the moisture tube would cause an error of 5 percent if only 1 ml of water is obtained and of 1.25 percent if 4 ml of water is obtained. At the boiling point of toluene, 110 to 111°, the moisture in the drug is volatilized. The moisture distills with the toluene, and as the vapors condense and drop into the moisture tube, the water, which is immiscible with and heavier than toluene, separates and collects at the bottom of the moisture tube.

2. "When the water has apparently all distilled over, rinse the inside of the condenser tube with toluene while brushing down the tube with a tube brush attached to a copper wire and saturated with toluene. Continue the distillation for 5 minutes, then remove the heat and allow the receiving tube to cool to room temperature. If any droplets of water adhere to the walls of the receiving tube, force them down with a rubber band wrapped around a copper wire and wet with toluene. When the water and toluene have separated completely, read the volume of water and calculate the percentage that was present in the substance."

Practically all the moisture distills over during the first 4 or 5 min of distillation. The condenser tube is washed down with toluene to remove water that may adhere to its inner wall. If the condenser is cleaned as directed, little difficulty will be experienced with water adhering to the wall of the tube. Some drugs hold their moisture quite tenaciously. Consequently, it is always well to continue distillation for 2 or 3 min after it is apparent from the test that practically all the moisture has distilled. Droplets adhering to the wall of a properly cleaned moisture tube can be dislodged by rotating the tube between the fingers or by tapping the tube gently on wood or some other base. Volatile organic substances, such as volatile oils and amines, remain dissolved in the toluene and so are not calculated as moisture. Since the result is read directly in milliliters, it is apparent that the percent of moisture would be 2 × number of milliliters of water when a 50-Gm sample is used.

Questions and Problems

1. Consult one of the official compendiums for the method of sampling vegetable drugs. What quantity constitutes an official sample when the total weight of drug exceeds 100 kg? How is the test sample prepared from the official sample?

2. Less than 1 part of water is soluble in 10,000 parts of toluene at 25°. Would the water dissolved in the toluene when moisture is determined in a drug by the toluene distillation method be appreciable if 100 ml of toluene unsaturated with water is employed?

Chapter 12

EXTRACTIVE AND CRUDE FIBER CONTENT

The amount of extractive that a drug yields to a given solvent is often an approximate measure of the amount of a certain constituent or group of related constituents that the drug contains. In cases where this is the object sought, it is important that a solvent be used which will not dissolve appreciable quantities of substances other than those sought in the extraction. In some cases, the amount of a drug soluble in a given solvent is an index of its purity; e.g., acacia must yield not more than 1 percent of water-insoluble residue to meet the official requirement.

For efficient and complete extraction, the solvent should be divided into a number of relatively small portions and used successively. Each portion should be brought into intimate contact with all parts of the sample and then be removed before the addition of the next portion. The use of fresh portions of the solvent that are not saturated with extractive favors the rapid and thorough removal of the solute with the minimum quantity of solvent.

The complete removal of extractive matter from a drug by this method, however, requires a large quantity of solvent in most cases, and most volatile solvents are expensive. Consequently, it is convenient and economical to be able to use the same portion of solvent repeatedly. To do so necessitates the separation of the solvent and solute after the extraction, the separated solvent being employed in a second extraction. This is readily accomplished by means of a special form of extraction apparatus of which the Soxhlet is a good example.

The Soxhlet (Fig. 19) is the apparatus generally used for extraction with volatile solvents where small quantities of drug are extracted, as in quantitative work. The solvent is placed in a weighed flask *A*, which is heated over an electric hot plate or over a water bath in the absence of a gas flame when the solvent is highly inflammable. The weighed sample of drug is placed in a porous paper extraction thimble, which is then fitted into the extraction tube *B*. A reflux condenser *C* is attached to the top of the extraction tube. The solvent vapors pass upward from the flask *A* through the side tube *D* of the extractor to the condenser. The condensate drips into the thimble until a sufficient amount has accumulated to raise the level of liquid in the extraction tube to the top of the siphon tube *E*. The solution that has been standing in contact with the drug is then discharged into the flask where the solvent is vaporized again. The solvent passes through this cycle repeatedly, and the solute is collected in the flask. The heat should be so adjusted that the solvent passes around the cycle about once every 10 min. The top of the thimble should be above the level of the top of the siphon side tube. The solvent must not be volatilized more rapidly than it will filter through the porous thimble, or the latter will overflow and carry insoluble matter over mechanically. After the extraction is complete, the flask is removed, the solvent is driven off until the flask and its contents are of constant weight, and the percentage of extractive in the drug is calculated from the weight of the residue.

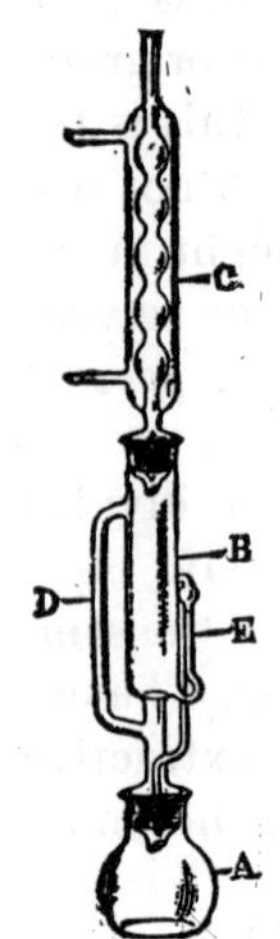

Fig. 19. Soxhlet extraction apparatus

In some cases, it is advantageous to determine the insoluble residue after extraction. The extraction thimble should be dried and weighed accurately before placing the sample in it. After the extraction is complete, it is dried with the residue and weighed accurately again. *Caution:* Do not place thimbles and residues saturated with ether or other inflammable or explosive solvents in an electric oven until they have been air-dried.

Volatile and Nonvolatile Ether-soluble Extractive. The determination of the total ether-soluble volatile constituents is applied to those drugs which contain volatile oils, while the determination of the nonvolatile ether-soluble constituents is applied to those drugs the active constituents of which are associated with volatile matter. In either case, the determination is only approximate.

Exercise 74

Object. To Determine the Nonvolatile Ether-soluble Content of Cocoa.

Materials Required. 10 Gm of Cocoa
50 to 100 ml of absolute ether
Soxhlet apparatus

Procedure. 1. "Extract about 10 Gm. of Cocoa, accurately weighed, with absolute ether in a continuous-extraction apparatus for 8 hours."

The extraction is continued for 8 hr to ensure complete removal of the ether-soluble constituents from the powdered cocoa. Absolute ether is used, since ordinary ether, which contains small, variable quantities of water, dissolves some tannin, sugar, etc.

2. "Allow the ether solution to evaporate spontaneously in a suitable tared container, dry at 105° for 1 hour, and weigh the nonvolatile, ether-soluble extractive. Retain the ether-insoluble residue for the test for Ether-insoluble residue."

Since ether boils at 35°, it evaporates rapidly at ordinary temperatures without carrying off more than negligible amounts of the volatile extractive. The residue represents the ether-soluble extractive of cocoa. This residue consists chiefly of fixed oil and resin. When the residue is heated at 105° until of constant weight, the volatile substances are volatilized, leaving resin, coloring matter, and fixed oil as the nonvolatile ether-soluble extractive.

The ether-insoluble residue is dried at 105° for 1 hr, and tests for total ash, crude fiber, etc., may be made.

To meet the official requirement set forth in the U.S.P. XVII, Cocoa must yield not less than 10 percent and more than 22 percent of nonvolatile, ether-soluble extractive. Purified Cotton, which is also official in the U.S.P. XVII, must yield not more than 0.7 percent total ether-soluble extractive.

Three substances official in the N.F. XII which have ether-soluble extractive limits are Ginger, Pancreatin, and Thimerosal. Ginger must yield not less than 4.5 percent total ether-soluble extractive to meet the official requirement, whereas Pancreatin and Thimerosal must yield residues not greater than 3.0 and 0.8 percent, respectively.

Alcohol-soluble Extractive. Alcohol is a good solvent for resinous matter. Consequently, the determination of the alcohol-soluble extractive is most frequently employed to determine approximately the amount of resin in those drugs in which the resinous matter is the important constituent. For some drugs, the determination of the alcohol-soluble extractive is considered officially as an assay process.

Exercise 75

Object. Assay of Benzoin.

Materials Required. 2 Gm of Benzoin
100 ml of alcohol, NaOH

Procedure. 1. Weigh 2 Gm of the prepared drug into a dried and tared paper extraction thimble, using a glass-stoppered weighing bottle as the container, and place 0.1 Gm of sodium hydroxide in the receiving flask. Extract the drug in a continuous extraction apparatus with alcohol, during 5 hr.

The alcohol dissolves the resins, benzoic acid, cinnamic acid, styrene, benzaldehyde, vanillin, esters, etc., contained in the balsamic resin. The NaOH in the receiving flask forms nonvolatile salts of the volatile acids.

2. Dry the insoluble residue at 105° for 2 hr and weigh. Determine the moisture in the drug by the toluene distillation method (see Exercise 73, Chap. 11), calculate the weight of moisture in the quantity of drug taken for the test and subtract it from the original weight of the drug taken for the test. The difference between this result and the weight of the residue determined as directed above represents the amount of alcohol-soluble extractive.

TABLE 37. SUBSTANCES WITH ALCOHOL-SOLUBLE EXTRACTIVE LIMITS

Substance	Alcohol used, percent	Extractive, percent
U.S.P.		
Benzoin, Siam	94.9	90.0
Benzoin, Sumatra	94.9	75.0
Storax	94.9	70.0
N.F.		
Vanilla	49.5	12.0

The residue should be dried and weighed in the extraction thimble. When the extraction is complete, the thimble should be removed from the apparatus and allowed to drain and dry in air for about 5 min to remove most of the alcohol. The residue and thimble can be dried conveniently at 105° in an electric oven. Since moisture remaining in the marc as well as the alcohol is volatilized at 105° it is necessary to correct for the loss of moisture by a moisture determination.

Calculate the percent of alcohol-soluble extractive assuming that the benzoin contained 3.20 percent moisture.

Diluted Alcohol Extractive. The diluted alcohol extractive determination is carried out as follows: Macerate about 2 Gm of the prepared drug, accurately weighed, in about 70 ml of diluted alcohol in a suitable flask. Shake the mixture during 8 hr at 30-min intervals and then allow it to stand for 16 hr without shaking. Filter and wash the flask and residue with small portions of diluted alcohol, passing the washings through the filter, until the filtrate measures 100 ml. Evaporate a 50-ml aliquot portion of this filtrate to dryness in a suitable tared dish on a water bath and dry to constant weight at 110°. Calculate the percentage of this extractive from the weight of drug taken.

Water-soluble Extractive and Water-insoluble Residue. The determination of the water-soluble extractive content is applied to those drugs of which one or more of the most important constituents are soluble in water, e.g., aloe. The determination is performed in the same way as the determination of the diluted alcohol-soluble extractive.

Exercise 76

Object. Assay of Aloe.

Materials Required. 2 Gm of Aloe

Procedure. Proceed as directed above for the determination of diluted alcohol extractive, using distilled water instead of diluted alcohol as the solvent.

Barbaloin, isobarbaloin, and other active constituents of a glycosidal nature dissolve in the water, leaving resinous matter, etc., undissolved.

The U.S.P. requires that aloe contain not less than 50 percent of water-soluble extractive. Calculate the percentage of water-soluble extractive contained in the sample examined, and compare the result with the official requirement. A water-soluble extractive limit is established for the following substances; minimum percent is given in parentheses: U.S.P. Aloe (50); N.F. Ginger (12).

In addition to the official definitions of the *water-soluble extractive* of the above substances, there are official limitations on the amounts of *water-soluble substances* permissible in certain other official drugs; e.g., the maximum percent of water-soluble substances in the following are given in parentheses: U.S.P. chlorophenothane (0.25), cotton, purified (0.25), pyroxylin (0.30), titanium dioxide (0.15), magnesium trisilicate (1.5), talc (0.1); N.F. bismuth magma (0.1) W/V, calcium phosphate, tribasic (0.5), magnesium phosphate, tribasic (1.5), pumice (0.2).

Solvent Hexane-soluble Extractive. Solvent Hexane is a good solvent for fats and fatty oils. The determination of the amount of extractive with this solvent is performed as follows:

Extract completely about 2 Gm of the prepared drug, accurately weighed, by subjecting it during 20 hours to the action of solvent hexane in a continuous extraction apparatus. Transfer the hexane solution to a tared porcelain dish and allow it to evaporate spontaneously. Then dry it over sulfuric acid for 18 hours and weigh. Calculate the percentage of anhydrous extractive from the weight of drug taken.

The hexane extractive is an index to the amount of fat present and serves as a check on the amount of seeds contained in colocynth pulp, since the seeds contain a large amount of fatty matter.

Crude Fiber. The crude fiber content of a drug is the residue, consisting chiefly of cellulose, that remains undissolved after successive treatment with boiling acid and alkali. The determination of crude fiber is of considerable importance in the examination of certain drugs and particularly of spices, since the commonly used adulterants consist of waste or refuse material derived from the drugs or spices themselves or from other food products. Frequently, this material is the outer cellular layer or protective coating, which contains a larger proportion of lignified tissues and, consequently, more crude fiber than the part official.

Exercise 77

Object. To Determine the Crude Fiber Content of Cocoa.

Materials Required. Ether-insoluble Residue of Cocoa
200 ml of 1.25 percent H_2SO_4
200 ml of 1.25 percent NaOH

Procedure. 1. Exhaust a weighed quantity of the prepared drug representing about 2 Gm of the drug, with ether or use the residue from the determination of the volatile ether-soluble extractive.

The drug should be powdered as finely as possible so that the acid and alkali will penetrate the innermost tissues and thus ensure complete removal of soluble material. It is exhausted first with ether to remove fats and waxes, which, being immiscible with aqueous solutions, tend to prevent the penetration of the acid and alkali solution into the drug particles.

2. Add 200 ml of boiling sulfuric acid solution, adjusted to exactly 1.25 percent by titration, to the ether-exhausted drug, in a 500-ml flask, and connect the flask with a reflux condenser, the tube of which passes only a short distance below the rubber stopper, into the flask. Heat the mixture to boiling and continue the boiling exactly 30 min. Then filter through a linen or hardened-paper filter and wash the residue on the filter with boiling water until no longer acid.

By diluting exactly 51 ml of 1 *N* sulfuric acid or an equivalent amount of a standard acid of known normality to 200 ml at 25°, 200 ml of 1.25 percent sulfuric acid may conveniently be prepared. The prepared acid solution should be measured and heated to boiling before it is added to the drug. When the acid solution is added to the powdered, ether-exhausted drug contained in a cold flask, the solution is cooled below its boiling point. The mixture should be heated rapidly to bring it to the boiling point again, and thereafter the flame should be cut down to the minimum required to maintain slow, steady boiling. Since the amount of material dissolved by the acid solution is partly dependent upon the length of time that the latter is heated in contact with the sample, it is important that the mixture should be boiled exactly 30 min, the time period being taken from the time the mixture starts to boil and not from the time the hot acid solution is added. The flask should be rotated gently from time to time while the mixture is being boiled to bring particles of the drug that adhere to the walls of the flask into contact with the acid. A current of air, passed into and beneath the surface of the mixture, through a capillary tube, will frequently prevent excessive frothing. The acid-insoluble residue is collected on a filter and washed until free of acid, since the latter, if present, would partially neutralize and, consequently, reduce the concentration of the alkali used in the next step of the procedure.

3. Rinse the residue back into the flask with 200 ml of boiling sodium hydroxide solution, adjusted to exactly 1.25 percent by titration and free from sodium carbonate. Again heat the mixture to boiling and continue the boiling exactly 30 min under the reflux condenser as described under the treatment with acid, then rapidly filter through a tared filter, wash the residue with boiling water until the last washing is neutral, dry it at 110° until of constant weight, and note the weight.

The alkali solution may be prepared conveniently by diluting 70 ml of recently standardized, carbonate-free, 1 *N* sodium hydroxide or an equivalent quantity of standard alkali solution of known normality to exactly 200 ml. The residue from the acid treatment in step 2 should be boiled with alkali in a manner duplicating as nearly as possible that used in the

TABLE 38. SUBSTANCES WITH INSOLUBLE RESIDUE LIMITS

Substance	Solvent	Insoluble residue limit, percent
U.S.P.		
Acacia	Dilute hydrochloric acid	1.0
Agar	Water	1.0
Aloe	Alcohol	10.0
Aluminum	Dilute hydrochloric acid	5.0
Amaranth	Water	0.5
Aspirin	Sodium carbonate T.S.	n.
Barium sulfate	Dilute hydrochloric acid	0.3
Boric acid	Water or alcohol	n.
Calamine	Dilute hydrochloric acid	2.0
Calcium carbonate, precipitated	Dilute hydrochloric acid	0.2
Calcium hydroxide	Dilute hydrochloric acid	0.5
Calcium phosphate, dibasic	Dilute hydrochloric acid	0.1
Cocoa	Ether	7.0
Evans blue	Water	0.05
Gentian violet	Alcohol	1.0
Hexachlorophene liquid soap	Alcohol	3.0
Magnesia, milk of	Dilute hydrochloric acid	n.
Magnesium oxide	Dilute hydrochloric acid	0.1
Phenolsulfonphthalein	Sodium bicarbonate	0.2
Potassium hydroxide	Water	n.
Pyroxylin	Water	0.3
Sodium bicarbonate	Water	n.
Sodium hydroxide	Water	n.
Sodium indigotindisulfonate	Water	0.5
N.F.		
Alum, exsiccated	Water	2.5
Calcium phosphate, tribasic	Dilute hydrochloric acid	0.2
Fuchsin, basic	Alcohol	1.0
Glycyrrhiza extract	Water	25.0
Green soap	Alcohol	3.0
Lime	Dilute hydrochloric acid	1.0
Magnesium carbonate	Dilute hydrochloric acid	0.5
Magnesium phosphate, tribasic	Water	0.2
Mercuric oxide, yellow	Dilute hydrochloric acid	n.
Mercury bichloride	Ether	0.3
Potassium bitartrate	Ammonia	n.
Sodium alginate	Water	0.02
Stannous fluoride	Water	0.2

n. = negligible.

boiling with acid. The residue is then collected on a filter, preferably in a Gooch crucible, washed free of alkali, dried, and weighed accurately. The dried residue consists chiefly of cellulose and small amounts of hemicelluloses, pentoses, inorganic matter, etc.

4. Incinerate the dried residue, and weigh the ash.

The fibrous matter should be incinerated cautiously at first and then at a low red heat until the carbon is completely burned off, as indicated by the formation of a white ash. The cellulose, etc., is burned when the fiber is incinerated, leaving the inorganic matter. The weight of the ash obtained subtracted from the weight of the residue found in step 3 gives the weight of the crude fiber. The crude fiber limit for cocoa is 7 percent of ether-insoluble residue.

Questions and Problems

1. What is meant by the term *crude fiber?*
2. Why should the sample used be finely ground and exhausted with ether before it is boiled with acid and alkali?
3. Why should the boiling with 1.25 percent acid and alkali be continued for exactly 30 min?
4. What substances remain undissolved by the acid and alkali?

In certain monographs limitations have been placed on the amount of substance that is insoluble in a given solvent: these limitations serve primarily to check the purity and identity of the substance.

Residues of official substances obtained by using solvents such as water, alcohol, ether, acids, and bases are also a means of establishing purity standards. Table 38 lists substances with insoluble residue limits.

Chapter 13

CONSTANTS OF FATS, FATTY OILS, WAXES, BALSAMS, RESINS, ETC.

The methods of analysis of fatty substances, waxes, resins, etc., usually consist of the determination of a number of physical and chemical properties or values commonly known as *constants*, although they are constant only within certain limits. These constants, when taken in conjunction with the color, odor, taste, and special identity tests for the given substance and for common adulterants, are the basis upon which the purity and quality of these substances are judged. Although the following determinations are not required for all the official substances that fall under the above caption, they are generally used in the examination of such substances.

The specific gravity, solubility, melting point, congealing point, refractive index, and optical activity determinations are performed according to the methods which are discussed later (see Chaps. 27 to 30).

Preparation of the Sample of Fats and Oils. "If a sample of oil shows turbidity owing to separated stearin, warm the container in a bath of water at 50° until the turbidity has disappeared and the oil is clear. Thoroughly mix the clarified oil before weighing the samples. If the oil does not become clear on warming, filter it through dry filter paper in a funnel contained in a hot water jacket. Weigh at one time as many portions as are needed for the various determinations, using preferably a bottle having a pipet dropper, or a weighing buret. Keep the sample melted, if solid at room temperature, until the desired samples are withdrawn."

Acid Value. The *acid value*, also known as the *acid number* and as the *acidity index*, is defined as *the number of milligrams of potassium hydroxide necessary to neutralize the free acids in* 1 *Gm of oil, fat, wax, resin, balsam, or similar organic substance of complex composition.* In other words, it gives the amount of potassium hydroxide, expressed in tenths of 1 percent, required to neutralize the free acids in a substance. The acidity may also be expressed as the number of milliliters of 0.1 *N* NaOH required to neutralize the free acid in 10 Gm of substance. This value is determined by titrating a weighed sample of the substance, contained in an alcoholic or in an alcohol-ether solution, with standard alkali solution, using phenolphthalein as the indicator. For the titration, 0.5 *N*, 0.1 *N*, or 0.02 *N* alkali may be employed, but a solution of 0.1 *N* concentration is most suitable in the majority of the official determinations. For substances like the balsams and resins, a mixed solvent, such as alcohol and ether, may be used advantageously, since the coloring matter that interferes with the observation of the end point is dissolved in the ethereal layer when the alkali is added. Solid fats and waxes are usually melted on a water bath and titrated while hot.

The presence of free acids in the oils, fats, and waxes is due chiefly to the hydrolysis of the esters composing them and is caused by chemical treatment, by bacterial action, or by the catalytic action of light and heat. As a rule, fresh or recently prepared fatty substances contain little or no free acids. Upon aging, the acid value increases slowly at first and more rapidly later, especially if the substance is not well protected from the simultaneous action of light and air. The acid values of the official fatty substances are quite variable; consequently, the official standards fix maximum limits, in most cases, which if exceeded indicate that the substances have undergone hydrolytic decomposition in their preparation or purification, or during the period of storage. High acid values are not necessarily an indication of rancidity, since the latter is a result of the action of the air, or possibly bacteria, on the liberated fatty acids.

A minimum acid value (or a minimum and a maximum acid value) is generally given in the official standards for substances that are balsamic or resinous in character. The value of these substances is usually based, in part at least, upon their content of free acids; e.g., the acid value of tolu is required to be not less than 112 and not more than 168. Since a good grade of tolu usually contains from 12 to 15 percent of cinnamic and benzoic acid in the free state, an acid value of less than 112 would indicate that the acid content of the balsam was low and that it was of inferior quality or adulterated. On the other hand, an acid value greater than 168 would indicate adulteration with some substance having high acid value, such as certain resins.

Exercise 78

Object. To Determine the Acid Value of Rosin.

Materials Required. About 1 Gm of Rosin
0.1 *N* sodium hydroxide
About 50 ml of a mixture of equal volumes of alcohol and ether, phenolphthalein T.S.

Procedure. 1. Pulverize the rosin in a mortar and dissolve about 1 Gm of the powder, accurately weighed, in from 40 to 50 ml of the neutral alcohol-ether mixture.

The rosin is powdered to facilitate its solution. Commercial alcohol and ether are frequently acid in reaction, and the mixture should always be tested before the sample is dissolved. This test may conveniently be

TABLE 39. ACIDITY LIMITS OF FATTY SUBSTANCES

Substance	Amount used, Gm	Alkali used	Official requirement, acid value
Expressed in Terms of Acid Values			
U.S.P.			
Oleic acid	2	0.1 *N* NaOH	195 to 204
Polyoxyl 40 stearate	10	0.1 *N* NaOH	n.m.t. 1
Stearyl alcohol	10	0.1 *N* NaOH	n.m.t. 2
Storax, American	1	0.5 *N* NaOH	36 to 85
Storax, Levant	1	0.5 *N* NaOH	50 to 85
Tolu balsam	1	0.5 *N* NaOH	112 to 168
White wax	3	0.5 *N* KOH (alcoholic)	17 to 24
N.F.			
Cetyl alcohol	10	0.1 *N* NaOH	n.m.t. 2
Glyceryl monostearate	10	0.1 *N* NaOH	n.m.t. 18
Rosin	1	0.1 *N* NaOH	n.l.t. 150
Undecylenic acid	10	0.1 *N* NaOH	n.l.t. 285
Yellow wax	3	0.5 *N* KOH (alcoholic)	18 to 24
Expressed in Terms of the Amount of Standard Alkali Required for Neutralization			
U.S.P.			
Castor oil	10	0.1 *N* NaOH	n.m.t. 7
Corn oil	10	0.02 *N* NaOH	n.m.t. 2
Cottonseed oil	10	0.02 *N* NaOH	n.m.t. 2
Lanolin, anhydrous	10	0.1 *N* NaOH	n.m.t. 2
Lanolin, hydrous	12.5	0.1 *N* NaOH	n.m.t. 2
Olive oil	10	0.1 *N* NaOH	n.m.t. 5
Peanut oil	10	0.02 *N* NaOH	n.m.t. 2
Polysorbate 80	10	0.1 *N* NaOH	n.m.t. 4
Sesame oil	10	0.02 *N* NaOH	n.m.t. 3
Sodium morrhuate injection	5	0.1 *N* NaOH	n.m.t. 0.5
Theobroma oil	10	0.1 *N* NaOH	n.m.t. 5
N.F.			
Almond oil	10	0.1 *N* NaOH	n.m.t. 5
Cod-liver oil	2	0.1 *N* NaOH	n.m.t. 1
Cod-liver oil, nondestearinated	2	0.1 *N* NaOH	n.m.t. 1
Iodized oil	1	0.1 *N* NaOH	n.m.t. 0.3
Persic oil	10	0.02 *N* NaOH	n.m.t. 2
Petrolatum	10	0.1 *N* NaOH	n.m.t. 0.4
Polyethylene glycol 1540	5	0.1 *N* NaOH	n.m.t. 0.2
Tocopherols concentrate, mixed	10	0.1 *N* NaOH	n.m.t. 10
Tocopheryl acetate concentrate, *d*-alpha	10	0.1 *N* NaOH	n.m.t. 10

n.m.t. = not more than; n.l.t. = not less than.

carried out in this case by adding 1 ml of phenolphthalein indicator solution to the alcohol-ether mixture before the rosin is dissolved. If the mixture is found to be acidic, it should be rendered neutral to phenolphthalein by the addition of the standard alkali solution.

In the titration of substances that form colored solutions, the end point may frequently be made observable by dilution with alcohol to a volume of from 100 to 200 ml.

2. Add 1 ml of phenolphthalein indicator solution (if it has not already been added in testing the neutrality of the alcohol-ether mixture) and titrate the solution with 0.1 *N* NaOH, agitating the mixture continuously.

In the titration of fatty substances, it is necessary to shake the mixture thoroughly after the addition of each portion of alkali to secure complete extraction of the fatty acids from the immiscible oily layer. This precaution is not of great importance in the titration of rosin, however, since rosin is soluble in alcohol and ether, and also in alkalies.

The alkali reacts with the dibasic abietic acid formed from the abietic acid anhydride and with the other acids contained in the rosin, with the formation of so-called "rosin soaps" and water. It has been suggested that the acid value of rosin can be more accurately determined in alcoholic solution, using alcoholic KOH in the titration, since the presence of water in the above determination causes dissociation of the rosin soap with consequent low results.

In addition to the listed substances for which definite acid value limits are given in the official standards, there are a number of substances for which the maximum content of free-fatty acids is fixed by the volume of standard alkali solution required in their titration.

The calculation of the acid value is illustrated as follows:

Problem. Find the acid number of a rosin sample weighing 1.100 Gm which required 28.00 ml of 0.1100 *N* NaOH to bring about the end point.

Solution

1. Determine the number of milliequivalents of NaOH consumed by the sample: 28.00 ml × 0.1100 *N* = 3.08 mEq.
2. Multiply 3.08 mEq by 56.11 mg, the weight expressed in milligrams of 1 mEq of KOH: 3.08 mEq × 56.11 mg/mEq = 172.8 mg of KOH required to neutralize the fatty acids in 1.100 Gm of sample.
3. Divide the sample weight into milligrams of KOH in order to find milligrams of KOH required to neutralize the fatty acids in exactly 1 Gm of sample: 172.8 mg/1.100 Gm = 157.1 mg/1 Gm sample.
4. Combining all mathematical operations (steps 1, 2, and 3):

$$\frac{28.00 \text{ ml} \times 0.1100\ N \times 56.11 \text{ mg/mEq}}{1.100 \text{ Gm}} = 157.1 \text{ mg/Gm}$$

or simply 157.1, the acid value of rosin.

Alternative Solution

1. Determine the number of milliliters of exactly 0.1 *N* NaOH which is equivalent to 28.00 ml of 0.1100 *N* NaOH solution:

$$\frac{28.00 \text{ ml} \times 0.1100\ N}{0.1000\ N} = 30.80 \text{ ml}$$

2. Multiply 30.80 ml by 5.611 mg, the number of milligrams of KOH equivalent to 1.00 ml of exactly 0.1 *N* NaOH solution: 30.80 ml × 5.611 mg = 172.8 mg KOH.

3. Continue calculations as in step 3 of previous example.

Questions and Problems

1. Define the term *acid value.*
2. Why must the alcohol or ether used as a solvent for fats, resins, etc., in acid value determinations be neutral?
3. Why do the official standards set up maximum acid values for most of the official fats and oils and minimum values for most of the resinous substances?
4. If a 2-Gm sample of cod-liver oil required 4.5 ml of 0.02 *N* NaOH in the titration of the free fatty acids, would the oil conform with the official purity requirement? What would be the acid number of the oil?

Saponification Value. *The saponification value, saponification number,* or *Koettsdorfer number,* as it is sometimes called from the originator of the process, is defined as *the number of milligrams of potassium hydroxide required to neutralize the free acids and saponify the esters contained in 1 Gm of fat, fatty or volatile oil, wax, resin, balsam, or other substance of similar composition.* This value represents the amount of potassium hydroxide, expressed in tenths of 1 percent, required to neutralize the total free and combined acids in 1 Gm of the substance, or, in other words, it is ten times the percentage of potassium hydroxide required to neutralize all the acids contained in the sample after saponification. Since the natural fats and oils consist of mixtures of glyceryl esters of the higher acids, their saponification values do not differ greatly. The determination of the saponification value, however, serves to aid in the detection of the presence of the glycerides of acids containing less than 16 or more than 18 carbon atoms, since the value of this constant is inversely proportional to the mean molecular weights of the acids present. In some cases, it may also indicate adulteration of the sample with unsaponifiable matter, such as mineral oil.

Exercise 79

Object. To Determine the Saponification Value of Cottonseed Oil.

Materials Required. 2 Gm of cottonseed oil
50 ml of 0.5 *N* alcoholic potassium hydroxide (see Exercise 82, Chap. 14)
About 50 ml 0.5 *N* hydrochloric acid

"If the oil has been saturated with carbon dioxide for the purpose of preservation, it should be exposed in a shallow dish in a vacuum desiccator for twenty-four hours before the portions are weighed for this determination."

Procedure. 1. Place from 1.5 to 2 Gm of the sample, accurately weighed, in a flask of from 200- to 250-ml capacity, and add to it exactly 25 ml of alcoholic 0.5 *N* potassium hydroxide. Insert into the neck of the flask, by means of a perforated stopper, an air condenser consisting of a glass tube from 70 to 80 cm in length and from 5 to 8 mm in diameter, and heat the flask on a water bath for 30 min, frequently rotating the contents.

If the sample is clear, it need not, as a rule, be filtered. When the presence of extraneous matter, however, makes it necessary to filter the oil, the operation may be carried out easily by means of a suction filter. The sample of oil should be introduced into the weighed flask by means of a dropper to avoid the possibility of a portion adhering to the neck of the flask, and the whole reweighed. The long glass tube serves as an air condenser to prevent the escape of alcohol vapor. An ordinary water-jacketed condenser may be used with equally satisfactory results. The mixture of oil and 0.5 *N* alcoholic KOH is heated on a water bath and mixed by imparting a rotatory motion to the flask from time to time, in order to accelerate the saponification. The alcoholic KOH solution used should not be deeply colored (see Exercise 82), since the color would interfere with the observation of the end point in the titration of the saponification mixture. Alcoholic KOH, rather than an aqueous solution, is employed, because the oils are more soluble in alcohol than in water, and because the products of saponification are completely soluble in alcohol, whereas when aqueous solutions are used, the unsaponifiable matter remains insoluble.

When saponification is complete, the mixture should be clear and free from oil globules. Although the glyceryl esters present in fatty oils vary greatly with respect to their acid components, the character of the saponification reaction may be illustrated as follows:

$$\underset{\text{Glycerol oleate}}{C_3H_5(C_{17}H_{33}CO_2)_3} + 3KOH \longrightarrow \underset{\text{Glycerin}}{C_3H_5(OH)_3} + \underset{\text{Potassium oleate}}{3C_{17}H_{33}CO_2K}$$

2. Then add 1 ml of phenolphthalein T.S. and titrate the excess of potassium hydroxide with 0.5 *N* hydrochloric acid.

Sulfuric acid should not be substituted for hydrochloric acid, since potassium sulfate would be precipitated from the alcoholic solution and this would interfere with the observation of the end point.

3. Make a blank test at the same time, using exactly the same amount of alcoholic 0.5 *N* potassium hydroxide. The difference in the number of milliliters of 0.5 *N* hydrochloric acid consumed in the actual test and the blank, multiplied by 28.05 and divided by the weight of the sample taken, gives the saponification value.

The blank test should be carried out at the same time as the sample is run, using similar flasks, boiling for the same length of time and under similar conditions, except that the oil should be omitted. This is done to eliminate as far as possible errors from every source, such as those which would be introduced by the absorption of CO_2 by the alkali or by the alkalinity of the glass.

The calculation of the saponification value is illustrated as follows:

Problem. Find the saponification value of cottonseed oil if a 1.532-Gm sample, refluxed with 25.00 ml of about 0.5 *N* alcoholic KOH, required 15.70 ml of 0.5100 *N* HCl for the residual titration. The blank was run using the same volume of 0.5 *N* alcoholic KOH and required 26.00 ml of 0.5100 *N* HCl to bring about the end point.

Solution

1. Determine the number of milliequivalents of KOH consumed by the sample:

$$(26.00 \text{ ml} \times 0.5100\ N) = 13.26 \text{ mEq of HCl (also KOH)}$$
$$(15.70 \text{ ml} \times 0.5100\ N) = 8.01 \text{ mEq of HCl (also KOH)}$$
$$13.26 \text{ mEq} - 8.01 \text{ mEq} = 5.25 \text{ mEq of KOH}$$

the amount required to saponify the esters and neutralize the free fatty acids in the sample.

2. Multiply 5.25 mEq of KOH so consumed by 56.11 mg, the weight in milligrams of 1 mEq of KOH: 5.25 mEq × 56.11 mg/mEq = 294.6 mg KOH consumed by 1.532 Gm of sample.

3. Divide sample weight into milligrams of KOH in order to determine milligrams of KOH consumed by exactly 1 Gm of sample. 294.6 mg/1.532 Gm = 192.3 mg KOH/Gm of sample.

4. Combining all mathematical operations (steps 1, 2, and 3) and factoring:

$$\frac{(26.00 \text{ ml} - 15.70 \text{ ml})0.5100\ N \times 56.11 \text{ mg/mEq}}{1.532 \text{ Gm}} = 192.3 \text{ mg KOH/Gm sample}$$

or simply 192.3, the saponification value of the oil.

Alternative Solution

1. Determine the number of milliliters of exactly 0.5 *N* alcoholic KOH consumed by the sample:

$$\frac{26.00 \text{ ml} \times 0.5100\ N}{0.5000\ N} = 26.52 \text{ ml HCl (also KOH) blank titration}$$

$$\frac{15.70 \text{ ml} \times 0.5100\ N}{0.5000\ N} = 16.01 \text{ ml HCl (also KOH) residual titration}$$

$$26.52 \text{ ml} - 16.01 \text{ ml} = 10.51 \text{ ml}$$

or the amount of exactly 0.5 *N* alcoholic KOH consumed by the sample.

2. Multiply 10.51 ml by 28.05 mg, the number of milligrams of KOH present in 1.00 ml of exactly 0.5 *N* alcoholic KOH solution. 10.51 ml × 28.05 mg = 294.6 mg.

3. Continue calculations as in step 3 of the previous example.

TABLE 40. SAPONIFICATION VALUE LIMITS OF FATTY SUBSTANCES

Substance	Amount used, Gm	Saponification value
U.S.P.		
Castor oil	1.5 to 2	179 to 185
Corn oil	1.5 to 2	187 to 193
Cottonseed oil	1.5 to 2	190 to 198
Olive oil	1.5 to 2	190 to 195
Peanut oil	1.5 to 2	185 to 195
Polyoxyl 40 stearate	1.5 to 2	25 to 35
Sesame oil	1.5 to 2	188 to 195
Storax	2	160 to 200
Theobroma oil	1.5 to 2	188 to 195
Tolu balsam	1	154 to 220
N.F.		
Almond oil	1.5 to 2	190 to 200
Cod-liver oil	1.5 to 2	180 to 192
Cod-liver oil, nondestearinated	1.5 to 2	180 to 192
Glyceryl monostearate	1.5 to 2	164 to 170
Persic oil	1.5 to 2	185 to 195

The saponification values of the official fats, fatty oils, waxes, resins, etc., are determined in the same way.

Questions and Problems

1. Define the term *saponification value.*
2. Why should a blank determination be run parallel with the determination of the saponification value of the sample?
3. Consult the following table and enumerate several oils that, if used to adulterate sesame oil, might lower its saponification value.
4. A sample of corn oil weighing 1.600 Gm saponified with 25 ml of 0.4 *N* KOH required 9.0 ml of 0.5 *N* HCl to titrate the excess KOH. In the blank determination, 20 ml of 0.5 *N* HCl was required to titrate the alkali. Calculate the saponification value of the sample. Does the value found correspond with the U.S.P. requirement?
5. Need the alcoholic KOH used in a saponification value determination be of an exact known normality? Explain.
6. How is the saponification value of balsam of tolu determined in the U.S.P.?

Ester Value. The *ester value,* or *ester number,* is defined as *the number of milligrams of potassium hydroxide required to saponify the esters in* 1 *Gm of a fatty or volatile oil, fat, wax, balsam, resin, or similar organic substance.* In those substances that do not contain free acids, the ester number is equal to the saponification value. When free acids are present, however, the ester value is the difference between the acid and saponification values.

The ester value is of particular importance in the analysis of yellow and white wax, since it serves, in many cases, to indicate the presence of adulterants, such as paraffin. Considered in conjunction with the acid value, the ester value may aid in the detection of such adulterants as rosin and stearic acid in the wax.

The United States Pharmacopeia gives the following directions for the determination of the ester value:

"Shake from 1.5 to 2 Gm. of the substance, accurately weighed in a 200- to 250-ml. tared flask, with from 20 to 30 ml. of neutralized alcohol, add 1 ml. of phenolphthalein T.S., and titrate with 0.5 *N* alcoholic potassium hydroxide until the free acid is neutralized. Add 25.0 ml. of 0.5 *N* alcoholic potassium hydroxide and proceed as directed under *Saponification Value*, beginning with 'Insert in the neck of the flask' and omitting the further addition of phenolphthalein T.S. The difference between the number of milliliters of 0.5 *N* hydrochloric acid consumed in the actual test and in the blank, multiplied by 28.05 and divided by the weight in Grams of the sample taken, is the Ester Value."

The ester number may also be calculated in the same manner as that used for the saponification value.

Questions and Problems

1. Define the term *ester number.*

2. If a sample of beeswax is found to have an acid number of 20.4 and a saponification value of 89.8, what would be the ester number of the sample?

3. If a sample of white beeswax has an acid value of 18.45 and an ester value of 74, what would be the saponification value of the sample?

Official substances with their required ester numbers include white wax 72 to 79 and yellow wax 72 to 77.

Unsaponifiable Matter. The term *unsaponifiable matter* is defined in the Pharmacopeia as *the substances present in oils or fats that are not saponified by alkali hydroxides but are soluble in ordinary fat solvents.*

When oils and fats are saponified, there remains a small amount of residue that may consist of phytosterol in the vegetable oils and fats, of cholesterol in the animal oils and fats, and in some cases, of unsaponifiable substances added for the purpose of adulteration. The normal unsaponifiable matter present in most oils and fats dissolves in the alcoholic soap solution obtained upon saponification. When a large amount of unsaponifiable matter is present, however, it may separate out and float on the surface of the mixture.

The determination of the amount of unsaponifiable matter present in certain oils, such as linseed oil and cod-liver oil, may be indicative of the quality and purity of the oil. Thus, the addition of petroleum oils, a common adulterant, to linseed oil would greatly increase the unsaponifiable-matter content of the latter.

The official compendiums give the following procedure for the determination of the amount of unsaponifiable matter in a fat or wax:

Weigh 5 Gm of the oil or fat into a 250-ml Erlenmeyer flask, add a solution of 2 Gm of potassium hydroxide in 40 ml of alcohol, and heat under a reflux condenser for 2 hr, keeping the alcohol gently boiling. Evaporate the alcohol on a water bath, dissolve the residue in 50 ml of hot water, and transfer the solution to a separator, rinsing the flask with two 25-ml portions of hot water which are added to the separator. Cool to room temperature, and extract with two successive portions of 50 ml each of ether, adding a few drops of alcohol to facilitate the separation of the two liquids. Combine the ether extracts in another separator and wash the ether solution first with 20 ml of a solution of sodium hydroxide (4 in 1,000), then with 20 ml of a solution of sodium hydroxide (8 in 1,000), and finally with 15-ml portions of water until the last washing is not reddened by the addition of 2 drops of phenolphthalein T.S. Transfer the ethereal solution to a tared beaker, and rinse the separator with 10 ml of ether, adding the rinsings to the beaker. Evaporate the ether just to dryness on a water bath, and dry the residue for 30 min at 100°. Cool the beaker in a desiccator for 30 min, and weigh the residue of Unsaponifiable Matter.

The Pharmacopeia and N.F. require that corn oil contain not more than 1.5 percent and cod-liver oil and nondestearinated cod-liver oil shall contain not more than 1.3 percent of unsaponificable matter.

Iodine Value. The *iodine value*, or *number, is the number of grams of iodine absorbed under specified conditions by* 100 *Gm of oil, fat, wax, or other substance.* This value is a quantitative measure of the proportion of unsaturated fatty acids present, both free and combined as esters, that have the property of absorbing iodine.

The determination of the iodine number of fats and oils is important, since it serves to characterize them and to indicate whether they are pure or admixtures. The so-called "drying oils," such as linseed oil, and the fish oils, such as cod-liver oil, have very high iodine numbers, usually above 120, since they contain a large proportion of unsaturated fatty acids; the nondrying oils, such as olive oil and almond oil, have relatively low iodine numbers, below 100; and the semidrying oils, such as cottonseed oil and sesame oil, have intermediate iodine values, that is, between 100 and 120. In the case of the animal fats, the iodine number is not very high, usually being less than 90. The determination of the iodine number, therefore, not only serves as an aid to the identification of known oils, but it also serves to indicate in a definite manner the class to which an unknown fat or oil belongs. Furthermore, when the iodine number is considered in conjunction with the saponification value of a fat or oil, it serves as a means of detecting adulteration, and frequently it indicates the nature of the adulterant; e.g., olive oil might be adulterated with cottonseed oil without changing the saponification value appreciably, but the iodine number of the olive oil would be increased. Again, castor oil might be adulterated with olive oil without changing the iodine number greatly, but the saponification value of the castor oil would be increased.

Several methods have been developed for the determination of the iodine number of fats and oils. These methods are generally designated by the name of their originators, as, for example, the Hübl, Hanus, and Wijs methods. The method given in the Pharmacopeia is essentially that of Hanus. In this method, the following solutions are required:

1. *Iodobromide Test Solution.* This solution is prepared as follows:

Dissolve 13.2 Gm of reagent iodine in 1,000 ml of glacial acetic acid with the aid of gentle heat, if necessary. Cool the solution to 25° and determine the iodine content in 20 ml by titration with 0.1 *N* sodium thiosulfate. Add to the remainder of the solution a quantity of bromine equivalent to that of the iodine present. Preserve in glass-stoppered bottles, protected from light.

No difficulty will be experienced in dissolving the iodine, as a rule, if the glacial acetic acid contains less than 0.5 percent of water. The concentration of this solution is determined by titration with 0.1 *N* sodium thiosulfate using the following procedure: Carefully measure 20 ml of the solution into a flask. Then add gradually and cautiously from a buret 0.1 *N* sodium thiosulfate, shaking constantly until the color of the solution is but slightly yellow. Then add about 3 ml of starch T.S. and continue the addition of the 0.1 *N* sodium thiosulfate until the blue color is just discharged. Thus, if 20.5 ml of the iodine solution requires 20 ml of 0.1 *N* sodium thiosulfate, each milliliter of the iodine solution must contain $\frac{20}{20.5} \times 0.01269 = 0.01233$ Gm of iodine, and the remaining 980 ml of iodine test solution would contain $980 \times 0.01233 = 12.05$ Gm of iodine. Since 126.90 Gm of iodine is equivalent to 79.91 Gm of bromine, the quantity of bromine equivalent to the iodine contained in the solution may be calculated:

$$X:12.05::79.91:126.90$$

$$X = 7.59 \text{ Gm of bromine}$$

Since bromine has a specific gravity of about 3.1, 7.64 Gm of bromine is equivalent to $7.64/3.1 = 2.5$ ml in round numbers. Inasmuch as a slight excess of iodine does no harm, the addition of 2.25 ml of bromine to the iodine solution would be sufficient in this case.

The bromine may be drawn into a measuring pipet by means of a rubber bulb and then added to the cool iodine solution. (Never draw bromine into a pipet by means of the mouth.)

2. *Potassium iodide test solution*, prepared by dissolving 16.5 Gm of potassium iodide in sufficient water to make 100 ml.

3. 0.1 *N sodium thiosulfate solution*, prepared and standardized as directed in Exercise 55, Chap. 9.

4. *Starch test solution*, prepared as follows:

Triturate 1 Gm of arrowroot starch with 10 ml of cold water, and pour slowly, with constant stirring, into 200 ml of boiling water. Boil the mixture until a thin, translucent fluid is obtained. (Longer boiling than necessary renders the solution less sensitive.) Allow to settle and use only the clear supernatant liquid. This test solution must be freshly prepared.

Exercise 80

Object. To Determine the Iodine Value of Olive Oil.

Materials Required. About 1 Gm of olive oil
20 ml of chloroform
50 ml of iodobromide test solution
60 ml of potassium iodide test solution
About 100 ml of 0.1 *N* sodium thiosulfate solution
Starch test solution

Procedure. 1. Introduce about 0.3 Gm of the olive oil, accurately weighed, into a glass-stoppered flask or bottle of 250-ml capacity, dissolve it in 10 ml of chloroform, add 25 ml of iodobromide T.S., accurately measured from a buret or pipet, stopper the vessel securely, and allow it to stand for 30 min[1] in a cool place protected from light.

Only 0.3 Gm of olive oil is used in this determination, as its iodine value is comparatively high (79 to 88).[2]

The Pharmacopeia directs that 0.8 Gm be taken for solid fats, as their iodine values are low as compared with those of the fatty oils, the iodine value of oil of theobroma, for instance, being 35 to 43.

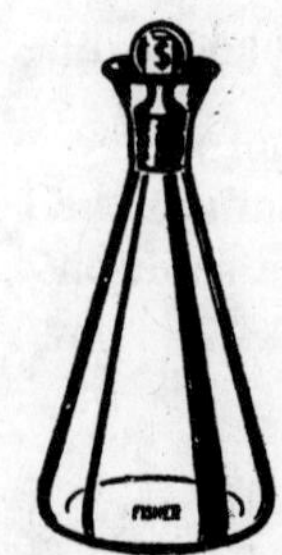

FIG. 20. Iodine titration flask.

The oil is soluble in chloroform, and since iodine and iodine bromide are also soluble in this solvent, it serves as a good reaction medium.

The oil, upon standing in contact with the iodobromide test solution, absorbs iodine. Iodine, itself, is absorbed very slowly by the oil, but iodine is absorbed readily from bromine-containing solutions, probably through a reaction of iodine bromide, IBr, wherein the bromine functions as a catalytic agent. The unsaturated acids of the oleic and linoleic series present in the olive oil, as well as their glyceryl esters, absorb iodine to form addition products. Thus oleic acid $C_{17}H_{33}COOH$ takes up 2 atoms of iodine and forms the addition product diiodostearic acid $C_{17}H_{33}I_2COOH$. The mixture, which is to be placed in a glass-stoppered flask (Fig. 20) to prevent the escape of iodine

[1] Allow castor oil, cod-liver oil, or linseed oil to stand for 1 hr.
[2] 0.12 to 0.15 Gm of linseed oil or of cod-liver oil, 0.8 to 1.0 Gm of theobroma oil.

and bromine vapors, should also be protected from light to prevent other side reactions from taking place in which iodine is consumed. If after standing for ½ hr the solution is no longer colored brown, the amount of iodine present was insufficient to saturate the oil.

2. Then add, in the order named, 30 ml of potassium iodide T.S. and 100 ml of water, and titrate the liberated iodine with 0.1 *N* sodium thiosulfate, shaking thoroughly after each addition of thiosulfate. When the iodine color becomes quite pale, add 1 ml of starch T.S. and continue the titration with thiosulfate until the blue color is discharged.

Upon the addition of the potassium iodide solution, the iodine bromide reacts with the KI, liberating iodine until all the bromine is used up:

$$KI + IBr \longrightarrow KBr + I_2$$

An excess of KI is added to ensure the complete removal of the free bromine and to prevent the precipitation of the iodine in the aqueous solution. The sodium thiosulfate reacts with the excess iodine as follows:

$$2Na_2S_2O_3 + I_2 \longrightarrow 2NaI + Na_2S_4O_6$$

3. Carry out a blank test at the same time with the same quantities of the same reagents and in the same manner and titrating as directed. The difference between the number of milliliters of thiosulfate consumed by the blank test and the actual test, multiplied by 1.269 and divided by the weight in Grams of the sample taken, is the Iodine Value.

"NOTE—If more than half of the iodobromide T.S. is absorbed by the portion of the substance taken, the determination must be repeated, a smaller portion of the substance under examination being used."

The blank test, when carried out under the same conditions as the actual test, corrects for the presence of impurities in the reagents, changes in volume at different temperatures, etc., and makes it unnecessary to know the exact normality of the iodobromide test solution. The blank test should be carried out in a manner exactly duplicating that used in the actual test insofar as possible, especially in the measurement of the iodobromide test solution, which has a high coefficient of expansion. It is required that the determination be repeated if more than half of the iodobromide T.S. is absorbed by the sample to ensure the presence of an excess of iodine.

The calculation of the iodine value is illustrated as follows:

Problem. Determine the iodine value of a sample of olive oil weighing 0.3100 Gm, if 24.15 ml and 5.25 ml of 0.1200 *N* sodium thiosulfate solution are required for the blank and residual titrations, respectively.

Solution

1. Determine the number of milliequivalents of Iodine consumed by the sample:

$$(24.15 \text{ ml} \times 0.1200\ N) = 2.898 \text{ mEq } Na_2S_2O_3 \text{ (also Iodine)}$$
$$(5.25 \text{ ml} \times 0.1200\ N) = 0.630 \text{ mEq } Na_2S_2O_3 \text{ (also Iodine)}$$
$$2.898 \text{ mEq} - 0.630 \text{ mEq} = 2.268 \text{ mEq Iodine}$$

or the amount consumed by the oil.

2. Multiply 2.268 mEq of Iodine by 0.1269 Gm, the weight of 1 mEq of iodine: 2.268 mEq × 0.1269 Gm/mEq = 0.2878 Gm of Iodine consumed by 0.3100 Gm of oil.

3. Divide sample weight into grams of Iodine in order to determine grams of iodine consumed by exactly 1 Gm of oil. Multiply this value by 100 to obtain the iodine number:

$$\frac{0.2878 \text{ Gm} \times 100}{0.3100 \text{ Gm}} = 92.8 \text{ Gm of Iodine consumed by 100 Gm of oil}$$

4. Combining all mathematical operations (steps 1, 2, and 3) and factoring:

$$\frac{(24.15 \text{ ml} - 5.25 \text{ ml}) \times 0.1200\ N \times 0.1269 \text{ Gm/mEq}}{0.3100 \text{ Gm}} \times 100 = 92.8$$

the iodine value of the oil.

Alternative Solution

1. Determine the number of milliliters of exactly 0.1 N iodine absorbed by the sample:

$$\frac{24.15 \text{ ml} \times 0.1200\ N}{0.1000\ N} = 28.98 \text{ ml } Na_2S_2O_3 \text{ (also iodine) blank titration}$$
$$\frac{5.25 \text{ ml} \times 0.1200\ N}{0.1000\ N} = 6.30 \text{ ml } Na_2S_2O_3 \text{ (also iodine) residual titration}$$
$$28.98 \text{ ml} - 6.30 \text{ ml} = 22.68 \text{ ml of exactly } 0.1000\ N \text{ iodine}$$

or the amount absorbed by the sample.

2. Multiply 22.68 ml × 0.01269 Gm, the number of grams of iodine present in 1.00 ml of exactly 0.1 N iodine solution:

$$22.68 \text{ ml} \times 0.01269 \text{ Gm} = 0.2878 \text{ Gm of iodine.}$$

3. Continue calculations as in step 3 of previous example.

TABLE 41. FATTY SUBSTANCES WITH THEIR IODINE NUMBERS

Substance	Iodine number official requirement
U.S.P.	
Castor oil	83 to 88
Corn oil	102 to 128
Cottonseed oil	109 to 116
Lanolin, hydrous	18 to 36
Lanolin, anhydrous	18 to 36
Oleic acid	85 to 95
Olive oil	79 to 88
Peanut oil	84 to 100
Sesame oil	103 to 116
Sodium morrhuate injection	n.l.t. 130
Stearic acid	n.m.t. 4
Stearyl alcohol	n.m.t. 2
Theobroma oil	35 to 43
N.F.	
Almond oil	95 to 105
Cetyl alcohol	n.m.t. 5
Cod-liver oil	145 to 180
Cod-liver oil, nondestearinated	128 to 180
Glyceryl monostearate	n.m.t. 6
Persic oil	90 to 108
Undecylenic acid	131 to 138

n.m.t. = not more than; n.l.t. = not less than.

Questions and Problems

1. Define the term *iodine value.*

2. Indicate by structural formulas how iodine may add on to an unsaturated fatty acid.

3. How may a determination of the iodine number be used to supplement the information relative to the nature and purity of an oil gained through a determination of the saponification value?

4. What functions do the bromine, acetic acid, chloroform, and potassium iodide serve in the determination of the iodine number of an oil?

5. Given an unknown sample of a fatty oil, what observations, tests, and determinations would you perform in order to establish the identity and purity of the oil?

Hydroxyl Value. The *hydroxyl value,* or *hydroxyl number, is the number of milligrams of potassium hydroxide equivalent to the hydroxyl content of* 1 *Gm of the substance.* This constant gives an indication of the identity and purity of fatty substances possessing alcoholic hydroxyl groups.

The hydroxyl value is inversely proportional to the molecular weight. Therefore, an abnormally low value is indicative of adulteration with higher molecular weight alcohols or with nonalcoholic fatty substances (paraffin, petroleum oil, etc.).

Exercise 81

Object. To Determine the Hydroxyl Number of Cetyl Alcohol.

Materials Required. 2 Gm of cetyl alcohol
2 ml of pyridine
10 ml of toluene
10 ml of 1.5 *M* acetyl chloride in toluene
About 30 ml of 1 *N* sodium hydroxide

Procedure. 1. Transfer into a dry 250-ml iodine flask about 2 Gm of Cetyl Alcohol, accurately weighed, and add exactly 2 ml of pyridine followed by 10 ml of toluene, using the latter to rinse the neck of the flask. To the mixture add exactly 10 ml of approximately 1.5 *M* acetyl chloride in toluene, stopper the flask and immerse in a water bath heated to 60 to 65° for 20 min.

The determination can be carried out in an iodometric flask or in an ordinary glass-stoppered 250-ml flask. The approximately 1.5 *M* acetyl chloride solution can be conveniently prepared by dissolving 10 ml of acetyl chloride in 90 ml of toluene. The toluene serves as a mutual solvent for the solid cetyl alcohol, the acetyl chloride, and the cetyl acetate formed. Pyridine acts as a condensing agent by reacting with the hydrogen chloride formed during the acetylation, which proceeds according to the following equation:

$$CH_3(CH_2)_{14}CH_2OH + CH_3COCl \longrightarrow CH_3(CH_2)_{14}CH_2O_2CCH_3 + HCl$$

At 60 to 65°, the acetylation is complete in a very short time.

2. Add 25 ml of distilled water, stopper the flask and shake it vigorously for 1 or 2 min and allow to stand for 5 min to decompose the excess acetyl chloride. Titrate to a permanent pink end point, shaking the flask vigorously toward the end of the titration to maintain the contents in an emulsified condition, with 1 *N* sodium hydroxide, using 1 ml of phenolphthalein T.S. as the indicator.

The excess acetyl chloride is hydrolyzed to acetic acid by shaking vigorously with water to ensure contact between the water and the acetyl chloride that is dissolved in the oily layer. During the titration of the acetic acid with 1 *N* sodium hydroxide, it is again necessary to shake vigorously to ensure complete neutralization of the acetic acid.

3. Run a blank under identical conditions, using the same amounts of reagents as in the determination. Subtract the number of milliliters of 1 *N* sodium hydroxide used for the sample from the number of milliliters used for the blank. Multiply the difference by 56.11 and divide by the weight of sample to obtain the hydroxyl number which is the number of milligrams of potassium hydroxide equivalent to the hydroxyl content of 1 Gm of the sample.

The amount of acetyl chloride that has reacted with the fatty alcohol can be determined from the difference between the amount of 1 *N* sodium hydroxide consumed by the excess acetyl chloride in the blank and in the sample. Each milliliter of 1 *N* sodium hydroxide is equivalent to 0.001

mole or 56.11 mg of potassium hydroxide. Therefore, the expression

$$\frac{\text{ml of 1 } N \text{ NaOH} \times 56.11}{\text{Wt of sample}}$$

gives the hydroxyl number, or the number of milligrams of potassium hydroxide that is equivalent chemically to the acetyl chloride required for the acetylation of the hydroxyl content of 1 Gm of sample of the fatty alcohol.

The hydroxyl values of official substances are as follows: stearyl alcohol, U.S.P. (200 to 220); and cetyl alcohol, N.F. (218 to 238).

Acetyl Value of Fatty Acids. The acetyl value of fatty acids corresponds closely to the hydroxyl value of fatty alcohols, and the two constants have much the same significance with respect to identity and purity of the substances. The acetyl value is found by acetylating hydroxy fatty acids, e.g., hydroxystearic acid or hydrogenated castor oil, and determining the saponification value of the product. From the saponification value of the acetylated fatty acid and the acid value of the original fatty acid, the acetyl value is calculated and expressed as *the number of milligrams of potassium hydroxide required to neutralize the acetic acid obtained by the saponification of* 1 *Gm of acetylated fatty acids.*

The official procedure for the determination of the acetyl value of fatty acids is as follows:

Boil about 30 Gm of the fatty acids, obtained as described in the monograph, with 30 ml of acetic anhydride under a reflux condenser for 2 hr. Pour the mixture into 500 ml of distilled water contained in an 800-ml beaker and boil gently for 15 min while bubbling carbon dioxide through the mixture. Siphon off the greater part of the water and repeat the boiling operation with 2 successive 500-ml portions of distilled water. Transfer the acetylated oil into a pear-shaped 500-ml separator and wash with two 200-ml portions of distilled water warmed to about 50°. Separate as much of the water as possible and add 5 Gm of anhydrous sodium sulfate, shake well, allow to stand 30 min, then filter through a folded filter, carrying out the filtration in a drying oven at 105°, if necessary.

A 2-Gm sample of the acetylated fatty acid is then used for the determination of the saponification value (see Saponification Value, Chap. 13), and the acid value of the original fatty acid is determined (see Acid Value, Chap. 13). The acetyl value can be calculated using the following formula:

$$A = \frac{S - F}{1 - 0.00075S}$$

where A = acetyl value of the free fatty acids
S = saponification value of the acetylated fatty acids
F = acid value of the original fatty acids

The complete derivation of this formula for calculation of the acetyl value is somewhat involved and the original literature should be consulted.[3] The factor 0.00075 gives the number of grams of acetyl group that corresponds to 1 mg of potassium hydroxide. The denominator $1 - 0.00075S$ corrects for the discrepancy between the molecular weight of the acetylated oil (used for determination of the saponification value) and that of the original fatty acid (used for determination of the acid value).

Questions and Problems

1. Define *hydroxyl value* and *acetyl value*. For what types of fatty substances are these constants significant?
2. Give equations for all reactions involved in the determination of the acetyl value of the fatty acids of hydroxystearin sulfate, starting with the hydrolysis of hydroxystearin sulfate.
3. A sample of stearyl alcohol has been adulterated with paraffin wax to the extent that it is 90 percent pure. Calculate its hydroxyl value.

Water and Sediment in Fatty Oils. Certain fatty oils, particularly those of animal origin, which are not highly refined, frequently contain some moisture and nonfatty tissue residues. Official limits for moisture and sediment have been fixed by the Pharmacopeia for oils of this type. For example, Nondestearinated Cod-liver Oil must contain not more than 0.5 percent by volume of water and sediment when treated according to the following directions.

Place exactly 50 ml of benzene in each of two centrifuge tubes and to each tube add exactly 50 ml of the oil, warmed if necessary to reincorporate separated stearin, and thoroughly mixed at 25°. Tightly stopper the tubes and shake them vigorously until the contents are thoroughly mixed, then immerse the tubes in a water bath at 50° for 10 min. Place the tubes on opposite sides of the centrifuge and whirl for 10 min. Read the combined volume of water and sediment at the bottom of each tube. Repeat the whirling for 10-min periods until the combined volume of water and sediment remains constant for three consecutive readings. The sum of the volumes of combined water and sediment in the two tubes represents the percentage, by volume, of water and sediment in the oil.

The determination is carried out in pear-shaped centrifuge tubes that are graduated in such a way that small volumes can be more accurately read.

Range, ml	Scale Division, ml
0 to 3	0.1
3 to 5	0.5
5 to 10	1.0
10 to 25	5.0
25 to 50	25.0
50 to 100	50.0

[3] L. W. Cook, *J. Am. Chem. Soc.*, **44**, 392, 1922.

A centrifuge that has a diameter of swing of 38 to 43 cm (measured from tip to tip of the whirling tubes) is operated at 1,500 rpm. If a centrifuge of different dimensions be used, it should be operated at a rate calculated by using the following formula:

$$\text{rpm} = 1{,}500\sqrt{\frac{40.6}{d}}$$

where d = diameter, cm

Chapter 14

ASSAY OF VOLATILE OILS

Volatile oils, also known as *ethereal oils* or *essential oils* and, in some cases, as *essences*, are generally complex products composed of mixtures of compounds of widely variant chemical characteristics. The most important chemical components of the official volatile oils are:

1. *Hydrocarbons*, occasionally of the acyclic series, such as heptane and myrcene, but more often of the isocyclic series, e.g., pinene, camphene, limonene, bornylene, fenchene, dipentene, sylvestrene, and phellandrene.
2. *Alcohols*, present both in the free state and in combination with acids as esters. The alcohols most generally found in the official volatile oils are linalol, geraniool, citronellol, terpineol, borneol, menthol, and santalol.
3. *Aldehydes*, such as benzaldehyde, cinnamic aldehyde, salicyl aldehyde, citral, and citronellal.
4. *Ketones*, the most important being camphor, carvone, fenchone, thujone, and menthone.
5. *Phenols*, such as anethol, eugenol, carvacrol, safrol, chavicol, and thymol.
6. *Acids*, sometimes present in the free state in small quantities, those occurring most commonly being acetic, propionic, butyric, valeric, benzoic, cinnamic, and hydrocyanic acids; but more often occurring in combination with the alcohols mentioned under 2 as esters, e.g., linalyl acetate, bornyl acetate, and menthyl acetate.
7. *Sulfur compounds*, such as the allyl isothiocyanate (mustard oil).

The analysis of volatile oils for the purposes of determining their purity and value is based on the measurement of certain physical characteristics, on the quantitative estimation of certain components such as alcohols, phenols, esters, and aldehydes, and on the qualitative tests for the various substances that are commonly employed as adulterants. Only the official quantitative procedures will be discussed here.

Methods of General Application. The purity and quality of volatile oils may be judged to some extent by their appearance, odor, color, etc., but the information gained from the determination of the specific gravity, rotatory power, refractive index, solidifying point, solubility, and behavior on distillation is of much greater importance.

Specific Gravity. The specific gravity of a volatile oil may be determined with the Westphal balance or pycnometer, the latter being the official method and the more accurate of the two, and expressed as the ratio of the weight of the volume of oil to that of an equal volume of pure water when both are determined at 25° (see Methods Used to Determine the Specific Gravity of Liquids, Chap. 30).

The specific gravities of the official volatile oils vary approximately between 0.84 and 1.2. Those oils which are lighter than water, such as orange, caraway, coriander, lemon, turpentine, and rosemary oils, are usually rich in hydrocarbons, alcohols, esters, and ketones. Oils with specific gravities that approach or exceed 1.0, e.g., anise, cinnamon, clove, and sassafras oils, usually contain chiefly aldehydes, phenols, phenolic derivatives, or certain esters.

The specific gravity of any volatile oil is not absolutely constant, since it is influenced by such factors as the maturity of the plant from which the oil is obtained and the age of the oil, as well as by the methods of preparation and purification.

Rotatory Power. The rotatory power of a volatile oil is generally measured with a Laurent half-shadow polarimeter, according to the procedure described in Chap. 27, using sodium light and a tube 10 cm long, but for highly colored oils, tubes 5 or even 2.5 cm long may be used. The observation of the optical activities of the official essential oils should be made at 25°. Slight deviations from this temperature do not greatly affect the rotatory power of a volatile oil, except in the cases of lemon oil and orange oil.

The rotatory power of some of the volatile oils varies within relatively wide limits. This determination should never be omitted in their examination, however, since it frequently serves as a valuable means of detecting adulteration with inactive substances, such as alcohol, or with substances of different rotatory power from that of the oil being examined, e.g., lemon oil (+57 to +65.6 deg) adulterated with turpentine oil (about +25 to −40 deg).

Refractive Index. The refractive index of a volatile oil is most commonly determined by means of an Abbé refractometer, as described in Chap. 28. The measurement of the refractive indices of the official volatile oils is required to be performed at 20°, except in the case of rose oil, which is measured at 30°. The index of refraction does not vary greatly with different official volatile oils, the values being between about 1.46 and 1.61 at 20°. In some cases, however, this determination may serve for the detection of extraneous matter.

Congealing Temperature or Point. The congealing temperature or point of a volatile oil is determined by the method described under Congealing Temperature, Chap. 29. Most essential oils solidify only at low temperatures. Consequently, in practice, this determination is carried out with only a few oils, such as anise oil, eucalyptus oil, and fennel oil, which contain large amounts of the readily crystallizable constituents anethol and eucalyptol. The higher the congealing temperature of these oils, the more they are valued. An abnormally low congealing temperature of an essential oil indicates the partial removal of the characteristic constituent for which the oil is valued or the addition of extraneous matter, such as alcohol.

Distillation Range or Limits. The distillation range of a volatile oil is determined by the general method described under Boiling and Distilling Ranges, Chap. 29. Volatile oils that are composed of mixtures of hydrocarbons, alcohols, esters, etc., boil between certain limits of temperature, frequently separated widely. Consequently, the official standards usually designate the temperature or range of temperature at which a definite percentage of the oil distills; e.g., not less than 90 percent of turpentine oil should distill between 154 and 170°; not less than 95 percent of pine oil should distill between 200 and 225°, and less than 10 percent of dwarf pine needle oil should distill below 165°.

Fractional Distillation. Fractional distillation is resorted to occasionally in the official tests for purity of volatile oils. It serves either to separate the various components of volatile oils or to detect adulteration. Thus, alcohol and petroleum ether, both of which boil below 100°, may be separated and identified in this way. The addition of turpentine oil and other adulterants may be detected by collecting various fractions of the oil and determining the rotatory power and refractive index of each fraction; e.g., see Orange Oil and Lemon Oil, U.S.P.

Solubility. The volatile oils are generally soluble in organic solvents, such as absolute alcohol, ether, chloroform, benzene, carbon disulfide, etc. They dissolve more or less readily in dilute alcohol according to the nature of their components. Oils containing a large percentage of oxygenated substances often produce turbid solutions with petroleum ether or carbon disulfide because of the separation of water, small quantities of which are dissolved in such oils.

Practically all the volatile oils exhibit almost constant solubility in 90, 80, or 70 percent alcohol. Consequently, the test of the solubility of a volatile oil in dilute alcohol frequently gives valuable data relative to its purity, since the commonly used adulterants, turpentine oil, petroleum oils, and fatty oils are but slightly soluble in 80 or 70 percent alcohol.

Assay for Ester Content. A number of volatile oils are evaluated on the basis of their ester content. The esters found are mostly the acetates of alcohols of the formula $C_{10}H_{17}OH$, e.g., borneol, geraniol, linalool, and terpineol; $C_{10}H_{19}OH$, e.g., menthol or citronellol; or $C_{15}H_{23}OH$, e.g., santalol. When these esters are refluxed with alcoholic potassium hydroxide, they are saponified and yield the free alcohol and the potassium salt corresponding to the acid component of the ester.

The number of milligrams of KOH required to saponify 1 Gm of the oil represents the *ester value* or *ester number* of the oil. The ester value for any given oil may be subject to considerable variation, but the minimum requirement set by the official standards must be complied with when the oil is represented as a product conforming to the U.S.P. or N.F. standards. The determination of the total esters when taken in conjunction with the official tests for purity serves to detect adulteration and to establish the quality of those oils valued for their ester content.

The ester values of oils that contain appreciable amounts of aldehydes or phenols cannot be estimated accurately by saponification with an alkali, since variable amounts of the latter are consumed by reaction with the aldehydes and phenols, or their decomposition products.

Exercise 82

Object. To Prepare 0.5 *N* Alcoholic Potassium Hydroxide.

Materials Required. 2.5 Gm of lead acetate
1,500 ml of alcohol
40 Gm of potassium hydroxide

Procedure. 1. Dissolve 2.5 Gm of lead acetate in 5 ml of water, add the solution to 1,000 ml of alcohol contained in a glass-stoppered bottle, and mix thoroughly. Dissolve 5 Gm of potassium hydroxide in 25 ml of warm alcohol, cool the solution, and add it slowly, without stirring, to the alcohol solution of lead acetate. After 1 hr shake the mixture vigorously, allow it to stand overnight, decant the clear liquid, and recover the alcohol by distillation.

The potassium hydroxide converts any acids present into nonvolatile salts and causes the polymerization of aldehydes. The lead acetate is decomposed by the KOH with the formation of lead oxide and potassium acetate. The lead forms insoluble and nonvolatile compounds with any sulfur compounds present. The lead oxide carries down the polymerized aldehydes when it is precipitated so that it is easy to decant the clear supernatant liquid.

2. Dissolve about 35 Gm of potassium hydroxide in 20 ml of water and add sufficient aldehyde-free alcohol to make 1,000 ml. Allow the solution to stand in a tightly stoppered bottle (using either a glass or a rubber stopper) for 24 hr. Then quickly decant the clear supernatant liquid into a bottle provided with a well-fitting rubber stopper and standardize as follows:

Traces of aldehydes remaining in the alcoholic potassium hydroxide solution are polymerized and settle out when the mixture is allowed to stand. The rubber stopper used should be washed with a small quantity of the alkaline alcoholic solution before it is inserted into the mouth of the bottle, to dissolve sulfur and any other material that might affect the concentration of the solution.

3. Measure accurately, from a buret, about 25 ml of 0.5 *N* hydrochloric acid. Dilute with 50 ml of water, add 2 drops of phenolphthalein T.S., and titrate with the alcoholic potassium hydroxide solution until a permanent, pale pink color is produced. Calculate the normality.

The titration is carried out as described in Exercise 11, Chap. 3. The prepared solution should be preserved in tightly stoppered bottles to prevent changes in the concentration of alkali through absorption of CO_2. Well-fitting rubber stoppers may be used. It should also be protected from light to prevent the development of reddish compounds, which interfere with the correct determination of the end point during titration. The solution should be standardized whenever it is used, and it should always be measured at the same temperature at which it was standardized, since alcohol has a high coefficient of expansion.

Exercise 83

Object. Assay of Peppermint Oil for Total Esters.

Materials Required. 10 ml of peppermint oil
25 ml of 0.5 *N* alcoholic KOH
About 25 ml of 0.5 *N* H_2SO_4

Procedure. 1. "Place about 10 Gm. of Peppermint Oil, accurately weighed, in a 250-ml. Erlenmeyer flask, add 10 ml. of neutralized alcohol and 2 drops of phenolphthalein T.S., then add, dropwise, 0.1 *N* sodium hydroxide until a faint pink color appears. Add 25 ml. of 0.5 *N* alcoholic potassium hydroxide, connect the flask to a reflux condenser, and heat on a boiling water bath for 1 hour."

Instead of weighing the oil directly in a previously tared Erlenmeyer flask, it may be weighed in a weighing bottle, and the weighing bottle with its contents may then be transferred into the Erlenmeyer flask. The esters of peppermint oil, which consist chiefly of menthyl acetate, are saponified by the alkali when the mixture is refluxed, forming free

TABLE 42. OFFICIAL VOLATILE OILS ASSAYED FOR THEIR ESTER CONTENT

Oil	Amount used, Gm	Equivalent of 1 ml of 0.5 *N* KOH, Gm	Official requirement, percent
U.S.P.			
Methyl salicylate..........	2	0.07608	Methyl salicylate $C_6H_4(OH)CO_2CH_3$ = n.l.t. 98
Peppermint oil............	10	0.09915	Menthyl acetate $C_{10}H_{19}C_2H_3O_2$ = n.l.t. 5
N.F.			
Lavender oil..............	5	0.09815	Linalyl acetate $C_{10}H_{17}C_2H_3O_2$ = n.l.t. 35
Pine needle oil...........	10	0.09815	Bornyl acetate $C_{10}H_{17}C_2H_3O_2$ = 3 to 10
Rosemary oil..............	10	0.09815	Bornyl acetate $C_{10}H_{17}C_2H_3O_2$ = n.l.t. 1.5

n.l.t. = not less than.

menthol and potassium acetate. Alcoholic potassium hydroxide is used, because the oil is soluble in strong alcoholic solution.

$$-O-\overset{O}{\overset{\|}{C}}-CH_3 + KOH \longrightarrow -OH + CH_3COOK$$

2. "Allow the mixture to cool, add 20 ml. of water and 3 drops of phenolphthalein T.S., and titrate the excess alkali with 0.5 *N* hydrochloric acid. Perform a blank determination with the same quantities of the same reagents and in the same manner, disregarding the 0.1 *N* sodium hydroxide. Each milliliter of 0.5 *N* alcoholic potassium hydroxide consumed in the saponification is equivalent to 99.15 mg. of total esters calculated as menthyl acetate ($C_{10}H_{19}\cdot C_2H_3O_2$)."

The calculation for the determination of percent menthyl acetate is illustrated as follows:

Problem. Calculate the percent menthyl acetate, if a 9.120-Gm sample of peppermint oil, refluxed with 25.00 ml of about 0.5 *N* alcoholic KOH, required 21.52 ml of 0.4900 *N* H_2SO_4 for the residual titration. The blank was run using the same volume of 0.5 *N* alcoholic KOH and required 26.00 ml of 0.4900 *N* H_2SO_4 to bring about the end point.

Solution

1. Determine the number of milliequivalents of KOH required to saponify the esters in the sample:

$$(26.00 \text{ ml} \times 0.4900\ N) = 12.740 \text{ mEq } H_2SO_4 \text{ (also KOH)}$$
$$(21.52 \text{ ml} \times 0.4900\ N) = 10.545 \text{ mEq } H_2SO_4 \text{ (also KOH)}$$
$$12.740 \text{ mEq} - 10.545 \text{ mEq} = 2.195 \text{ mEq of KOH}$$

the amount required to saponify the total esters calculated as menthyl acetate.

2. Multiply 2.195 mEq of KOH so consumed by 0.19831, the weight in grams of 1 mEq of menthyl acetate. Observe that in the balanced chemical reaction, 1 mole of KOH saponifies 1 mole of menthyl acetate and the number of milliequivalents of KOH consumed represents the number of milliequivalents of menthyl acetate present in the sample: 2.195 mEq × 0.19831 Gm/mEq = 0.4353 Gm menthyl acetate.

3. Divide the weight of menthyl acetate by the weight of the sample and multiply by 100 in order to obtain the percentage of total esters:

$$\frac{0.4353 \text{ Gm menthyl acetate} \times 100}{9.120 \text{ Gm sample}} = 4.77\%\ W/W$$

4. Combining all mathematical operations (steps 1, 2, and 3) and factoring:

$$\frac{(26.00\text{ ml} - 21.52\text{ ml}) \times 0.4900\ N \times 0.19831\text{ Gm/mEq} \times 100}{9.120\text{ Gm}}$$

$$= 4.77\%\ W/W$$

Alternative Solution

1. Determine the number of milliliters of exactly 0.5 N alcoholic KOH consumed by the sample:

$$\frac{26.00\text{ ml} \times 0.4900\ N}{0.5000\ N} = 25.49\text{ ml HCl (also KOH) blank titration}$$

$$\frac{21.52\text{ ml} \times 0.4900\ N}{0.5000\ N} = 21.09\text{ ml HCl (also KOH) residual titration}$$

$$25.48\text{ ml} - 21.09\text{ ml} = 4.39\text{ ml } 0.5\ N\text{ alcoholic KOH}$$

or the amount consumed in the saponification reaction.

2. Multiply 4.38 ml by 0.09915, the number of grams of menthyl acetate equivalent to 1.00 ml of 0.5 N alcoholic KOH consumed: 4.39 × 0.09915 Gm = 0.4353 Gm menthyl acetate.

3. Continue calculations as in step 3 of previous example.

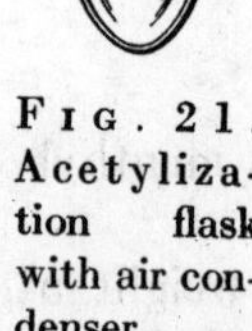

FIG. 21. Acetylization flask with air condenser.

Questions and Problems

1. What effect would the presence of free acids in the oil have upon the result of the determination of total esters?

2. Why is an alcoholic rather than an aqueous solution of KOH used to saponify the esters?

3. What factor would be necessary for the ester in order to obtain the percent of total esters directly as described above, if the chief ester present in an oil was bornyl acetate; linalyl acetate? Show how these factors are derived.

Assay for Alcohol Content. The alcohols present in volatile oils, such as menthol and borneol, occur both free and combined as esters. Usually, the determination of the total alcohol, occurring free and combined, is sufficient to establish the purity and value of an oil with respect to its content of alcoholic constituents.

The total alcohols present in any given oil are determined by transforming the free alcohols into the corresponding acetates by boiling the oil with acetic anhydride in an acetylization flask (Fig. 21) and then determining the saponification value of the acetylized product. When the alcohol occurs partly free and partly combined in the form of an ester, a correction factor must be employed. The method of determining the total alcohol content of volatile oils is illustrated in the following exercise.

Exercise 84

Object. Assay of Peppermint Oil for Total Menthol.

Materials Required. 10 ml of peppermint oil
10 ml of acetic anhydride
1 Gm of anhydrous sodium acetate
3 Gm of monohydrated sodium carbonate
About 0.2 Gm of fused calcium chloride
50 ml of 0.5 *N* alcoholic KOH
50 ml of 0.5 *N* H_2SO_4

An acetylization flask of about 100-ml capacity with an air condenser about 100 cm long

Procedure. 1. Place 10 ml of Peppermint Oil in an acetylization flask of 100-ml capacity, and add 10 ml of acetic anhydride and 1 Gm of anhydrous sodium acetate. Boil the mixture gently for exactly 1 hr, cool, disconnect the flask from the condenser, transfer the mixture to a small separator, rinsing the acetylization flask with three successive 5-ml portions of warm water, and add the rinsings to the separator.

The acetylization flask should be supported on an asbestos board while heating to prevent decomposition of the oil. The acetic anhydride reacts with the menthol, forming menthyl acetate and acetic acid. The acetylization is promoted by the anhydrous sodium acetate.

$$R{-}OH + O(C(=O){-}CH_3)_2 \longrightarrow R{-}O{-}C(=O){-}CH_3 + CH_3COOH$$

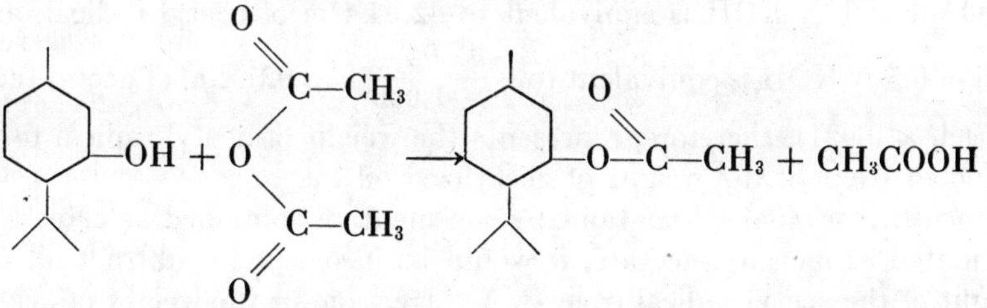

2. When the liquids have completely separated, reject the water layer, and wash the remaining oil with successive portions of sodium carbonate T.S., diluted with an equal volume of water, until the last washing is alkaline to 2 drops of phenolphthalein T.S. Dry the resulting oil with anhydrous sodium sulfate and filter.

The oil is washed with sodium carbonate test solution (12.5 Gm of $Na_2CO_3 \cdot H_2O$ in sufficient distilled water to make 100 ml) to neutralize the excess acetic acid. The washed oil is then dried over anhydrous sodium sulfate to remove the small amounts of alkaline sodium carbonate solution that it contains.

3. Transfer 5 ml of the dry acetylized oil to a tared, 100-ml Erlenmeyer flask, note its exact weight, add 50 ml of 0.5 *N* alcoholic potassium hydroxide, connect the flask with a reflux condenser, and boil the mixture on a water bath for exactly 1 hr.

The 0.5 N KOH saponifies the menthyl acetate, forming menthol and potassium acetate in the same manner as in the assay for total esters. The liberated menthol dissolves in the alcoholic solution.

4. Allow the mixture to cool, add 10 drops of phenolphthalein T.S., and titrate the excess alkali with 0.5 N sulfuric acid. Perform a blank determination with the same quantities of the same reagents and in the same manner. Calculate the percentage of total menthol in the oil tested by the formula

$$\frac{A \times 7.813}{B - (A \times 0.021)} \times [1 - (E \times 0.0021)]$$

in which A is the result obtained by subtracting the number of milliliters of 0.5 N sulfuric acid required in the above titration from the number of milliliters of 0.5 N sulfuric acid required in the residual titration blank, B is the weight of acetylated oil taken, and E is the percentage of esters calculated as menthyl acetate ($C_{10}H_{19}\cdot C_2H_3O_2$).

The formula used in the calculation is derived as follows: The molecular weight of menthol is 156.27. One milliliter of 0.5 N KOH is equivalent to $\frac{156.27}{2 \times 1{,}000} = 0.07813$ Gm of menthol. In order that the result may be read directly in terms of percent, the factor 0.07813 is multiplied by 100, giving 7.813. $B - (A \times 0.021)$ corrects for the weight of acetyl radical in the sample of acetylized oil, the factor 0.021 being the amount of acetyl radical equivalent to 1 ml of 0.5 N KOH. The latter is obtained from 198.31 (molecular weight of menthyl acetate) $-$ 156.27 (molecular weight of menthol) $=$ 42.04 (molecular weight of the acetyl radical). Therefore, 1,000 ml of 1 N KOH is equivalent to 42.04 Gm of acetyl radical, and 1 ml of 0.5 N KOH is equivalent to $\frac{42.04}{2 \times 1{,}000} = 0.021$ Gm of acetyl radical. $(A \times 0.021)$, therefore, represents the weight of acetyl radical to be deducted from B, the weight of acetylized oil taken.

Since the original oil contained some menthol combined as ester and calculated as menthyl acetate, it would be incorrect to subtract all the weight of the acetyl radical from B. A correction for the weight of acetyl radical in the original oil is consequently applied by multiplying the percent found from the formula $\frac{A \times 7.813}{B - (A \times 0.021)}$ by the correction formula $[1 - (E \times 0.0021)]$. The factor 0.0021 is the ratio of the weight of acetyl radical to that of menthyl acetate $42.04/198.31 = 0.21$. If E, the percent of total esters determined by the preceding assay, is found to be 5, then the formula will be $[1 - (0.05 \times 0.21)]$ or $[1 - (5 \times 0.0021)]$. Applying this correction, the formula becomes

$$\frac{A \times 7.813}{B - (A \times 0.021)} \times [1 - (5 \times 0.0021)] = \text{percent of total menthol}$$

The importance of the correction $[1 - (0.05 \times 0.21)]$ may be seen from the following example: A sample of peppermint oil found to contain 5 percent of total esters, calculated as menthyl acetate, and 50 percent of total menthol, as calculated from the formula, percent of total mënthol = $\frac{A \times 7.813}{B - (A \times 0.021)}$, gave the following percentage of total menthol when the correction was applied: $50 \times [1 - (5 \times 0.0021)] = 50 \times (1 - 0.0105) = 50 \times 0.9895 = 49.475$ percent of menthol free and as esters. The correction involves about one-half of 1 percent menthol or $50 - 49.475 = 0.525$ percent.

Peppermint Oil is official in the U.S.P. XVII and must contain not less than 50 percent total menthol ($C_{10}H_{19}OH$) in order to meet official requirements. Rosemary Oil is the only other official volatile oil assayed for its alicyclic alcohol content. It is official in the N.F. and is assayed in a manner analogous to that used for Peppermint Oil. Rosemary Oil must contain not less than 8 percent total borneol ($C_{10}H_{17}OH$).

Questions and Problems

1. Having determined the ester value and the percentage of total borneol present in rosemary oil, how could the percentage of free borneol be found? Give an example.
2. Write the reactions that take place in the assay of oil of rosemary for total borneol.
3. If peppermint oil containing 7.2 percent of total esters is found to contain 51.5 percent of total menthol (uncorrected), what change in the result would the correction $1 - (E \times 0.0021)$ cause? Is this correction important?
4. Show by calculations that the equivalent of 1 ml of 0.5 N KOH is 0.07712 Gm for borneol, the alicyclic alcohol present in Rosemary Oil.

Assay for Aldehyde Content. Aldehydes, like ketones, form addition products with certain reagents, and this property is utilized in their assay. When no other constituents are present which react with the reagent, the assay for aldehydes may be performed by the bisulfite method described later in this chapter under the assay for ketones. Thus, the cinnamic aldehyde content of the official cinnamon oil is determined by this method, the basic reaction being

$$C_6H_5{-}CH{=}CH{-}C({=}O){-}H + 2NaO{-}S(\rightarrow O){-}OH \longrightarrow C_6H_5{-}CH(SO_3Na){-}CH_2{-}C(OH)(H)(SO_3Na)$$

The bisulfite addition product dissolves in water, leaving the nonaldehyde constituents as a water-insoluble layer. The volume of this water-insoluble layer is then measured in a cassia flask (Fig. 22).

The cassia flask is a glass flask of about 100-ml capacity with a long, narrow neck graduated to 10 ml in 0.1-ml divisions. In the assay for aldehydes and ketones by the bisulfite method, the portion of the oil that does not react to form a water-soluble addition product rises to the surface, and when the flask is filled, this oily layer rises into the graduated neck of the cassia flask, where its volume can be measured. The volume of the residual oily layer subtracted from the volume of the sample of oil used represents the volume of the aldehyde constituents that formed a water-soluble addition product with the bisulfite.

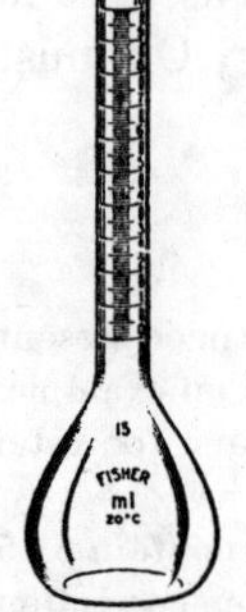

Fig. 22. Cassia flask.

Those volatile oils which contain aldehydes in very small amounts, such as lemon oil, or which contain other constituents that form water-soluble addition products with sodium bisulfite cannot be assayed accurately by this method. In these cases, the aldehydes are generally determined by the hydroxylamine method described in the following exercise.

Exercise 85

Object. Assay of Benzaldehyde.

Materials Required. 1 ml of Benzaldehyde
100 ml of hydroxylamine hydrochloride solution
Bromophenol blue T.S.
1 *N* NaOH

Procedure. 1. Transfer about 1 ml of Benzaldehyde to a tared glass-stoppered weighing bottle and weigh accurately. Loosen the stopper and transfer the weighing bottle and contents to a 250-ml Erlenmeyer flask containing 25 ml of a reagent prepared by dissolving 34.7 Gm of hydroxylamine hydrochloride in 160 ml of water and diluting this solution to 1,000 ml with alcohol. Rinse down the sides of the flask with an additional 50 ml of the reagent.

The hydroxylamine hydrochloride reacts with the benzaldehyde in the oil to form benzaldoxime, liberating hydrochloric acid.

$$C_6H_5\text{-}CH{=}O + H_2NOH{\cdot}HCl \longrightarrow C_6H_5\text{-}CH{=}NOH + H_2O + HCl$$

A standing period of at least 10 min is necessary for completion of this reaction. A somewhat longer period might, in fact, be desirable, if results are consistently too low.

2. Allow to stand for 10 min, add 1 ml of bromophenol blue T.S., and titrate the liberated hydrochloric acid to a light green end point with 1 *N* sodium hydroxide. Run a blank using the same quantities of reagents but omitting the Benzaldehyde. Match the color of this end point in the above assay. To determine the required number of milliliters of 1 *N* sodium hydroxide to neutralize the liberated hydrochloric acid, subtract the number of milliliters of 1 *N* sodium hydroxide used in the blank from the number of milliliters required in the assay. Each milliliter of 1 *N* sodium hydroxide is equivalent to 106.1 mg of C_7H_6O.

Since benzaldehyde reacts mole for mole with hydroxylamine, each milliliter of 1 *N* NaOH used by the sample but not by the blank, i.e., the difference, is equivalent to 0.001 mole of benzaldehyde or to 0.1061 Gm.

TABLE 43. VOLATILE OILS ASSAYED FOR ALDEHYDE CONTENT

Oil	Amount used	Method	Official requirement, percent
U.S.P.			
Cinnamon oil	10 ml	Bisulfite	Total aldehyde, n.l.t. 80
Lemon oil, California	5 Gm	Hydroxylamine	Citral, 2.2 to 3.8
Lemon oil, Italian	5 Gm	Hydroxylamine	Citral, 3.0 to 5.5
N.F.			
Benzaldehyde	1 Gm	Hydroxylamine	Benzaldehyde, n.l.t. 98

Questions and Problems

1. Show how the equivalent factor for benzaldehyde is derived.
2. Why is the titration carried out in a hydroalcoholic solution?
3. Why is it not necessary to measure the hydroxylamine HCl accurately?

Assay for Ketone Content. Only two of the official volatile oils are assayed for their ketone content. The determination is made by the "bisulfite addition reaction" or by the hydroxylamine method.

Exercise 86

Object. Assay of Caraway Oil.

Materials Required. A 100-ml cassia flask
10 ml of caraway oil
100 ml of saturated sodium sulfite solution (40 Gm in 100 ml)
100 ml of sodium bisulfite solution (5 percent)

Procedure. 1. "Introduce 10 ml. of Caraway Oil, accurately measured, into a cassia flask, and add 50 ml. of a saturated solution of sodium sulfite which has been carefully rendered neutral to 2 drops of phenolphthalein T.S. by means of a saturated sodium bisulfite solution. Heat the flask in boiling water and shake it repeatedly, neutralizing the mixture from time to time by the addition of a few drops of saturated sodium bisulfite solution."

Carvone, the principal ketone present in the oil, reacts with the sodium bisulfite, forming a water-soluble addition product as follows:

$$C_{10}H_{14}{=}O + NaO{-}\overset{\overset{O}{\uparrow}}{S}{-}OH \longrightarrow C_{10}H_{14}(OH)(SO_3Na)$$

Although the chemical reaction is run in sodium sulfite, the final addition product results from the reaction of the ketone with sodium bisulfite.

The N.F. directs that the reaction is to take place in a saturated solution of sodium sulfite. This saturated solution acts to "salt-out" from the aqueous phase those volatile oil constituents which did not react with the bisulfite. Since the volume of the water-insoluble constituents is an experimentally determined value, it is important that as little as possible remain in the aqueous solution. The concomitant use of Na_2SO_3 with $NaHSO_3$ in this determination also provides a convenient device whereby the end of the addition reaction is readily determined when no coloration appears upon adding a few drops of phenolphthalein to the reaction mixture and heating for 15 mins.

The sodium sulfite is hydrolyzed in part by the water with the formation of sodium hydroxide, viz.,

$$Na_2SO_3 + H_2O \rightleftharpoons NaHSO_3 + NaOH$$

The presence of free alkali in the aqueous layer renders other components of the oil, such as phenols, water-soluble and also tends to reverse the addition reaction. To prevent this the sodium hydroxide is neutralized by the addition of sodium bisulfite

$$NaHSO_3 + NaOH \longrightarrow Na_2SO_3 + H_2O$$

2. "When no coloration appears upon adding a few drops of phenolphthalein T.S. and heating for 15 minutes, cool the mixture to room temperature, and when the liquids have separated completely, add sufficient sodium sulfite solution to raise the lower limit of the oily layer within the graduated portion of the neck of the flask. Note the volume of the residual oily liquid. This volume does not exceed 5 ml., indicating the presence in caraway oil of not less than 50 per cent, by volume, of carvone ($C_{10}H_{14}O$)."

The difference between the volume of oil used as the sample and the volume of the oily layer that remains insoluble represents the carvone which dissolve in the aqueous layer. If drops of oil adhere to the walls of the flask, they may be made to rise into the neck by gently tapping and rotating it. If the residual liquid measures 4.5 ml, the percentage of carvone by volume in the oil would be $\frac{10 - 4.5}{10} \times 100 = 55$ percent.

Another official oil evaluated for its ketone content is spearmint oil. It is assayed in exactly the same way as is caraway oil, and it is required to contain not less than 55 percent by volume of carvone.

Questions and Problems

1. Ketones may also be estimated in volatile oils by the use of sodium bisulfite alone. Write the reaction between carvone and sodium bisulfite.
2. What substances, if present in the oil, would cause an error in the results of the sulfite method of assay for ketones?
3. How could the carvone be isolated from the aqueous solution?

Assay for Phenol Content. Volatile oils that contain phenols when shaken with solutions of sodium or potassium hydroxide diminish in volume because of the ready solubility of the phenol constituents in alkali; the nonphenolic portion of the oil remains undissolved.

Use is made of this property of phenols in the official assay methods. In these determinations a flask with a graduated neck, like that employed in the estimation of aldehydes (see Assay for Aldehyde Content, above), is used. In this flask, an accurately measured volume of the oil is shaken vigorously with a portion of the caustic alkali, and a further quantity of alkali is then added to bring the undissolved oil into the graduated neck of the flask. When the liquids have separated and become clear, the volume of the oil insoluble in alkali is read off, the volume of the phenol being determined by difference.

Exercise 87

Object. Assay of Clove Oil.

Materials Required. 10 ml of Clove Oil
150 ml of KOH, T.S.
A cassia flask.

Procedure. 1. "Pipet 10 ml. of Clove Oil into a 150-ml. cassia flask the neck of which is graduated from 0 to 6 ml. at intervals of 0.01 ml. Add 75 ml. of potassium hydroxide T.S., shake the mixture for 5 minutes, and heat for 10 minutes in boiling water, shaking the flask at least 3 times during the heating. Remove the flask from the bath, and cool to room temperature."

The KOH reacts with the phenol, eugenol, as follows:

$$C_6H_3(OH)(OCH_3)(CH_2{-}CH{=}CH_2) + KOH \longrightarrow C_6H_3(OK)(OCH_3)(CH_2{-}CH{=}CH_2) + HOH$$

The mixture is heated on a water bath for 10 min to saponify any eugenyl acetate present.

2. "When the liquids have separated completely add sufficient potassium hydroxide T.S. to raise the lower limit of the oily layer within the graduated portion of the neck. After standing for 18 hours the volume of the oily layer does not exceed 1.5 ml., indicating the presence in the Oil of not less than 85 percent by volume of total phenolic substances."

The potassium eugenolate dissolves in the water, leaving undissolved hydrocarbons, ketones, etc., which rise into the graduated neck of the flask and are measured. If stronger solutions are employed, the results obtained are too high, since the strong alkali exercises some solvent action on the nonphenolic constituents, particularly on oxygenated compounds. The volume of the undissolved liquid subtracted from the volume of oil used as the sample gives the volume of eugenol that dissolved in the alkali; e.g., if a 10-ml sample of oil yields 1.6 ml of residual liquid, then $10 - 1.6 = 8.4$ ml of eugenol dissolved, and the percentage of eugenol in the sample is $\frac{8.4}{10} \times 100 = 84$ percent. Clove Oil is official in the U.S.P. XVII.

Thyme Oil is the only other official volatile oil assayed for phenol content. It is official in the N.F. XII and is assayed in a manner analogous to that used for Clove Oil.

Questions and Problems

1. Is the assay of phenols as given above an exact method?
2. How could the eugenol be liberated from the alkaline solution?
3. If a part of the eugenol is present combined as esters, is that part estimated in the determination?

Determination of Volatile Oil Content of Crude Drugs and Oleoresins. Certain crude drugs and oleoresins, used as medicinal or flavoring agents, owe their virtues primarily to volatile oil constituents. Official assay methods for drugs and preparations of these types have been developed, and the limits for volatile oil content have been fixed in a number of instances.

The official methods prescribed for these determinations consist essentially of the separation of the oil from the other components of the drug or oleoresin by means of steam distillation, and the accurate measurement of the volume of oil obtained. These operations are carried out in a distillation apparatus designed especially for the purpose (see Fig. 23). All parts that come in contact with the material to be assayed, the volatile oil or steam, are made of glass. The condenser and receiver are combined and are designed to minimize the loss of volatile oil through the escape of vapors and to facilitate the accurate measurement of the volume of oil in the distillate. Receivers that differ slightly from one another in design are employed, depending on whether the oil is lighter or heavier than water (see Fig. 23). Distillation for 4 to 8 hr is required to ensure complete isolation of the oil.

The chief precaution to be observed in carrying out the determination is not to boil the contents of the flask too rapidly during distillation, as this causes vapors to escape around the condenser and results in a loss of volatile oil.

If the oil in the sample to be assayed is lighter than water, it is measured directly in the receiver, the lower part of which is shaped like a pipet and graduated in tenths of a milliliter. If it is heavier than water, the oil is transferred to a graduated cylinder after distillation is complete, and the

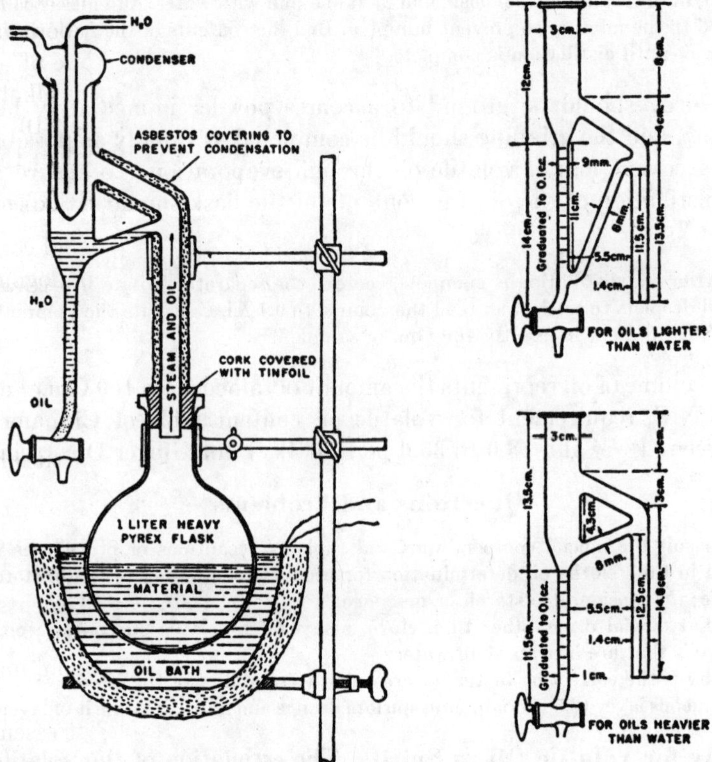

FIG. 23. Volatile oil distillation apparatus.

water with any remaining oil is drawn off into a small separator. The oil trap is washed with 10 ml of ether, the ether washings are added to the separator, the mixture is shaken and allowed to separate. The aqueous layer is drawn off and discarded. The ethereal layer is evaporated until the odor of ether is no longer perceptible, the residual oil is drained into the graduated cylinder, and the volume is measured at 25°.

If the oil content is to be determined by weight, a small quantity of anhydrous sodium sulfate is added, the mixture is agitated gently, the clear oil is decanted, and its specific gravity is determined at 25°.

Exercise 88

Object. To Determine the Volati e Oil Content of Cinnamon.

Materials Required. 100 Gm Cinnamon, preferablv coarsely comminuted
Distilled Water
Volatile oil determination apparatus for oils heavier than water

Procedure. 1. "Place 100 Gm. of coarsely comminuted cinnamon, accurately weighed, in the round bottom flask, and fill it one-half with water. Add glass beads or a glass rod to the mixture to prevent bumping. Boil the contents of the flask slowly for 2 hours, or until distillation is complete."

The drug should be ground to a coarse powder immediately before weighing, and the grinding should be completed as rapidly as possible to avoid excessive loss of volatile oil through evaporation. To ensure complete distillation of the oil, the contents of the flask should be boiled for at least 2 hr.

2. "After the distillation is complete, remove the separator, rotate if necessary to cause oil droplets to settle, and read the volume to 0.1 ml. Calculate the volume of oil in milliliters present in exactly 100 Gm. of sample."

The volume of oil represents the amount obtained from 100 Gm of drug.

The N.F. requirement for volatile oil content limit of Cinnamon is 2.5 percent *W/V* and 18.0 to 35.0 percent *W/V* for Ginger Oleoresin.

Questions and Problems

1. Consult the official compendiums and explain precautions or procedures to be followed in the volatile oil determination for products containing (*a*) little starch or mucilage; (*b*) considerable starch or mucilage.
2. What official drugs other than clove, assayed for volatile oil content, contain volatile oils that are heavier than water?
3. Why is the correction factor for error due to solubility of the oil-kerosene layer in the aqueous layer 10.5 in compound spirit of orange and only 2.2 in spirit of lavender?

Assay for Volatile Oil in Spirits. The estimation of the volatile oil content of certain official spirits is based upon the separation of the volatile oil by means of an immiscible solvent and measurement of the volume of the oil as illustrated in the following exercise.

Exercise 89

Object. Assay of Peppermint Spirit.

Materials Required. 5 ml of peppermint spirit
A Babcock bottle and centrifuge
A 1-ml pipet
1 ml of kerosene
50 ml of a saturated solution of calcium chloride (about 30 Gm in sufficient water to make 50 ml) acidified with HCl

Procedure. 1. Transfer exactly 5 ml of peppermint spirit to a Babcock bottle, graduated to 8 percent, add exactly 1 ml of kerosene from a pipet calibrated to deliver that amount, and mix well. Then add sufficient saturated calcium chloride solution, acidified with hydrochloric acid, almost to fill the bulb of the bottle. Rotate the bottle vigorously to ensure thorough mixing, then add sufficient of the calcium chloride solution to bring the separated oil into the neck of the bottle.

The acidified calcium chloride solution, when added to the alcoholic spirit, throws the volatile oil out of solution, and the oil dissolves in the kerosene when the mixture is shaken. The calcium chloride serves to "salt out" any volatile oil from the diluted alcoholic mixture. The acid ensures the absence of alkali which would render acids, traces of aldehydes, etc., water-soluble. The determination is carried out in a Babcock bottle (Fig. 24), the neck of which is calibrated into eight main divisions of 0.2 ml each.

Fig. 24. Babcock bottle.

2. Centrifuge for 5 min at about 1,500 rpm, and then read the volume of oil in the stem. Subtract five divisions for the kerosene added, and multiply the remaining number of divisions by 4.2 to obtain the volume of peppermint oil in 100 ml of the spirit.

The oil-kerosene layer, being lighter than the aqueous layer, rises into the neck of the flask when the mixture is centrifuged. The length of the oil-kerosene layer supplies the information needed for the calculation of the percent of oil in the spirit; e.g., if 7.4 divisions represents the length of the column of oil-kerosene in the graduated neck of the flask, the oil measures $7.4 - 5.0 = 2.4$ divisions, because each 5 divisions are equal to 1 ml (5×0.2 ml) and 1 ml of kerosene was added. The 2.4 divisions, each of which represents 0.2 ml, are equal to 0.48 ml of oil. Since a 5-ml sample is employed, 100 ml of the spirit yielded $0.48 \times 20 = 9.60$ ml of oil. The procedure is subject to slight error, which has been experimentally determined. Consequently, the volume of oil in scale divisions is multiplied by 4.2 (20×0.21 ml) to correct for the error due to solubility of the oil-kerosene layer in the aqueous layer and also due to contraction of volume of the immiscible solvents when mixed. Thus in the example given, $2.4 \times 4.2 = 10.08$ ml of oil in 100 ml of the spirit. Peppermint Spirit is the only spirit official in the U.S.P. XVII or N.F. XII.

Chapter 15

PRINCIPLES OF ALKALOIDAL ASSAYS

Proximate and Ultimate Assays. Assays of crude drugs may be classified according to the manner in which the percent of extractive is determined. If the percent of extractive represents the total of a class of plant principles, such as alkaloids or glycosides, etc., the assay is referred to as a *proximate assay*. If on the other hand, the percent of extractive from a crude drug represents a single chemical species, such as morphine, the assay is referred to as an *ultimate assay*.

The assay of Ipecac is an example of a proximate assay, because the total ether-soluble alkaloidal content is determined. In the assay of Opium, however, the percent morphine, a single chemical species, is determined, and the assay is therefore referred to as an ultimate assay.

General Principles. The alkaloidal drugs and preparations derived from them constitute a relatively important group of the official substances that are employed frequently in modern therapy. As a class of medicinal agents, they are characterized by their high potency. A slight deficiency of alkaloid in a preparation may cause a marked decrease in physiological effect; on the other hand, a slight excess may cause toxic effects when the preparation is administered. It therefore follows that the accurate estimation of the quantity of alkaloids present in a medicinal substance is an important subdivision of pharmaceutical analysis.

The assay of alkaloidal drugs and preparations is generally performed for purposes of standardization, proof of purity, commercial evaluation, or pharmacolegal purposes. Methods of various types have been developed for the quantitative estimation of these principles, e.g., gravimetric, volumetric, colorimetric, potentiometric, and physiological.

The amounts of alkaloids that occur in crude drugs are subject to considerable variation in different samples of the same drug. The variations may be caused by several factors: (1) the age of the plant when it is collected; (2) the season of the year when the drug is harvested; (3) the soil and climate in which the drug is grown; (4) the conditions under which the drug is collected, dried, and stored. The quantity of alkaloid present in galenical preparations is also subject to variation because of a number of factors, some of which are (*a*) the quality of drug employed; (*b*) the menstruum used in the extraction of the alkaloid; (*c*) the amount of decomposition of the alkaloid during the process of extraction and during the period of storage.

Many alkaloidal preparations deteriorate comparatively rapidly, the rate of deterioration being markedly affected by the nature of the alkaloid, the pH value of the preparation, heat, and light. Frequent restandardization of the alkaloidal drugs and their preparations is therefore essential.

In view of the fact that the alkaloids may comprise only a fraction of 1 percent of the substance assayed and that this small amount must be separated from numerous other constituents present in the crude drug or preparation, such as resins, volatile and fatty oils, coloring matter, glycosides, fatty acids, gums, and proteins, it is evident that the exact technique involved in any given method must be carefully adhered to in order to estimate the variations in alkaloidal content. It is this special technique that characterizes the chemical assay of alkaloidal drugs, rather than the gravimetric or volumetric nature of the procedure employed.

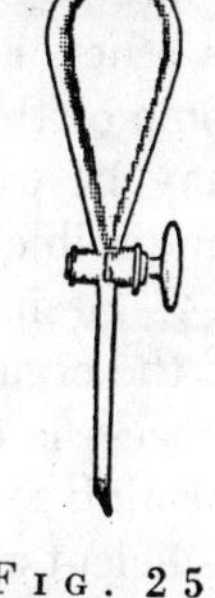

FIG. 25. Separatory funnel.

The principles employed in the quantitative determination of the alkaloids by chemical methods are based upon certain characteristic properties of these substances. The following general properties are possessed by most of the members of this class of compounds: (1) Alkaloids are usually sparingly soluble in water but readily soluble in certain organic solvents that are immiscible with water, such as chloroform, ether, amyl alcohol, benzene, petroleum benzin, or mixtures of these solvents. (2) Alkaloids combine directly with acids to form salts that are usually soluble in water but insoluble in certain organic solvents, such as chloroform and ether. (3) Alkaloids are liberated and usually precipitated from aqueous solutions of their salts by alkalies. (4) Alkaloids form highly insoluble precipitates with a considerable number of reagents, especially with the salts of some of the heavy metals such as mercury, gold, and platinum.

In general, the methods employed to separate the alkaloids from the other constituents present in a drug involve the use of the so-called "immiscible solvents," such as water with ether, water with chloroform, or water with ether-chloroform mixture. The process is generally carried out as follows, the separation of the immiscible solvents being accomplished by means of a separatory funnel (Fig. 25): An accurately weighed portion of the drug is treated with a suitable alkali, whereupon the alkaloids, which usually occur in the drug in combination with acids as salts, are set free. An organic solvent, such as ether, is then added, which dissolves the free alkaloids. An aliquot portion or all of the ethereal solution corresponding to a definite weight of the sample taken is separated and shaken successively with several portions of acidulated water. The complete extraction of the alkaloids from the ethereal solution is determined by testing a portion of the latter with a suitable alkaloidal reagent. The alkaloidal salts formed with the acid dissolve in the water, leaving most of the resins, coloring matter, fatty substances, etc., in the ethereal portion. The acid extractions obtained are combined, made alkaline, and shaken with successive portions of ether. The alkaloids liberated from their salts by the alkali pass into solution in the ether, leaving water-soluble impurities behind. The separated ethereal extractives may then be combined, evaporated to dryness, and weighed; or the residual alkaloid may be dissolved in an excess of acid of known normality, and the excess of acid may be determined by titration with a standard solution of alkali, using a suitable indicator to show when the end point is reached.

When the above general procedures are applied in assaying, numerous difficulties may be encountered that will lead to serious error, unless proper precautions are observed, e.g.:

1. The organic solvents employed are sometimes appreciably soluble in water; thus, ether is soluble in twelve times its volume of water at 25°. Consequently, when an organic solvent partially soluble in water, such as ether, is employed to extract the alkaloid from an aqueous solution, some of the alkaloid will remain in the aqueous solution. This difficulty may be overcome by extraction with small successive portions of the immiscible solvent.

2. Resins, fats, and many other organic substances remain in solution in the organic solvent if the aqueous liquid employed in the shaking-out process is acid, but they dissolve in the aqueous liquid if it is alkaline, forming soaps, resinates, etc. Care should be exercised, therefore, to have sufficient acid in the aqueous solution used to react with the alkaloid from the organic solvent and also to neutralize all the alkali and leave a slight excess of acid.

3. When the liquids are shaken in the separatory funnel, emulsification may occur between the organic solvent and the aqueous solution, making it exceedingly difficult or impossible to separate the liquids quantitatively. The formation of emulsions may frequently be avoided by rotating the separatory funnel containing the liquids slowly and occasionally inverting it in place of shaking vigorously. Since the rapidity with which the alkaloids are extracted from a liquid in which they are insoluble, by an immiscible liquid in which they are soluble, is largely dependent on the rate of change of the interface between the liquids, it is necessary that gentle agitation be continued longer than is required when strong agitation is employed. If emulsification occurs, it is often possible to break the emulsion by some one or more of the following procedures: (*a*) addition of more of the aqueous or organic solvent; (*b*) gentle stirring with a glass rod; (*c*) drawing off the separated solvent as rapidly as it separates; (*d*) addition of a few drops of alcohol; (*e*) addition of a few milliliters of salt solution; (*f*) addition of acid, if alkaline; (*g*) addition of alkali, if acid;

(*h*) filtration through a pledget of cotton; (*i*) drawing off the emulsion and treating it with anhydrous sodium sulfate. If none of these methods suffices to break the emulsion, it may be necessary to start the assay over. Time is always saved by taking every precaution to prevent the formation of emulsions. Emulsions formed in the assay of drugs containing a large proportion of fatty matter should be treated according to the following directions:

"In such cases add sufficient sulfuric acid to acidify, and evaporate the volatile solvent, while stirring with a glass rod. When the resinous and fatty matter has been agglutinated, cool the acid solution and filter it through a small, wetted filter into a separator. Redissolve the residue in 15 ml. of ether, add 5 to 10 ml. of 0.1 *N* acid, evaporate the ether as before, with continued stirring, and pour the acid solution through the filter into the separator. Repeat the extraction of the fatty residue with dilute acid two or three times and finally wash the filter free from alkaloids."

4. When an acid solution, contained in a separatory funnel to which ether has been added, is made alkaline, sufficient heat is frequently generated to volatilize some of the ether and cause considerable pressure in the funnel. Often the pressure generated is sufficient to force the stopper out of the funnel with loss of some of its contents, particularly when the funnel is shaken. For this reason, it is advisable to permit the mixture to stand and cool for a short time after adding the alkali and to agitate it gently at first.

Sources of Error. Numerous possibilities of error are encountered in alkaloidal assaying other than those common to all analytical work, such as those caused by inaccurate weighing, faulty calibration and reading of measuring apparatus, incorrect strength of volumetric solutions, dirty apparatus, etc. Some of the more common errors are caused by (1) failure to secure complete extraction of the alkaloids from the drug or from solution; (2) loss of volatile solvent during maceration of the drug before the aliquot portion is taken; (3) imperfect separation of the immiscible liquids; (4) failure to wash the stems of separators and funnels, graduates, and other apparatus with solvent to recapture small amounts of alkaloidal residues; (5) decomposition of the alkaloids, especially those which have an ester structure, upon prolonged contact with strong acids and alkalies; (6) use of the wrong indicator.

Other sources of error, such as variation in the moisture content of the sample, frequently cause different analysts to obtain nonconcordant results in the assay of the same drug. The most common cause of variations in results obtained by different workers, however, is referable directly to a lack of familiarity with the procedures involved, since considerable experience in making assays of this type is essential to acquire proficiency.

Theory of Distribution Coefficient. According to Nernst's law, if two practically immiscible solvents are in contact and a substance that is soluble in both liquids is added to them, the substance will distribute itself between the liquids in such a way that the ratio of the concentrations of the two solutions is a constant irrespective of the quantity of solute. This constant is variously termed the *distribution ratio, distribution constant, distribution coefficient,* or *partition coefficient.*

If C_1 is used to designate the concentration of an alkaloid in ether, and C_2 to designate the concentration of the same alkaloid in an equal volume of an aqueous acid solution, Nernst's law may be formulated $C_2/C_1 = K$, where K is the distribution constant or coefficient. If there are several solutes, such as alkaloids, resins, etc., present in the ether when it is shaken with the acid solution, the distribution of each solute is the same as if it were present alone.

Measurable deviations from Nernst's law take place in alkaloidal assaying owing to molecular association and dissociation, ionization, chemical reaction, etc., but the fundamental principle that it expresses may be used to explain why it is better to extract the alkaloid from a solvent with several small portions instead of one large portion of an immiscible solvent; e.g., assume that 1 Gm of atropine in 100 ml of ether solution when shaken with an equal volume of dilute sulfuric acid distributes itself so that 0.1 Gm remains in the ether layer and 0.9 Gm in the acid layer of the solvents, $C_{\text{acid}}/C_{\text{ether}} = 9$ (the distribution coefficient). Obviously, only 90 percent of the atropine present would be extracted from the ethereal solution by extracting it once with 100 ml of dilute sulfuric acid solution. If the extraction is performed with 200 ml of the acid solution, the amount of atropine extracted should be

$$\frac{x/200}{(1 - x)/100} = 9 \quad \text{and} \quad x = 0.947 \text{ Gm of atropine}$$

where x = the amount of atropine extracted from the ether by the acid solution

$$C_{\text{acid}} = \frac{x}{200}$$

$$C_{\text{ether}} = \frac{1 - x}{100}$$

If the original solution is extracted with several small portions of acid solution, i.e., 50 ml, the amount of alkaloid extracted will be $C_{\text{acid}} = x/50$ and $C_{\text{ether}} = \frac{1 - x}{100}$; and since $\frac{x/50}{(1 - x)/100} = 9$, $x = 0.818$ Gm of atropine removed by the first extraction with acid, leaving 0.182 Gm in the ethereal layer. A second extraction with a 50-ml portion of acid solution will remove $\frac{x_1/50}{(0.182 - x_1)/100} = 9$, and $x_1 = 0.1489$ Gm of atropine, where x_1 equals the amount of atropine that passes into the acid solution. Thus, in two extractions, $0.818 + 0.1489 = 0.9669$ Gm of atropine removed from the ethereal solution by 100 ml of acid solution, whereas a single extraction with 200 ml of acid solution removed only 0.947 Gm of the alkaloid.

In making successive extractions using equal volumes of extracting solvent, the following equation is useful in determining the quantity of solute remaining after n extractions.

$$W_n = W\left(\frac{V_1}{KV_2 + V_1}\right)^n \tag{1}$$

where $K = C_2/C_1$

C_2 = concentration of solute in the extracting solvent (2)

C_1 = concentration of solute in the original solvent (1)

W = weight of solute originally taken

W_n = weight of solute remaining in the original solvent after n extractions

V_2 and V_1 = volume of solvents (2) and (1), respectively

n = number of extractions

The following calculation illustrates an application of this equation.

Problem:

1. Determine the number of grams of atropine which remain in 100 ml of ether after extracting with three 50-ml portions of dilute sulfuric acid. The original quantity of atropine in the ether solvent is 1 Gm.

$$K = C_{\text{acid}}/C_{\text{ether}} = 9$$

2. Calculate the percent error resulting from incomplete extraction.

Solution:

1. Substitute the appropriate values into Eq. (1).

$$W_3 = 1\left(\frac{100}{(9)(50) + 100}\right)^3 = 0.006 \text{ Gm}$$

the quantity of atropine remaining in the ether solution after three successive 50-ml extractions with dilute H_2SO_4.

2. 1.000 Gm is the initial quantity of atropine in the ether solution. 0.006 Gm. is the quantity of atropine remaining in the ether after three extractions:

$$\frac{0.006 \times 100}{1.000} = 0.6\%$$

the percent of error resulting from incomplete extraction.

From the examples given, it can be seen that the error resulting from incomplete extraction using one 200-ml portion of dilute H_2SO_4 is 5.3 percent, while the error involved using two 50-ml portions of dilute H_2SO_4 is 3.3 percent and the error using three 50-ml portions of dilute H_2SO_4 is only 0.6 percent. These results indicate that the use of several small portions of solvent is recommended, rather than one large portion. In the official assay processes, 10- to 20-ml portions of the immiscible solvent are used, because it has been found that these quantities serve best to extract the alkaloids quickly, economically, and with less error.

Choice of Indicators for Alkaloidal Titrations. In the official alkaloidal assays by volumetric methods, the use of methyl red solution is recommended in all the titrations. The same lot of indicator used in the standardization of the volumetric solutions should be employed in titrating the alkaloids. Numerous investigators have found that an indicator other than that employed in the official assays gives more accurate results. Table 44, showing the indicators best suited to the titration of some of the more important alkaloids, is taken with slight modification from a paper by H. Wales.[1]

Materials Required. In the statement of the materials required under the following respective exercises, it is assumed that the student's locker is supplied with the four separatory funnels necessary to conduct alkaloidal assays in duplicate, as well as measuring cylinders, burets, flasks, etc. It is also assumed that the following substances are available in the laboratory: reagent sulfuric acid, hydrochloric acid, Valser's reagent, ammonia water, distilled water, alcohol, mercuric iodide test solution, iodine test solution, and methyl red T.S. The quantities of immiscible organic solvent required are approximate.

Alkaloidal Test Solutions. A number of reagents form insoluble compounds with many of the alkaloids, precipitating the latter from aqueous solution. In many cases, the presence of a mere trace of alkaloid can be detected by adding a suitable reagent to its solution. Some of the frequently used reagents are auric chloride, platinic chloride, mercuric chloride, gallotannic acid, picric acid, phosphomolybdic acid, potassium mercuric iodide, mercuric iodide, and iodine in potassium iodide. Although most of these reagents are employed in qualitative tests for the alkaloids, only the last two are of importance in the official assays, since they are employed to test for the complete extraction of the alkaloids by immiscible solvents.

Mercuric Iodide Test Solution, commonly known as *Valser's reagent*, is prepared by slowly adding a 10 percent solution of potassium iodide to red mercuric iodide until almost all the red mercuric iodide is dissolved. The excess mercuric iodide is then removed by filtration. This reagent forms white precipitates with minute traces of many of the alkaloids, when it is

[1] Wales, H., *J. Ind. Eng. Chem.*, **18,** 390, 1926.

added to their acidified aqueous solutions. The composition of the precipitates usually varies somewhat depending upon the conditions under which the precipitation is conducted.

Iodine test solution, commonly known as *Wagner's reagent*, is a solution containing approximately 1.275 Gm of iodine and 3.6 Gm of potassium

TABLE 44. INDICATORS FOR ALKALOIDAL TITRATIONS

Salt of	Average pH	pH range for indicator	Indicator
Aconitine	5.04	4.2 to 5.8	Methyl red
Arecoline	4.81	3.8 to 5.8	Methyl red
Atropine	5.56	3.8 to 7.2	Methyl red
Brucine	4.85	3.9 to 6.0	Methyl red
Cephaeline	4.81	4.2 to 5.4	Methyl red
Cinchonidine	5.90	5.4 to 6.4	Bromocresol purple
Cinchonine	6.02	5.5 to 6.5	Bromocresol purple
Cocaine	5.20	4.0 to 6.5	Methyl red
Codeine	4.86	3.6 to 6.3	Methyl red
Cotarnine	5.97	4.9 to 7.0	Bromocresol purple
Diacetyl morphine	4.89	4.2 to 5.7	Methyl red
Emetine	4.90	4.2 to 5.6	Methyl red
Ethyl hydrocupreine	6.33	5.6 to 7.0	Bromocresol purple
Ethyl morphine	4.99	4.2 to 5.8	Methyl red
Homatropine	5.74	3.9 to 7.6	Methyl red, propyl red, bromocresol purple
Hydrastine	4.45	3.8 to 5.0	No end point with methyl red or bromophenol blue
Hyoscine	4.83	3.6 to 5.6	Bromophenol blue
Hyoscyamine	5.83	3.8 to 7.6	Methyl red, propyl red, bromocresol purple
Morphine	4.68	4.0 to 5.2	Methyl red
Narcotine	4.43	3.9 to 4.9	Bromophenol blue
Nicotine	5.26	4.4 to 6.1	Methyl red
Papaverine	4.23	3.8 to 4.6	Indistinct, cannot be titrated
Physostigmine	4.85	3.8 to 6.0	Methyl red
Pilocarpine	4.31	3.6 to 5.0	Bromophenol blue
Quinidine	6.10	5.5 to 6.5	Bromocresol purple
Quinine	6.12	5.5 to 6.5	Bromocresol purple
Strychnine	4.81	3.8 to 6.0	Methyl red
Thebaine	5.08	4.0 to 6.2	Methyl red
Yohimbine	4.72	4.0 to 5.3	Methyl red

iodide dissolved in sufficient distilled water to make 100 ml; 0.1 *N* iodine solution may be employed. Wagner's reagent yields reddish or red-brown precipitates of complex composition with nearly all the alkaloids, even in dilute solutions, when it is added to an aqueous solution slightly acidified with sulfuric acid. The precipitates are usually amorphous. The quantity of the reagent used should not be sufficient to color the solution more than a faint yellow.

Mercuric Potassium Iodide Test Solution, known as *Mayer's reagent*, is a solution of 1.358 Gm of mercuric chloride in 60 ml of water mixed with 5 Gm of potassium iodide in 10 ml of water and the mixture diluted to 100 ml with water. Mayer's reagent yields white or slightly yellow precipitates when mixed with dilute solutions of many alkaloids.

General Procedures. 1. *Selection of the Sample.* The sample of a crude drug taken for assay must represent the entire drug and be powdered to a specified fineness. Usually the drug must be ground to a fine powder (No. 60), but finer powders may be employed. When the whole crude drug, such as leaves or seeds, is powdered for assay, care must be exercised not to exclude leaf stems, seed coats, etc., since the alkaloid may be localized in amounts varying in the different morphological parts of the drug. Care should be taken to avoid the loss of water during the powdering of the drug. If it is impossible to avoid this loss, the drug should be dried at a low temperature before powdering, the loss of water noted, and a correction made in the final calculations.

2. *Weighing of Portions for Assay.* In the preparation of drugs for assay, all portions directed to be weighed should be weighed accurately, but with crude drugs, accuracy to within 10 mg for quantities of 5 Gm and over is sufficient. Portions of pilular extracts may be weighed on a piece of waxed or parchmentized paper, the surplus paper cut away, and the extract and adhering paper dropped into a separator, breaker, or dish containing the solvent, and the extract dissolved. In transferring weighed portions to a separator, the dish containing the material to be assayed should be thoroughly rinsed and the rinsings should be added to the separator.

3. *Methods of Extraction.* The alkaloids are extracted from drugs by the process of maceration, by the process of percolation, and by the process of continuous extraction. The maceration process is adapted to those drugs which contain a relatively large amount of alkaloids and which, because of their nature or the nature of the drug, are difficult to extract completely by the process of percolation. In the maceration process, it is assumed that all the alkaloids are dissolved in the organic solvent and that an aliquot part of the solvent contains an amount of the extracted alkaloids in definite proportion to the weight of drug taken as sample. Many analysts believe that the aliquot portion of solvent obtained by the maceration process does not contain exactly the assumed proportion of the total alkaloids. Consequently, methods for the total extraction of the alkaloids are regarded as more satisfactory.

a. By maceration. An accurately weighed portion of the ground drug is treated with the specified solvent or mixture of solvents, made alkaline with ammonia T.S., and thoroughly mixed. It is then allowed to macerate for 12 to 24 hr with occasional agitation, or for a shorter period with continuous agitation. At the end of this period, the drug is allowed to settle; an aliquot portion of the solvent is decanted and treated as described for the purification of the alkaloids.

The organic solvent, accurately measured from a calibrated cylindrical graduate, is added and allowed to stand in contact with the powdered drug so that the small particles of the drug will be impregnated with the organic solvent before the ammonia is added. If the ammonia test solution is added first, the drug particles become wetted by the water with which the organic solvent is immiscible, and the latter will not permeate the drug cells to extract the alkaloids completely. The alkaloids generally occur in combination with various acids as salts in the crude drug. The ammonia reacts with these alkaloidal salts, forming the free alkaloids and the ammonium salts of the acids, and the liberated alkaloids dissolve in the organic solvent. The process of extraction of the alkaloid from the cellular structure of the drug proceeds slowly, probably because it is an osmotic process. Consequently, a period of maceration is required to permit the complete transference of the alkaloids from the drug cells to the organic solvent. The rate of extraction is accelerated by grinding the drug to a fine powder, thereby breaking down its cellular structure, and by frequent agitation of the drug-solvent mixture, so that the solvent in contact with the particles is changed frequently. The small particles of drug, which sometimes remain in suspension in the organic solvent, become hydrated when the latter is shaken with a little water and settle out. The flask in which the maceration and agitation of the drug with the solvents are carried out should be tightly stoppered throughout the entire period of maceration to prevent the escape of volatile organic solvent, since the volatilization of some of the solvent might cause a considerable error; e.g., if 100 ml of solvent is taken for the extraction of the alkaloids from a drug and 10 ml is lost by volatilization, and an aliquot portion of 50 ml, representing one-half of the weight of the sample is measured for assay, the aliquot portion will contain five-ninths instead of one-half of the alkaloids contained in the drug, and an error of approximately 5 percent will be caused in the assay result.

An aliquot portion of the supernatant organic solvent is poured off from the mixture and measured accurately in a calibrated cylinder; e.g., if 4 Gm of a drug is macerated with 100 ml of solvent and a 50-ml portion of the solvent is decanted, the 50-ml portion of solvent will contain an amount of extractive equivalent to $^{50}/_{100} \times 4 = 2$ Gm of the drug, provided that none of the solvent originally added is lost by volatilization. The aliquot portion of solvent should be measured at the same temperature at which the original solvent was measured, since the volume of liquids varies somewhat with changes in temperature.

b. By percolation. An accurately weighed quantity of the ground drug is placed in a suitable container and completely saturated with the specified solvent or mixture of solvents, and allowed to stand for 5 min. A quantity of ammonia T.S. sufficient to make the mixture distinctly alkaline is added and thoroughly mixed with the drug. The mixture is then transferred to a cylindrical percolator, previously prepared by packing the outlet with purified cotton. A small amount of the solvent may be used to rinse the container and the rinsing added to the percolator. The drug is allowed to macerate for a suitable period of time (from 1 to 12 hr or overnight, depending upon the drug to be assayed). Then the drug is firmly packed, a pledget of purified cotton placed above it, and percolated slowly with the solvent until it is completely exhausted of its alkaloid contents. Complete extraction of the alkaloid is determined by evaporating about 4 ml of the last percolate to dryness, dissolving the residue in 0.5 ml of approximately 0.5 *N* acid and adding a drop of mercuric iodide T.S. (Valser's reagent), or, when testing for caffeine or colchicine, a drop of iodine T.S.; not more than a slight turbidity should be produced. The percolate is then treated for the purification of the alkaloids.

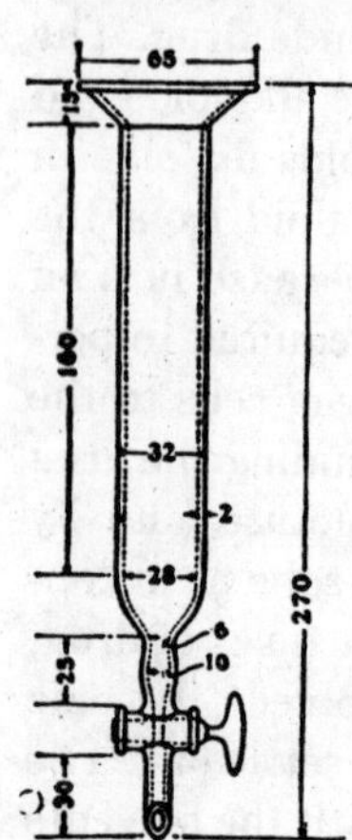

FIG. 26. Percolator recommended (measurements are given in millimeters).

The extraction of the alkaloids from a drug by cold percolation has the disadvantage that it requires a relatively large amount of solvent to effect complete extraction. The cold percolation process of extraction is best suited to the extraction of drugs that contain unstable alkaloids or that yield a large amount of nonalkaloidal extractive by hot percolation. The cold percolation process is well suited to the extraction of those drugs which contain a comparatively small percentage of alkaloids and which can be readily extracted by percolation. The importance of the total extraction of the alkaloid from a drug such as hyoscyamus becomes apparent when it is recalled that the official drug is required to contain only 0.040 percent of alkaloids.

The percolator employed in the extraction of the drug should be of a special type similar to that illustrated in Fig. 26. The barrel of the percolator should be about 15 or 16 cm long and about 3 or 4 cm in diameter. A percolator of these dimensions will accommodate from 10 to 25 Gm of drug, and, because of the small diameter, a long column of the drug will be provided through which the menstruum must pass. The glass stopcock is necessary to eliminate the use of rubber connections, since rubber is dissolved to some extent by chloroform and ether.

The alkaloids are liberated by the ammonia, and they dissolve in the organic solvent as described under (*a*).

c. By continuous extraction. An accurately weighed portion of the ground drug is placed in an extraction thimble and the thimble inserted into a suitable extractor (a Soxhlet extractor of appropriate size is satisfactory). The drug is moistened with the specified solvent, mixed by means of a stirring rod, and allowed to stand about 5 min. It is then made alkaline with the specified quantity of ammonia T.S. and thoroughly mixed. The stirring rod is rinsed with a small portion of the solvent, and the drug macerated for 6 to 12 hr or overnight. The drug is then packed in the thimble and covered with a pledget of purified cotton, a sufficient quantity of solvent is added, and the drug is extracted for a specified period of time or until extraction is complete. The solvent remaining in the extraction chamber is transferred to the receiving flask and the liquid extract treated for the extraction of alkaloids.

An extraction apparatus of the type illustrated in Fig. 19 and described in Chap. 12 is very satisfactory. The continuous extraction process requires a small amount of the organic solvent, and in most cases, complete extraction of the alkaloids is obtained in from 3 to 5 hr.

4. *Extraction of the Alkaloids from Organic Solvents with Acid.* The alkaloids are extracted from ethereal or chloroformic solutions by shaking with dilute acids. The concentration and quantity of the acid employed are left to the discretion of the analyst, but the use of 10 ml of a 1 *N* or 5 percent acid or sufficient to render the mixture distinctly acid for the first extraction and of 10-ml portions of 0.5 *N* or 2 percent acid for succeeding extractions will be found to be practical and convenient. When the drug contains a large amount of fat, the use of a smaller volume of more concentrated acid for the first extraction may be advantageous in preventing emulsions. The shaking of the immiscible solvents should not be violent, but it should be sufficient to mix them thoroughly without emulsification. Repeated successive extractions with small portions of the acid must be continued until the alkaloid is completely removed from the ethereal or chloroformic liquid, as shown by tests with the designated reagents. Each portion of the acid solution should be carefully separated from the organic solvent and run directly into another separatory funnel. In all assays, the extractions should be continued until 0.5 ml of the last acid washing shows no more than a slight turbidity on the addition of a drop of mercuric iodide T.S., or, in the case of caffeine and colchicine, on the addition of a drop of iodine T.S. If traces of organic solvent are drawn off with the acid solution, they may be removed by washing the combined acid extractives with 2 to 3 ml of ether or chloroform, which is separated and washed in turn with about 5 ml of distilled water, the aqueous washings being separated and added to the acid solution.

If sulfuric acid is used, the acid converts the free alkaloids dissolved in the organic solvent into their sulfates, which pass into solution in the aqueous acid layer of the immiscible solvents when the mixture is shaken, leaving chlorophyll, resins, fixed and volatile oils, and other impurities dissolved in the organic solvent. The rate at which the alkaloids pass from solution in the organic solvent into the acid solution is increased greatly by shaking the mixture, since a greater area of contact surface between the liquids is produced. Sulfuric acid solution is employed to extract the alkaloids, because the ammonium sulfate formed in the next step in the procedure is not appreciably soluble in the organic solvent used for the final extraction of the alkaloid.

5. *Extraction of the Alkaloids from Acid Solution with Organic Solvents.* The acid solution of the alkaloids contained in a separatory funnel is made alkaline, in most cases, with ammonia T.S., and then shaken with several successive portions of organic solvent, such as ether, chloroform, or ether-chloroform mixture. The volume of organic solvent used in each extraction should be at least one-half that of the aqueous solution. The extraction with portions of organic solvent must be repeated as long as any alkaloid is extracted by the solvent. To determine the completion of extraction, evaporate 1 ml of the last washing and dissolve the residue in 0.5 ml of approximately 0.5 N hydrochloric acid; the resulting solution should show no more than a slight turbidity on the addition of a drop of mercuric iodide T.S., or, in the case of caffeine and colchicine, on the addition of a drop of iodine T.S. The number of extractions required depends largely on the character of the alkaloid. With most alkaloids, it is advisable to extract three or four times before testing. Physostigmine and pilocarpine require about twice as many extractions as other alkaloids.

The ammonia reacts with the alkaloidal sulfates contained in the aqueous solution to form the free alkaloids and ammonium sulfate. If considerable quantities of alkaloid are contained in the acid solution, the alkaloids may be precipitated when the alkali is added. Consequently, the first portion of organic solvent is added prior to the addition of the ammonia water to dissolve any alkaloidal precipitate formed. If this procedure is not followed, the precipitate frequently agglomerates into large particles which dissolve slowly when the organic solvent is added.

6. *Evaporation of Organic Solvents.* The final alkaloidal solution in organic solvents may be evaporated by distillation with recovery of the solvent, by use of a steam bath or sand bath, or in a current of warm air. When the evaporation is conducted on a steam bath, precautions must be observed to prevent the ignition of the vaporized solvent. The operation may be performed by heating a prepared water bath until the water boils, turning out the flame, and placing the alkaloidal solution, in a suitable container, on the bath. If a sand bath is employed, the steam-heated type is best, and the temperature of the bath should be about 60°.

Chloroform tends to form unstable compounds with many alkaloids; consequently, when chloroformic solutions of the alkaloids are evaporated, the residue may retain traces of chloroform rather tenaciously. These traces of chloroform are best removed by treating the alkaloidal residue with 2 to 3 ml of ether or alcohol and evaporating the solution to dryness again. This is a wise precaution to employ in the evaporation of all solutions of alkaloids in organic solvents, when the final determination is made by volumetric methods in order to remove small traces of ammonia that may be occluded in the alkaloidal residue, particularly if the latter is gummy.

Those alkaloids which have the structure of esters, such as atropine, hyoscyamine, and cocaine, and some others, are decomposed more or less by heat and by prolonged contact with water and strong acids and alkalies. Consequently, their solutions should be evaporated at a low temperature, usually at about 60°; and, if it is necessary to allow the alkaloid in solution to stand from one laboratory period to the next before the assay can be completed, it should be shaken out of acid or alkali solutions into an organic solvent and kept in a tightly stoppered flask.

7. *Gravimetric Determinations of Alkaloids.* If the alkaloid is to be estimated gravimetrically, the solvent is evaporated in the usual manner from a previously weighed evaporating dish, wide-mouth flask, or beaker, and the residue is dried to constant weight in an electric or hot-air oven at 100 to 110° and weighed. The same procedure should be followed in drying the flask or beaker used as is employed in drying the container with its residue. When the residue is to be weighed, if the final solvent has been chloroform, the last traces of that solvent should be removed by the addition of a little neutral ether or alcohol, followed by evaporation. Care must be taken to avoid loss by decrepitation, especially when evaporating chloroformic solutions of nux vomica or cinchona alkaloids. Decrepitation may usually be prevented by the addition of a little alcohol after the solution has been reduced to a volume of 1 or 2 ml, evaporating at a low temperature, and rotating the container during the process.

8. *Volumetric Determinations of Alkaloids.* If the alkaloid is directed to be assayed volumetrically, the residue should be softened by the addition of about 1 ml of neutral alcohol or ether, the required amount of standard acid added, and the mixture gently warmed to ensure the complete solution of the alkaloid. If preferred, the alkaloidal residue may be dissolved in chloroform, the standard acid added, and the chloroform removed by evaporation. Before titrating, add a sufficient quantity of distilled water to make the volume of the mixture measure about 25 ml. The amount of acid in excess of that which combines with the alkaloid is then determined at once by titration with standard alkali.

As a rule, the volumetric method of determining the quantity of alkaloid obtained in a given assay is more accurate than the gravimetric method, since many impurities that cannot be removed are present in the alkaloidal residue and are computed as alkaloid when the residue is weighed. If the residue is titrated, however, only those impurities which are acidic or basic affect the result.

Absorbents. In assaying fluidextracts, tinctures, and other preparations of alkaloid-bearing drugs, it is often necessary to evaporate them to dryness and, to avoid loss and to aid in the evaporation, they are usually added to some absorbent material. Paper pulp or asbestos fiber should be used for this purpose. Such absorbent material must be acid- and alkali-washed, then rendered neutral by being washed with distilled water, and dried before being used.

Apparatus for Proximate Assays. When a container of definite size and shape is recommended in a proximate assay process, it is understood that this is advisory and not obligatory, except when volumetric flasks, measuring burets, or other exact measuring apparatus are specified.

Problems

1. (*a*) Calculate the amount, in milligrams, of atropine which remains in 20 ml of ether after extracting with three 20-ml portions of dilute sulfuric acid. The ethereal solution originally contained 500 mg of atropine. $K = C_{acid}/C_{ether} = 9$. (*b*) What is the percent error resulting from incomplete extraction?

2. In the assay of Meperidine HCl, the N.F. XII directs that five 10-ml portions of chloroform be used for the extraction process. It further directs that excess NaOH solution be added to an accurately measured volume of the injection before extraction is undertaken. Assuming $K = C_{CHCl_3}/C_{basified\,H_2O} = 12$, and 4 ml of injection containing 400 mg of the salt was taken, calculate the extraction error after three, four, and five extractions using 10-ml portions of chloroform in each case. Discuss results.

Chapter 16

ALKALOIDAL ASSAYS BY OFFICIAL TYPE METHODS

The official alkaloidal drugs and the galenical preparations containing alkaloids are assayed by two type methods and by special methods. The type methods are (1) the total extraction method for drugs and galenical preparations, and (2) the aliquot method for drugs and galenical preparations. Belladonna Leaf and Belladonna Tincture, official in the U.S.P., and Belladonna Leaf Fluidextract, Belladonna Pillular Extract, and Powdered Belladonna Extract, official in the N.F., are assayed by the total extraction method. Ipecac and Powdered Ipecac are assayed by either method, and Areca is assayed only by the aliquot part method.

TOTAL EXTRACTION METHOD

Exercise 90

Object. Assay of Belladonna Leaf.

Materials Required. 10 Gm of Belladonna Leaf in Moderately Coarse Powder
Mercuric Iodide T.S.
0.5 *N* Sulfuric acid
Methyl red T.S.
10 ml of alcohol
0.02 *N* Sulfuric acid
0.02 *N* NaOH
Ether
Ammonia T.S.
Chloroform

Procedure. "Moisten 10 Gm, previously reduced to a moderately coarse powder and accurately weighed, with a mixture of 8 ml of stronger ammonia, 10 ml of alcohol, and 20 ml of ether, and extract the alkaloids by either of the methods given in the following two paragraphs. If necessary, reduce the volume of the extract to 100 ml. by evaporation on a steam bath."

I. "Place the moistened drug in the thimble of a Soxhlet or similar extractor, and allow maceration to proceed overnight, then extract with ether for 3 hours, or longer if necessary to effect complete extraction."

II. 1. "Place the moistened drug in a small percolator, and allow maceration to proceed overnight. Percolate slowly with a mixture of 3 volumes of ether and 1 volume of chloroform. Continue the percolation until the residue from 3 to 4 ml. of percolate last passed, when dissolved in 0.5 *N* sulfuric acid and treated with mercuric iodide T.S., shows only a faint turbidity."

l-Hyoscyamine, the chief alkaloid of belladonna, and small quantities of other alkaloids are set free from the acids with which they are combined in the crude drug by the ammonia, and dissolve in the organic solvent. The *l*-hyoscyamine is largely converted into its racemic isomer, atropine, in the ammoniacal solution.

The large volume of liquid obtained, especially by cold percolation, is concentrated to a small volume to secure a convenient amount. It should not be evaporated to dryness because the chlorophyll and other nonalkaloidal residue make it difficult to dissolve completely the alkaloids in the acid. The alkaloids hyoscyamine and atropine, being esters, undergo hydrolysis readily in aqueous acid and alkaline solution, especially when heated. Consequently, the extraction process should be conducted at a low temperature. The evaporation may be carried out advantageously in a vacuum desiccator under reduced pressure.

2. "Transfer the extract quantitatively to a separator with the aid of ether. Extract with successive portions of approximately 0.5 *N* sulfuric acid, filtering each portion drawn off."

Upon filtration, the nonalkaloidal residue consisting mostly of chlorophyll is retained on the filter. Quite stable emulsions are particularly likely to be encountered in shaking out alkaloids in the assay of hyoscyamus, belladonna, and stramonium, owing to the presence of small amounts of saponins which act as emulsifying agents.

The sulfates of the alkaloids are formed and dissolved in the acid solution, leaving most of the coloring matter, etc., in the acid-insoluble residue.

3. "Render the combined acid solutions distinctly alkaline with ammonia T.S., and completely extract the alkaloids at once with successive portions of chloroform."

The alkaloids are set free by the ammonia water and dissolve in the chloroform.

4. "Evaporate the combined extracts on a steam bath to dryness, and then heat in a bath of boiling water for 15 minutes."

The solution of the alkaloids in organic solvent is evaporated to drive off the solvent and ammonia. The residue is repeatedly dissolved, evaporated, and dried to drive off small amounts of volatile amines which are basic and which, unless removed, would neutralize some of the acid in the next step of the procedure, with subsequent high results.

5. "Dissolve the residue in a small volume of chloroform, again evaporate to dryness, and heat in a few milliliters of chloroform, add 15.0 ml. 0.02 *N* sulfuric acid, remove the chloroform by evaporation, cool, add methyl red T.S., and titrate the excess acid with 0.02 *N* sodium hydroxide. Each milliliter of 0.02 *N* acid is equivalent

to 5.788 mg. of the alkaloids of Belladonna Leaf, calculated as hyoscyamine (or atropine)."

The residual alkaloids, which consist chiefly of the *d-l* isomers of hyoscyamine, react with sulfuric acid as follows:

$$\underset{2 \times 289.38}{2C_{17}H_{23}O_3N} + \underset{98.08}{H_2SO_4} \longrightarrow (C_{17}H_{23}O_3N)_2 \cdot H_2SO_4$$

The results for this assay may be calculated by using either the gram-milliequivalent weight of atropine or the U.S.P. titer value. Both these values will be determined and used in the sample calculations. It should be pointed that these calculations are generally applicable to most assays of alkaloids and other related substances.

The gram-milliequivalent weight of atropine is calculated as follows: From the balanced chemical reaction it can be seen that 2 millimoles of atropine react with 1 millimole or 2 mEq of sulfuric acid. Consequently 1 millimole of atropine is equal to 1 mEq of atropine and the weight of 1 millimole of atropine, which equals 0.28938 Gm, is also equal to the weight in grams of 1 mEq of atropine (0.2894 Gm per mEq).

The U.S.P. titer value may be calculated by using the gram-milliequivalent weight as follows: Each milliliter of 0.02 N H_2SO_4 contains 0.02 mEq (ml $\times$ N = mEq) of H_2SO_4 and accordingly, reacts with 0.02 mEq of atropine. The weight in grams of 0.02 mEq of atropine is found by multiplying the gram-milliequivalent weight by 0.02. Therefore,

$$0.2894 \text{ Gm/mEq} \times 0.02 \text{ mEq} = 0.005788 \text{ Gm or } 5.788 \text{ mg}$$

and the atropine titer value for 0.02 N sulfuric acid is 5.788 mg per ml.

The calculations for the assay of belladonna leaf using these values are illustrated in the following sample calculations.

Problem. Determine the percent W/W of alkaloids in belladonna leaf if 15.00 ml of 0.0210 N sulfuric acid were added to the extracted alkaloids and 7.30 ml of 0.0198 N NaOH were required for the residual titration. The weight of the powder of belladonna leaf taken was 10.21 Gm.

Solution (Using the Milliequivalent Value of Atropine):

1. Determine the number of milliequivalents of sulfuric acid which reacted with atropine:

$$15.00 \text{ ml} \times 0.0210\ N = 0.3150 \text{ mEq } H_2SO_4 \text{ initially present}$$
$$7.30 \text{ ml} \times 0.0198\ N = 0.1445 \text{ mEq NaOH (residual } H_2SO_4)$$
$$0.3150 \text{ mEq} - 0.1445 \text{ mEq} = 0.1705 \text{ mEq of } H_2SO_4$$

reacted with 0.1705 mEq of atropine.

2. Determine the weight in grams of 0.1705 mEq of atropine.

$$0.2894 \text{ Gm/mEq} \times 0.1705 \text{ mEq} = 0.04934 \text{ Gm}$$

This is the quantity present in 10.21 Gm of sample.

3. Convert to percent *W/W*:

$$\frac{0.04934 \text{ Gm atropine}}{10.21 \text{ Gm sample}} \times 100 = 0.483 \text{ percent}$$

Solution (Using the U.S.P. Titer Value):

1. Determine the number of milliliters of 0.02 *N* H_2SO_4 consumed by the alkaloids.

$$\frac{15.00 \text{ ml} \times 0.0210\ N}{0.02\ N} = 15.75 \text{ ml } H_2SO_4 \text{ initially present}$$

$$\frac{7.30 \text{ ml} \times 0.0198\ N}{0.02\ N} = 7.23 \text{ ml NaOH (residual } H_2SO_4)$$

$$15.75 \text{ ml} - 7.23 \text{ ml} = 8.52 \text{ ml of } 0.02\ N \text{ sulfuric acid}$$

the amount consumed by the alkaloids.

2. Multiply 8.52 ml by 5.788 mg, the number of milligrams of atropine equivalent to 1 ml of 0.02 *N* H_2SO_4:

$$8.52 \text{ ml} \times 5.788 \text{ mg} = 49.31 \text{ mg}$$

3. Divide the weight of sample in milligrams into the weight of atropine as calculated in (2) and multiply by 100:

$$\frac{49.31 \text{ mg}}{10{,}210 \text{ mg}} \times 100 = 0.483 \text{ percent}$$

Fluidextracts. Belladonna Leaf Fluidextract is assayed in a manner similar to the assay of the tincture. The volumes of fluidextract taken as samples are designated so that approximately 300 mg of the alkaloids will be obtained in the assay residue.

Exercise 90

Object. Assay of Belladonna Leaf Fluidextract.

Materials Required. 10 ml of Belladonna Leaf Fluidextract
Chloroform
Ammonia T.S.
0.5 *N* sulfuric acid
0.02 *N* sulfuric acid
0.02 *N* sodium hydroxide
Methyl red T.S.

Procedure. "Transfer 10.0 ml of Belladonna Leaf Fluidextract to a separator containing 25 ml of chloroform. Add 25 ml. of water and render the mixture alkaline by the addition of ammonia T.S. Shake the mixture vigorously for about 1 minute, separate the chloroform layer, and complete the extraction of the alkaloids with successive portions of chloroform. Completely remove the alkaloids from the immiscible solvent by extracting with successive portions of approximately 0.5 *N* sulfuric acid filtering each portion drawn off. Render the combined acid solutions distinctly alkaline with ammonia T.S., and completely remove the alkaloids at once by extracting with successive portions of chloroform. Evaporate the combined chloroform extracts to dryness on a steam bath and keep the dry residue at this temperature for 15 minutes. Again dissolve the residue in chloroform, evaporate to dryness on a steam bath, and continue the heating for 15 minutes. Dissolve, evaporate, and heat for the third time. Dissolve the resulting residue in a few milliliters of chloroform, add 15.0 ml. of 0.02 *N* sulfuric acid, remove the chloroform by evaporation, cool, add methyl red T.S., and titrate the excess acid with 0.02 *N* sodium hydroxide. Each milliliter of 0.02 *N* sulfuric acid is equivalent to 5.788 mg. of the alkaloids of belladonna leaf."

Tinctures. The tincture of belladonna leaf is assayed by the same method except that the amount of the tincture employed as sample is designated so that approximately 0.030 Gm of the alkaloids will be obtained in the assay residue. The assay method is the same as that for belladonna leaf, except that the extraction of the alkaloids is omitted, since they are already in solution. The assay process is illustrated in the following exercise.

Exercise 91

Object. Assay of Belladonna Tincture.

Materials Required. 100 ml of Belladonna Tincture
Ammonia T.S.
About 10 ml of Diluted Alcohol
Approximately 0.5 *N* sulfuric acid
0.02 *N* sulfuric acid
0.02 *N* NaOH
Chloroform
Methyl red T.S.

Procedure. "Measure accurately 100 ml., and evaporate it, at a temperature not exceeding 100°, to a volume of about 10 ml. Transfer the concentrated liquid to a separator containing 25 ml. of chloroform, rinse the container with small portions of diluted alcohol, and add the rinsings to the separator. Add 25 ml. of water, and render the mixture alkaline by the addition of ammonia T.S. Completely extract the alkaloids with successive portions of chloroform. Transfer the extract quantitatively to a separator with the aid of ether. Extract with successive portions of approximately 0.5 *N* sulfuric acid, filtering each portion drawn off. Render the combined acid solutions distinctly alkaline with ammonia T.S., and completely extract the alkaloids at once with successive portions of chloroform. Evaporate the combined extracts on a steam bath to dryness, and then heat in a bath of boiling water for 15 minutes. Dissolve the residue in a small volume of chloroform, again evaporate to dryness, and heat in a bath of boiling water for 15 minutes. Dissolve the residue in a few milliliters of chloroform, add 15.0 ml. of 0.02 *N* sulfuric acid, remove the chloroform by evaporation, cool, add methyl red T.S., and titrate the excess acid with 0.02 *N* NaOH. Each milliliter of 0.02 *N* acid is equivalent to 5.788 mg of the alkaloids of belladonna leaf, calculated as hyoscyamine (or atropine)."

Questions and Problems

1. What isomeric change does the chief alkaloid of belladonna undergo in the assay process?

2. The alkaloids of belladonna have the structure of esters. What type of decomposition might they undergo if allowed to stand in contact for a long time with a strong alkali?

3. How much 0.1015 *N* H_2SO_4 will be required to form the sulfates of the alkaloids from 25 Gm of hyoscyamus leaves that contain 0.070 percent alkaloids?

4. Ten milliliters of Belladonna Leaf Fluidextract was extracted following the procedure given in the N.F. XII. After adding 15.00 ml of 0.0220 *N* sulfuric acid to the extracted alkaloids, the residual acid was titrated and 8.40 ml of 0.0210 *N* sodium hydroxide was required to bring about the methyl red end point. Calculate the milligrams of alkaloids present in 100 ml of Belladonna Leaf Fluidextract.

ALIQUOT PART METHOD: VOLUMETRIC ASSAYS

Exercise 92

Object. Assay of Ipecac for Ether-soluble Alkaloids.

Materials Required. Ipecac, finely powdered (No. 60)
Peroxide-free Ether
Dilute Sulfuric acid (1 in 20)
Stronger Ammonia T.S.
Ammonia T.S.
0.1 *N* Sulfuric acid (standardized)
0.1 *N* NaOH (standardized)
Methyl red T.S.

Procedure. 1. I. "To 10 Gm. of finely powdered Ipecac, in a suitable container, add 100 ml. of ether, accurately measured at 25°, stopper the container tightly, shake the mixture thoroughly, and allow it to stand for 5 minutes. Then add 10 ml. of ammonia, T.S., close again tightly, shake it for 1 hour in a mechanical shaker or intermittently during 2 hours, and allow to stand overnight at a temperature not exceeding 25°. Again shake the mixture intermittently during 30 minutes, and allow the drug to settle at 25°. Transfer to a separator a 50 ml. aliquot of the clear, supernatant liquid, representing 5 Gm. of Ipecac."

II. "Place 5 Gm. of the finely powdered Ipecac in an extraction thimble, in a Soxhlet or similar extractor. Add enough ether to cover the powder, and allow to stand for 10 minutes with occasional stirring. Add 3 ml. of stronger ammonia T.S., mix, and allow to stand overnight. Cover the drug with a pledget of cotton, pack well, and extract with ether for 5 hours. Transfer the ether extract to a separator."

"Either method I or II may be used to extract the alkaloids from the powdered Ipecac. The ether used in this assay must be shown by U.S.P. test to be free of peroxides within 24 hours prior to use."

Ipecac contains a number of alkaloids, the most important of which are emetine, cephaeline, and psychotrine. These three alkaloids occur in varying proportions in different varieties of the root. Psychotrine constitutes a relatively small proportion of the total alkaloids and is not known to be of importance therapeutically. Consequently, this alkaloid is eliminated in the assay process, since it is practically insoluble in ether, while the liberated emetine and cephaeline, being soluble in ether, are determined. The ipecac must be finely powdered, since it is not penetrated readily by the solvent because of its high starch content.

The aliquot portion of the ethereal solution contains an amount of emetine and cephaeline equivalent to one-half the weight of sample taken.

2. "Extract the alkaloids from the ether with dilute sulfuric acid (1 in 20), using at first 15 ml., or more, if necessary, to insure an acid reaction, then successive 10 ml. portions until extraction is complete, and filtering all extracts through the same filter into a second separator."

The sulfate salts of emetine and cephaeline are formed and dissolve in the aqueous solution.

3. "To the combined acid solutions add about an equal volume of peroxide-free ether, render the mixture alkaline with ammonia T.S., and completely extract with successive portions of peroxide-free ether. Filter each portion of the ether extract into a flask or beaker, and carefully evaporate the combined ether extracts on a steam bath until nearly but not quite dry. Add 5 ml. of peroxide-free ether and exactly 10 ml. of 0.1 *N* sulfuric acid, and heat on a steam bath to effect complete solution of the alkaloids and to remove all the ether. Cool, add 15 ml. of water, then add methyl red T.S. and titrate the excess acid with 0.1 *N* sodium hydroxide. Each milliliter of 0.1 *N* sulfuric acid is equivalent to 24.0 mg. of the ether-soluble alkaloids of ipecac."

The ammonia sets the emetine and cephaeline free, and the liberated alkaloids dissolve in the ether.

The residue consists almost entirely of emetine and cephaeline which dissolve in the standard acid solution, forming emetine and cephaeline sulfates, as illustrated by the following reaction for emetine:

$$\underset{480.65}{C_{29}H_{40}N_2O_4} + \underset{98.08}{H_2SO_4} \longrightarrow C_{29}H_{40}N_2O_4 \cdot H_2SO_4$$

The gram-milliequivalent weight of emetine is calculated as follows: From the balanced chemical reaction it can be seen that 1 millimole of emetine reacts with 1 millimole or 2 mEq of sulfuric acid. Consequently 1 millimole of emetine is equal to 2 mEq and 0.5 millimole is equal to 1 mEq. The weight of 0.5 millimole of emetine is 0.48065 Gm/2 or 0.24031 Gm. The weight of 1 mEq of emetine therefore is equal to 0.48065 Gm/2 or 0.24032 Gm.

The U.S.P. titer value may be calculated in a manner analogous to that used for atropine in the assay of Belladonna Leaf.

Since cephaeline has nearly the same pharmacological properties, molecular structure (differing by only one methyl group), and chemical properties as emetine, it is included in the calculation of total ether-soluble alkaloids of ipecac, calculated as emetine.

Questions and Problems

1. Calculate the alkaloidal equivalent for each milliliter of 0.02 *N* sulfuric acid.
2. Would the error in the assay result be very considerable if the ether-soluble alkaloids of the ipecac consisted almost entirely of emetine?

3. Enumerate several possible sources of error that might be encountered in the assay of ipecac.

Exercises 93

Object. Assay of Areca.

Materials Required. 8 Gm of areca
80 ml of ether
10 Gm of anhydrous sodium sulfate
0.5 Gm of talc

Procedure. "Place 8 Gm. of Areca, in moderately coarse powder and accurately weighed, into a glass-stoppered flask, add 80 ml. of ether, shake well, add 4 ml. of ammonia T.S., and shake during 10 minutes."

The most important alkaloid in areca is arecoline. Several other alkaloids, viz., arecaidine, guavacine, guavacoline, and arecolidine, also occur in the drug in small amounts. These alkaloids, which are largely present in the drug as salts, are liberated by the NH_3 in the ammonia T.S. The arecoline, guavacoline, and arecolidine, although soluble in both water and ether, dissolve in large part in the latter; the arecaidine and guavacine dissolve mainly in the alkaline aqueous liquid, because they contain a carboxy group and form water-soluble salts with alkalies. Since arecoline is the methyl ester of arecaidine, and guavacoline is the methyl ester of guavacine, the carboxy groups are blocked, and these alkaloids do not form salts with alkalies.

2. "Add 10 Gm. of anhydrous sodium sulfate and shake during 5 minutes. Allow to settle and quickly decant the ether into another glass-stoppered flask."

Since arecoline, guavacoline, and arecolidine are soluble in both water and ether, the water present as a result of the addition of ammonia T.S. is removed by shaking with anhydrous sodium sulfate. After this treatment, all the arecoline, guavacoline, and arecolidine are dissolved by the ether, and the other alkaloids, arecaidine and guavacine, which are almost insoluble in organic solvents, remain with the sodium sulfate and other constituents insoluble in ether.

3. "Add 0.5 Gm. of purified talc to the decanted ether solution and shake during 3 minutes, then add 2.5 ml. of distilled water and shake during 3 minutes more. Allow to stand until clear, quickly decant 50 ml. of the ether, equivalent to 5 Gm. of Areca, and . . ."

The decanted ether layer, which contains fat and suspended matter, is freed from fat and clarified by shaking it with talc and a small amount of water and allowing it to settle. Exactly 50 ml of the clear supernatant ether solution is decanted, so that the aliquot will represent 5 Gm of the drug.

4. "Evaporate off about two-thirds of the ether."

About two-thirds of the ether is distilled off to reduce the volume to about 15 ml and thus facilitate the shaking out of the alkaloids. The arecoline, guavacoline, and arecolidine, which are soluble in all proportions in ether and water, form salts with the acid which are readily soluble in water but only slightly soluble in ether.

5. "Extract the remaining ether solution in a separator with 15 ml. of 0.02 *N* sulfuric acid and then with 3 portions of distilled water using 5 ml. each time. The 15 ml. of 0.02 *N* sulfuric acid may be added directly to the ether in the glass-stoppered flask and the flask shaken for 10 minutes, after which its contents are then transferred to the separator. The flask is then washed with the three 5-ml. portions of distilled water, each being passed successively through the separator. To the combined acid and washings, add methyl red T.S. and titrate the excess acid with 0.02 *N* sodium hydroxide solution. Each milliliter of 0.02 *N* sulfuric acid is equivalent to 3.104 mg. of alkaloids calculated as arecoline."

The acid used is 0.02 *N* H_2SO_4, and the amount consumed is determined by residual titrations with 0.02 *N* NaOH. Calculate the arecoline titer value for 0.02 *N* sulfuric acid using the following chemical reaction.

$$\underset{2 \times 155.20}{2C_8H_{13}O_2N} + H_2SO_4 \longrightarrow (C_8H_{13}O_2N)_2 \cdot H_2SO_4$$

Chapter 17

ALKALOIDAL ASSAYS BY SPECIAL METHODS

A number of the official alkaloidal drugs and their preparations cannot be assayed by any of the type methods previously discussed because of the nature of the drug or of the properties of the alkaloids to be determined. Drugs of this type are assayed by special methods.

The assays of opium and its preparations differ from those of the other alkaloidal drugs, because morphine exhibits exceptional solubility properties; i.e., it is practically insoluble in most organic solvents.

Exercise 94

Object. Assay of Opium.

Materials Required. 6 Gm of opium
3 Gm of calcium hydroxide
1 Gm of ammonium chloride
100 ml of ether
25 ml of water saturated with morphine
0.1 *N* H_2SO_4
0.1 *N* NaOH

Procedure. 1. Weigh accurately 6 Gm of Opium which, if fresh, should be in very small pieces and, if dry, in a fine powder. Place the Opium in a mortar and triturate it with about 40 ml of warm water for 15 min. Transfer the contents of the mortar completely with the aid of 30 ml of water into a flask, stopper the flask, and shake it every 10 min, or continuously in a mechanical shaker during 1 hr. The extraction of the opium may also be made in the following manner: Place the comminuted opium as described above in a flask with 70 ml of water, add a few glass beads, if desired, and shake every 10 min, or continuously in a mechanical shaker during 2 hr, and allow to stand overnight, then on the following morning shake the flask for 30 min.

"Pour the contents of the flask as evenly as possible upon a wetted filter of 10 to 11 cm. diameter, and, when the liquid has drained off well, wash the residue with about 20 ml. of water, carefully dropped upon the edges of the filter and its contents. Transfer the residue on the filter as completely as possible into the flask with the aid of about 40 ml. of water, shake it thoroughly during 10 minutes, and filter through the original filter. Filtration and washing of the residue may be facilitated by the use of a centrifuge. When the liquid has drained off, wash the flask and residue with small portions of water until the washings are nearly colorless. (The total filtrate will usually measure 300 to 350 ml.)"

The alkaloids of opium, about 25 of which are known, occur in the drug as meconates and sulfates. When the opium is agitated and macerated with distilled water, most of the meconates and sulfates of the alkaloids are dissolved. The alkaloids are completely extracted when the residual opium is treated with a second portion of distilled water. Rubbing to a smooth paste breaks up any lumps, so that the water can come in contact with all the particles of the drug. The residue is washed with distilled water after each extraction to displace all the solvent absorbed by the opium.

2. Evaporate the filtrate and washings in a tared dish to about 30 Gm, and allow to cool. Now add 3 Gm of calcium hydroxide, triturate for 15 min, then transfer the mixture completely to a tared flask with the aid of small portions of water. Add sufficient water to make 54 Gm and mix well. Filter through a dry filter of 10 to 11 cm diameter, collecting the filtrate in a dry cylinder or a small flask. During the filtration keep the funnel covered with a glass plate and its stem well within the neck of the receiving vessel.

The combined aqueous filtrates are evaporated on a water bath to a volume of about 30 ml, since the alkaloids that are quite stable at 100° might be decomposed if the evaporation were carried out over a direct flame. The purpose of concentrating the aqueous alkaloidal extractive is to reduce the volume to a convenient working quantity. The concentrated extract is made up to a weight of exactly 54 Gm, so that an aliquot portion representing a definite weight of opium may be taken. When the aqueous opium extract is triturated with calcium hydroxide, the alkaloids are liberated from their salts and the morphine, codeine, and narcotine are dissolved in the alkaline calcium hydroxide solution, most of the other alkaloids of opium remaining as insoluble matter along with the calcium meconate, calcium sulfate, resins, etc. The mixture may be transferred to a flask and shaken for 15 min to allow the reactions in which the alkaloids are liberated to go to completion, the flask being stoppered to prevent evaporation, so that the volume of liquid will not change appreciably, since an aliquot portion of it is employed in the assay.

3. Place 34 Gm of the filtrate, corresponding to 4 Gm of Opium, in an Erlenmeyer flask of suitable capacity, add 2 ml of alcohol and 15 ml of ether, and, after shaking, add 1 Gm of ammonium chloride. Stopper the flask, shaking it frequently during 10 min, and set aside overnight at a temperature of 5 to 10°. Remove the stopper and brush any adhering crystals back into the flask.

Since the total aqueous extract containing the alkaloids weighed 54 Gm after adding the lime, 54 − 3 = the weight of liquid, and an aliquot portion of two-thirds of this liquid or 34 Gm contains an amount of the alkaloids corresponding to ⅔ of 6 = 4 Gm of opium.

Ammonia is liberated from ammonium chloride:

$$2NH_4Cl + Ca(OH)_2 \longrightarrow 2NH_4OH + CaCl_2$$

and morphine is precipitated, since it is insoluble in ammoniacal solution and in ether. The alcohol causes the morphine to separate out in the form of relatively large crystals; whereas, when alcohol is not used, the alkaloid comes down in the form of a fine powder. The crystals are easier to collect and less rapidly soluble in ether than the powder. The mixture is allowed to stand in a cool place for 12 hr to allow the larger crystals to grow at the expense of the smaller ones. The codeine and narcotine remain in solution in the ether.

4. Decant the ether layer through a small filter paper or through a sintered-glass crucible, rinse the flask and contents with 15 ml of ether, and pass these washings through the filter. Again wash the filter and precipitate with an additional 10 ml of ether. When all the ether has passed through the filter, pour the water layer on the filter without trying to remove the crystals from the flask. Wash the crystals in the flask and the contents of the filter with 5-ml portions of water saturated with morphine, using a total of 40 ml of the morphinated water. Then add, dropwise, 1 ml of cold water to displace the morphinated water.

The crystals are washed with ether to remove any traces of codeine or narcotine, since these alkaloids are insoluble in ammonia water but soluble in ether. Water saturated with morphine is used to wash the crystals to prevent solution of the crystalline morphine, which is very slightly soluble in water. This aqueous solution removes any ammonia present, as well as any other water-soluble impurity adhering to the surface of the crystals, such as gums, coloring matter, etc. The morphinated water must be displaced from the filter paper with distilled water to avoid including in the final result some of the morphine from the morphinated water.

5. To the flask containing the crystals add about 15 ml of boiling reagent methanol, agitate to dissolve as much of the morphine as possible, and pour the boiling solution over the morphine on the filter, receiving the filtrate in a suitable dry flask or beaker. Repeat this treatment with boiling reagent methanol eight to ten times, using 5 to 15 ml each time, until all the morphine has been dissolved. Cool the methanol solution, and add to it exactly 25 ml of 0.1 *N* sulfuric acid. Dilute with 75 ml of water, and boil carefully, or evaporate on a steam bath, to a volume of about 100 ml. Cool, and titrate the excess of acid with 0.1 *N* sodium hydroxide, using 0.25 ml of methyl red T.S. as the indicator. Each milliliter of 0.1 *N* sulfuric acid is equivalent to 28.53 mg of anhydrous morphine.

The morphine crystals dissolve in the 0.1 *N* sulfuric acid with the formation of morphine sulfate.

$$\underset{2 \times 285.35}{2C_{17}H_{19}O_3N} + H_2SO_4 \longrightarrow (C_{17}H_{19}O_3N)_2 \cdot H_2SO_4$$

The gram-milliequivalent weight of morphine is calculated as follows: From the balanced chemical reaction it can be seen that 2 millimoles (mM) of morphine react with 1 mM or 2 mEq of sulfuric acid. Therefore, 1 mM of morphine is equal to 1 mEq of morphine, and the weight of 1 mM of morphine, which equals 0.28535 Gm, is also equal to the weight in grams of 1 mEq of morphine (0.2853 Gm per mEq).

The U.S.P. titer value may be calculated in a manner analogous to that used for atropine in the assay of Belladonna Leaf.

The Pharmacopeia requires that Opium contain not less than 9.5 percent of anhydrous morphine. Powdered Opium, which is assayed in exactly the same manner as described for Opium, must contain not less than 10.0 and not more than 10.5 percent of anhydrous morphine. Paregoric, which is assayed by a related method, must contain, in each 100 ml, not less than 35 mg and not more than 45 mg of anhydrous morphine.

Questions and Problems

1. What functions do the calcium oxide and ammonium chloride serve in the above assay?
2. One gram of morphine is soluble in about 7.632 ml of ether at 25°. Look up the solubility of codeine in the U.S.P. and explain how ether may be used to separate the alkaloids.
3. Why is water saturated with morphine used to wash the crystallized morphine?
4. Morphine contains a phenolic hydroxyl group. Explain why it is appreciably soluble in alkaline solutions.

Assay of Capsules and Tablets for Alkaloidal Content. The official assays of capsules and tablets for alkaloids are conducted in much the same way as the assays of the crude drugs in which the alkaloids occur (see General Procedures, Chap. 15 through end of Chap. 16), except that the methods of preparing the samples for assay differ and the procedures for the purification of the alkaloids are less involved, as a rule.

The usual procedure followed in taking the aliquot for an assay, in the case of capsules and tablets, is to empty about 20 capsules or to powder about 20 tablets and then weigh accurately the amount to be used in the assay. In some products, however, the aliquot is taken after the initial extraction of the alkaloid has been effected, e.g., aminophylline tablets. In the latter instance, a definite number of tablets are dissolved in distilled water and ammonia T.S.; the solution is diluted with distilled water to a definite volume and filtered. An accurately measured aliquot of the filtrate, equivalent to about 0.3 Gm of aminophylline, is then taken for assay.

Modifications of the usual methods of assay may be necessary for tablets and capsules when coated. The methods of obtaining the quantities of the drug contained in capsules and tablets to be used for assay and the procedures followed for isolating and purifying the alkaloids are illustrated in the following exercises.

Exercise 95

Object. Assay of Aminophylline Tablets.

Materials Required. 20 aminophylline tablets
20 ml of 0.1 *N* silver nitrate

Procedure. 1. Weigh not less than 20 Aminophylline Tablets and reduce them to a fine powder without appreciable loss. Transfer completely to a 200-ml flask with the aid of a mixture of 50 ml of water and 15 ml of ammonia T.S., and allow to stand for 30 min with frequent shaking. Dilute to 200 ml with water, mix well, and filter through a dry filter into a dry flask, rejecting the first 20 ml of the filtrate. Transfer an accurately measured aliquot, equivalent to about 250 mg of aminophylline, to a 250-ml Erlenmeyer flask and add water if necessary to make about 40 ml.

If the tablets are labeled to contain 3 grains each, 20 tablets, corresponding to about 4.0 Gm of aminophylline, should be dissolved in a total volume of 200 ml of distilled water and ammonia T.S. One milliliter of the solution will then contain approximately 0.02 Gm of aminophylline, and an aliquot of 155 ml of the solution will contain about 0.3 Gm.

2. Add 8 ml of ammonia T.S. and exactly 20 ml of 0.1 *N* silver nitrate, mix, and warm on the steam bath for 15 min. Cool the mixture to between 5 and 10° for 20 min, then filter, preferably through a filtering crucible under reduced pressure, and wash the precipitate three times with 10-ml portions of water. Acidify the combined filtrate and washings with nitric acid, and add an excess of 3 ml of the acid. Cool, add 2 ml of ferric ammonium sulfate T.S., and titrate the excess silver nitrate with 0.1 *N* ammonium thiocyanate. Each milliliter of 0.1 *N* silver nitrate is equivalent to 22.82 mg of $C_{16}H_{24}N_{10}O_4 \cdot 2H_2O$.

From the addition of the silver nitrate solution to the completion of the assay, the procedure is almost the same as that described under "Assay of Theophylline Sodium Acetate." The explanations given under the latter are also applicable (see Exercise 98, in this chapter).

Each milliliter of 0.1 *N* silver nitrate is equivalent to 0.02282 Gm of hydrated aminophylline. Calculate the percentage of the labeled amount of hydrated aminophylline in the tablets assayed.

Exercise 96

Object. Assay of Ephedrine Sulfate Capsules.

Materials Required. Twenty 25-mg ephedrine sulfate capsules
Ether
Distilled water saturated with sodium chloride
0.1 *N* NaOH
0.1 *N* sulfuric acid
Methyl red T.S.

Procedure. 1. "Transfer, as completely as possible, the contents of a counted number of not less than 20 Ephedrine Sulfate Capsules to a 100-ml. volumetric flask. Place the emptied capsules in a small beaker, add sufficient cold water to cover them, and agitate for 10 minutes. Filter into the flask, wash the beaker and the filter well with cold water, receiving the washings in the same flask, add water to make exactly 100 ml., mix well, and filter through a dry filter. Accurately measure an aliquot of the solution, equivalent to about 250 mg. of ephedrine sulfate, and concentrate it by evaporation to about 7 ml. Completely transfer to a continuous extraction apparatus or a separator with the aid of a few milliliters of water, saturate the solution with

sodium chloride (about 3 Gm.), add 5 ml. of sodium hydroxide T.S., and extract with ether in the continuous extraction apparatus or by shaking first with 25 ml. of ether, then with five or more 10-ml. portions of ether, until the alkaloid is completely extracted."

Ephedrine is liberated by the action of the sodium hydroxide upon ephedrine sulfate. The free base is soluble in water (about 1 in 20); therefore the solution is saturated with sodium chloride, in which it is practically insoluble. Ether is used instead of chloroform in extracting the ephedrine from the aqueous solution, because this alkaloid is a sufficiently strong base to decompose chloroform and form ephedrine hydrochloride.

2. "Wash the combined ether extracts with two 5-ml. portions of water saturated with sodium chloride, extract the washings with 10 ml. of ether, and add this extract to the main ether extract. To the ether solution add exactly 25 ml. of 0.1 *N* sulfuric acid, stir thoroughly, and warm until the ether is expelled. Cool, add methyl red T.S., and titrate the excess acid with 0.1 *N* sodium hydroxide. Each milliliter of 0.1 *N* sulfuric acid is equivalent to 21.43 mg. of $(C_{10}H_{15}NO)_2 \cdot H_2SO_4$."

The ephedrine combines quantitatively with the acid in accordance with the following reaction:

$$\underset{2 \times 165.24}{2C_{10}H_{15}ON} + H_2SO_4 \longrightarrow \underset{428.56}{(C_{10}H_{15}ON)_2 \cdot H_2SO_4}$$

The gram-milliequivalent weight of ephedrine sulfate is calculated as follows: From the balanced chemical equation it can be seen that 1 mM (2 mEq) of sulfuric acid is present in 1 mM of ephedrine sulfate. Consequently, 1 mM of ephedrine sulfate is equal to 2 mEq of ephedrine sulfate and 1 mEq of ephedrine sulfate is, therefore, equivalent to 0.5 mM of ephedrine sulfate. Since the weight of 1 mM of ephedrine sulfate is 0.42856 Gm, 0.5 mM (1 mEq) of ephedrine sulfate is equal to

$$0.42856 \text{ Gm/mM} \times 0.5 \text{ mM} = 0.2143 \text{ Gm}$$

and the gram-milliequivalent weight of ephedrine sulfate is 0.2143.

The U.S.P. states that each milliliter of 0.1 *N* sulfuric acid is equivalent to 21.43 mg of ephedrine sulfate. Show by calculation how this value is obtained.

Assay of Alkaloidal Salts. The determination of the amount of alkaloid present in an alkaloidal salt or in a simple preparation containing an alkaloid or an alkaloidal salt is generally a relatively simple process, since the alkaloid does not need to be freed from a number of organic impurities as is the case in the assay of crude drugs. The solvents used and the gravimetric or volumetric methods employed differ according to the characteristic properties of the alkaloid determined.

Exercise 97

Object. Assay of Citrated Caffeine.

Materials Required. 1 Gm of citrated caffeine
75 ml of chloroform

Procedure. "Accurately weigh about 1 Gm. of Citrated Caffeine, previously dried at 80° for 4 hours, and dissolve it in 10 ml. of hot water. Add 8 ml. of sodium hydroxide T.S., cool the solution, and shake it in a separator with three or more successive portions of 20 ml. each of chloroform to effect complete extraction of the caffeine. Filter the combined chloroform solutions through a small filter, previously moistened with chloroform, into a tared dish. Wash the stem of the separator, the filter, and the funnel with 10 ml. of hot chloroform, adding the washings to the dish, and evaporate the combined chloroform solutions on a water bath, adding 2 ml. of alcohol just before the last trace of chloroform is expelled. Complete the evaporation of the solvent, and dry the residue, consisting of $C_8H_{10}N_4O_2$, at 80° for 4 hours and weigh."

The citrated caffeine is dried before it is weighed to drive off moisture. Since caffeine is a very weak base and citric acid is feebly acidic, it is probable that the aqueous solution of citrated caffeine is dissociated almost completely into caffeine and citric acid. Citric acid is quite soluble in chloroform, but its sodium salt is insoluble in this solvent. The chloroform, therefore, dissolves the caffeine, leaving the sodium citrate in the aqueous solution.

Exercise 98

Object. Assay of Theophylline Sodium Acetate for Theophylline.

Materials Required. 0.3 Gm of theophylline sodium acetate
NH_3 T.S.
0.1 *N* $AgNO_3$
0.1 *N* NH_4SCN
HNO_3
$FeNH_4(SO_4)_2$ T.S.

Procedure. 1. "Place about 300 mg. of Theophylline and Sodium Acetate, accurately weighed, in a 250-ml. Erlenmeyer flask, add 50 ml. of water and 8 ml. of ammonia T.S., and gently warm the mixture on a steam bath until complete solution is effected."

The theophylline sodium acetate, which is a hydrated mixture of theophylline sodium and sodium acetate, dissolves completely in the warm aqueous liquid.

2. "Add exactly 20 ml. of 0.1 *N* silver nitrate, mix, and continue to warm on the bath for 15 minutes. Filter through a filtering crucible under reduced pressure while still warm, and wash the precipitate three times with 10-ml. portions of water."

The silver nitrate converts the theophylline sodium ($C_7H_7N_4O_2Na$) into the silver derivative ($C_7H_7N_4O_2Ag$), which is precipitated from hot ammonia solution in crystalline form. The precipitate is collected on a

TABLE 45. CAPSULES AND TABLETS ASSAYED FOR ALKALOIDS

Capsules or tablets	Weight of sample in aliquot, Gm	Official requirement, percent of labeled amount
U.S.P.		
Aminophylline	0.25	93 to 107
Atropine sulfate (20 mg or more)	0.06	93 to 107
Atropine sulfate (less than 20 mg)	0.04	90 to 110
Codeine phosphate	0.15	93 to 107
Colchicine	0.025	90 to 110
Ephedrine sulfate*	0.25	92 to 108
Ergonovine maleate†	0.002	90 to 110
Ergotamine tartrate†	0.005	90 to 110
Morphine sulfate	0.1	93 to 107
Neostigmine bromide	0.1	93 to 107
Quinidine sulfate	0.5	94 to 106
Reserpine‡	0.001	90 to 110
N.F.		
Apomorphine hydrochloride	0.05	90 to 110
Arecoline hydrobromide	0.3	91 to 109
Aspirin, phenacetin, and caffeine*	0.05	90 to 110
Aspirin, phenacetin, and caffeine	0.05	90 to 110
Caffeine, citrated	0.5	45 to 55
Cocaine hydrochloride	0.06	91 to 109
Codeine sulfate	0.15	93 to 107
Ephedrine hydrochloride*	0.25	92 to 108
Ephedrine hydrochloride	0.25	93 to 107
Ephedrine sulfate and phenobarbital*	0.3	91 to 109
Ephedrine sulfate	0.25	93 to 107
Mercurophylline	0.2	90 to 110
Morphine and atropine sulfate*	0.13	91 to 109
Papaverine hydrochloride	0.3	93 to 107
Quinidine sulfate*	0.3	94 to 106
Quinine sulfate*	0.3	94 to 106
Quinine sulfate	0.5	94 to 106
Rescinnamine	0.0025	90 to 115
Scopolamine hydrobromide	0.05	90 to 110
Theophylline	0.25	94 to 106
Theophylline sodium acetate	0.35	53 to 67
Theophylline sodium glycinate	0.25	47 to 54

* Capsules.
† Visible spectroscopy.
‡ Ultraviolet spectroscopy.

TABLE 46. ASSAYS OF ALKALOIDS, ALKALOIDAL SALTS, AND THEIR PREPARATIONS

Substance	Amount used, Gm	Official requirement, percent
U.S.P.		
Aminophylline	0.25	$C_7H_8N_4O_2$ = 84 to 86
Aminophylline injection	0.25	$C_{16}H_{24}N_{10}O_4 \cdot 2H_2O$ = 93 to 107 l.a.
Aminophylline suppositories	1.0	$C_{16}H_{24}N_{10}O_4 \cdot 2H_2O$ = 90 to 110 l.a.
Atropine sulfate injection	0.04	$(C_{17}H_{23}NO_3)_2 \cdot H_2SO_4 \cdot H_2O$ = 93 to 107 l.a.
Caffeine	0.1	$C_8H_{10}N_4O_2$ = 98.5 to 101
Caffeine and sodium benzoate	1.0	$C_8H_{10}N_4O_2$ = 47 to 50
Caffeine and sodium benzoate injection	0.4	$C_8H_{10}N_4O_2$ = 45 to 52
Codeine phosphate	1.0	$C_{18}H_{21}NO_3 \cdot H_3PO_4$ = 98 to 101.5
Codeine phosphate injection	0.075	$C_{18}H_{21}NO_3 \cdot H_3PO_4 \cdot \frac{1}{2}H_2O$ = 93 to 107 l.a.
Emetine hydrochloride	0.2	$C_{29}H_{40}N_2O_4 \cdot 2HCl$ = 98 to 101.5
Emetine hydrochloride injection	0.12	$C_{29}H_{40}N_2O_4 \cdot 2HCl$ = 84 to 91 l.a.
Ephedrine sulfate	0.4	$(C_{10}H_{15}NO)_2 \cdot H_2SO_4$ = 98 to 101
Ephedrine sulfate solution	0.25	$(C_{10}H_{15}NO)_2 \cdot H_2SO_4$ = 93 to 107 l.a.
Ergonovine maleate	0.1	$C_{19}H_{23}N_3O_2 \cdot C_4H_4O_4$ = 98 to 101
Ergonovine maleate injection	0.002	$C_{19}H_{23}N_3O_2 \cdot C_4H_4O_4$ = 90 to 110 l.a.
Ergotamine tartrate	0.01	$(C_{33}H_{35}N_5O_5)_2 \cdot C_4H_6O_6$ = n.l.t. 97
Ergotamine tartrate injection	0.005	$(C_{33}H_{35}N_5O_5)_2 \cdot C_4H_6O_6$ = 50 to 70 l.a.
Morphine injection	0.075	Morphine salt = 93 to 107 l.a.
Morphine sulfate	0.7	$(C_{17}H_{19}NO_3)_2 \cdot H_2SO_4$ = 98 to 101
Neostigmine bromide	0.1	$C_{12}H_{19}BrN_2O_2$ = 98 to 102
Neostigmine methylsulfate	0.1	$C_{13}H_{22}N_2O_6S$ = 98 to 102
Neostigmine methylsulfate injection	0.03	$C_{13}H_{22}N_2O_6S$ = 90 to 110 l.a.
Physostigmine salicylate	0.25	$C_{15}H_{21}N_3O_2 \cdot C_7H_6O_3$ = 97 to 102
Physostigmine sulfate	0.175	$(C_{15}H_{21}N_3O_2)_2 \cdot H_2SO_4$ = 97 to 102
Pilocarpine hydrochloride	0.5	$C_{11}H_{16}N_2O_2 \cdot HCl$ = n.l.t. 98.5
Pilocarpine nitrate	0.6	$C_{11}H_{16}N_2O_2 \cdot HNO_3$ = n.l.t. 98.5
Quinidine gluconate	0.5	$C_{20}H_{24}N_2O_2 \cdot C_6H_{12}O_7$ = n.l.t. 99
Quinidine gluconate injection	0.4	$C_{20}H_{24}N_2O_2 \cdot C_6H_{12}O_7$ = 95 to 105 l.a.
Quinidine sulfate	0.5	$(C_{20}H_{24}N_2O_2)_2 \cdot H_2SO_4$ = n.l.t. 98
Quinine and urea hydrochloride	0.5	$C_{20}H_{24}N_2O_2 \cdot HCl \cdot CH_4N_2O \cdot HCl \cdot 5H_2O$ = 98 to 110
Quinine and urea hydrochloride injection	0.5	$C_{20}H_{24}N_2O_2 \cdot HCl \cdot CH_4N_2O \cdot HCl \cdot 5H_2O$ = 93 to 107 l.a.
Reserpine	0.025	$C_{33}H_{40}N_2O_9$ = 96 to 101
Reserpine injection	0.01	$C_{33}H_{40}N_2O_9$ = 90 to 110 l.a.
Scopolamine hydrobromide	0.75	$C_{17}H_{21}NI_4 \cdot HBr$ = 98.5 to 102
Scopolamine hydrobromide injection	0.025	$C_{17}H_{21}NO_4 \cdot HBr$ = 90 to 110 l.a.

TABLE 46. ASSAYS OF ALKALOIDS, ALKALOIDAL SALTS, AND THEIR PREPARATIONS (*Continued*)

Substance	Amount used, Gm	Official requirement, percent
N.F.		
Caffeine, citrated	1.0	$C_8H_{10}N_4O_2$ = 48 to 52
Ephedrine, anhydrous	0.5	$C_{10}H_5NO$ = n.l.t. 98.5
Ephedrine, hydrated	0.5	$C_{10}H_5NO$ = n.l.t. 94
Ephedrine hydrochloride	0.4	$C_{10}H_5NO \cdot HCl$ = n.l.t. 98
Ephedrine sulfate injection	0.25	$(C_{10}H_5NO)_2 \cdot H_2SO_4$ = 95 to 105 l.a.
Ephedrine sulfate jelly	10.0	$(C_{10}H_5NO)_2 \cdot H_2SO_4$ = 0.8 to 1.2
Mercurophylline	0.7	$C_7H_8N_4O_2$ = 94 to 106
Mercurophylline injection	0.2	$C_7H_8N_4O_2 \cdot H_2O$ = 94 to 106
Noscapine hydrochloride	1.5	$C_{22}H_{23}NO_7 \cdot HCl$ = n.l.t. 99
Papaverine hydrochloride injection	0.12	$C_{20}H_{21}NO_4 \cdot HCl$ = 95 to 105 l.a.
Rescinnamine	0.02	$C_{35}H_{42}N_2O_9$ = n.l.t. 95
Theophylline sodium acetate	0.3	$C_7H_8N_4O_2$ = 55 to 65
Theophylline sodium glycinate	1.0	$C_7H_8N_4O_2 \cdot H_2O$ = 49 to 52

n.l.t. = not less than; l.a. = labeled amount.

filter and washed with successive portions of distilled water to remove completely any adhering silver nitrate solution.

3. "Acidify the combined filtrate and washings with nitric acid and add an excess of 3 ml. of the acid. Cool, add 2 ml. of ferric-ammonia sulfate T.S., and titrate the excess of silver nitrate with 0.1 *N* ammonium thiocyanate. Each milliliter of 0.1 *N* silver nitrate is equivalent to 18.02 mg. of $C_7H_8N_4O_2$."

The excess of 0.1 *N* silver nitrate in the filtrate is determined by titration with 0.1 *N* ammonium thiocyanate. In acid solution, ammonium thiocyanate reacts with silver nitrate to form insoluble silver thiocyanate, which is white. After all the silver is precipitated, any excess of 0.1 *N* ammonium thiocyanate will react with ferric ammonium sulfate to precipitate ferric thiocyanate, which is red (see Chap. 7).

$$\underbrace{C_7H_7N_4O_2Na}_{180.17 + 22.990} + \underbrace{AgNO_3}_{169.87} \longrightarrow \underbrace{C_7H_7N_4O_2Ag}_{180.17 + 107.87} + NaNO_3$$

The gram-milliequivalent weight of theophylline may be calculated as follows: The balanced chemical equation shows that 1 millimole (mM) of sodium theophylline reacts with 1 mM of silver nitrate. From the definition of gram-milliequivalent weight in precipitation methods of analysis, it is seen that 1 mEq of $AgNO_3$ is equal to 1 mM of $AgNO_3$. Since 1 mM (1 mEq) of silver nitrate reacts with 1 mM of sodium theophylline (also theophylline), 1 mM of sodium theophylline (also theophylline) is equivalent to 1 mEq. One mM of theophylline weighs 0.18017 Gm. Therefore, 0.1802 Gm per mEq is the gram-milliequivalent weight of theophylline in this chemical assay.

The U.S.P. titer value may be determined in a manner analogous to that used for atropine in the assay of Belladonna Leaf.

Problems

1. Calculate the theophylline titer value for 0.08 *N* silver nitrate solution.

2. Calculate the percent *W/W* of theophylline in Theophylline Sodium Acetate if a 0.3120-Gm sample, assayed following the procedure given in the N.F., required 12.10 ml of 0.1100 *N* NH_4SCN to bring about the end point in the residual titration of $AgNO_3$. Exactly 20 ml of 0.1200 *N* $AgNO_3$ was taken initially.

Chapter 18

OTHER OFFICIAL ASSAYS INVOLVING THE USE OF IMMISCIBLE SOLVENTS

Extraction with immiscible solvents is used in the proximate assay of a number of nonalkaloidal crude drugs such as podophyllum and in the analysis of certain synthetic medicinal products such as the barbiturates. Tables 3, 4, and 5 in Chap. 2 list products assayed gravimetrically following extraction with immiscible solvents.

Exercise 99

Object. Assay of Amobarbital Tablets.

Materials Required. 20 amobarbital tablets
10 ml of 2 percent solution of NaOH saturated with NaCl
About 50 ml of ether
About 75 ml of chloroform
Hydrochloric acid
Alcohol

Procedure. 1. "Weigh and finely powder not less than 20 Amobarbital Tablets. Weigh accurately a portion of the powder, equivalent to about 300 mg. of amobarbital, transfer completely to a glass mortar, and rub to a paste with a portion of an alkaline salt solution, prepared by saturating sodium hydroxide solution (1 in 50) with sodium chloride and filtering. Transfer the mixture completely to a separator with the aid of sufficient alkaline salt solution to make about 10 ml. If lubricants other than stearic acid are present, extract the solution with three 10 ml. portions of ether, and discard the ether extracts."

The amobarbital reacts with the NaOH to form amobarbital sodium, which is readily soluble in the aqueous layer. When the solution is shaken with portions of ether, organic matter, such as petrolatum used as a lubricant in the preparation of the tablets, dissolves in the ether, but the amobarbital sodium, being insoluble in ether, remains in the aqueous layer.

2. "To the alkaline solution add 5 ml. of water and 2 ml. of hydrochloric acid, and extract the liberated amobarbital with five 25 ml. portions of a mixture of 1 volume of alcohol and 4 volumes of chloroform, collecting these extracts in a second separator. Wash the combined alcohol-chloroform extracts with 25 ml. of 2% hydrochloric acid, and discard the washings. Filter the combined alcohol-chloroform extracts into a tared beaker, evaporate the solvent on a steam bath with the aid of a current of air, add 10 ml. of ether, again evaporate, dry the residue at 105° for 30 minutes, cool, and

weigh. The weight of the residue represents the weight of $C_{11}H_{18}N_2O_3$ in the portion of the Tablets taken.

"*When stearic acid or stearates are present:* Dissolve the weighed residue, obtained as directed in the preceding paragraph, in 10 ml. of alcohol, add 20 ml. of saturated barium hydroxide solution, and stir well. Filter into a separator, and wash the residue and the filter with three 10 ml. portions of the barium hydroxide solution. Acidify the filtrate with diluted hydrochloric acid, and proceed as directed in the preceding paragraph, beginning with 'and extract the liberated amobarbital.'"

The hydrochloric acid liberates the amobarbital from the amobarbital sodium, and the liberated amobarbital is extracted upon shaking with successive portions of chloroform.

Questions and Problems

1. Calculate the amount of amobarbital present per tablet.
2. Does the amount of amobarbital found fall within the tolerance limits established for these tablets in the Pharmacopeia?
3. Consult the U.S.P. for the formulas of amobarbital and soluble amobarbital and write the equations for the reactions that occur in the assay.

Exercise 100

Object. Assay of Podophyllum.

Materials Required. 10 Gm of podophyllum in fine powder
Percolator, small
200 ml of alcohol
50 ml of $CHCl_3$
50 ml of 0.6 percent HCl
5 ml of dehydrated alcohol

Procedure. 1. "Place 10 Gm. of Podophyllum, in fine powder, in a 125-ml conical flask, add 35 ml of alcohol, and reflux on a steam bath for 3 hours. Transfer the mixture to a small percolator, and percolate slowly with warm alcohol until the percolate measures 95 ml. Cool, add sufficient alcohol to the percolate to make it measure 100.0 ml, and mix."

The podophyllum is digested and percolated with alcohol to extract the resin completely.

2. "Transfer 10.0 ml of this percolate to a separator, and add 10 ml of chloroform and 10 ml of 0.6 per cent HCl. Shake the mixture, allow it to separate, draw off the alcohol-chloroform layer into a second separator, and wash the acid layer with three 15-ml portions of a mixture of 1 volume of alcohol and 2 volumes of chloroform, adding the washings to the second separator. Add 10 ml. of 0.6 per cent HCl to the combined extract and washings, again shake the mixture, allow it to separate, and draw off the alcohol-chloroform layer into a tared vessel. Wash the acid layer three times with 15 ml portions of the alcohol-chloroform mixture, adding the washings to the tared vessel."

The resin of podophyllum is insoluble in aqueous acid solution. It dissolves in the alcohol-chloroform layer, leaving nonresinous constituents in the aqueous layer.

3. "Evaporate the combined extracts on a steam bath to approximately 1 ml., add 5 ml. of dehydrated alcohol, again evaporate to dryness, and dry the residue at 80° for 4 hours; the weight of this residue is the weight of resin in 1 Gm. of the Podophyllum taken."

The lower aqueous layer, containing small amounts of sugars, gums, coloring matter, etc., is drawn off and discarded. The alcohol-chloroform solution when evaporated to constant weight leaves a residue that consists of practically pure resin of podophyllum.

Chapter 19

ASSAY OF ENZYME-CONTAINING SUBSTANCES

No methods are known whereby the quantity of enzymes in a substance can be accurately determined. Consequently, the assay of enzyme-containing substances is based on observations of their power to produce changes in other substances. In other words, the activity of the enzymes is measured. Thus, pepsin is evaluated on the basis of the quantity of egg albumen that it will digest, and extract of malt is assayed for its power to convert starch into soluble carbohydrates under definite conditions. Any condition that tends to increase or decrease the activity of the enzymes manifests itself in the assay results as an increase or decrease in the quantity of enzymes present.

Numerous factors tend to influence the activity of enzymes. Some of the more important of these factors are as follows:

1. Practically all enzymes are destroyed when they are heated to 100°. Many of them are gradually inactivated at comparatively low temperatures (35 to 40°), and many of them undergo a slow loss of activity at room temperature and even at 0°.
2. Enzymes have an *optimum temperature* of action; i.e., for each enzyme there is a definite range of temperature at which it exhibits its maximum activity under definite conditions. The optimum temperature for most enzymes lies between 30 and 45°.
3. The activity of the enzymes is affected by the mediums in which they react; e.g., pepsin is most active in acid mediums, and pancreatin is most active in alkaline mediums.
4. The enzymes are inactivated by many chemical agents, such as alcohol, salts of the heavy metals, tannin, and phosphotungstic acid, that precipitate the enzymes from solution.

Because the activity of enzymes is influenced by so many factors, it is necessary to adhere strictly to the procedures outlined in the various assay processes in order to secure concordant results.

Exercise 101

Object. Assay of Pepsin.

Materials Required. 0.1 Gm of pepsin

0.1 Gm of N.F. Reference Pepsin. The N.F. Reference Pepsin consists of a powdered pepsin, carefully selected, dried, and packaged and having a digestive power of

3,000 times its weight of egg albumen. It may be obtained from the Committee on National Formulary.

1 *N* hydrochloric acid
A hen's egg
A No. 40 sieve
A conical measure

Procedure. 1. Mix 35 ml of 1 *N* hydrochloric acid with 385 ml of distilled water. Dissolve 100 mg of pepsin in 150 ml of this dilute acid. Likewise dissolve 100 mg of Reference Pepsin in another 150-ml portion of the dilute acid.

When 35 ml of 1 *N* HCl is diluted to 420 ml, the concentration of acid in the solution is reduced to approximately 0.3 percent. This is about the normal acidity of the stomach. Investigation has shown that pepsin is most active as an albumen digestive in mediums containing this concentration of acid. Other strongly dissociated acids that supply the same hydrogen ion concentration as 0.3 percent HCl might be used, since the digestive activity of the pepsin is directly related to the hydrogen ion concentration of the reaction mixture.

2. Immerse one or more hen eggs in boiling water during 15 minutes. Cool them rapidly to room temperature by immersion in cold water; remove the shell and pellicle and all the yolk and at once rub the albumen through a clean, dry, No. 40 sieve, rejecting the first portion that passes through the sieve. Place 10 Gm of the succeeding well-mixed portion in each of three wide-mouth bottles of about 100-ml capacity.

The egg is immersed in boiling water for 15 min to coagulate thoroughly the albumen. The albumen is coagulated so that its digestion by the pepsin may be better observed. The coagulated albumen is passed through a sieve to break it into small particles, so that a greater surface area will be exposed to the digestive action of the pepsin enzymes. The first portion of the albumen that passes through the sieve is rejected, since it may contain dust particles and traces of metallic compounds from the materials of the sieve.

3. Immediately add 35 ml of the dilute acid at one time or in portions and, by suitable means, thoroughly disintegrate the particles of albumen. Place the bottles in a water bath at 52°. After the contents of the bottles have reached that temperature, add exactly 5 ml of the acidified solution of pepsin to one bottle, 4.30 ml of the same solution and 0.70 ml of the dilute acid to another bottle, and exactly 5 ml of the acidified solution of Reference Pepsin to the third bottle. At once stopper the bottles securely, invert them three times, and maintain them at a temperature of 52° for 2 hr and 30 min, agitating the contents equally every 10 min by inverting the bottles once.

The acid should be added at once, because the albumen loses moisture on exposure to the air and becomes tough and hard. The mixture is maintained at 52°, since pepsin exerts its optimum activity at this temperature. Exactly 5 ml of the pepsin solution being assayed and of the solution of Reference Pepsin is taken because this quantity introduces 1 part of pepsin into the mixture to each 3,000 parts of albumen. Thus 0.1 Gm of pepsin in 150 ml gives 0.1:150::X:5. X = 0.003333 Gm pepsin in 5 ml of the solution. 0.003333 × 3,000 = 9.999 (10 Gm), the weight of albumen used.

4. Remove the bottles from the bath, pour the contents into 100-ml conically shaped measuring vessels [Fig. 27], having diameters not exceeding 1 cm at the bottom and complying in other respects with the water and sediment tube ASTM Standard Method, D96-35, graduated from 0 to 0.5 ml in 0.05-ml graduations; from 0.5 to 3 ml in 0.1-ml graduations; from 3 to 5 ml in 0.5-ml graduations; from 5 to 10 ml in 1-ml graduations; from 10 to 25 ml in 5-ml graduations; and with graduation marks at 50-, 75-, and 100-ml points. Transfer the undigested egg albumen which adheres to the sides of the bottles to the respective measuring vessels with the aid of small portions of distilled water until 50 ml has been used for each. Mix the contents of each measuring vessel and allow them to stand for 30 min. The volume of the undissolved albumen in the measuring vessel corresponding to 5.0 ml of the solution of pepsin being assayed does not exceed the volume of the undissolved albumen in the measuring vessel corresponding to 5.0 ml of the Reference Pepsin solution, and the volume of the undissolved albumen in the measuring vessel corresponding to 4.30 ml of the solution of pepsin being assayed is not less than the volume of undissolved albumen in the measuring vessel corresponding to 5.0 ml of the Reference Pepsin solution.

FIG. 27. Sedimentation cone.

If the volume of egg albumen remaining undissolved from the sample containing 5.0 ml of the solution of pepsin being assayed is not greater than that remaining from the sample containing the solution of Reference Pepsin, it is evident that the pepsin being assayed must be at least as active as the Reference Pepsin and that it must conform to the minimum official requirement. If the volume of egg albumen remaining undissolved from the sample containing 4.3 ml of the solution of pepsin being assayed is not less than that remaining from the sample containing the Reference Pepsin, it is evident that the pepsin being assayed is a pepsin that will cause the solution of 3,500 times its weight of egg albumen. 4.3:5::3,000:X. X = 3,500. The relative proteolytic power of pepsin, stronger or weaker than that just described, may be determined by ascertaining through repeated trials the quantity of the pepsin solution, made as directed in the assay, required to digest under the prescribed conditions 10 Gm of boiled and disintegrated egg albumen.

Questions and Problems

1. Name several factors that affect the activity of pepsin.

2. Why is the proteolytic activity of pepsin determined in acid solution? What effect would the use of a neutral or alkaline medium have on the digestive power of pepsin?

3. Could proteins other than egg albumen be used to estimate the proteolytic activity of pepsin? Give examples.

4. What advantages does egg albumen possess as a source of protein for this assay?

Exercise 102

Object. Assay of Pancreatin for Starch Digestive Power.

Materials Required. 5 Gm of powdered potato starch
0.4 Gm of pancreatin
0.2 ml of 0.1 *N* iodine solution

Procedure. 1. "Determine the percentage of moisture in N.F. Potato Starch Reference Standard by drying about 500 mg., accurately weighed, at 120° for 4 hours. Boil a sufficient amount of water for 10 minutes, and cool to room temperature. Use this water for all dilutions hereinafter specifying water. Thoroughly mix a quantity of the N.F. Potato Starch Reference Standard, equivalent to 3.75 Gm. of dry Reference Standard, with 10 ml. of water. Add the mixture, with constant stirring, to 75 ml. of water, previously heated to about 55°, and contained in a tared, 250 ml. beaker. Rinse the remaining starch into the beaker with 10 ml. of water. Heat the mixture to boiling, and boil it gently, with constant stirring, for 5 minutes."

The percentage of moisture remaining in the starch is determined so that a sample equivalent to 3.75 Gm of dry starch can be used for digestion. Thus, if the sample dried at 50° is found to contain 10 percent of moisture, each gram of the sample is equivalent to 0.9 Gm of dry starch. Then $3.75/0.9 = 4.1666$ Gm of starch equivalent to 3.75 Gm of dry starch.

The starch is mixed with cold water to obtain a paste, so that it will not form lumpy masses when the hot water is added. When this paste is added to warm water and the resulting mixture is boiled, the starch grains burst, liberating their contents, which dissolve in the hot water to form a more or less mucilaginous mass. Until the starch grains are broken down in this way, they remain insoluble and are not acted upon by pancreatin.

2. "Add enough water to make the mixture weigh 100 Gm., cool the paste to 40°, and place the beaker in a water bath maintained at 40°. Suspend 150 mg. of Pancreatin in 5 ml. of water in a 250-ml. beaker and add the suspension to the starch paste, mixing it well by pouring the mixture from beaker to beaker for 30 seconds, noting the time when the Pancreatin suspension was first added to the paste. Maintain the mixture at a temperature of 40° for exactly 5 minutes."

The water lost by evaporation is replaced so that a definite concentration of starch will be present during the digestion with the pancreatin. The solution of pancreatin must be prepared immediately before it is added to the starch solution, since amylopsin, the starch-splitting enzyme of pancreatin, becomes inactivated quite rapidly in aqueous solutions. The mixture is maintained at 40° for 5 min, because 40° is the optimum temperature for amylopsin. The regulation of the temperature is very important in this assay, since a variation of 1° throughout the period of digestion has been found to cause a difference of about 10 percent in the activity of the amylopsin of pancreatin. The amylopsin converts the starch into soluble carbohydrates, chiefly maltose and isomaltose.

3. "Stir, and at once add 0.1 ml. of this mixture to a previously made mixture of 0.2 ml. of 0.1 *N* iodine and 60 ml. of water, at a temperature of from 23° to 25°: no blue or violet color is produced."

The presence of starch and primary hydrolytic products of starch is indicated by the development of a blue, red, or violet color.

Pancreatin must be capable of converting at least twenty-five times its weight of starch into soluble carbohydrates in order to meet the pharmacopeial requirements. Since 0.15 Gm of pancreatin is employed to digest 3.75 Gm of starch, each gram of pancreatin used corresponds to 3.75/0.15 = 25 Gm of starch. The relative amyloptic activity of a sample of pancreatin can be determined by repeated trials on varying quantities of starch.

Extract of malt of the N.F. is also assayed for its capacity to convert starch into soluble carbohydrates that do not give a color with iodine solution.

Questions and Problems

1. Define the terms *proteolytic enzyme* and *amylolytic enzyme.*
2. Why is the starch solution boiled before the digestion with pancreatin?
3. Look up the hydrolytic products of starch, and explain what products might give a red or violet color with iodine.

Exercise 103

Object. Assay of Pancreatin for Casein Digestive Power.

Materials Required. 0.2 Gm of casein
1 ml of 0.1 *N* sodium hydroxide
0.1 Gm of pancreatin
0.1 Gm Pancreatin Reference Standard
2 ml of glacial acetic acid
20 ml of alcohol

Procedure. 1. "Place 100 mg. of finely powdered casein in a 50-ml. volumetric flask, add 30 ml. of water, and shake well to bring the casein into suspension. Add exactly 1 ml. of 0.1 *N* sodium hydroxide, and heat the mixture at 40° until the casein is completely dissolved, which should not require more than 30 minutes. Cool, add sufficient water to make 50 ml., and mix well."

Casein is a complex compound, or mixture of compounds, that yields amino acids upon hydrolysis. Consequently, it is soluble in alkalies. The casein is dissolved in alkali, in this case, because trypsin, the proteolytic enzyme contained in pancreatin, functions best in alkaline mediums. Casein that requires more than 30 min to dissolve is usually of poor quality, containing an excess of fat.

2. "Dissolve 100 mg. of Pancreatin in 500 ml. of water. Also dissolve 100 mg. of N.F. Pancreatin Reference Standard for Casein Digestive Activity in another 500-ml. of water. Mix 1 ml. of glacial acetic acid with 9 ml. of water and 10 ml. of alcohol. Place 5 ml. of the casein solution in each of two test tubes. To one of the test tubes add 2 ml. of well-shaken Pancreatin solution and to the other test tube add 2 ml. of the well-shaken Reference Standard Solution. Add 3 ml. of water to both tubes, mix by

gentle agitation, and immediately immerse the test tubes in a water bath at 40° and keep it at this temperature for 1 hour. Then remove the tubes from the bath, and add 3 drops of the acetic acid mixture to each tube. The amount of haze in the test tube corresponding to 2 ml. of the Pancreatin being assayed is not greater than the amount of haze in the test tube containing 2 ml. of the Reference Standard Solution."

The concentrations of the solutions of casein and pancreatin are such that when they are mixed there will be 25 parts of casein present for each part of pancreatin. Thus, in 2 ml of the pancreatin solution there is $2 \times (0.1/500) = 0.0004$ Gm of pancreatin, and in 5 ml of casein solution there is $5 \times (0.1/50) = 0.01$ Gm of casein; $0.0004 \times 25 = 0.0100$.

The mixture is maintained at 40°, the optimum temperature for trypsin, for 1 hr to allow the proteolytic action to proceed. The casein is digested by the pancreatic enzymes with the formation of soluble proteoses and possibly small amounts of amino acids. When the alcoholic acetic acid solution is added, the alkali is neutralized and any undigested casein is precipitated, since casein is insoluble in alcohol and in acetic acid. The proteoses and other products of casein digestion, being soluble in alcohol and acetic acid, are not precipitated. Consequently, if a precipitate is formed, it indicates that the proteolytic activity of the pancreatin is below standard; i.e., that 1 part of the pancreatin is not capable of digesting 25 parts of casein.

Questions and Problems

1. Why is the protein digestive capacity of pancreatin determined in an alkaline medium?
2. Which of the enzymes of pancreatin take part in protein digestion?
3. Define the term *optimum temperature* as applied to enzymes.
4. How might the relative proteolytic power of a sample of pancreatin stronger or weaker than the official standard be determined?

TABLE 47. SUBSTANCES ASSAYED FOR ENZYME ACTIVITY

Substance	Substrate	Official requirement
N.F.		
Chymotrypsin	*p*-toluene-sulfonyl-L-arginine methyl ester HCl	Proteolytic activity n.l.t. 1,000 N.F. chymotrypsin units per mg
Pancreatin	Potato starch	Converts n.l.t. 25 × its weight of starch into soluble carbohydrates
	Casein	Converts n.l.t. 25 × its weight of casein into proteoses
Pepsin	Egg albumen	Digests 3,000 to 3,500 × its weight of egg albumen
Pepsin elixir, actated	Egg albumen	Proteolytic activity in each 100 ml = n.l.t. 1.75 Gm of Reference Pepsin
Trypsin, crystallized	*N*-acetyl-L-tyrosine ethyl ester	Proteolytic activity n.l.t. 2,500 N.F. trypsin units per mg

n.l.t. = not less than.

Table 17. Substrates Used for Enzyme Activity

PART 3

PHYSICOCHEMICAL AND INSTRUMENTAL METHODS USED IN OFFICIAL PHARMACEUTICAL ANALYSES

The analysis of many of the official substances for the purpose of identifying them and of determining their purity or therapeutic value is based on the measurement of certain physical and chemical characteristics or constants, considered in conjunction with the estimation of certain special components by assay, and qualitative tests for various substances commonly present as impurities or added for the purpose of adulteration.

The qualitative methods employed in the determination of the physical characteristics or physical constants of official substances are frequently applied to materials of widely divergent nature; e.g., valuable criteria relative to the purity of volatile oils, sugars, and alkaloids may be gained by the measurement of the optical activity. The fundamental principles involved in ascertaining the physical constants of these substances are the same regardless of the nature of the substance, although slight modifications of the technique employed are frequently necessary.

The Pharmacopeia and the National Formulary, as a rule, specify that certain methods and apparatus shall be employed in the determination of the physical constants of official substances. In such cases, the methods are known as *official methods*, and the apparatus as *official apparatus*.

Chapter 20

SPECTROMETRY

Introduction. When radiant energy such as light from an ordinary house or office lamp passes through a solution of copper sulfate of a given concentration, part of the light energy is absorbed by the cupric ions of this salt, and the remainder is transmitted. It is transmitted light that stimulates the eye of the human observer and registers a color sensation of blue. White light, as used in this example, is known to contain a mixture of the colors found in the rainbow. The colors—violet, indigo, blue, green, yellow, orange, and red—constitute the visible or color region of the electromagnetic spectrum. The question as to which of the colors in this mixture are absorbed and which are transmitted may be answered in a qualitative way as follows: Since the solution appears blue, the radiant energy corresponding to the green, yellow, orange, and red regions of the spectrum must have been absorbed by the cupric ions and the radiant energy corresponding to the blue region of the spectrum must have been transmitted. In other words, the cupric ions interacted with a portion of the incident (total) radiant energy and rejected the remainder of the radiant energy, which was then transmitted through the solution. If this same solution were now diluted with water, the intensity of the blue color would diminish. On the other hand, if the concentration of the copper sulfate were increased, one would observe that the intensity of the blue color increases. An important conclusion which can be made from these observations is that "the intensity of transmitted radiant energy is a function of (related to) the concentration of the chemical species which absorbs the radiant energy."

Those analytical methods which are based on the absorption of radiant energy are called *absorptimetric* methods, and those absorptimetric methods which utilize the visible region of the spectrum are called *colorimetric* methods. Some of the colorimetric methods of analysis which are of great importance in medicine and pharmacy are those used in the determination of glucose, ketone bodies, occult blood, pH, and protein in biological specimens. As an example, glucose, which is a colorless substance in solution, may be determined in urine by several different colorimetric methods. In one such method, the determination is conducted by immersing in the urine specimen a strip of paper previously impregnated with two enzymes and a color-producing reagent. In the presence of glucose, the enzyme glucose oxidase catalyzes the oxidation of glucose to gluconic acid and hydrogen peroxide. Hydrogen peroxide, in the presence of a peroxidase enzyme, then reacts with the color-producing reagent and brings about a color change on the strip of paper. The color so produced indicates the presence of glucose; the concentration of glucose in the specimen may be determined by matching the color of the paper strip with the color on a color chart previously prepared by using glucose solutions of known concentrations. In this colorimetric determination of glucose, the intensity of color reflected, rather than transmitted, is related to the presence or concentration of the glucose in the sample. Blood glucose concentration may be determined in a similar manner.

Glucose as well as many other substances of biological and medicinal importance may also be determined by colorimetric methods in which the color of the transmitted, rather than reflected, radiant energy is measured. The color and intensity of the transmitted radiant energy are a measure of the concentration and are determined by comparison with a color chart or, more accurately, by using an instrument called a *spectrophotometer*.

Interactions of matter and radiant energy also occur in the ultraviolet and infrared regions of the electromagnetic spectrum. Many qualitative and quantitative methods of pharmaceutical analysis are based on these interactions and will be discussed.

Before spectrometry can be considered further, it is necessary to define and discuss the terms most commonly used in this type of analysis. The definitions may not be immediately clear, but as the topic of spectrometry is developed the meaning of the terms should become apparent.

Terminology in Spectrometry. *Spectrometry* may be defined as methods of analysis which deal with the measurement of spectra. It may be thought of as a generic term which embraces the many aspects involved in these measurements. In pharmaceutical analysis the more common measurements which are made of spectra and which are aspects of spectrometry are (1) the measurements of the position (wavelength) in the electromagnetic spectrum where radiant energy has interacted with a chemical species, and (2) the measurement of the power of transmitted, fluorescent, reflected, or emitted energy.

Electromagnetic spectrum is the term used to define the complete system of energy which is propagated in wave form. Energy of this nature is referred to as *radiant energy*, and appears in such apparently different forms as sunlight, color, radio waves, etc. It must be understood, however, that these many apparently different forms of radiant energy have related properties. The equation which relates one form of radiant energy with another is

$$c = \nu\lambda$$

where c = velocity of light (3×10^{10} cm per sec)
ν = frequency, cps
λ = wavelength, cm

This equation shows (1) that the velocity of all forms of radiant energy in a given medium is constant, and (2) that the apparent differences lie in the frequency and wavelength.

Radiant energy in this discussion refers to energy in the ultraviolet, visible, and infrared regions of the electromagnetic spectrum.

Spectrophotometry is a branch of spectrometry which embraces the measurement of the absorption, by chemical species, of radiant energy of definite and narrow wavelength, approximating monochromatic radiation. Monochromatic radiation is radiation of a "single" wavelength. In actual practice monochromatic radiation is obtained by using a prism or diffraction grating and consists of more than a single wavelength. The usual spectral region isolated in the better commercial spectrophotometers ranges from 0.5 to 10 millimicrons (mμ), which approximates monochromatic radiant energy.

Colorimetry is a branch of spectrophotometry in which the absorption measurement is made in the visible region of the spectrum.

Wavelength. On the basis of the simple theory of wave motion, the length of a complete wave or cycle, from the peak of one wave to the peak of the next, is called the wavelength; it can be accurately measured. The units of wavelength measurement generally used in spectrometry are the *micron*, μ, which is equal to 10^{-4} cm, the millimicron, mμ, which is equal to 10^{-7} cm, and less frequently the Angstrom, A, which is equal to 10^{-8} cm.

The ranges of the wavelengths of radiant energy of importance in the practice of spectrophotometry are as follows:

Ultraviolet	185 to 380 mμ
Visible	380 to 780 mμ
Near infrared	780 to 3,000 mμ
Medium infrared	3.0 to 15 μ
Far infrared	15 to 40 μ

Wavenumber may be defined as the number of waves per centimeter and is equal to 1/wavelength (cm).

Frequency may be defined as the number of complete cycles which pass a given point per second.

Spectrophotometer and *spectrometer* are terms used to designate instruments which have a radiant energy–dispersing device, such as a prism or grating, and the associated electronics which permit the measurement of wavelength and radiant power. These instruments may be either manually operated or automatic recording instruments.

Colorimeter and *filter photometer* are terms used to designate instruments which have the necessary electronics to permit measurement of radiant power but which use a filter instead of a prism or diffraction grating for the purpose of increasing the sensitivity of the measurement.

Theories of Radiant Energy. Two theories are generally considered in a discussion of the interaction of radiant energy and matter. One theory states that energy in the electromagnetic spectrum occurs as bundles called *quanta* of photons which are related to the frequency of radiation by the equation

$$E = h\nu$$

where E = energy, ergs
ν = frequency, cps
h = a universal constant, known as Planck's constant, which equals 6.62×10^{-27} erg/sec

This equation relates the energy content of a quantum of energy to the frequency of radiation; the energy content possessed by a quantum of radiant energy determines whether a given chemical species absorbs or transmits radiant energy of a given frequency.

The other theory states that radiant energy is propagated in wave form and is expressed mathematically by $c = \nu\lambda$ (see "Electromagnetic spectrum," above, under Terminology in Spectrometry). This equation provides a means of determining the wavelength from frequency data. Both theories are of importance in understanding the principles of spectrometry.

FUNDAMENTAL LAWS AND RELATED TERMS OF SPECTROPHOTOMETRY

Beer's law states that the power[1] of a transmitted radiant beam decreases exponentially as the concentration of the solution containing the absorbing chemical species increases arithmetically.

Lambert's or Bouguer's law states that the power of a transmitted radiant beam decreases exponentially as the thickness of the solution containing the absorbing chemical species increases arithmetically.

Beer-Lambert or Beer-Bouguer law is a combination of the above laws and relates the power of the incident and transmitted radiant beams to the thickness and concentration of the solution containing the absorbing chemical species.

The mathematical relationship of the combined laws is:

$$P = P_o e^{-kcb}$$

$$T = \frac{P}{P_o} = e^{-kcb}$$

or

$$A = \log \frac{1}{T} = abc$$

where P = power of the transmitted radiant beam
P_o = power of the incident beam

[1] Power = intensity × time.

e = base of the natural logarithm = 2.7184
k = proportionality constant (characteristic for any one wavelength and chemical species under investigation)
c = concentration of chemical species in solution, Gm per liter
$-$ = negative sign indicates that the power of the radiant beam *decreases* as the concentration of the chemical species or thickness increases. (In pharmaceutical analysis the thickness is held constant and the concentration is varied.)
b = distance, cm, traversed by the radiant energy through the solution. In practice 1-cm cells (cuvettes) are usually used.

On the basis of these mathematical relationships, certain terms and symbols have found usage and are applicable to spectrophotometric analyses. The important ones, designated as official in the U.S.P. and N.F., are given below, with the symbols in parentheses.

Transmittance, P(solution)/P (solvent). (T). The ratio of the radiant power transmitted by the solution to the radiant power transmitted by the solvent.

Absorbance, log $1/T$ or log [P (solvent)/P(solution)]. (A). Also called optical density, absorbancy, extinction coefficient.

Concentration (c). Concentration expressed in grams of solute per liter of solution.

Absorptivity (a). The value obtained by dividing the absorbance (A) by the product of the concentration, expressed in grams per liter, and the path length, expressed in centimeters.

Molar absorptivity (ϵ). The value obtained by dividing the absorbance A by the product of the concentration of the solution, expressed in moles per liter, and the path length, expressed in centimeter.

Absorption spectrum. The graph obtained when absorbance or any function of absorbance is plotted against wavelength.

INSTRUMENTS IN SPECTROPHOTOMETRY[2]

For our purposes the terms *spectrophotometer* and *spectrometer* will be used synonymously to designate manually operated or automatic recording instruments. By our definition, both these types of instruments have (1) a radiant energy source, (2) a radiant energy–dispersing device, such

[2] Details regarding calibration of both the wavelength and photometer scales are found in the following publications of the National Bureau of Standards, Department of Commerce, Washington, and are available from the Superintendent of Documents, U.S. Government Printing Office: (1) Spectrophotometry (200 to 1,000 mμ), Circular 484; (2) Standards for Checking the Calibration of Spectrophotometers (200 to 1,000 mμ), Letter Circular LC-1017.

Additional information regarding calibration of spectrophotometers is available in the *Journal of Research*, N.B.S., **49,** 13, 1952; and *ibid.*, **58,** 195, 1957.

as a prism or diffraction grating, (3) a sample compartment, and (4) the associated electronics which permit the measurement of radiant power transmitted by the sample.

Automatic recording instruments, as the name implies, automatically scan the wavelength region of interest, and record absorbance or some function of absorbance versus wavelength. Many different types of recording spectrophotometers are available; the wavelength region beginning at 185 mμ (ultraviolet) and extending to 40 μ (infrared) is covered

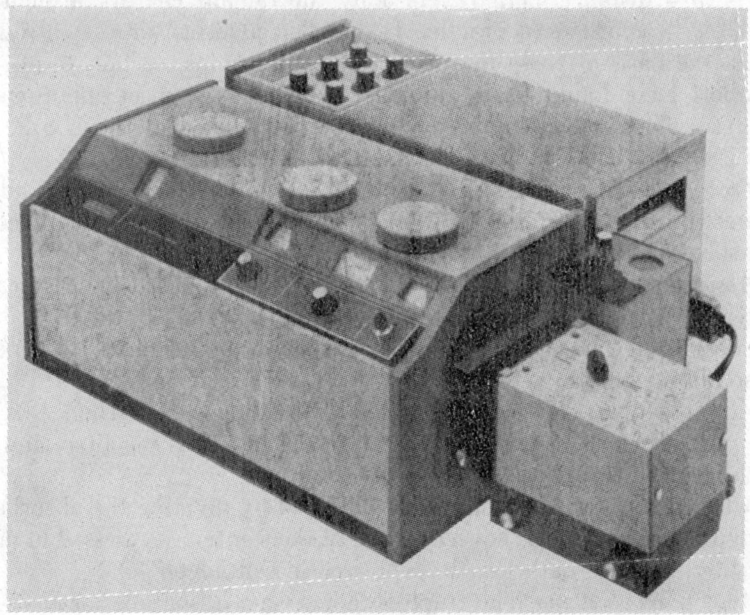

FIG. 28. Beckman DU-2 spectrophotometer. (*Courtesy of Beckman Instruments, Inc.*)

by several different instruments. These instruments are especially helpful in obtaining much useful analytical data quickly.

Manually operated spectrophotometers are either direct-reading or null-balance instruments. An example of a null-balance, manually operated instrument is the Beckman DU-2 Spectrophotomer (Fig. 28). This instrument, which covers the ultraviolet, visible, and near-infrared regions, has a continuous range from 190 to 1,000 mμ, isolating spectral regions from 0.5 to 1.5 mμ. No attempt will be made to give detailed instructions for the operation of this instrument, since they are supplied by the manufacturer. However, the essential steps required for normal operation may be summarized:

1. Set wavelength to desired value and turn absorbance-transmittance (A-T) knob to 100 percent T. (An alternative procedure is described in operating manual.)

2. With shutter closed, turn dark current knob until null position is reached on galvanometer.

3. Position cell containing blank in light path, open shutter, and turn slit-adjust knob until null position is reached.

4. Close shutter and position cell containing the sample solution. Open shutter and turn A-T knob until null position is again reached.

5. Read and record absorbance value.

An example of a direct-reading manually operated spectrophotometer is the Bausch and Lomb Spectronic 20 (Fig. 29). This instrument has an operating spectral range of 340 to 950 mμ. No attempt is made to give

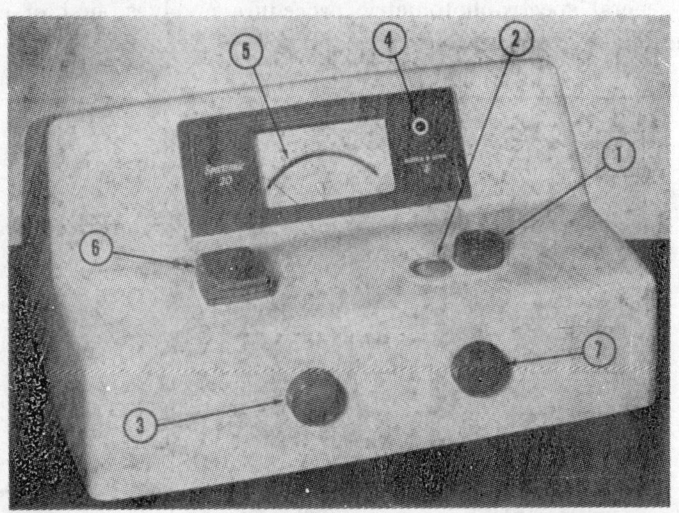

FIG. 29. **Bausch and Lomb Spectronic 20.** (*Courtesy of Bausch & Lomb Optical Co.*)

detailed instructions for the operation of this instrument as they are supplied by the manufacturer. However, the essential steps in its operation are presented. Refer to Fig. 29 and observe that there are only three knobs. Knob (1) is the wavelength control and is used to position the desired wavelength as indicated by the wavelength scale (2). Knob (3) combines the on-off switch with the zero adjust control. The instrument is turned on by turning this knob clockwise until a "click" is heard. Turning the knob more adjusts the zero setting of the transmittance-absorbance (T-A) scale (5). Knob (7), referred to as the light control, permits the operator to set 100 percent on the T-A scale (5).

To make spectrophotometric measurements using this instrument, knob (3) is turned clockwise to the "on" position. After the initial warm-up period, as designated in the instruction manual, knob (1) is turned to the desired wavelength as indicated on the wavelength scale (2). Knob (3) is then turned until the meter needle reads zero on the T-A scale (5). The blank solution is placed in a cell recommended by the manufacturer and then inserted in the sample holder (6). (Always set the index line on cell opposite the index line on the sample holder when positioning the cell in the sample holder.) Close the cover on the sample holder and rotate knob (7) until the meter reads 100 percent T on the T-A scale (5). Replace the blank with the unknown solution, close cover on sample holder, and read the absorbance directly from the meter (5).

GENERAL PROCEDURE

The general spectrophotometric procedure involves, first of all, the preparation of a spectral-absorbance curve (SA curve) of the solute.

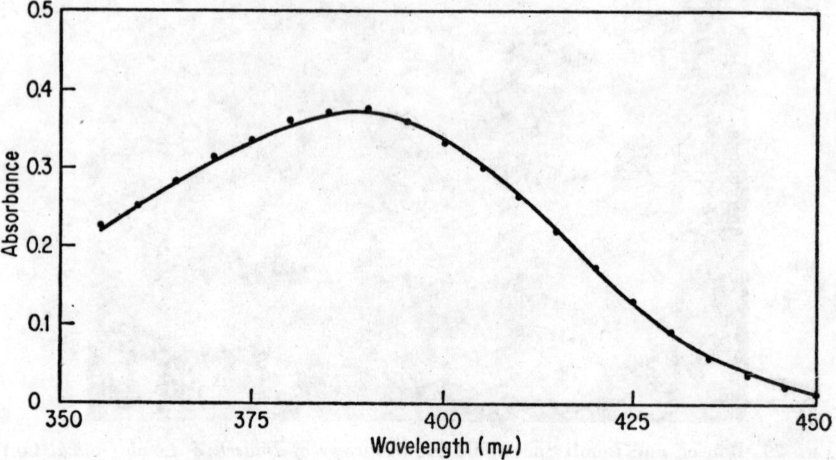

FIG. 30. Spectral-absorbance curve for the Schiff base of sulfisoxazole and vanillin.

This curve (see Fig. 30) is made by plotting the absorbance of the solute versus the wavelength. When a manually operated instrument is used, the absorbance value is usually determined at 5-mμ increments, except in the region of the maximum, where 2-mμ increments are used. The SA curve may also be obtained, of course, using an automatic recording spectrophotometer.

After the SA curve has been prepared, the wavelength at which maximum absorption occurs is chosen for the purpose of measuring absorbance values of a series of reference solute concentrations. The concentrations of the reference solutions should be prepared so that the transmittance readings fall between 20 and 80 percent, because the measurements are most accurate in this range. The wavelength at maximum absorption is selected, since the method is most sensitive at this wavelength and therefore will give the most accurate results.

A plot of the absorbance values against a series of known solute concentrations, called a "Beer's plot," should yield a straight line (Fig. 31). From this plot the unknown solute concentration is determined merely by measuring the absorbance value of the solution and finding the concentration value which corresponds to the measured absorbance value.

If a filter photometer is used, the filter which gives the greatest sensitivity should be taken. This filter can easily be determined by measuring the absorbance value of a given concentration using the different filters available. The filter which gives the highest absorbance value is then used to measure absorbance values of a series of known solute concentrations.

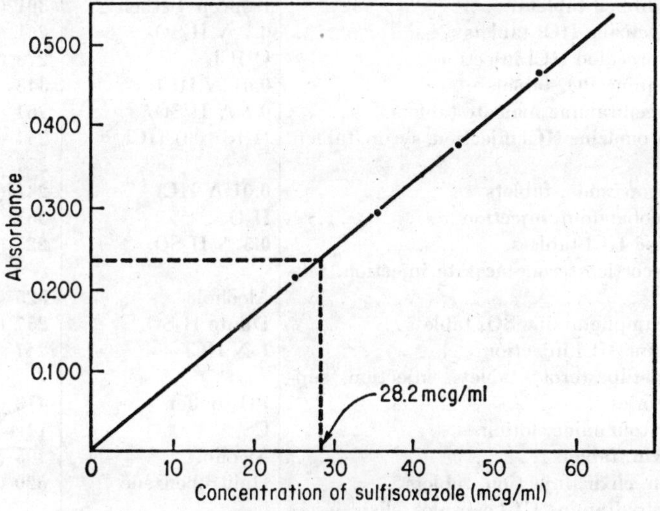

FIG. 31. Beer's plot.

The analysis of Sulfisoxazole Tablets will be used to illustrate the spectrophotometric method of analysis. It was chosen for the following reasons: (1) the procedure illustrates the fundamentals of a spectrophotometric assay by a colorimetric method in which a colorless chemical species is converted to a colored derivative, the color intensity of which is related to the concentration of the chemical species being analyzed; (2) the reagents required for this analysis are all readily available; (3) filter photometers or inexpensive or expensive spectrophotometers may be used to obtain the experimental absorbance values of the solutions.

Although the colorimetric region of the spectrum is used for this assay, the procedure is generally applicable to the ultraviolet and infrared regions of the spectrum.

Table 48 lists the official substances assayed using spectrometric methods.

TABLE 48. OFFICIAL SUBSTANCES ASSAYED USING SPECTROMETRIC METHODS

Substance	Solvent	Wavelength, mμ; spectral region
U.S.P.		
Acetazolamide	Pyridine	7.38 μ (IR)
Aspirin tablets	Glacial acetic acid in $CHCl_3$	280 (UV)
Bemegride injection	NH_3, H_2O	230 (UV)
Bishydroxycoumarin tablets	$CHCl_3$	309 (UV)
Carbachol ophthalmic solution	Acetone	525 (visible)
Chlorambucil tablets	Isopropyl ether	301 (UV)
Chlorcyclizine HCl tablets	0.5 *N* H_2SO_4	261 (UV)
Chloroprocaine HCl injection	$CHCl_3$	278 (UV)
Chloroquine PO_4 tablets	0.01 *N* HCl	343 (UV)
Chlorpheniramine maleate tablets	0.1 *N* H_2SO_4	264 (UV)
Chlorpromazine HCl injection, syrup, tablets	(1 to 100) HCl	254 (UV) 277
Chlorpropamide, tablets	0.01 *N* HCl	232 (UV)
Cyanocobalamin, injection	H_2O	361 (UV)
Cyclizine HCl tablets	0.5 *N* H_2SO_4	225 (UV)
Desoxycorticosterone acetate injection, tablets	Alcohol	525 (visible)
Dextroamphetamine SO_4 tablets	Dilute H_2SO_4	257 (UV)
Dibucaine HCl injection	1 *N* HCl	247 (UV)
Diethylstilbesterol* tablets, injection, suppositories	PO_4 buffer	418 (visible)
Diethyltoluamide, lotion	CS_2	14.4 μ (IR)
Digitoxin, tablets	Alcohol	495 (visible)
Digoxin, elixir, injection, tablets	Dinitrobenzene	620 (visible)
Diphenhydramine HCl capsules, elixir, injection	0.1 *N* H_2SO_4	258 (UV)
Dithiazanine iodide	DMF	652 (visible)
Edrophonium Cl injection	0.1 *N* HCl	273 (UV)
Ergocalciferol capsules, solution	Ethylene dichloride	500 (visible) 550
Ergonovine maleate injection, tablets	H_2O	550 (visible)
Ergotamine tartrate, injection, tablets	H_2O, tartaric acid	550 (visible)
Estradiol valerate injection	KOH, methanol	300 (UV)
Ethinyl estradiol tablets	Dilute H_2SO_4	420 (visible) 525
Evans blue, injection	H_2O	610 (visible)
Folic acid, tablets, injection	Isobutyl alcohol	550 (visible)
Guanethidine SO_4, tablets	H_2O	500 (visible)
Hexachlorophene liquid soap	KOH, methanol	312 (UV)
Histamine PO_4 injection	H_2O	460 (visible)
Hydrocortisone, lotion, ointment, tablets	Alcohol	525 (visible)
Hydrocortisone acetate, ointment, ophthalmic ointment, ophthalmic suspension	Alcohol	525 (visible)
Sterile hydrocortisone acetate suspension	Alcohol	525 (visible)

TABLE 48. OFFICIAL SUBSTANCES ASSAYED USING SPECTROMETRIC METHODS (*Continued*)

Substance	Solvent	Wavelength, mμ; spectral region
U.S.P.		
Hydrocortisone sodium succinate, for injection	Alcohol	525 (visible)
Hydroxychloroquine SO_4 tablets	0.01 *N* HCl	343 (UV)
Iodochlorhydroxyquin cream, ointment, suppositories	Glacial acetic acid	325 (UV)
Iron dextran injection	H_2O, bipyridine	510 (visible)
Isoproterenol HCl, inhalation, injection, tablets	H_2O	279 (UV)
Lucanthone, HCl, tablets	0.1 *N* HCl	256 (UV)
Medroxyprogesterone acetate, suspension, tablets	$CHCl_3$	380 (UV)
Mercaptopurine tablets	0.1 *N* HCl	325 (UV)
Metaraminol bitartrate injection	H_2O	272 (UV)
Methazolamide	Borate buffer	288 (UV)
Methotrexate	0.1 *N* HCl	306 (UV)
Methylene blue, injection	Dilute alcohol	663 (visible)
Methylergonovine maleate, injection, tablets	H_2O	555 (visible)
Metyrapone	(1 to 35) H_2SO_4	260 (UV)
Nalorphine HCl, injection	H_2O	285 (UV) 310
Niacinamide injection, tablets	H_2O	450 (visible)
Nitrofurantoin	H_2O	367 (UV)
Nitroglycerin tablets	CS_2	7.89 μ (IR)
Ouabain, injection	H_2O, alcohol	495 (visible)
Phenobarbital elixir, tablets	Borate buffer	240 (UV)
Phenolsulfonphthalein injection	(1 to 100) Na_2CO_3	559 (visible)
Phentolamine mesylate, for injection	H_2O	278 (UV)
Phenylephrine HCl ophthalmic solution	0.5 *N* HCl	273 (UV)
Sterile phytonadione emulsion	*n*-Hexane	248 (UV)
Phytonadione tablets	Isooctane	270 (UV)
Prednisolone, tablets, ointment	Alcohol	525 (visible)
Prednisolone acetate, suspension	Alcohol	525 (visible)
Prednisolone Na phosphate injection	CH_2Cl_2	241 (UV)
Prednisone, tablets	Alcohol	525 (visible)
Primidone, oral suspension, tablets	Alcohol	254 (UV) 257 261
Probenecid tablets	$CHCl_3$	257 (UV)
Procaine HCl injection	H_2O	530 (visible)
Prochlorperazine edisylate injection, syrup	(1 to 100) HCl	254 (UV) 278
Prochlorperazine maleate tablets	(1 to 100) HCl	254 (UV) 278

TABLE 48. OFFICIAL SUBSTANCES ASSAYED USING SPECTROMETRIC METHODS (*Continued*)

Substance	Solvent	Wavelength, mμ; spectral region
U.S.P.		
Promethazine HCl injection, syrup, tablets	(1 to 100) H_2SO_4	298 (UV)
Propoxyphene HCl capsules	$CHCl_3$	5.80 μ (IR)
Pyrimethamine tablets	0.5 *N* H_2SO_4	273 (UV)
Pyrvinium pamoate, oral suspension, tablets	$CHCl_3$, hexane	516 (visible)
Reserpine, injection, tablets	$CHCl_3$, CH_3OH, H_2O	390 (visible)
Sodium sulfobromophthalein, injection	(1 to 100) Na_2CO_3	580 (visible)
Sodium sulfoxone, tablets	H_2O	560 (visible)
Sodium thiopental, for injection	0.1 *N* NaOH	304 (UV)
Sodium warfarin, for injection, tablets	0.01 *N* NaOH	308 (UV)
Spironolactone, tablets	Methanol	238 (UV)
Testosterone cypionate, injection	$CHCl_3$	380 (visible)
Tetracaine ophthalmic ointment	H_2O	310 (UV)
Tetracaine HCl injection	H_2O	310 (UV)
Thiotepa for injection	$CHCl_3$	10.75 μ (IR)
Tolbutamide tablets	$CHCl_3$	263 (UV)
Trichlorethylene	0.01 *N* NaOH	590 (visible)
Trihexyphenidyl HCl tablets	$CHCl_3$	408 (visible)
Tripelennamine citrate, elixir	0.5 *N* H_2SO_4	313 (UV)
Tripelennamine HCl tablets	0.5 *N* H_2SO_4	313 (UV)
Vanillin	Methanol	308 (UV)
Vitamin A	Isopropyl alcohol	310 (UV) 325 334
Vitamin B_6	Isopropyl alcohol	650 (visible)
Vitamin D	$ClCH_2CH_2Cl$	500 (visible) 550
Vitamin E	Dehydrated alcohol	520 (visible)
N.F.		
Acetaminophen, tablets	Acidic methanol	249 (UV)
Aluminum aspirin	$CHCl_3$	278 (UV) 308
Amodiaquine HCl, tablets	0.1 *N* HCl	342 (UV)
Amphetamine PO_4, injection, tablets	(2 to 100) H_2SO_4 in $CHCl_3$	257 (UV)
Amphetamine SO_4, injection, tablets	(2 to 100) H_2SO_4 in $CHCl_3$	257 (UV)
Antazoline PO_4 ophthalmic solution	0.5 *N* H_2SO_4	242 (UV)
Anthralin, ointment	$CHCl_3$	354 (UV) 432 (visible)
Aspirin, phenacetin, and caffeine capsules, capsules, tablets	$CHCl_3$	276 (UV) 285 280 310
Azacyclonol HCl tablets	$CHCl_3$	260 (UV)

TABLE 48. OFFICIAL SUBSTANCES ASSAYED USING SPECTROMETRIC METHODS (*Continued*)

Substance	Solvent	Wavelength, mμ; spectral region
N.F.		
Azuresin	Acidic alcohol	690 (visible)
Benzestrol, tablets	Acidic alcohol	750 (visible)
Benzpyrinium Br injection	H_2O	269 (UV)
Benztropine mesylate tablets	2 *N* HCl	258 (UV)
Betamethasone, tablets	Acidic alcohol	525 (visible)
Bishydroxycoumarin capsules	$CHCl_3$	309 (UV)
Butethamine HCl	H_2O	290 (UV)
Butethamine HCl and epinephrine injection	H_2O	290 (UV)
Calcium cyclamate and calcium saccharin solution, tablets	H_2O	266 (UV)
Chlordiazepoxide HCl capsules, for injection	0.1 *N* alcohol H_2SO_4	245 (UV)
Chlorothen citrate tablets	Alcohol	243 (UV)
Chlorotrianisene capsules	$CHCl_3$	310 (UV)
Chlorpheniramine maleate injection	0.1 *N* H_2SO_4	264 (UV)
Cortisone acetate, ophthalmic suspension, suspension, tablets	Alcohol	525 (visible)
Cyclophosphamide, tablets	KBr	4.9 μ (IR) 9.5 μ
Cycrimine HCl tablets	$CHCl_3$	8.91 μ (IR)
Deslanoside injection	H_2O, methanol	490 (visible)
Desoxy-corticosterone pivalate suspension, sterile	Methanol	241 (UV)
Dexamethasone, elixir, tablets	Alcohol	525 (visible)
Dexamethasone Na phosphate	CH_2Cl_2	236 (UV)
Dexamethasone Na phosphate cream, ophthalmic ointment, ophthalmic solution	Reagent blank	410 (visible)
Dexchlorpheniramine maleate syrup, tablets	0.1 *N* HCl	265 (UV)
Dextroamphetamine PO_4, tablets	(2 to 100) H_2SO_4 in $CHCl_3$	257 (UV)
Dextromethorphan HBr syrup	0.1 *N* HCl	278 (UV)
Dibucaine ophthalmic ointment	1 *N* HCl	247 (UV)
Diethylstilbextrol dipropionate injection, tablets	Diluted alcohol	520 (visible)
Digitoxin injection	Alcohol	495 (visible)
Dimethisoquin HCl, lotion, ointment	Isopropyl alcohol	328 (UV) 360
Diphenadione, tablets	Alcohol, H_2SO_4	289 (UV)
Doxylamine succinate, tablets	H_2O	260 (UV)
Estradiol, tablets, pellets, injection	H_2O	420 (visible) 525
Estradiol benzoate injection	H_2O	420 (visible) 525
Estradiol cypionate injection	Alcohol	510 (visible)
Estradiol dipropionate injection	Isooctane, benzene	420 (visible) 525

TABLE 48. OFFICIAL SUBSTANCES ASSAYED USING SPECTROMETRIC METHODS (*Continued*)

Substance	Solvent	Wavelength, $m\mu$; spectral region
N.F.		
Furazolidone	H_2O	367 (UV)
Furazolidone and nifuroxime powder, suppositories	H_2O, DMF	341 (UV) 367
Glutethimide, tablets	Methanol	257 (UV)
Hexylcaine HCl, injection, solution	0.1 *N* HCl	232 (UV) 275
Homatropine methylbromide tablets	Acetone	525 (visible)
Hyaluronidase for injection	PO_4 buffer	640 (visible)
Isoflurophate ophthalmic ointment, ophthalmic solution	H_2O	710 (visible)
Lanatoside C tablets	Glacial acetic acid	570 (visible)
Levallorphan tartrate injection	$CHCl_3$	420 (visible)
Menadione capsules, injection	Reagent blank	635 (visible)
Mephenesin, capsules, tablets	Alcohol	272 (UV)
Methapyrilene HCl tablets	0.5 *N* H_2SO_4	313 (UV)
Metharbital, tablets	0.1 *N* NaOH	244 (UV)
Methscopolamine Br injection, tablets	H_2O	420 (visible)
Methsuximide, capsules	Alcohol	247 (UV)
Methylphenidate HCl tablets	$CHCl_3$	420 (visible)
Methylphenidate HCl	H_2O	500 (visible)
Methylprednisolone, tablets	Alcohol	525 (visible)
Methylprednisolone acetate, sterile suspension	Alcohol	525 (visible)
Methyl prednisolone sodium succinate, for injection	Alcohol	525 (visible)
Methyltestosterone tablets	Alcohol	241 (UV)
Naphazoline HCl solution	Methanol	280 (UV)
Nifuroxime	H_2O	341 (UV)
Nitrofurazone, ointment, solution	H_2O	375 (UV)
Penthienate Br tablets	0.02 *N* HCl	238 (UV)
Pentylenetetrazol injection	$CHCl_3$	10.05 μ (IR) 10.25
Phenindamine tartrate tablets	0.5 *N* H_2SO_4	260 (UV)
Phenmetrazine HCl tablets	4% HCl	256 (UV)
Phensuximide, capsules	Alcohol	258 (UV)
Pipradrol HCl tablets	$CHCl_3$	260 (UV)
Potassium guiacolsulfonate	PO_4 buffer	279.5 (UV)
Potassium warfarin, tablets	0.01 *N* NaOH	308 (UV)
Pramoxine HCl jelly	0.5 *N* H_2SO_4	286 (UV)
Procaine, tetracaine, and nordefrin HCls injection	H_2O	530 (visible)
Prochlorperazine suppositories	1% HCl	254 (UV) 278
Promazine HCl, injection, syrup, tablets	1% HCl	301 (UV)

TABLE 48. OFFICIAL SUBSTANCES ASSAYED USING SPECTROMETRIC METHODS (*Continued*)

Substance	Solvent	Wavelength, mμ; spectral region
N.F.		
Propoxycaine HCl injection	2 *N* HCl	303 (UV)
Propoxycaine and procaine HCls and levonordefrin injection	2 *N* HCl	300 (UV)
Pyrilamine maleate tablets	0.5 *N* H_2SO_4	315 (UV)
Pyrrobutamine PO_4 tablets	H_2O	226 (UV)
Rauwolfia serpentina, powder, tablets	Dilute alcohol	390 (visible)
Rescinnamine, tablets	$CHCl_3$, H_2O, alcohol	390 (visible)
Sodium methohexital for injection	(1 to 25,000) NaOH	247 (UV)
Syrosingopine, tablets	Methanol, $CHCl_3$	390 (visible)
Testosterone suspension, sterile	Methanol	241 (UV)
Thenyldiamine HCl tablets	0.5 *N* H_2SO_4	314 (UV)
Thimerosal ointment	$CHCl_3$	490 (visible)
Thonzylamine HCl tablets	0.5 *N* H_2SO_4	236 (UV)
Mixed tocopherols concentrate	Dehydrated alcohol	410 (visible) 520
d-Alpha tocopheryl acetate concentrate	Dehydrated alcohol	410 (visible) 520
Tridihexethyl Cl tablets	$CHCl_3$	408 (visible)
Trimethobenzamide HCl capsules, injection	0.1 *N* HCl	258 (UV)
Trimethobenzamide HCl and benzocaine suppositories	Dilute H_2SO_4	540 (visible)
Trypsin crystallized	PO_4 buffer	253 (UV)

* Assayed following irradiation with a germicidal lamp.
UV = ultraviolet; IR = infrared.

Exercise 104

Object. The Assay of Sulfisoxazole Tablets.

Materials Required. Ten 0.5-Gm Sulfisoxazole Tablets
100 mg U.S.P. Sulfisoxazole Reference Standard[3]
1 liter approximately 1 *N* HCl
200 ml of 3 percent Vanillin U.S.P. in Alcohol U.S.P.

Procedure. 1. *Preparation of Calibration Curve.*

a. Weigh accurately about 100 mg of Sulfisoxazole Reference Standard which has previously been dried at 105° for 4 hr. Transfer the powder quantitatively to a 100-ml beaker and add about 40 ml of 1 *N* HCl. Apply heat gently and carefully. Stir until the powder is in solution. Allow the solution to cool to room temperature and then transfer quanti-

[3] Available from U.S.P. Reference Standards, New York, N.Y.

tatively to a 100-ml volumetric flask using 1 *N* HCl to bring the solution to the mark. Mix thoroughly.

b. Using a buret, transfer exactly 25 ml of the solution prepared in part *a* to a 100-ml volumetric flask and add 1 *N* HCl to bring the solution to the mark. Designate this solution as the *Stock Solution.*

c. Using a buret, transfer accurately 10, 14, 18, and 22 ml of Stock Solution to separate 100-ml volumetric flasks appropriately marked 1 through 4, respectively. By means of a buret add exactly 20 ml of 3 percent vanillin solution (color developing solution) to each flask. Then accurately add 1 *N* HCl to volume. Prepare a blank solution by transferring exactly 20 ml of 3 percent vanillin solution to a 100-ml volumetric flask and adding 1 *N* HCl to the mark. Mix all solutions thoroughly.

d. Following the instructions given earlier in this chapter under General Procedure, prepare a spectral-absorbance (SA) curve of the solution containing the most concentrated dilution. As for all spectrophotometric measurements, use matched cells if possible. If the cells are mismatched, appropriate corrections must be applied to the absorbance readings.[4]

e. Observe from the SA curve that the absorbance maximum occurs at 390 mμ. Determine the absorbance of each solution at this wavelength, using matched cells, a blank as prepared above, and a suitable spectrophotometer or a filter photometer with a blue filter. Record the data and plot the absorbance of the solutions against concentration. Note that the points of the plot form a straight line in accordance with the equation $A = abc$ (Fig. 31).

This assay depends on the following acid-catalyzed equilibrium reaction which takes place between an aldehyde (vanillin) and an aryl amine (sulfisoxazole). The chemical species which forms is called a Schiff base and is yellow in color.

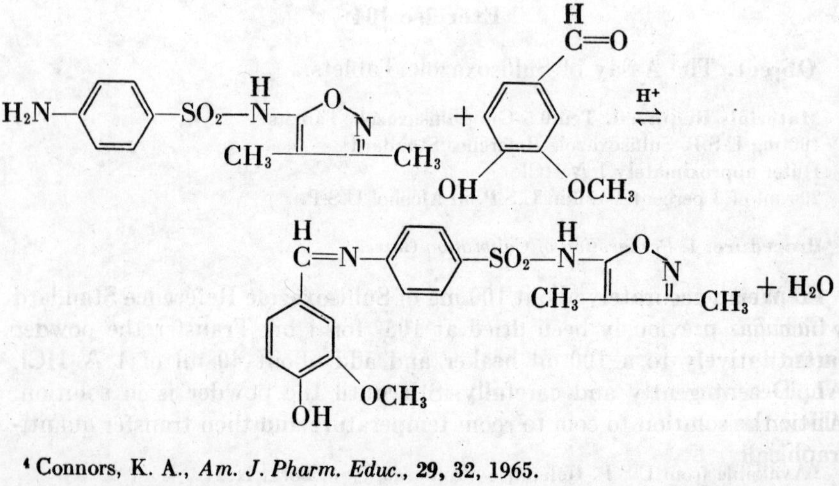

H₂N—C₆H₄—SO₂—N(H)—(isoxazole: O, N, CH₃, CH₃) + (benzene ring: H—C=O, OH, OCH₃) $\overset{H^+}{\rightleftharpoons}$

H—C=N—C₆H₄—SO₂—N(H)—(isoxazole: O, N, CH₃, CH₃) (benzene ring: OH, OCH₃) + H_2O

[4] Connors, K. A., *Am. J. Pharm. Educ.*, **29**, 32, 1965.

An absorption spectrum of the Schiff base is shown in Fig. 30. Note that the maximum absorption occurs at 390 mμ.

2. *Assay of Sulfisoxazole Tablets*

a. Determine the exact weight of 10 tablets. Using a mortar and pestle reduce these tablets to a fine powder. Weigh accurately about 100 mg and transfer to a 100-ml beaker. Add about 30 ml of 1 *N* HCl. Heat the solution gently while stirring for about 5 min. Filter the solution through a small filter paper wetted with 1 *N* HCl in order to remove any insoluble tablet excipients which may be present.

Collect the filtrate in a 100-ml volumetric flask and add sufficient 1 *N* HCl through the filter paper to bring the solution to the mark. Mix thoroughly.

b. Using a buret, transfer exactly 25 ml of the solution prepared in part *a* to a 100-ml volumetric flask and add 1 *N* HCl to bring the solution to the mark. Designate this solution as the *Tablet Stock Solution.*

c. Using a buret, transfer exactly 15 ml of the Tablet Stock Solution to a 100-ml volumetric flask and designate as solution 5. Add exactly 20 ml of the 3 percent vanillin solution and sufficient 1 *N* HCl to measure exactly 100 ml. Mix thoroughly. Determine the absorbance of this solution using the same blank which was used above, and from the calibration curve determine the concentration of sulfisoxazole in solution. Calculate the amount in milligrams of sulfisoxazole per tablet.

A sample calculation follows.

Illustration

Experimental Data

Weight of U.S.P. Reference Standard Sulfisoxazole, mg	100.6
Weight of 10 Sulfisoxazole Tablets, mg	6,490
Weight of powdered tablet, mg	98.5

Solution No.	Concentration, μg/ml	Absorbance at 390 mμ
1	25.2	0.210
2	35.3	0.295
3	45.4	0.375
4	55.4	0.465
5	(Tablets)	0.235

Prepare a Beer's plot (Fig. 31).

Calculation. The calculation may conveniently be divided into three parts:

1. Determination of the concentration of active constituent in the final dilution using spectrophotometric data. In this example it was determined graphically. From the Beer's plot (Fig. 31), the concentration of sulfi-

soxazole which corresponds to an absorbance reading of 0.235 was found to equal 28.2 μg per ml. In other words, the concentration of drug in the final dilution is 28.2 μg per ml.

The concentration could also have been determined from the following mathematical relationship:

$$C_1 = \frac{A_1}{A_2} \times C_2$$

where C_1 and A_1 = the concentration and absorbance, respectively, of the Sulfisoxazole Solution of unknown concentration

C_2 and A_2 = the concentration and absorbance, respectively, of a U.S.P. Reference Standard solution of Sulfisoxazole of known concentration

2. Determination of the concentration of the sample taken in the final dilution. In this example the sample was powdered tablets. The powdered tablet sample weighing 98.5 mg was diluted to 100 ml. Exactly 25 ml of this solution was diluted to 100 ml.

$$\frac{98.5 \text{ mg}}{100 \text{ ml}} \times 25 \text{ ml} = 24.63 \text{ mg in } 100 \text{ ml}$$

Exactly 15 ml of this solution was diluted to 100 ml.

$$\frac{24.63 \text{ mg}}{100 \text{ ml}} \times 15 \text{ ml} = 3.695 \text{ mg in } 100 \text{ ml or } 0.03695 \text{ mg/ml}$$

Convert to micrograms per milliliter:

$$\frac{0.03695 \text{ mg}}{1.00 \text{ ml}} \times \frac{1{,}000\ \mu\text{g}}{1 \text{ mg}} = 36.95\ \mu\text{g/ml}$$

in final dilution, or combining the above mathematical operations gives:

$$\frac{98.5 \text{ mg}}{100 \text{ ml}} \times \frac{25 \text{ ml}}{100 \text{ ml}} \times \frac{15 \text{ ml}}{100 \text{ ml}} \times \frac{1{,}000\ \mu\text{g}}{1 \text{ mg}} = 36.95\ \mu\text{g}$$

which is the amount of powdered sample per milliliter in the final dilution.

3. Determination of the quantity of active constituent in a given quantity of sample. In this example the quantity of sulfisoxazole present per tablet, or 649.0 mg, is the result desired. Therefore,

$$\frac{28.2\ \mu\text{g/ml sulfisoxazole}}{36.95\ \mu\text{g/ml powdered sample}} \times \frac{649.0 \text{ mg}}{\text{tablet}}$$

$$= 495 \text{ mg sulfisoxazole per tablet}$$

U.S.P. Trisulfapyrimidines Tablets and Oral Suspension are also assayed colorimetrically. For the assay of these dosage forms, accurately measured quantities of the mixture, U.S.P. reference standard sulfadiazine, sulfamethazine, and sulfamerazine, are spotted on chromatography paper and then developed. The designated areas on the chromatography paper containing the drugs are eluted and analyzed by the Bratton-Marshall method. The analysis by this method is specific for aryl (aromatic) amines and is based upon the diazotization of the aryl amine using nitrous acid followed by coupling with a color-producing reagent called *N*-(1-naphthyl)ethylenediaminedihydrochloride. In actual practice, the excess, unreacted nitrous acid is first destroyed using the ammonium salt of sulfamic acid, before the color-producing reagent is added. The power of the color as measured at 545 mμ is proportional to the concentration of the sulfonamide being analyzed. The procedure used for the analysis of these sulfa drugs is completely analogous to the analysis of sulfa by Schiff base formation.[5]

$$Ar{-}NH_2 + HNO_2 + HCl \longrightarrow Ar{-}N_2^+Cl^- + 2H_2O$$

$$HNO_2(\text{excess}) + H_2NSO_3H \longrightarrow H_2SO_4 + N_2 + H_2O$$

$$Ar{-}N_2^+Cl^- + C_{10}H_7{-}\overset{H}{N}{-}CH_2CH_2{-}NH_2 \rightleftharpoons$$

$$Ar{-}N{=}N{-}C_{10}H_6{-}\overset{H}{N}{-}CH_2CH_2{-}NH_2 + HCl$$

The analysis of other aryl amine compounds such as procaine and other sulfonamides may also be adapted to this and the Schiff base method.

Questions and Problems

1. Define the terms: (*a*) light; (*b*) radiant energy; (*c*) infrared; (*d*) absorptivity; (*e*) SA curve.

2. Prepare a table listing the names of the proprietary products, the biochemical agent for which the test was devised, and a concise description of the various tests which are based on the colorimetric method available in kit form from the hospital and community pharmacy.

3. How does colorimetry differ from spectrophotometry? Which gives rise to more precise measurements? Why?

4. Outline the general procedure for determining a substance spectrophotometrically.

5. Calculate the percent transmittance value for a solution of *N*-allylmorphine HCl having an absorbance value of 0.482.

6. Calculate the absorptivity value of a medicinal agent if a solution having a concentration of 8.0 mg per 100 ml has a transmittance reading of 54.0 percent.

[5] $Ar = R{-}\overset{H}{N}{-}SO_2{-}C_6H_4{-}$

7. Calculate the milligrams of mercaptopurine present per tablet from the following data and procedure. Twenty tablets of mercaptopurine, weighing 1.200 Gm, were reduced to a fine powder. Exactly 65.0 mg of the powder was transferred to a 100-ml volumetric flask, and 0.1 *N* HCl was added to the mark. Exactly 5 ml of this solution was transferred to a 500-ml flask and diluted to the mark with 0.1 *N* HCl. The absorbance value of this solution was found to equal 0.542. Assume the absorptivity value of mercaptopurine to be 108.4.

Exercise 105

Object. Assay of Vitamin A Content of Cod-Liver Oil.

Materials Required. 1 Gm of Cod-Liver Oil. Isopropyl Alcohol (USP), redistill if necessary. This reagent must meet the following requirements for spectral purity: When measured in a 1-cm silica cell against distilled water, it must show an absorbance not greater than 0.01 between 350 and 320 mμ and not greater than 0.05 at 300 mμ.

Phenolphthalein T.S.

Freshly redistilled ether in which the first and last 10 percent portions were discarded.

Potassium hydroxide solution, 50 percent (*W/W*).

Spectrophotometer capable of measuring radiant energy in the ultraviolet region of the spectrum.

Procedure. 1. "Transfer to a saponification flask an amount of the sample accurately measured and expected to contain not less than the equivalent of 0.15 mg. of vitamin A but not more than 1 Gm. of fat. Reflux under a condenser, in an all-glass apparatus with 30 ml. of alcohol and 3 ml. of potassium hydroxide solution (1 in 2) for 30 minutes."

To prevent destruction of vitamin A by light during the sampling and analysis, amber or nonactinic glassware must be used unless the assay is carried out under subdued light. Throughout the procedure care must be taken to avoid undue exposure to oxidizing agents; peroxide-free reagents and promptness in completing the assay are important. Rubber stoppers and corks may not be used. Other fat-soluble substances occurring with vitamin A interfere with its determination, making some purification necessary. This is accomplished by saponification and extraction.

2. "Cool the solution, add 30 ml. of water, and transfer the resulting solution to a conical separator. Add 2 Gm. of finely powdered sodium sulfate. Extract with one 150 ml. portion of ether, or, if an emulsion occurs, with three additional 25 ml. portions of ether. Combine the ether extracts, if necessary, and wash by swirling gently with 50 ml. of water. Repeat the washing more vigorously until the last washing gives no color with phenolphthalein T.S. Transfer the washed ether extract to a 250 ml. volumetric flask, and make to volume with ether."

The unsaponifiable fraction of the oil, containing the vitamin A, is extracted in the ether.

3. "Evaporate a 25.0 ml. aliquot of the ether solution of the unsaponifiable extract to about 5 ml. Without the application of heat, remove the remaining ether in a stream of inert gas or by vacuum. Dissolve the residue in sufficient isopropyl alcohol

to give an expected concentration of the equivalent of 3 to 5 mcg. of vitamin A per ml. or such that it will give an absorbance in the range 0.4 to 0.8 at 325 mμ. Determine the absorbances of the resulting solution at the wave lengths 310 mμ, 325 mμ and 334 mμ, using a suitable spectrophotometer and matched silica cells."

4. Calculate the vitamin A content as follows:

$$A_{325\,(\text{corrected})} = 6.815\ A_{325} - 2.555\ A_{310} - 4.260\ A_{334}$$

The absorbances at 325, 310, and 334 mμ are substituted in the formula to determine the corrected absorbance at 325 mμ, taking into consideration the *irrelevant* absorption. In this calculation the irrelevant absorption is considered to be linear and the corrections are based on the procedure originally reported by Morton and Stubbs.[6]

The corrected absorbance is substituted into the following formula if $A_{325(\text{corrected})}$ has a value less than $A_{325}/1.030$:

$$\frac{0.549\ A_{325(\text{corrected})}}{L \times C} = \text{vitamin A content, mg}$$

where L = length of light path, cm
C = amount of sample expressed as grams, capsule or tablet, in 100 ml of final isopropyl alcohol solution

0.549 is a factor for conversion from spectrophotometric to gravimetric absorbance, in milligram units. If A_{325} has a value not less than $A_{325(\text{corrected})}/1.03$ and not more than $A_{325(\text{corrected})}/0.970$, substitute A_{325} for $A_{325(\text{corrected})}$ in the above formula and complete the calculation.

Example. Sample weight of cod-liver oil is 0.9980 Gm. Sample following saponification is taken up in 250 ml of ether, and a 25.0-ml aliquot is evaporated and diluted to 10 ml with isopropyl alcohol for spectrophotometric reading. Concentration of sample in final isopropyl alcohol solution is 0.9980 Gm per 100 ml. Light path = 1 cm. Absorbances found are $A_{310} = 0.498$; $A_{325} = 0.582$; $A_{334} = 0.510$.

Calculation. Substitute into correction formula:

$$A_{(\text{corrected})} = 6.815\ (0.582) - 2.555\ (0.498) - 4.260\ (0.510)$$
$$0.522 = 3.966 - 1.272 - 2.172$$

Determine which equation to use for final calculation.

$$\frac{A_{325}}{1.030} = \frac{0.582}{1.030} = 0.565$$

Since $A_{(\text{corrected})}$, 0.522, is less than 0.565, this value is substituted into the original equation and number of milligrams of vitamin A is determined.

$$\frac{0.549 \times 0.522}{1 \times 0.9980} = 0.287 \text{ mg of vitamin A per Gm of oil}$$

[6] Morton, R. A., and A. L. Stubbs, *Analyst*, 71, 348, 1946.

OFFICIAL IDENTIFICATION TESTS USING SPECTROMETRIC ABSORPTION METHODS

Spectrometric absorption methods are frequently used as identification tests for pharmaceutical substances. Although all three regions of the electromagnetic spectrum are used, the ultraviolet and infrared regions are used most frequently. When utilizing the *ultraviolet* region for the purpose of identification, the test usually consists of one or more of the following three procedures: (1) the determination of absorbance values at two wavelengths followed by the calculation of the ratio and comparison with the value given in the official compendiums; (2) the determination of the absorptivity values of the official substance and a corresponding reference standard and comparison of the results; or (3) the comparison of the reference standard spectrum with a sample spectrum.

Of the three *infrared* regions, the medium infrared region[7] (3.0 to 15.0) is most often applied for identification purposes. In most cases, spectra of the medicinal agent under investigation and the corresponding U.S.P. or N.F. reference standard are prepared and then compared with one another. To meet the requirements set forth in the official compendiums, the spectrum of the medicinal agent must exhibit absorption maximums and minimums only at the same wavelengths as that of a similar preparation of the reference standard.

In the infrared region of the electromagnetic spectrum, water cannot be used as a solvent because it strongly absorbs most of the infrared radiation, thereby obscuring absorption by the solute. Another good reason why water cannot be used is that it would destroy the sodium chloride cells which hold the sample. Therefore, the sample is prepared by one of the following three methods: (1) dispersing the finely ground solid in liquid petrolatum (mull technique), (2) dissolving the solid in a suitable organic solvent, or (3) incorporating the solid in a potassium bromide pellet.

When using the mull technique, a small sample (about 100 mg) of medicinal agent is placed in an agate mortar and ground vigorously with an agate pestle. Several drops of liquid petrolatum is then added, and the mixture is again ground. The mull, so formed, is then transferred to a NaCl plate and covered with a second NaCl plate and mounted. The absorption spectrum is then run. The procedure is repeated, using the corresponding official reference standard. It should be noted that the preparation of "mulls" is an art which is easily mastered with a little practice.

When using solutions for spectral analysis, a concentration of solute sample and reference standard as specified in the official compendiums is

[7] For brevity, this region is simply called the infrared region.

prepared using a designated organic spectral grade solvent, $CHCl_3$ or CS_2. The solutions are transferred to NaCl liquid cells, and absorption spectra are run.

When preparing the sample for infrared analysis using KBr, the following procedure is recommended. Place about 5 mg of medicinal agent and 500 mg of infrared-quality KBr in an agate mortar and mix rapidly with an agate pestle. Transfer about 200 mg of the mixture to the compression chamber of an evacuable die and evacuate the chamber for 5 min. Compress the mixture under vacuum at 20,000 lb for 10 min. Then remove the pellet, transfer carefully to the holder, and obtain an absorption spectrum. The same procedure is repeated using the corresponding reference standard. Occasionally a difference appears in the spectra of the medicinal agent and the reference standard which may be due to polymorphism

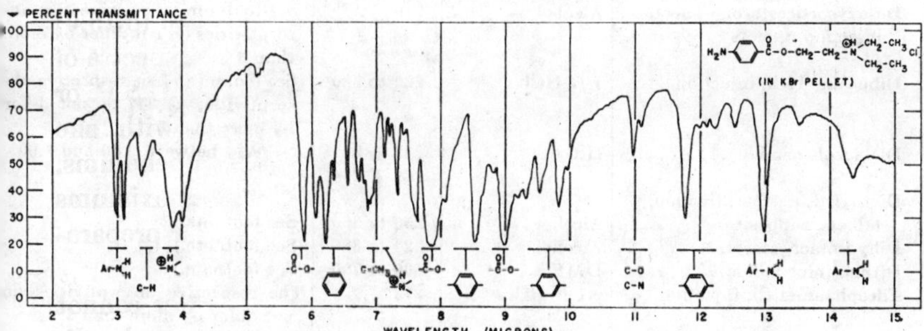

FIG. 32. An infrared spectrum of procaine HCl prepared as a KBr dispersion.

(a property some substances have of existing in two or more crystalline forms). To correct for this difference, portions of both the sample and reference standard are dissolved in methanol, the solutions are evaporated to dryness, and KBr pellets are prepared using the residues. If differences in spectra still persist, the sample may be assumed to be contaminated.

An infrared absorption spectrum of U.S.P. Procaine HCl prepared as a dispersion in KBr is shown in Fig. 32. The groups responsible for absorption of infrared radiant energy at various wavelengths are designated on the spectrum.

Identification tests based on spectrometric methods and used for medicinal agents and products official in the U.S.P. and N.F. are listed in Tables 49 and 50.

Exercise 106

Object. Determine Identification Test A for Aminosalicylic Acid.

Materials Required. U.S.P. Aminosalicylic Acid
Sodium Hydroxide T.S.
pH 7 phosphate buffer

TABLE 49. OFFICIAL IDENTIFICATION TESTS USING ULTRAVIOLET ABSORPTION METHODS

Substance	Solvent	Wavelength, mμ	Remarks
U.S.P.			
Aminosalicylic acid	PO_4 buffer	265 ± 2 299 ± 2	A_{265}/A_{299} lies between 1.50 and 1.56.
Bishydroxycoumarin tablets	$CHCl_3$	....	Exhibits absorbance maximum @ 309 ± 2.
Chlorambucil	Isopropyl ether	256 301	The respective absorptivities do not differ by more than 3 %.
Chlorambucil tablets	Isopropyl ether	220 to 380	(See footnote.)
Chloroprocaine HCl, injection	Buffer No. 1	220 to 380	(See footnote.)
Cholecalciferol	(3 to 200) alcohol	265	The respective absorptivities do not differ by more than 3 %.
Cyanocobalamin, injection	H_2O	278 ± 1 361 ± 1 550 ± 2	A_{361}/A_{278} between 1.70 and 1.90. A_{361}/A_{550} between 3.15 and 3.40.
Dapsone, tablets	Methanol	220 to 380	(See footnote.)
Desoxycorticosterone acetate, injection, tablets	Alcohol	240	(See footnote.) The respective absorptivities do not differ by more than 3 %.
Dibucaine HCl, injection	1 *N* HCl	220 to 380	(See footnote.) The respective absorptivities @ 247 do not differ by more than 3 %.
Dichlorphenamide	H_2O	295 ± 2 286 ± 2	A_{295}/A_{286} between 0.90 and 1.00.
Diethylstilbestrol, injection, tablets, suppositories	Alcohol	250 to 350	(See footnote.)
Dihydrotachysterol	Alcohol	220 to 380	(See footnote.)
Dithiazanine iodide	DMF	400 to 700	(See footnote.)
Edrophonium Cl, injection	0.1 *N* HCl	273	The respective absorptivities do not differ by more than 2 %.
Ergocalciferol	Alcohol	265	The respective absorptivities do not differ by more than 3 %.
Estradiol valerate	Alcohol	281	The respective absorptivities do not differ by more than 3 %.
Ethinyl estradiol	Alcohol	281●	(See Estradiol valerate).
Evans blue, injection	H_2O	....	Exhibits absorbance maximum @ 610.
Folic acid, tablets	0.1 *N* NaOH	256 365	A_{256}/A_{365} between 2.80 and 3.00.
Hydrochlorothiazide	Methanol	220 to 380	(See footnote.)
Hydrocortisone, acetate	Methanol	242	The respective absorptivities do not differ by more than 2.5 %.
Hydrocortisone Na succinate	Methanol	242	The respective absorptivities do not differ by more than 3 %.
Hydroxychloroquine SO_4, tablets	0.1 *N* HCl	343	The respective absorptivities do not differ by more than 3 %.
Hydroxystilbamidine isethionate	0.01 *N* HCl	220 to 380	(See footnote.)
Iodochlorhydroxyquin, cream, ointment, suppositories	3 *N* HCl	267	(See footnote.) The respective absorptivities do not differ by more than 3 %.
Isoproterenol HCl, inhalation, tablets	H_2O	220 to 380	(See footnote.)
Lucanthone HCl, tablets	0.1 *N* HCl	220 to 380	(See footnote.)
Meclizine HCl, tablets	0.1 *N* HCl	232	The respective absorptivities do not differ by more than 3 %.

TABLE 49. OFFICIAL IDENTIFICATION TESTS USING ULTRAVIOLET ABSORPTION METHODS (*Continued*)

Substance	Solvent	Wavelength, mμ	Remarks
U.S.P.			
Medroxyprogesterone acetate..	Alcohol	241	The respective absorptivities do not differ by more than 2.5 %
Meglumine diatrizoate injection	0.1 *N* NaOH	220 to 380	(See footnote.)
Meglumine iodipamide injection	0.1 *N* NaOH	220 to 380	(See footnote.)
Mercaptopurine............	0.1 *N* HCl	255 325	A_{255}/A_{325} does not exceed 0.06.
Mercaptopurine tablets.......	0.1 *N* HCl	255 325	A_{255}/A_{325} does not exceed 0.08.
Methazolamide.............	0.1 *N* NaOH	220 to 380	(See footnote.)
Methotrexate..............	0.1 *N* HCl	220 to 380	(See footnote.)
Methylergonovine maleate.....	0.4 *N* H_2SO_4	310	The respective absorptivities do not differ by more than 3 %.
Metyrapone...............	1 *N* H_2SO_4	260	The respective absorptivities do not differ by more than 3 %.
Niacinamide, tablets..........	H_2O	245 262	A_{245}/A_{262} between 0.63 and 0.67.
Niacinamide injection........	H_2O	237 262	A_{237}/A_{262} between 0.35 and 0.39.
Nitrofurantoin.............	H_2O	220 to 380	(See footnote.)
Noscapine................	Methanol	290 ± 1 310 ± 1	A_{310}/A_{290} is 1.23.
Phentolamine mesylate, for injection	H_2O	255 278	A_{278}/A_{255} between 1.30 and 1.50.
Phytonadione, sterile emulsion, tablets	Isopropyl alcohol	270	(See footnote.) The respective absorptivities do not differ by more than 3 %.
Prednisolone, acetate.........	Methanol	242	The respective absorptivities do not differ by more than 2.5 %.
Prednisone...............	Methanol	239	The respective absorptivities do not differ by more than 2.5 %.
Primidone, tablets, oral suspension	Alcohol	258	The respective absorptivities do not differ by more than 3 %.
Probenecid, tablets..........	Alcohol	248	The respective absorptivities do not differ by more than 3 %.
Pyrimethamine tablets........	Glacial acetic acid	220 to 380	(See footnote.)
Pyrvinium pamoate..........	Methoxyethanol	358 508	A_{508}/A_{358} between 1.75 and 2.05.
Pyrvinium pamoate oral suspension, tablets...........	Mobile solvent	300 to 600	(See footnote.)
Reserpine................	$CHCl_3$, CH_3OH, H_2O	268 295	The respective absorptivities do not differ by more than 3 %.
Sodium diatrizoate injection...	Alcohol	220 to 380	(See footnote.)
Sodium iodipamide injection...	0.1 *N* NaOH	220 to 380	(See footnote.)
Sodium liothyronine	0.2 *N* HCl in alcohol	297	The respective absorptivities do not differ by more than 5 %.
Sodium warfarin tablets.......	0.01 *N* NaOH	220 to 380	(See footnote.)
Spironolactone.............	Methanol	238	The respective absorptivities do not differ by more than 3 %.
Sulfisoxazole..............	PO_4 buffer	220 to 380	(See footnote.)
Testosterone cypionate........	Alcohol	242	The respective absorptivities do not differ by more than 3 %.
Testosterone propionate.......	Alcohol	241	The respective absorptivities do not differ by more than 3 %.
Tetracaine, HCl, ophthalmic ointment	PO_4 buffer	310	The respective absorptivities do not differ by more than 2 %.
Thiotepa.................	$CHCl_3$	220 to 380	(See footnote.)
Vanillin..................	Methanol	220 to 380	(See footnote.)

TABLE 49. OFFICIAL IDENTIFICATION TESTS USING ULTRAVIOLET ABSORPTION METHODS (*Continued*)

Substance	Solvent	Wavelength, $m\mu$	Remarks
N.F.			
Acetyl sulfisoxazole, oral suspension	Alcohol	220 to 380	(See footnote.)
Amodiaquine HCl, tablets	0.1 *N* HCl	220 to 380	(See footnote.)
Anileridine HCl	PO_4 buffer	234 ± 1 285 ± 2	$5A_{234}/A_{285}$ is 8.8.
Azacyclonol HCl	H_2O	258	The respective absorptivities do not differ by more than 2 %.
Betamethasone	Methanol	239	The respective absorptivities do not differ by more than 3 %.
Bishydroxycoumarin capsules	$CHCl_3$	309 ± 2	Exhibits an absorbance maximum @ 309 ± 2.
Brompheniramine maleate	Methanol	262	The respective absorptivities do not differ by more than 3 %.
Calcium aminosalicylate	NaOH T.S.	265 ± 2 299 ± 2	A_{265}/A_{299} between 1.50 and 1.56.
Calcium cyclamate and calcium saccharin solution, tablets	H_2O	266 ± 2 247 ± 2	Exhibits a maximum @ 266 ± 2 and a minimum @ 247 ± 2.
Carbinoxamine maleate	Methanol	260 ± 2	The respective absorptivities do not differ by more than 3 %.
Cetylpyridinium Cl	H_2O	259	The respective absorptivities do not differ by more than 3 %.
Chlordiazepoxide HCl	0.1 *N* alcohol H_2SO_4	245 ± 2	The respective absorptivities do not differ by more than 3 %.
Chlorothiazide	0.1 *N* NaOH	292	The respective absorptivities do not differ by more than 3 %.
Chlorotrianisene	$CHCl_3$	310 ± 2	The respective absorptivities do not differ by more than 2 %.
Cortisone acetate	Methanol	238	The respective absorptivities do not differ by more than 3 %.
Danthron	Benzene	432 ± 2	The respective absorptivities do not differ by more than 5 %.
Desoxycorticosterone acetate pellets	Alcohol	240	The respective absorptivities do not differ by more than 3 %.
Desoxycorticosterone pivalate	Methanol	241	The respective absorptivities do not differ by more than 4 %.
Dexamethasone	Methanol	239	The respective absorptivities do not differ by more than 3 %.
Dexchlorpheniramine maleate	H_2O	261	The respective absorptivities do not differ by more than 3%.
Dibucaine	1 *N* HCl	247	The respective absorptivities do not differ by more than 3 %.
Dimethisoquin HCl	Isopropyl alcohol	328 ± 2 316 ± 2 285 ± 2	A_{316}/A_{328} between 1.05 and 1.12.
Diphenadione	0.1 *N* H_2SO_4	289	Exhibits a maximum @ 289.
Estradiol cypionate	Alcohol	280 ± 2	The respective absorptivities do not differ by more than 3 %.
Estrone	Alcohol	280	The respective absorptivities do not differ by more than 3 %.
Ethisterone	Methanol	241	The respective absorptivities do not differ by more than 2 %.
Furazolidone	H_2O-DMF	367 ± 2	Exhibits a maximum @ 367 ± 2.
Glutethimide	Methanol	....	Exhibits maximums @ 251 ± 2, 257 ± 2, 263 ± 2, and minimums @ 249 ± 2, 255 ± 2, and 262 ± 2.

Table 49. Official Identification Tests Using Ultraviolet Absorption Methods (*Continued*)

Substance	Solvent	Wavelength, $m\mu$	Remarks
N.F.			
Hexylcaine HCl	0.1 *N* HCl	232 275	The respective absorptivities do not differ by more than 3%.
Hydroxyzine HCl	Alcohol	230 ± 2	The respective absorptivities do not differ by more than 3%.
Levallorphan tartrate	0.1 *N* H_2SO_4	279	The respective absorptivities do not differ by more than 2%.
Meperidine HCl	H_2O	....	Exhibits maximums @ 251, 257, 263, and minimums @ 248, 253.5, 260.5.
Metharbital	0.1 *N* NaOH	....	Exhibits maximum @ 244 ± 2.
Methylprednisolone	Alcohol	243	The respective absorptivities do not differ by more than 3%.
Methylprednisolone acetate	Alcohol	243	(See Methylprednisolone.)
Methylprednisolone sodium succinate	Alcohol	243	(See Methylprednisolone.)
Methyltestosterone	Alcohol	241	The respective absorptivities do not differ by more than 2%.
Methyprylon	Isopropanol	295 ± 5	The respective absorptivities do not differ by more than 3%.
Niacin	H_2O	237 262	A_{237}/A_{262} is 0.37 ± 0.02.
Nitrofurazone	H_2O-alcohol	....	Exhibits maximum @ 375 ± 2.
Phenmetrazine HCl	4% HCl	256 ± 2	The respective absorptivities do not differ by more than 2%.
Potassium warfarin	0.01 *N* KOH	220 to 380	(See footnote.)
Pramoxine HCl	0.1 *N* HCl	224 ± 2	The respective absorptivities do not differ by more than 3%.
Progesterone	Methanol	241	The respective absorptivities do not differ by more than 2%.
Promazine HCl	(1 to 100) HCl	251.5 301	$A_{251.5}/A_{301}$ between 7.1 and 7.9.
Pyrrobutamine PO_4	0.1 *N* HCl	....	Exhibits a maximum @ 226 ± 2.
Rescinnamine	$CHCl_3$-alcohol	305	The respective absorptivities do not differ by more than 3%.
Sodium vinbarbital injection	0.1 *N* NaOH	....	Exhibits maximums @ 252 ± 2 and a minimum @ 233 ± 2.
Sulfadimethoxine	Alcohol	....	Exhibits maximums @ 272 ± 2 and a minimum @ 234 ± 2.
Syrosingopine	Methanol	258 298	The respective absorptivities do not differ by more than 3%.
Testosterone enanthate	Alcohol	240	The respective absorptivities do not differ by more than 3%.
Tetrahydrozoline HCl	H_2O	264 271.5	The respective absorptivities do not differ by more than 4%.
Thenyldiamine HCl	Alcohol	246	The respective absorptivities do not differ by more than 3%.
Thiabendazole	0.1 *N* HCl	302 ± 2 243 ± 2 258 ± 2	A_{243}/A_{302} between 0.47 and 0.53. A_{258}/A_{302} is 0.15.
Triethylenemelamine	0.1 *N* HCl	222 240	A_{222}/A_{240} between 1.40 and 1.50.
Trimethobenzamide HCl	0.1 *N* HCl	220 to 380	(See footnote.)
Vinbarbital	0.1 *N* NaOH	....	Exhibits a maximum @ 252 ± 2 and a minimum @ 233 ± 2.

The sample solution exhibits maximums and minimums at the same wavelengths as a similar solution of reference standard, and 220 to 380 represents the practical and approximate wavelength range from which the absorption spectrum is determined.

TABLE 50. OFFICIAL IDENTIFICATION TESTS USING INFRARED ABSORPTION METHODS*

Substance	Solvent or dispersion	Wavelength, μ
U.S.P.		
Acetazolamide	KBr	2.0 to 15.0
γ-benzene hexachloride	KBr	2.0 to 15.0
Betazole HCl	Liquid pet.	2.0 to 15.0
Bishydroxycoumarin	KBr	2.0 to 15.0
Calcium disodium edetate	Liquid pet.	2.0 to 15.0
Calcium pantothenate	KBr	2.0 to 15.0
Chlorcyclizine HCl, tablets	CS_2	7.0 to 15.0
Chlorpheniramine maleate	CS_2	7.0 to 15.0
Chlorpropamide	KBr	2.0 to 15.0
Chlorpropamide tablets	$CHCl_3$	2.0 to 15.0
Cholecalciferol	KBr	2.0 to 12.0
Cyclizine HCl	KBr	2.0 to 12.0
Cyclizine HCl tablets	CS_2	7.0 to 15.0
Desoxycorticosterone acetate	KBr	2.0 to 15.0
Diethyltoluamide, lotion	CS_2	8.0 to 15.0
Dihydrotachysterol	KBr	2.0 to 15.0
Dimenhydrinate, syrup, tablets	CS_2	7.0 to 15.0
Ergocalciferol	KBr	2.0 to 12.0
Estradiol valerate	KBr	2.0 to 15.0
Ethinyl estradiol	KBr	2.0 to 15.0
Guanethidine sulfate	Liquid pet.	2.0 to 15.0
Hydrochlorothiazide	KBr	2.0 to 15.0
Hydrocortisone, tablets	KBr	2.0 to 15.0
Hydrocortisone acetate, suspension	KBr	2.0 to 15.0
Hydrocortisone Na succinate, for injection	KBr	2.0 to 15.0
Hydroxystilbamidine isethionate	KBr	2.0 to 15.0
Mecamylamine HCl, tablets	KBr	2.0 to 15.0
Meclizine HCl	KBr	2.0 to 15.0
Meglumine iothalamate injection	KBr	2.0 to 15.0
Methazolamide, tablets	KBr	2.0 to 15.0
Methotrexate	KBr	2.0 to 15.0
Metyrapone	Liquid pet.	2.0 to 15.0
Phenobarbital, elixir, tablets	KBr	2.0 to 15.0
Phentolamine mesylate, for injection	KBr	2.0 to 15.0
Prednisolone, tablets	KBr	2.0 to 15.0
Prednisolone acetate, suspension	KBr	2.0 to 15.0
Prednisone, tablets	KBr	2.0 to 15.0
Primidone	KBr	2.0 to 15.0
Procaine HCl, injection	KBr	2.0 to 15.0
Prochlorperazine edisylate, injection	CS_2	7.0 to 15.0
Prochlorperazine maleate, tablets	CS_2	7.0 to 15.0
Promethazine HCl, injection, syrup, tablets	CS_2	7.0 to 15.0
Propoxyphene HCl, capsules	$CHCl_3$	2.0 to 15.0
Pyrimethamine	KBr	2.0 to 15.0

TABLE 50. OFFICIAL IDENTIFICATION TESTS USING INFRARED ABSORPTION METHODS* (*Continued*)

Substance	Solvent or dispersion	Wavelength, μ
U.S.P.		
Pyrvinium pamoate	KBr	2.0 to 15.0
Sodium acetazolamide	KBr	2.0 to 15.0
Sodium iothalamate injection	KBr	2.0 to 15.0
Sodium phenobarbital, injection	KBr	2.0 to 15.0
Sodium warfarin	KBr	2.0 to 15.0
Spironolactone, tablets	KBr	2.0 to 15.0
Sulfamerazine	KBr	2.0 to 15.0
Sulfamethazine	KBr	2.0 to 15.0
Sulfamethoxypyridazine, tablets	KBr	2.0 to 15.0
Testosterone cypionate	KBr	2.0 to 15.0
Testosterone propionate	KBr	2.0 to 15.0
Thiotepa, for injection	$CHCl_3$	2.0 to 15.0
Tolbutamide, tablets	Liquid pet.	2.0 to 12.0
Tripelennamine citrate	CS_2	7.0 to 15.0
Tripelennamine HCl, tablets	CS_2	7.0 to 15.0
Tubocurarine Cl	KBr	2.0 to 15.0
N.F.		
Azacyclonol HCl	KBr	2.0 to 15.0
Betamethasone	KBr	2.0 to 15.0
Brompheniramine maleate, elixir, tablets	CS_2	7.0 to 15.0
Carbetapentane citrate	KBr	2.0 to 15.0
Chlormerodrin	Liquid pet.	2.0 to 15.0
Chlorothiazide	KBr	2.0 to 15.0
Cortisone acetate, ophthalmic suspension, suspension, tablets	KBr	2.0 to 15.0
Cyclophosphamide, tablets	KBr	2.0 to 15.0
Desoxycorticosterone acetate pellets	KBr	2.0 to 15.0
Desoxycorticosterone pivalate	KBr	2.0 to 15.0
Dexamethasone	KBr	2.0 to 15.0
Dexchlorpheniramine maleate, tablets	CS_2	7.0 to 15.0
Doxylamine succinate, tablets	CS_2	7.0 to 15.0
Estradiol benzoate	KBr	2.0 to 15.0
Estradiol cypionate	Liquid pet.	2.0 to 12.0
Estradiol dipropionate	KBr	2.0 to 15.0
Estrone	KBr	2.0 to 15.0
Gallamine triethiodide	KBr	2.0 to 15.0
Glutethimide tablets	KBr	2.0 to 15.0
Mepivacaine HCl	KBr	2.0 to 15.0
Meprobamate, tablets	KBr	2.0 to 15.0
Methscopolamine Br	KBr	2.0 to 15.0
Methsuximide, capsules	KBr	2.0 to 15.0
Methylphenidate HCl	Liquid pet.	2.0 to 15.0
Methylprednisolone, tablets	KBr	2.0 to 15.0
Methylprednisolone acetate, suspension	KBr	2.0 to 15.0

TABLE 50. OFFICIAL IDENTIFICATION TESTS USING INFRARED ABSORPTION METHODS* (*Continued*)

Substance	Solvent or dispersion	Wavelength, μ
N.F.		
Methylprednisolone Na succinate, for injection	KBr	2.0 to 15.0
Methyltestosterone, tablets	KBr	2.0 to 15.0
Pentylenetetrazol	KBr	2.0 to 15.0
Phenindamine tartrate, syrup, tablets	CS_2	7.0 to 15.0
Phensuximide, capsules, oral suspension	KBr	2.0 to 15.0
Pipradol HCl	KBr	2.0 to 15.0
Povidone-Iodine, solution	KBr	6.0, 7.0, 7.7
Progesterone, suspension, tablets	KBr	2.0 to 15.0
Pyrilamine maleate, tablets	CS_2	7.0 to 15.0
Sitosterols, suspension	CS_2	3.0 to 12.0
Sodium methohexital for injection	$CHCl_3$	2.0 to 15.0
Syrosingopine	Liquid pet.	2.0 to 15.0
Testosterone, pellets, suspension	KBr	2.0 to 15.0
Testosterone enanthate	KBr	2.0 to 15.0
Triethylenemelamine	KBr	3.3 to 13.28
Urethan	$CHCl_3$	2.0 to 15.0

* Unless otherwise designated, the sample preparations exhibit maximums only at the same wavelength as that of a similar preparation of a Reference Standard.

Procedure. "Dissolve 250 mg. in 3 ml. of sodium hydroxide, T.S., transfer to a 500 ml. volumetric flask, dilute to volume, and mix. Transfer a 5 ml. aliquot to a 250 ml. volumetric flask containing 12.5 ml. of pH 7 phosphate buffer, dilute to volume with water and mix. This solution, when compared in a suitable spectrophotometer against a blank of the same buffer in the same concentration, exhibits absorbance maxima at 265 ± 2 mμ and 299 ± 2 mμ, and the ratio A_{265}/A_{299} is between 1.50 and 1.56."

Exercise 107

Object. Determine Identification Test *A* for Acetazolamide.

Materials Required. U.S.P. Acetazolamide
U.S.P. Reference Standard Acetazolamide
Spectral grade potassium bromide

Procedure. Prepare two potassium bromide pellets of U.S.P. and U.S.P. Reference Standard Acetazolamide following the directions given on page 349. Obtain spectrum of each, and compare. If differences occur, repeat the procedure after first correcting for polymorphism (p. 349).

QUANTITATIVE ANALYSIS IN THE INFRARED REGION

The fundamental laws and principles of spectrometry which have been previously discussed with regard to quantitative methods in the visible and ultraviolet regions apply also to quantitative methods in the infrared regions. As in the ultraviolet and visible regions, the absorbance value of the solute in a given concentration of solution is the experimental value desired. In some infrared analyses, the absorbance value of the solute is most simply obtained by subtracting the absorbance value of the blank (organic solvent) from the absorbance value of the solution (solute plus solvent). In other infrared analyses, the "base-line" technique is used to

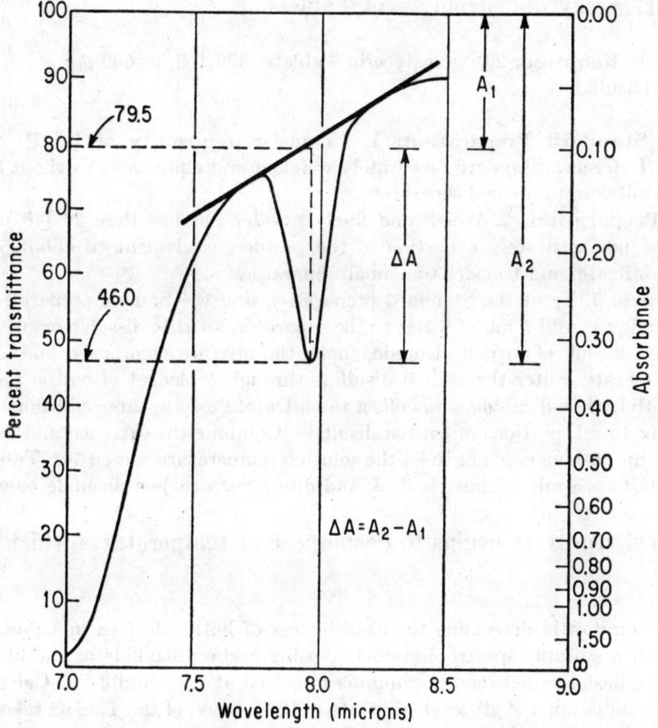

FIG. 33. Base-line technique for absorbance determination of nitroglycerin.

obtain the absorbance value of the solute. This technique will be illustrated by discussing the assay of Nitroglycerin Tablets, U.S.P. In this assay spectra of accurately weighed samples of powdered tablets and U.S.P. Reference Standard of nitroglycerin are obtained by scanning the region from 7.0 to 9.0 μ using an automatic recording infrared spectrophotometer. Each spectrum is then treated in the following manner: First a line is drawn tangent to the minimums at 7.5 and 8.3 μ. Then a line is drawn between the apex of the maximum and a point on the X axis corresponding to the wavelength value of the maximum. The transmittance values at the apex and at the intersection of the two lines are read and converted to absorbance values. The absorbances, as measured from the apex of the maximum to the intersection of the two lines, are then determined on each spectrum (Fig. 33). From these data the potency of nitroglycerin tablets is calculated in a manner analogous to that used for the assay of sulfisoxazole tablets.

Exercise 108

Object. Assay of Nitroglycerin Tablets.

Materials Required. 20 Nitroglycerin Tablets (300,400, or 600 μg)
Carbon Disulfide

Assay. Standard Preparation. 1. "Transfer a quantity of U.S.P. Glyceryl Trinitrate Reference Standard, accurately weighed and equivalent to about 5 mg. of glyceryl trinitrate, to a small separator.

Assay Preparation. 2. Weigh and finely powder not less than 20 Nitroglycerin Tablets. Weigh accurately a portion of the powder, equivalent to about 5 mg. of glyceryl trinitrate, and transfer to a small separator.

Procedure. 3. Treat the Standard preparation and the Sample preparation separately as follows: Add 5 ml. of water to the separator, swirl to dissolve or suspend the sample, add 20 ml. of carbon disulfide, shake the mixture vigorously, and allow the layer to separate. Filter the carbon disulfide through a pledget of cotton previously washed with carbon disulfide, and collect the filtrate. Perform three additional extractions, using 10 ml. portions of carbon disulfide. Combine the extracts, and evaporate to about 3 ml., taking care not to let the solution temperature exceed 50°. Transfer the concentrate to a 5 ml. volumetric flask and dilute with carbon disulfide to volume."

Glyceryl trinitrate begins to decompose at temperatures which exceed 50°.

4. "Concomitantly determine the absorbances of both solutions in 1-mm. cells at 7.89 μ, with a suitable spectrophotometer, using carbon disulfide as the blank, and drawing the base line between the minima observed at 7.5 μ and 8.3 μ. Calculate the quantity in milligrams of glyceryl trinitrate in the portion of the Tablets taken by the formula $W(A_u/A_s)$, in which W is the weight, in mg., of glyceryl trinitrate represented in the quantity of U.S.P. Glyceryl Trinitrate Reference Standard taken, and A_u and A_s are the absorbances of the solutions from the *Assay preparation* and *Standard preparation*, respectively."

Absorption of infrared radiation by nitroglycerin is due to the presence of nitrogen-oxygen bonds ($O—NO_2$).

To calculate the absorbance A_u using the base-line technique, it is first necessary to convert the transmittance values 79.5 and 46.0 percent (Fig. 33) to corresponding absorbance values. Recall that

$$A = \log \frac{1}{T} = \log \frac{100}{\%T} \qquad A_1 = \log \frac{100}{79.5} = \log 100 - \log 79.5$$

$$= 2.000 - 1.9004 = 0.0996$$

Similarly

$$A_2 = \log \frac{100}{46.0} \qquad \log 100 - \log 46.0 = 2.000 - 1.6628 = 0.3372$$

$$A_u = \Delta A = A_2 - A_1 \qquad A_u = 0.3372 - 0.0996 = 0.2376 \text{ or } 0.238$$

(The Greek letter delta Δ signifies a difference between two values.)

Calculate A_s of Reference Standard in a like manner.

Questions and Problems

1. Why must the reagents like the isopropyl alcohol and the ether used in the vitamin A assay be purified?
2. Why are the oils containing vitamin A saponified before determining absorbance with the spectrophotometer?
3. Matched quartz glass absorption cells are specified in the U.S.P. assay for vitamin A. Why?
4. Could a filter photometer or an infrared spectrophotometer be used in the vitamin A assay? Why?
5. Look up the reference for the general formula for calculation of irrelevant absorbancies, and explain the derivation of the correction equation.

FLUOROMETRY

A number of materials can be determined with greater sensitivity and specificity by fluorometric methods than by other optical methods such as spectrophotometry. However, the method is limited to the analysis of those substances or their derivatives which fluoresce. Fluorometry consists of the measurement of the intensity of fluorescence of the solute when fluorescence is excited by radiant energy, usually in the ultraviolet region. Fluorometric measurements may be made by using a spectrophotometer, such as the Beckman DU-2, fitted with appropriate attachments, or by using commercially available fluorometers.

The general procedure consists first of illuminating a sample with monochromatic ultraviolet radiant energy and measuring photoelectrically, at right angles to the incident radiation, the fluorescent light produced in the sample. An official reference standard is then illuminated in the same manner, and the fluorescent light produced is measured. Quantities of reference standard and sample taken for fluorometric analysis are such that the transmittance readings on the fluorometers are directly proportional to the concentrations taken. As in other spectrophotometric methods of analysis, corrections must be taken into consideration for the solvent and other interfering substances.

The calculation for the assay of thiamine will be discussed briefly. The U.S.P. states that the quantity in micrograms of thiamine in each 5 ml of the assay preparation taken is given by the formula: $(A - b)/(S - d)$, in which A and S are the average fluorometer readings of the assay preparation and standard preparation treated with oxidizing reagent, respectively, and b and d are the readings for the blanks. The concentration of the reference standard which the U.S.P. states must be taken is 0.2 μg per ml. This is also equal to 1.0 μg per 5 ml. Hence, dividing the corrected fluorescence value of the assay preparation automatically gives the concentration of thiamine in micrograms in 5 ml of solution.

The fluorometric method for the assay of riboflavin consists briefly of extraction of the riboflavin from the sample with sulfuric acid (0.1 N) by autoclaving at 121 to 123° for 30 min, removing interfering substances by precipitation through pH adjustment, followed by oxidation with potassium permanganate, and finally measurement of the fluorescence of the riboflavin in a suitable fluorometer as compared with the fluorescence of U.S.P. riboflavin reference standard, taking into consideration the fluorescence of a blank and other interfering substances.

The method has found most application in the analysis of the vitamins and in particular in the analysis of thiamine and riboflavin, which methods are official.

Questions

1. To what types of materials is fluorometric analysis limited?
2. Explain the phenomenon of fluorescence.
3. How do fluorophotometers differ from spectrophotometers?
4. List the factors on which the intensity of fluorescence depends.

TURBIDIMETRY AND NEPHELOMETRY

Turbidimetry and nephelometry may be considered as branches of spectrometry in which transmitted or reflected light, respectively, is measured after radiant energy passes through a turbid solution or suspension. Since light transmittance is used as a measure of turbidity, instruments such as colorimeters and spectrophotometers may be employed in turbidimetric measurements. Nephelometry is based on the measurement of the brightness of the light reflected by a cloud of finely divided particles suspended in a liquid. Again colorimeters may be used. However, attachments must be provided in order that the measurement of the reflected light in the medium can be made at right angles to the direction of illumination.

Turbidimetric methods are used in official assays of the majority of antibiotics, calcium pantothenate, vitamin B_{12}, and other medicinal agents. In these methods the activity of the biological agent is determined by measuring the turbidity produced in a series of antibiotic dilutions containing a microbiological culture. The greater the turbidity (due to microbial growth), the less is the activity of the antibiotic. This procedure is also used to standardize the bacterial concentration of the inoculum used in the microbial assays, the transmittance being measured with a suitable spectrophotometer at 580 mμ or with a colorimeter fitted with the proper filter. The procedures for the antibiotics listed in the Pharmacopeia are published in the Federal Register of the United States by the Food and Drug Administration. A summary of these procedures is given in the U.S.P.

The term used by the U.S.P. to designate solution turbidity and light-scattering power is *turbidance*, *S*, and the equation which can be employed to obtain the turbidance value of a given suspension is:

$$S = \log \frac{P_o}{P} = k \frac{bd^3}{d^4 + \lambda^4} C$$

where P_o = power of incident radiation
P = transmitted radiation
b = thickness of the layer of sample
C = concentration
d = average diameter of particles
λ = wavelength
k = a proportionality constant which depends on the nature of the suspension and on the method of measurement

For a given suspension measured in a given apparatus using monochromatic light, the turbidance value S is equal to the product of a constant K and concentration C, where K is a proportionality constant which combines all the other fixed values, b, k, d, and λ. Symbolically, $S = KC$. This relationship is analogous to absorbance in spectrophotometry, and the data may be treated in an identical manner. However, it is important to realize that this relationship holds only for very dilute suspensions. In more concentrated suspensions, secondary scattering brings about effects which result in deviation from linearity. It has been found, however, that transmittance values plotted against concentration give good results. In practice, therefore, transmittance values of each suspension of known concentration are usually obtained for the purpose of preparing a calibration curve. The following procedure is recommended for nephelometric measurements: (1) Attach nephelometric accessories to a suitable colorimeter. (2) Set wavelength (or use filter) as directed in U.S.P. If no specific wavelength is recommended, use a wavelength setting or filter which gives blue illumination. (3) Place the most concentrated suspension in the instrument and turn the absorbance-transmittance knob to the 100 percent T position. Then open or close slits until null position is reached. (4) Measure the T of remaining known concentrations and prepare a calibration curve by plotting T versus concentration of suspension. (5) Measure the T of the unknown and determine the concentration using the calibration curve.

Turbidimetric measurements are made in a manner analogous to colorimetric measurements. If no specific wavelength is recommended for the determination, use a wavelength setting or filter which gives blue illumination. Also recall that transmittance values plotted against concentration of suspension give greater linearity than absorbance values.

Turbidimetric tests are also applied to certain official chemicals to ensure the absence of excessive amounts of chloride and sulfate. The following directions are given in the U.S.P. for the performance of these tests:

"Perform the tests in glass cylinders of the same diameter and matched as closely as practicable in other respects. Use the same quantities of the same reagents for both the solution under test and the control solution containing the specified volume of chloride or sulfate. If after acidification the solution is not perfectly clear, filter it through a filter paper that gives negative tests for chloride and sulfate. Add the precipitant, silver nitrate T.S. or barium chloride T.S. as required, to both the test solution and the control solution in immediate sequence.

"Where the individual monograph calls for applying the test to a specific volume of a solution of the substance, and the limit for chloride or sulfate corresponds to 0.2 ml. or less of 0.02 *N* hydrochloric or sulfuric acid, respectively, apply the test to the solution without further dilution. In such cases maintain the same volume relationships for the control solution as specified for the solution under test. In applying the test to the salts of heavy metals, which normally show an acid reaction, omit the acidification and do not neutralize the solution. Dissolve bismuth salts in a few milliliters of water and 2 ml. of nitric acid before treating with the precipitant.

"**Chloride.** Dissolve the specified quantity of the substance under test in 30 to 40 ml. of water and, if necessary, neutralize the solutions to litmus with nitric acid. Add 1 ml. each of nitric acid and of silver nitrate T.S., and sufficient water to make 50 ml. Mix well, allow to stand for 5 minutes protected from direct sunlight, and compare the turbidity, if any, with that produced in a solution containing the volume of 0.02 *N* hydrochloric acid specified in the monograph.

"**Sulfate.** Dissolve the specified quantity of the substance under test in 30 to 40 ml. of water and, if necessary, neutralize the solution to litmus with hydrochloric acid. Add 1 ml. of diluted hydrochloric acid and 2 ml. of barium chloride T.S., and sufficient water to make 50 ml. Mix well, allow to stand for 10 minutes, and compare the turbidity, if any, with that produced in a solution containing the volume of 0.02 *N* sulfuric acid specified in the monograph."

Exercise 109

Object. Limit Test for Chloride and Sulfate in Calcium Gluconate.

Materials Required. 3 Gm of calcium gluconate
0.02 *N* HCl
0.02 *N* H_2SO_4
Silver nitrate T.S. (0.1 *N* silver nitrate)
Barium chloride T.S. (12 Gm of barium chloride in sufficient water to make 100 ml)
Matched glass cylinders
Spectrophotometer or colorimeter fitted with appropriate filter

Procedure. 1. Proceed as directed above, using 1 Gm of calcium gluconate and 1 ml of 0.02 *N* HCl in the test for chloride and 2 Gm of calcium gluconate and 1 ml of 0.02 *N* sulfuric acid in the test for sulfate.

2. Using the solutions prepared in part 1, determine the transmittance of each sample with a spectrophotometer or a colorimeter at about 480 mμ. Make dilutions of the solutions if necessary. From the data calculate the concentration of chloride in the sample.

Questions and Problems

1. Calculate the percents of chloride and of sulfate permitted in calcium gluconate in the official test for purity.
2. List several official substances that are required to show no turbidity when tested for chlorides and sulfates.
3. How might minute quantities of each of the following ions be determined: iodide, sulfate, phosphate, and calcium?
4. Could minute quantities of alkaloids in solution be estimated nephelometrically? Suggest possible precipitants.

References

Those who desire to make a more extended study of spectrometric methods of analysis will find valuable information in the following book references:

1. DAY, R. A., Jr., and A. L. UNDERWOOD, "Quantitative Analysis," Prentice-Hall, Inc., Englewood Cliffs, N.J., 1958.
2. EWING, G. L., "Instrumental Methods of Chemical Analysis," McGraw-Hill Book Company, New York, 1960.
3. MEITES, L., and H. C. THOMAS, "Advanced Analytical Chemistry," McGraw-Hill Book Company, New York, 1958.
4. Mellon, M. G., "Analytical Absorption Spectroscopy," John Wiley & Sons, Inc., New York, 1950.
5. Rao, C. N. R., "Chemical Applications of Infrared Spectroscopy," Academic Press Inc., New York, 1963.

Chapter 21

POTENTIOMETRY

Introduction. Potentiometry may be defined as a branch of electrochemistry which deals with the study and measurement of electrode potentials. Measurements of electrode potentials find wide application in pharmaceutical sciences, especially in pH determinations and potentiometric titrations.

Before we consider the applications of potentiometry in pharmaceutical sciences it is necessary to have an understanding of electrode potentials. This may best be accomplished by determining the relative tendencies which different metals, ions, or molecules have to gain or lose electrons under controlled conditions. It is convenient to arrange these chemical species in the order of their affinities for electrons. The partial list shown in Table 51 contains such an arrangement and is referred to as a table of

TABLE 51. STANDARD ELECTRODE POTENTIALS AT 25°

Half-cell reaction	$E°$, volts
$Na^+ + e = Na°$	−2.710
$Zn^{++} + 2e = Zn°$	−0.763
$Fe^{++} + 2e = Fe°$	−0.440
$Cd^{++} + 2e = Cd°$	−0.403
$2H^+ + 2e = H_2$	0.000
$S_4O_6^= + 2e = 2S_2O_3^=$	+0.080
$Cu^{++} + 2e = Cu°$	+0.340
$Hg_2Cl_2 + 2e = 2Hg° + Cl^-$ (saturated)	+0.246
$I_3^- + 2e = 3I^-$	+0.536
$I_2° + 2e = 2I^-$	+0.536
$C_6H_4O_2 + 2H^+ + 2e = C_6H_4(OH)_2$	+0.699
$Fe^{3+} + e = Fe^{++}$	+0.771
$Br_2 + 2e = 2Br^-$	+1.065
$OP' + e = OP$ (*o*-phenanthroline)	+1.110
$Cr_2O_7^= + 14H^+ + 6e = 2Cr^{3+} + 7H_2O$	+1.330
$MnO_4^- + 8H^+ + 5e = Mn^{++} + 4H_2O$	+1.500
$Ce^{4+} + e = Ce^{3+}$	+1.610

"standard electrode potentials." The contribution of the numerical values found in this table to the understanding of the principles underlying potentiometry becomes apparent as the discussion unfolds.

When a bar of zinc metal is immersed in a solution of cupric sulfate, a shiny deposit of metallic copper is seen to coat the bar of zinc almost instantaneously. In order that copper deposit on the zinc, it is necessary for the cupric ions in solution to pick up two electrons. These electrons come from the zinc metal as it goes into solution. The zinc ions so produced and the cupric ions already in the solution then compete for the electrons. Because the cupric ions have a greater affinity for the electrons, metallic copper forms and deposits on the zinc bar. The chemical changes which take place can be represented by two chemical equations. First, when Zn metal is placed in solution, some of it is dissolved:

$$Zn^{\circ} \rightleftharpoons Zn^{++} + 2e$$

Only a very small amount of zinc goes into solution, because the zinc ions (which are positive) have a strong attraction for the electrons (which are negative) formed and hence return to metallic zinc. If an ionic species is present which has a greater affinity for electrons than the zinc ions, then these ions will react with the electrons and form the reduced species. In

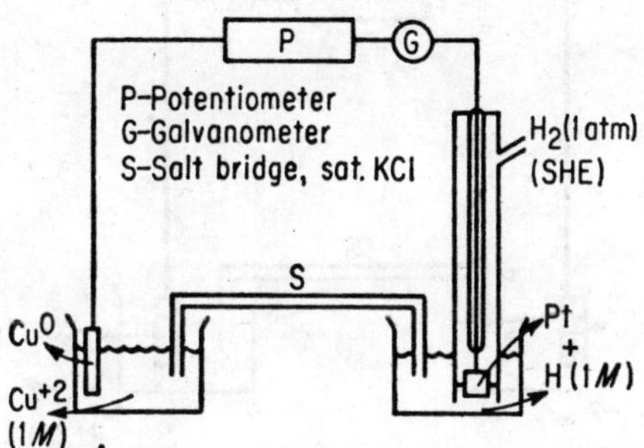

FIG. 34. Schematic of apparatus used to measure standard electrode potentials.

the example cited, cupric ions react with electrons to form metallic copper, as shown in the following equation:

$$Cu^{++} + 2e \longrightarrow Cu^{\circ}$$

The relative tendency of metals, ions, or molecules to gain or lose electrons can be studied quantitatively; one approach is to determine the so-called "standard electrode potentials" (abbreviated E°) of these species. This may be accomplished by (1) choosing an electrode which is used as a standard with which to compare all other electrode measurements, (2) choosing a suitable single concentration for all solutions measured, and (3) choosing a suitable temperature. By international agreement the hydrogen electrode has been arbitrarily chosen as the standard and assigned a potential value of zero. The concentration (more strictly, the activity) taken is 1 M, and the temperature is 25°. To measure the potential of the copper electrode immersed in a 1 M solution of its ions relative to the standard hydrogen electrode (abbreviated SHE), a schematic of the apparatus as shown in Fig. 34 is used. The potentiometer is adjusted until no current flows through the galvanometer. The voltage (0.34) is then measured and assigned to the copper, cupric ion electrode (copper electrode, for brevity), since by agreement the potential of the hydrogen electrode is taken as zero. Also by convention it has been agreed to assign a positive value to those electrode systems in which electrons flow from the hydrogen electrode to the test electrode, and negative values if electrons flow in the opposite direction. When the electrodes in this example are short-circuited, electrons flow from the hydrogen electrode to the copper electrode. Therefore, the copper electrode is assigned a positive $E°$ of $+0.34$ volt.

In a like manner we can determine the $E°$ for other systems, such as zinc in contact with a 1 M concentration of its ions. The potential reading obtained for this electrode system versus the SHE is 0.76 volt. When the

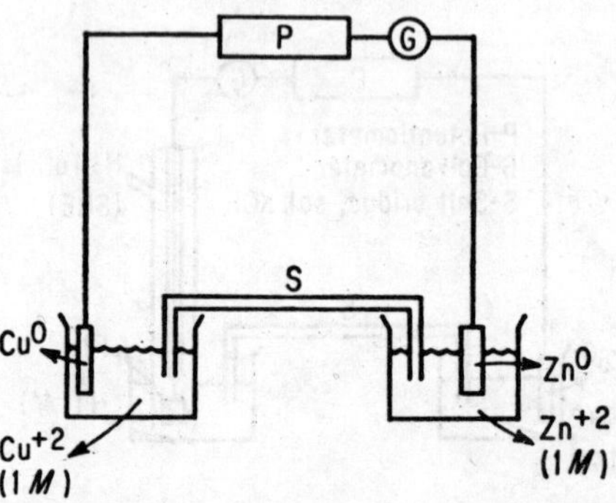

FIG. 35. Schematic of apparatus used to measure cell potentials.

cell is short-circuited, electrons flow from the zinc electrode to the SHE. The zinc electrode is therefore assigned a negative $E°$ of -0.76 volt. If we now connect the zinc half-cell to a copper half-cell (Fig. 35) in which the molar concentrations of each species are 1 M and adjust the potentiometer until no current flows through the galvanometer, we obtain an emf reading of 1.10 volts. Actually it is not necessary to make experimental measurements every time emf data are wanted. In the example just cited, the value (1.10 volts) of the cell could have been calculated by simply subtracting the potential of the zinc half-cell from the potential of the copper half-cell: $E_{cell} = E_1 - E_2 = 0.34 - (-0.76) = 1.10$ volts.

Oxidized and reduced species in different oxidation states such as iron (Fe^{3+}, Fe^{++}) or cerium (Ce^{4+}, Ce^{3+}) when placed in contact with a platinum electrode also develop a potential which can be compared to the SHE. If the concentrations of the species are kept at 1 M when the measurements are made, the potential of the electrode system (Fe^{3+}, Fe^{++}, Pt) developed may also be included in the standard electrode potential table. For the electrode system under consideration, the $E°$ value is equal to

-0.77 volt. It is interesting to note that strong oxidizing agents have high positive potentials and appear at the bottom of the table and that strong reducing agents have high negative potentials and appear at the top of the table.

Since it is not often convenient to draw a schematic of the potential measuring assembly as shown in Fig. 34, the information is represented as follows:

$$\text{Pt} \left| \begin{matrix} \text{H}^+ (1\ M) \\ \text{H}_2 (1 \text{ atm}) \end{matrix} \right\| (1\ M)\ \text{Cu}^{++} \Big| \text{Cu}$$

where the single vertical line represents an interface, i.e., a solid-solution interface, and the double or parallel lines signify connection of the two half-cells with a salt bridge whose liquid junction potential is negligible. When the concentrations of the oxidized and reduced species of a half-cell are not unity (equal to 1), the potential of the half-cell may be calculated using the Nernst equation:

$$E = E^\circ + \frac{0.0591}{n} \log \frac{[\text{ox}]}{[\text{red}]} \text{ at } 25^\circ$$

where E = calculated electrode potential of the half-cell
E° = standard electrode potential
[ox] = molar concentration of the oxidized species
[red] = molar concentration of the reduced species
n = the difference in the number of electrons found in the oxidized and reduced states

When calculating emf values from half-cell data it is necessary to follow carefully a set of conventions or rules in order to assure accurate results. The following conventions are presented for this purpose.

1. When representing the cell, the electrode which has the more positive E° as listed in Table 51 is written on the left.
2. To determine the emf of the cell, the potential of the right electrode is always subtracted from the left electrode: $E_c = E_l - E_r$.
3. To determine the chemical reaction which takes place spontaneously in the half-cells, the chemical species in the left half-cell appears in its higher valence state on the left side of the chemical equation. The other chemical species are written accordingly. For example, the chemical reaction for the copper-zinc cell would be written as:

$$\text{Cu}^{++} + \text{Zn}^\circ \longrightarrow \text{Zn}^{++} + \text{Cu}^\circ$$

4. A positive emf value of the cell means that electrons flow spontaneously from right to left when the cell is short-circuited and the chemical reaction proceeds as written in convention 3. If the value is negative, electrons flow from left to right and the chemical reaction proceeds in the reverse direction.

5. Solids and pure liquids have values of 1 (unity).

The following example illustrates the use of the conventions listed above:

Example. Calculate the emf developed by a cell consisting of a copper bar immersed in a 1 *M* solution of its ions, and a cadmium bar immersed in a 1 *M* solution of its ions.

Solution. By convention 1, the copper electrode is placed in the left-hand compartment of the cell, since it has a higher positive $E°$ value than Cd.

$$Cu|Cu^{++}\ (1\ M)\|(1\ M)\ Cd^{++}|Cd$$

By convention 2, the emf of the cell is $E_l - E_r$.

$$0.34 - (-0.403) = +0.743$$

By convention 3, the chemical reaction is written so that Cu^{++} ions appear on the left side of the equation. The other species appear accordingly:

$$Cu^{++} + Cd° \longrightarrow Cu° + Cd^{++}$$

By convention 4, the positive emf of the cell (+0.743) means that the chemical reaction proceeds spontaneously as written in convention 3. This information is important when predicting whether a given chemical reaction will proceed spontaneously or not.

In addition to calculating the potential value of cells, emf data may also be used to calculate equilibrium constants for chemical reactions. The student interested in this type of calculation is referred for details to one of the many texts on quantitative analysis or physical chemistry listed in the front of the text.

Oxidation-Reduction Titration Curve. Titration curves may be prepared from emf data obtained experimentally, or the data for the graph may be calculated. For purposes of considering principles underlying titrimetric analysis based on redox reactions, data obtained from calculations are satisfactory. Consider the titration of 100 ml of 0.1 *M* ferrous sulfate solution in 1 *N* sulfuric acid using 0.1 *M* ceric sulfate as the titrant. Calculate the emf of a cell consisting of a platinum electrode, connected to a SHE by a salt bridge after the addition of 25 ml, 100 ml, and 130 ml of 0.1 *M* ceric sulfate. It is important to realize that during the titration, the chemical species Fe^{++}, Ce^{3+}, Fe^{3+}, and Ce^{4+} are in a state of equilibrium in the titration beaker at all times.

$$Fe^{++} + Ce^{4+} \rightleftharpoons Fe^{3+} + Ce^{3+} \qquad K = 10^{14} \text{ at } 25°$$

Since these species are in the same container, $E_{Fe} = E_{Ce}$ at all times during the titration and either Eq. (1) or (3) may be used to calculate the

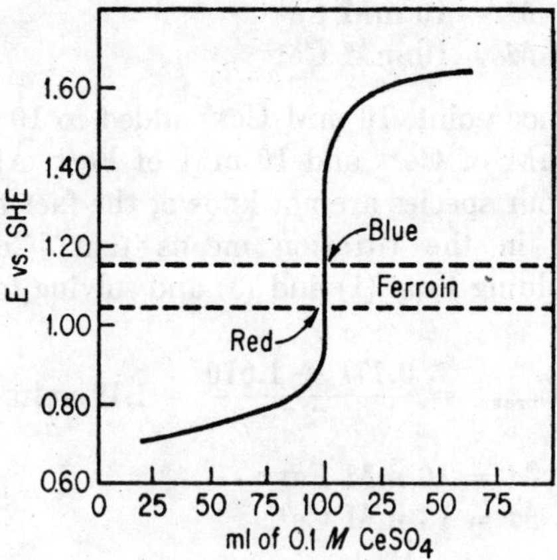

FIG. 36. Redox titration curve.

potential of the half-cell E_c (beaker). It is important to recognize also at this time that $E_c = E_{Fe} - E_H$ or $E_c = E_{Ce} - E_H$. Since $E_H = 0$, $E_c = E_{Fe} = E_{Ce} = E_{beaker}$.

Only three types of points need to be considered for the preparation of the redox titration curve (Fig. 36), and only three different equations are needed for the calculations:

1. Before the equivalence point is reached:

$$E_{Fe} = E^{\circ}_{Fe} + \frac{0.0591}{n} \log \frac{[Fe^{3+}]}{[Fe^{++}]} \quad (1)$$

2. At the equivalence point:

$$E_{beaker} = \frac{E^{\circ}_{Fe} + E^{\circ}_{Ce}}{2} \quad (2)$$

3. Beyond the equivalence point:

$$E_{Ce} = E^{\circ}_{Ce} + \frac{0.0591}{n} \log \frac{[Ce^{4+}]}{[Ce^{3+}]} \quad (3)$$

A sample calculation for each equation follows:

1. 100 ml $\times$ 0.1 M = 10 mM Fe^{++}
   25 ml $\times$ 0.1 M = 2.5 mM Ce^{4+}

When 2.5 mM of Ce^{4+} is added to 10 mM of Fe^{++}, nearly 2.5 mM of Fe^{3+} and 2.5 mM of Ce^{3+} are produced. The solution contains, to a very close approximation, 2.5 mM of Fe^{3+} and 2.5 mM of Ce^{3+}, 7.5 mM of Fe^{++}, and a trace of Ce^{4+}. At this point in the titration it is most convenient to use the Fe^{3+} and Fe^{++} ions to calculate the emf of the cell, since these concentrations are more accurately known.

$$E_{beaker} = E_{Fe} = 0.771 + \frac{0.0591}{1} \log \frac{2.5/125}{7.5/125} = 0.743 \text{ volt}$$

2. 100 ml × 0.1 M = 10 mM Fe^{++}
   100 ml × 0.1 M = 10 mM Ce^{4+}

At the equivalence point, 10 mM Ce^{4+} added to 10 mM of Fe^{++} produces almost 10 mM of Ce^{3+} and 10 mM of Fe^{3+}. Although the exact quantities of the four species are not known, the fact that we are at the equivalence point in the titration means that $[Ce^{4+}] = [Fe^{++}]$ and $[Ce^{3+}] = [Fe^{3+}]$. Adding Eqs. (1) and (3) and solving for E_{Fe} or E_{Ce} gives Eq. (2):

$$E_{beaker} = \frac{0.771 + 1.610}{2} = 1.19 \text{ volts}$$

3. 100 ml × 0.1 M = 10 mM Fe^{++}
   130 ml × 0.1 M = 13 mM Ce^{4+}

When 13 mM of Ce^{4+} is added to 10 mM of Fe^{++}, 10 mM of Fe^{3+} and 10 mM of Ce^{3+} are produced. The solution contains, to a very close approximation, 3.0 mM of Ce^{4+}, 10 mM of Ce^{3+}, 10 mM of Fe^{3+}, and a trace of Fe^{++}. At this point in the titration it is convenient to use Ce^{4+} and Ce^{3+} to calculate the emf, since their concentrations are more accurately known.

$$E_{beaker} = 1.61 + \frac{0.0591}{1} \log \frac{3/230}{10/230} = 1.58 \text{ volts}$$

Redox Indicators. Certain chemicals have the property of changing color with a change of potential. Since the color change comes about as a result of a redox phenomenon, $[ox] + e = [red]$, the concentrations of oxidized and reduced forms can be expressed in the Nernst equation.

$$E = E^\circ_{Ind} + \frac{0.0591}{n} \log \frac{[ox]}{[red]}$$

It has been shown that for most redox indicators perceptible color changes occur when the ratio of oxidized and reduced species is 10:1 or 1:10. This is the case with indicators used in neutralization reactions also. For redox indicators a range of about 0.06 volt on each side of the E° of the indicator usually brings about a suitable color change.

The indicator ferroin (1,10-phenanthroline ferrous sulfate) has an E° value of 1.11 volts, with the reduced form red and the oxidized form blue. The indicator range is about 1.11 ± 0.06 volt. Note that on the titration curve the equivalence point of the titration passes through this range using a negligible amount of titrant.

Potentiometric Considerations of pH. The hydronium ion concentration of an aqueous solution may be determined by measuring the voltage developed between two electrodes of special character immersed in a solution using a potentiometer or a modified vacuum tube voltmeter called a pH meter. The electrodes employed in the determination of

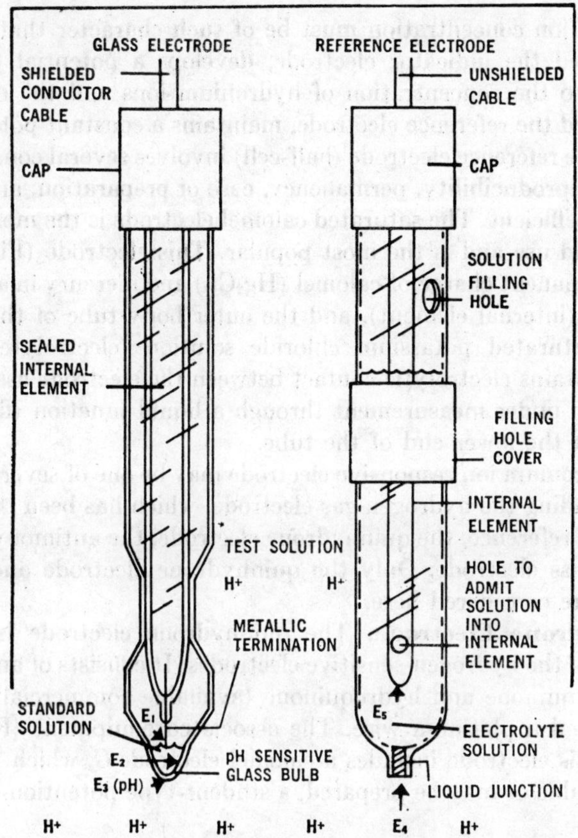

E_1 – potential between internal element and standard solution.

E_2 – potential between standard solution and inner surface of glass membrane.

E_3 – potential between outer surface of pH-sensitive glass bulb and test solution; this is the only potential in electrode system that varies in accordance with change in hydrogen-ion concentration, or pH, of test solution.

E_4 – potential of liquid junction between electrolyte and test solutions.

E_5 – potential developed at internal element in electrolyte solution.

FIG. 37. Glass and reference electrodes. (*Courtesy of Beckman Instruments, Inc.*)

hydronium ion concentration must be of such character that one, commonly called the indicator electrode, develops a potential that varies according to the concentration of hydronium ions and the other, commonly called the reference electrode, maintains a constant potential. The choice of the reference electrode (half-cell) involves several considerations, including reproducibility, permanency, ease of preparation, and the temperature coefficient. The saturated calomel electrode is the most practical for all-round use and is the most popular. This electrode (Fig. 37) contains a permanent charge of calomel (Hg_2Cl_2) and mercury in a protective inner tube (internal element), and the outer body tube of the electrode contains saturated potassium chloride solution (electrolyte solution), which maintains electrolytic contact between the electrode assembly and the solution under measurement through a liquid junction (fiber) which is placed at the lower end of the tube.

The hydronium ion responsive electrode may be one of several different kinds, including the hydrogen gas electrode, which has been the ultimate standard of reference, the quinhydrone electrode, the antimony electrode, and the glass electrode. Only the quinhydrone electrode and the glass electrode are considered here.

Quinhydrone Electrode. The quinhydrone electrode is the least expensive of the hydrogen-sensitive electrodes. It consists of an equimolar mixture of quinone and hydroquinone (available commercially as quinhydrone) and a platinum wire. The associated equipment (Fig. 38) for use with this electrode includes a calomel electrode *G*, which is commercially available or may be prepared, a student-type potentiometer, which

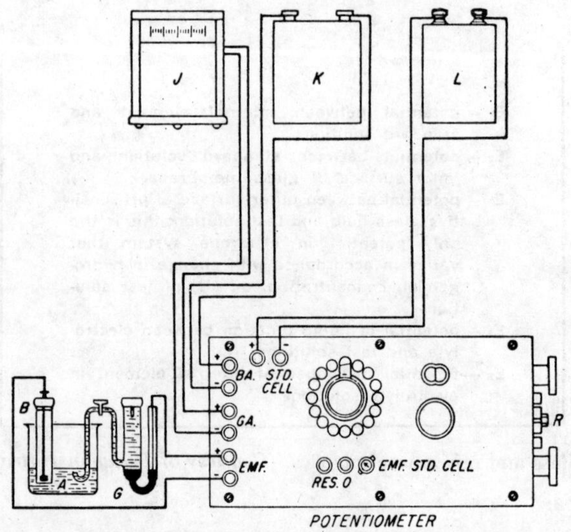

FIG. 38. Scheme of assembly for hydrogen ion measurements.

may be purchased or assembled using appropriate electrical components, two 1.5-volt batteries K, a standard cell L, and a galvanometer J. To make pH measurements, a small visible excess of quinhydrone is placed directly in the test solution the pH of which is to be determined. A platinum wire or platinum electrode B and the calomel electrode are immersed in the test solution A, and the emf developed by the cell is measured with a standardized potentiometer. The quinhydrone electrode cannot be used for pH measurements above 8 to 9. Neither can it be used in the presence of strong oxidizing or reducing agents.

When quinhydrone is dissolved in an aqueous solution, it dissociates, forming quinone and hydroquinone. The equilibrium established between the dissociation products may be indicated by the following equation.

$$Q + 2H^+ + 2e \rightleftharpoons H_2Q$$

A schematic representation of the quinhydrone and calomel electrode cell is:

$$\text{Pt}|\text{Q,H}_2\text{Q solution of unknown pH}|| \underset{\text{(saturated solution)}}{\text{KCl}} \quad \text{Hg}_2\text{Cl}_2|\text{Hg}$$

The potential developed by the platinum electrode at 25° immersed in the test solution containing the quinhydrone is given by:

$$E_Q = E_Q{}^\circ + \frac{0.0591}{2} \log \frac{[Q][H^+]^2}{[H_2Q]} \tag{4}$$

Taking into consideration the fact that (1) the activity values of Q and H_2Q are unity at all acidic concentrations, and (2) the standard electrode potential of the quinhydrone electrode is +0.699 volt, Eq. (4) may be written as:

$$E_Q = 0.699 + 0.0591 \log [H^+]$$

or changing the signs of both sides of the equation and substituting pH as indicated:

$$-E_Q = -0.699 + 0.0591 \text{ pH}$$

Recall that $E_c = E_l - E_r$; $E_l = E_Q$; and $E_r = E_{Hg}{}^\circ = +0.246$.[1] Substituting and rearranging gives:

$$-E_c = -0.699 + 0.0591 \text{ pH} + 0.246 = -0.453 + 0.0591 \text{ pH}$$

$$\text{pH} = \frac{0.453 - E_c}{0.0591}$$

The quinhydrone electrode is very convenient and easy to use. It gives very accurate results in acid medium but, as previously stated, cannot be used to measure pH values above 8 to 9.

[1] E_c is read on the potentiometer.

Glass Electrode. The glass electrode is the most popular of all electrodes for pH determinations. Because of the high resistance of the glass, pH measurements must be made using a modified vacuum tube voltmeter (VTVM) called a pH meter.

A glass electrode (Fig. 37) consists of an outer glass tube containing about 0.1 *N* HCl (not necessary to be exact) which makes contact with a silver, silver chloride electrode (sealed internal element). The half-cell may be shown schematically as follows and is designated as E'_g for convenience.

Ag, AgCl|HCl (0.1 *N*)|glass membrane|H^+ (test solution)

Theoretically, this half-cell has a standard electrode potential $E_g°$, and because it is sensitive to hydronium ion, changes in the potential may be shown mathematically as:

$$E'_g = E_g° + 0.0591 \log [H^+] = E_g° - 0.0591 \text{ pH} \tag{5}$$

The entire cell assembly E_c must of course include a reference electrode such as a calomel electrode. It may be expressed as:

$$E_c = E_{Hg}° - E'_g \tag{6}$$

Substituting Eq. (5) into (6) and rearranging gives

$$E_c = E_{Hg}° - E_g° + 0.0591 \text{ pH} \tag{7}$$

For any given measurement, $E_{Hg}° - E_g°$ remains constant. Therefore, these two values may be combined and designated E_g, the glass electrode assembly. Substituting into Eq. (7) gives:

$$E_c = E_g + 0.0591 \text{ pH}$$

Rearranging:

$$\text{pH} = \frac{E_c - E_g}{0.0591} \tag{8}$$

To use Eq. (8) in combination with a glass electrode, it is first necessary to standardize or determine E_g experimentally, since this value depends on the reference electrodes (internal and external), the asymmetry potential which develops because of unequal strain in the glass, and junction potentials, all of which were included in E_g. This can be done by using a buffer solution of known pH_s and determining the potential of the cell E_s.

$$\text{pH}_s = \frac{E_s - E_g}{0.0591} \tag{9}$$

Then substitute these values into Eq. (9) and solve for E_g. When E_g is known, the *E* of the unknown solution is measured, both values are substituted in Eq. (8), and the pH is calculated.

TABLE 52. OFFICIAL SUBSTANCES ASSAYED BY THE POTENTIOMETRIC METHOD*

Substance	Titration medium	Titrant	Indicator
U.S.P.			
Caffeine	Glacial acetic acid	0.05 *N* $HClO_4$	Glass
Chlorcyclizine HCl	Glacial acetic acid	0.1 *N* $HClO_4$	Glass
Chloroprocaine HCl	Glacial acetic acid	0.1 *N* $HClO_4$	Glass
Chlorpromazine HCl	Glacial acetic acid	0.1 *N* $HClO_4$	Glass
Codeine PO_4	Glacial acetic acid	0.1 *N* $HClO_4$	Glass
Cyclizine HCl	Glacial acetic acid	0.1 *N* $HClO_4$	Glass
Hexachlorophene	Alcohol	0.1 *N* NaOH	Glass
Homatropine HBr ophthalmic solution	$CHCl_3$	0.02 *N* $HClO_4$	Glass
Iodochlorhydroxyquin	Acetic acid	0.1 *N* $AgNO_3$	Ag
Meglumine iothalamate injection	H_2SO_4	0.05 *N* $AgNO_3$	Ag
Menadiol sodium diphosphate and injection	Acetic acid-HCl	0.02 *N* $Ce(SO_4)_2$	Pt
Phenobarbital	Alcohol	0.1 *N* NaOH	Glass
Physostigmine	Acetic acid	0.02 *N* $HClO_4$	Glass
Physostigmine salicylate	$CHCl_3$-acetic acid	0.02 *N* $HClO_4$ in dioxane	Glass
Physostigmine sulfate	$CHCl_3$-acetic acid	0.02 *N* $HClO_4$	Glass
Pilocarpine nitrate	Glacial acetic acid	0.1 *N* $HClO_4$	Glass
Prochlorperazine maleate	$CHCl_3$-acetic acid	0.05 *N* $HClO_4$	Glass
Sodium carboxymethylcellulose	Glacial acetic acid	0.1 *N* $HClO_4$	Glass
Sodium citrate	Glacial acetic acid	0.1 *N* $HClO_4$	Glass
Anticoagulant sodium citrate solution	Glacial acetic acid	0.1 *N* $HClO_4$	Glass
N.F.			
Acetyl sulfisoxazole and oral suspension	H_2O	0.1 *M* $NaNO_2$	Pt
Aminobenzoic acid and tablets	H_2O	0.1 *M* $NaNO_2$	Pt
Anileridine ICl	Acetic acid	0.1 *N* $HClO_4$	Glass
Antazoline PO_4	Acetic acid	0.1 *N* $HClO_4$	Glass
Bithionol	Acetone	0.1 *N* NaOH	Glass
Calcium aminosalicylate and tablets	H_2O	0.1 *M* $NaNO_2$	Pt
Calcium cyclamate	H_2O	0.1 *M* $NaNO_2$	Pt
Chlorothen citrate	Glacial acetic acid	0.1 *N* $HClO_4$	Glass
Ethinamate and tablets	H_2O	0.05 *N* NaOH	Glass
Glycobiarsol and tablets	H_2O	0.1 *M* $NaNO_2$	Pt
Levallorphan tartrate	Glacial acetic acid	0.02 *N* $HClO_4$ in dioxane	Glass
Menadiol sodium diphosphate tablets	H_2O	0.01 *N* $Ce(SO_4)_2$	Pt
Methapyrilene HCl	Acetic acid	0.1 *N* $HClO_4$	Glass
Methyprylon, elixir, and tablets	H_2O	0.1 *N* $K_3Fe(CN)_6$	Pt
Phenmetrazine HCl	H_2O	0.1 *N* $AgNO_3$	Ag
Phthalylsulfacetamide and tablets	H_2O	0.1 *M* $NaNO_2$	Pt
Pipradol HCl	Glacial acetic acid	0.05 *N* $HClO_4$	Glass
Potassium PO_4	H_2O	0.1 *N* NaOH	Glass

TABLE 52. OFFICIAL SUBSTANCES ASSAYED BY THE POTENTIOMETRIC METHOD* (*Continued*)

Substance	Titration medium	Titrant	Indicator
N.F.			
Propoxycaine HCl	H_2O	0.1 *M* $NaNO_2$	Pt
Sodium carboxymethylcellulose tablets	Glacial acetic acid	0.1 *N* $HClO_4$	Glass
Sodium cyclamate	H_2O	0.1 *M* $NaNO_2$	Pt
Trisodium PO_4	H_2O	0.1 *N* NaOH	Glass
Trisodium PO_4, exsiccated	H_2O	0.1 *N* NaOH	Glass
Sodium sulfabromomethazine	H_2O	0.1 *M* $NaNO_2$	Pt
Sodium sulfamerazine injection	H_2O	0.1 *M* $NaNO_2$	Pt
Sulfacetamide and tablets	H_2O	0.1 *M* $NaNO_2$	Pt
Sulfacetamide, sulfadiazine, and sulfamerazine suspension and tablets	H_2O	0.1 *M* $NaNO_2$	Pt
Sulfadimethoxine, suspension, and tablets	H_2O	0.1 *M* $NaNO_2$	Pt
Sulfamerazine tablets	H_2O	0.1 *M* $NaNO_2$	Pt
Sulfamethizole, suspension, and tablets	H_2O	0.1 *M* $NaNO_2$	Pt
Tetrahydrozoline HCl	Glacial acetic acid	0.1 *N* $HClO_4$	Glass
Thenyldiamine HCl	Glacial acetic acid	0.1 *N* $HClO_4$	Glass
Thiabendazole	Glacial acetic acid	0.1 *N* $HClO_4$	Glass
Thonzylamine HCl	Glacial acetic acid	0.1 *N* $HClO_4$	Glass

* The calomel electrode is recommended as the reference electrode in each assay except Iodochlorhydroxyquin, where a mercuric sulfate–potassium sulfate electrode is recommended.

Another approach is to subtract Eq. (9) from Eq. (8).

$$\mathrm{pH} - \mathrm{pH}_s = \frac{E - E_g - (E_s - E_g)}{0.0591} = \frac{E_c - E_s}{0.0591}$$

$$\mathrm{pH} = \mathrm{pH}_s + \frac{E_c - E_s}{0.0591}$$

The U.S.P. and N.F. define pH according to this equation. Chapter 22, dealing with pH and hydronium ion concentration, describes in detail the recommended procedure for experimental determination of pH by the potentiometric method.

Potentiometric Titrations. The analyst for many years has depended on his ability to perceive indicator color change as the end point in titrimetric procedures. Several factors influence the accuracy of such indicator titrations, including (1) lighting conditions; (2) color or turbidity of the solution; (3) color blindness; (4) fatigue of the operator. For many assays, a suitable indicator is not available, and other end-point techniques must be used. Of the several methods which have been devised, the potentiometric method is very important and widely used.

The monographs in the U.S.P. and N.F. usually designate the electrode pair which is to be used for a particular potentiometric titration. For example, Phenmetrazine HCl, N.F., is titrated using 0.1 *N* $AgNO_3$ and the end point is determined potentiometrically using a silver electrode as the indicator electrode and the calomel electrode as the reference electrode.

For most assays conducted in nonaqueous media, however, the monographs simply state that the titration is conducted potentiometrically, without specifying the electrodes. The analyst then refers to a table listed in the official compendiums under "Titrimetry" and chooses an electrode pair according to the nonaqueous solvent used in the titration medium. Table 52 lists the official compounds which may be assayed potentiometrically, along with the electrode pair which may be used for the titration.

Exercise 110

Object. To Determine the End Point of Niacin (Nicotinic Acid) Titrated with Sodium Hydroxide Potentiometrically.

Materials Required. A pH meter fitted with glass and calomel electrodes
Magnetic stirrer and stirring bar
Standard buffers
0.1 *N* sodium hydroxide
Niacin, N.F., previously dried at 105° for 4 hr

Procedure. Weigh accurately, two 250-mg samples of Niacin and transfer each sample to a 250-ml beaker. Add 50 ml of distilled water to each sample and, if available, gently drop a magnetic stirring bar into the first sample. Place the beaker containing the sample on a magnetic stirrer and carefully immerse the electrodes of a previously standardized pH meter into the niacin solution. Before starting the magnetic stirrer be absolutely certain that the magnetic stirring bar will not strike the electrodes.

Determine the pH of the solution before adding any 0.1 *N* NaOH titrant, and record this value. Then add 3-ml increments of NaOH and record both milliliters of titrant and pH of the solution. Continue the titration until the pH of the solution reads between 11 and 12. Examine the data, and note the increments which gave the greatest difference in pH. This pH contains the equivalence point. Discard the sample and replace with a fresh sample.

Determine and record the pH of the solution before adding any 0.1 *N* NaOH titrant. Then add 0.1 *N* NaOH in 0.5 ml-increments within 3 ml of the predicted end point and record milliliters of each increment of titrant added along with the corresponding pH of the solution. Continue to add titrant dropwise, recording milliliters of titrant and pH after each addition, to 3 ml beyond the end point. Prepare a graph by plotting milliliters of 0.1 *N* NaOH solution along the abscissa and the corresponding pH values along the ordinate. The end point of the titration may be determined graphically by drawing a vertical line from the midpoint of the steepest part of the titration curve to the X axis. The point where this line intersects the X axis is then taken as the volume of titrant consumed at the equivalence point.

In actual practice the end point is determined more conveniently by first running an exploratory titration as described previously, in order to determine the approximate end point. Then the titration is repeated using a fresh sample. However, in this titration the initial quantity of titrant is added in one volume to within 2 ml of the end point, after which 0.1 ml-increments are added to 2 ml beyond the end point. The pH of the solution and the volume of titrant added are recorded, and the data are treated in the manner shown in Table 53. The value $\Delta pH/\Delta V$ is referred

TABLE 53. "FIRST" AND "SECOND" DERIVATIVE VALUES USED TO DETERMINE END POINT IN A POTENTIOMETRIC TITRATION

V, ml	pH	$V_2 - V_1$ (ΔV)	$pH_2 - pH_1$ (ΔpH)	($\Delta pH/\Delta V$)	($\Delta^2 pH/\Delta V^2$)
19.90	7.34				
		0.10	0.18	1.8	
20.00	7.52				+10
		0.10	0.28	2.8	
20.10	7.80				+102
		0.10	1.30	13.0	
20.20	9.10				−90
		0.10	0.40	4.0	
20.30	9.50				−15
		0.10	0.25	2.5	
20.40	9.75				

to as the "first derivative" and can be shown to equal a maximum at the end point. The value of $\Delta^2 pH/\Delta V^2$ is called the "second derivative" and can be shown to equal zero at the end point. Examining the data in Table 53 one sees that the end point must be near 20.10 ml because the first derivative is a maximum at this value. Also it can be seen from the table that the second derivative value of zero lies between +102 and −90 corresponding to 20.10 ml and 20.20 ml of titrant, respectively. To find the end point in the titration, simply calculate as shown:

$$20.10 + \left(\frac{102}{102 + 90} \times 0.10\right) = 20.15 \text{ ml}$$

or the amount of sodium hydroxide titrant consumed at the end point.

Chapter 22

pH AND HYDRONIUM ION CONCENTRATION IN AQUEOUS SYSTEMS

Introduction. Many compounds of pharmaceutical importance may be classified chemically, as acids, bases, and salts. In aqueous or hydroalcoholic solutions these compounds tend, in great or small degree, to dissociate into their respective ions. The concentration of hydronium ions influences the concentrations of the anions, cations, and undissociated molecules present in solution. In turn, these chemical species often affect the stability, therapeutic activity (through drug absorption), and pharmaceutical elegance of medicinal agents in aqueous or hydroalcoholic preparations. pH and hydronium ion concentration also play an important role in the study of drug receptor-site interactions, an area of research which has gained considerable impetus during recent years. In pharmaceutical analyses pH and hydronium ion concentration are also of considerable importance. Many chemical reactions require optimum pH conditions if the reaction is to proceed to completion. In analyses based on neutralization reactions the pH of the solution at the equivalence point determines which indicator is to be used for the analytical determination. Other examples could be cited to illustrate the importance of pH and hydronium ion concentration in pharmaceutical sciences, but it is hoped that these few will emphasize to the student the need of gaining a sound understanding of pH and related topics for use not only in pharmaceutical analysis but also in other areas of science yet to be encountered.

pH and Hydrogen Ion Concentration of Water. In Chap. 3, Principles of Titrimetric Analysis, the stoichiometric relationship existing between titrant and medicinal agent being assayed and the calculations based on the milliequivalent factor were discussed in detail. Since the present chapter deals with pH and its effects on solutions, a discussion of the manner in which pH changes during titration will be considered. Before we take up the subject of these changes (as illustrated with titration curves), equations which may be applied to the pH calculation of various ionic species must be developed and discussed. Recall from Chap. 3 that acids and bases are defined, according to the Brönsted-Lowry concept, as proton donors and proton acceptors, respectively; and that water acts either as an acid (donates a proton to ammonia) or as a base (accepts a proton from hydrochloric acid).

The equilibrium constant for water may be expressed as

$$\frac{a_{H_3O^+} \times a_{OH^-}}{a^2_{H_2O}} = K \quad (1)$$

Since water is very slightly ionized, the activity of unionized water may be considered constant, and therefore incorporated into the *equilibrium constant*. The expression then becomes:

$$a_{H_3O^+} \times a_{OH^-} = Ka^2_{H_2O} = K_w \quad (2)$$

where a = activity

K_w = ion-product constant of water, the ionization constant for water, or the protolytic constant of water

These terms are used synonymously. Activity in the above expression is the term used to designate effective concentration and is related to the actual concentration (i.e., molar) by the equation

$$a = fC \quad (3)$$

where f = a correction factor known as the activity coefficient

This factor is a number without units and very nearly equals 1 in very dilute solutions of electrolytes. In many analytical procedures in pharmacy and pharmaceutical sciences the solutions encountered are of low concentrations, and therefore, as a close approximation, the activity coefficient value may be taken to equal 1. Substituting the value 1 into Eq. (3) sets activity equal to concentration, and Eq. (2) is then expressed as:

$$[H_3O^+][OH^-] = K_w \quad (4)$$

where $[H_3O^+]$, or $[H^+]$, for the sake of brevity, and $[OH^-]$ equal molar concentrations. The numerical value of K_w at 22° is equal to 1×10^{-14}. An important but relatively simple fact about Eq. (4) is that $[H^+]$ may be determined in any aqueous solution regardless of the ionic species (salts, acids, or bases) present, if the $[OH^-]$ is known.

In many calculations the hydronium ion concentration becomes a cumbersome value whose magnitudes vary enormously, i.e., from one to one ten-millionth squared or 10^{-14}. The desirability of an exponential or logarithmic scale of expression is at once apparent; the system generally adopted is that devised by Sørenson, who defined the $[H^+]$ in terms of the logarithm of the reciprocal of $[H^+]$, called pH. The mathematical expression is given as:

$$\text{pH} = \log \frac{1}{[H^+]} = -\log [H^+] \quad (5)$$

The relation of pH to $[H^+]$ may be seen in Table 54.

TABLE 54

pH		Normality, hydrogen ions	Normality, hydroxyl ions
	0	1	10^{-14}
	1	10^{-1}	10^{-13}
	2	10^{-2}	10^{-12}
	3	10^{-3}	10^{-11}
	4	10^{-4}	10^{-10}
	5	10^{-5}	10^{-9}
	6	10^{-6}	10^{-8}
Neutral point	7	10^{-7}	10^{-7}
	8	10^{-8}	10^{-6}
	9	10^{-9}	10^{-5}
	10	10^{-10}	10^{-4}
	11	10^{-11}	10^{-3}
	12	10^{-12}	10^{-2}
	13	10^{-13}	10^{-1}
	14	10^{-14}	1

The concentration of other chemical species present in solution may also be expressed in terms of the negative logarithm of their concentrations. For example, hydroxyl ion concentration, may be defined as the logarithm of the reciprocal of $[OH^-]$, called pOH. The mathematical expression for this term is given as:

$$pOH = \log \frac{1}{[OH^-]} = -\log [OH^-] \tag{6}$$

In a similar manner K_w may be defined in terms of the logarithm of the reciprocal of ion-product constant of water, called pK_w. Mathematically this term is defined as:

$$pK_w = \log \frac{1}{K_w} = -\log 10^{-14} = 14 \tag{7}$$

Taking the logarithm of both sides of Eq. (4) and changing the signs of both sides of the equation, the following relation results:

$$-\log [H^+] - \log [OH^-] = -\log K_w \tag{8}$$

substituting Eqs. (5), (6), and (7) into Eq. (8) gives a very useful relationship which will be referred to on many occasions.

$$pH + pOH = pK_w \tag{9}$$

pH Calculation of Strong Acids. Strong acids may be defined as proton donors whose chemical reaction with water proceeds to virtual completion with the formation of hydronium ions. Hydrochloric acid is an example of a strong monoprotic[1] acid. In an aqueous medium, HCl

[1] See Chap. 3 for definition of *monoprotic acid*.

molecules react completely with water molecules to form hydronium and chloride ions. This reaction is represented by the following equation:

$$HCl + H_2O \longrightarrow H_3O^+ + Cl^-$$

Since HCl is completely reacted, the hydronium ion concentration which forms as a result of this reaction enters the equilibrium reaction of water.

$$[H_3O^+][OH^-] = K_w$$

Because of this phenomenon, the total hydronium ion concentration in an aqueous solution of a strong monoprotic acid is equal to the sum of the molar concentration of the acid plus the molar concentration of hydronium ions contributed by the protolysis (dissociation) of water.

To determine the total hydronium ion concentration in any solution, it is most exact to employ the electroneutrality rule. This rule simply states that all solutions must contain an equal number of positive and negative charges. For example, dilute solutions of HCl contain only H_3O^+, OH^-, and Cl^- ions. By the electroneutrality rule, then, the total number of positive charges $[H_3O^+]$ must equal the total number of negative charges, $[Cl^-]$ and $[OH^-]$. This statement, expressed symbolically as

$$[H_3O^+] = [Cl^-] + [OH^-] \qquad (10)$$

permits the calculation of the hydronium ion concentration in terms of the total chloride and hydroxyl ion concentration. To illustrate the use of this approach in the determination of hydronium ion concentration of a strong acid, consider a 10^{-3} M HCl solution. First recognize that the chloride ion concentration is 10^{-3} M and the contribution to the total hydronium ion concentration from this source is also 10^{-3} M. It must be realized that, as in any chemical system at equilibrium, all equilibrium and ionization constant expressions, such as $[H_3O^+][OH^-] = K_w$, must be obeyed at all times, and therefore $[OH^-] = K_w/[H_3O^+]$. Substituting into Eq. (10) gives $[H_3O^+] = 10^{-3} + K_w/[H_3O^+]$. As a close approximation, $K_w/[H_3O^+] = 10^{-14}/10^{-3} = 10^{-11}$. Since 10^{-11} is very small compared with 10^{-3}, it is dropped from Eq. (10), and $[H_3O^+] = 10^{-3}$ M.

For all practical purposes it may be stated that the hydronium ion concentration of a strong monoprotic acid in aqueous solution is equal to the molar concentration of the acid. However, it must again be emphasized that the hydronium ion concentration is equal to the sum of all contributing chemical reactions and that it is valid to drop the contributory concentrations only under certain conditions. An example in which the hydronium ion contribution of water may not be dropped is illustrated in the following: "Calculate the pH of a 10^{-8} M HCl solution." The answer obviously cannot be 8, since a pH of 8 indicates an excess of $[OH^-]$, and this cannot be true in an acid solution. In this example the hydronium ion contribution by water must be taken into account. The problem is best solved by employing the electroneutrality rule. Since [HCl] = $[Cl^-] = 10^{-8}$ *M* and $[OH^-] = K_w/[H_3O^+]$, substitution into Eq. (10) gives $[H_3O^+] = 10^{-8} + K_w/[H_3O^+]$. Multiplying both sides of the equation by $[H_3O^+]$ and rearranging gives:

$$[H_3O^+]^2 - 10^{-8}[H_3O^+] - K_w = 0$$

This equation is solved for hydronium ion concentration using the quadratic formula,[2] from which the pH is found to equal 6.98.

The calculation of the hydronium ion concentration of a strong diprotic acid such as H_2SO_4 is best illustrated by example.

"Calculate the pH of a 10^{-2} *M* sulfuric acid solution." In aqueous solutions in which the concentration is less than 1 *M*, sulfuric acid is completely dissociated to $[H_3O^+]$ and $[HSO_4^-]$ ions. The $[HSO_4^-]$ ions undergo further dissociation to form H_3O^+ and $SO_4^=$ with a $K_a = 1.02 \times 10^{-2}$. The hydronium ions which are formed by these two sources affect the equilibrium reaction of water in a similar manner to that discussed for monoprotic acids. The following equations summarize the chemical reactions which take place:

$$H_2SO_4 + H_2O \longrightarrow H_3O^+ + HSO_4^-$$

$$HSO_4^- + H_2O \rightleftharpoons H_3O^+ + SO_4^= \qquad K_a = \frac{[H_3O^+][SO_4^=]}{[HSO_4^-]} \tag{11}$$

$$2H_2O \rightleftharpoons H_3O^+ + OH^- \qquad K_w = [H_3O^+][OH^-] \tag{12}$$

Employing the electroneutrality rule:[3]

$$[H_3O^+] = [HSO_4^-] + 2[SO_4^=] + [OH^-]$$

[2] Quadratic formula:

$$x = \frac{-b \pm \sqrt{b^2 - 4ac}}{2a}$$

The quadratic equation from which the quadratic formula is derived is $ax^2 + bx + c = 0$. In the above example, $a = 1$; $b = 10^{-8}$; $c = K_w = 10^{-14}$, and $x = [H_3O^+]$. Only the positive value of x has physical significance.

[3] Consider a hypothetical solution prepared by adding 1,000 molecules of sulfuric acid to water. Assume that the following conditions prevail *at equilibrium:*

$$\underset{(10^8)}{H_2O} \rightleftharpoons \underset{(10)}{H_3O^+} + \underset{(10)}{OH^-}$$

$$H_2SO_4 + H_2O \longrightarrow \underset{(1{,}000)}{H_3O^+} + \underset{(1{,}000\,-\,200)}{HSO_4^-}$$

$$\underset{(800)}{HSO_4^-} + \underset{(200)}{H_2O} \rightleftharpoons \underset{(200)}{H_3O^+} + \underset{(200)}{SO_4^=}$$

Adding positive and negative charges in the solution yields:

$$(+)\text{ charges} = 10^+ + 1{,}000^+ + 200^+ = 1{,}210\ (+)\text{ charges}$$

$$(-)\text{ charges} = \underset{(OH^-)}{10^-} + \underset{(HSO_4^-)}{800^-} + \underset{2(SO_4^=)}{2(200^=)} = 1{,}210\ (-)\text{ charges}$$

Conclusion: The hydronium ion concentration is equal to the sum of the negative charges in solution.

For all practical purposes, however:

$$[H_3O^+] = [HSO_4^-] + 2[SO_4^=] \tag{13}$$

Let $x = [SO_4^=]$ in Eq. (11). Then $[10^{-2} - x]$ equals bisulfate concentration and $[10^{-2} + x]$ equals the hydronium ion concentration in the solution.

$$\underset{(10^{-2}-x)}{HSO_4^-} + H_2O \rightleftharpoons \underset{(10^{-2}+x)}{H_3O^+} + \underset{(x)}{SO_4^=}$$

Substituting into the equilibrium expression (11) gives:

$$\frac{(10^{-2} + x)(x)}{(10^{-2} - x)} = 1.02 \times 10^{-2}$$

Collecting terms and rearranging:

$$x^2 + 0.0202x - 1.02 \times 10^{-4} = 0$$

Using the quadratic formula, x is found to equal 4.18×10^{-3}. Substituting into Eq. (13) gives:

$$[H_3O^+] = [0.01 - 0.00418] + 2[0.00418]$$
$$[H_3O^+] = 1.418 \times 10^{-2}$$

or more simply:

$$[H_3O^+] = [10^{-2} + x] = [0.01 + 0.00418] = 1.418 \times 10^{-2}$$
$$pH = 1.85$$

pH Calculation of Strong Bases. Strong bases may be defined as proton acceptors whose chemical reaction with water proceeds to virtual completion with the formation of hydroxyl ions. Sodium hydroxide and potassium hydroxide are examples of strong inorganic bases, because their reaction with water leads to the formation of equal molar quantities of hydroxyl ion. In other words 10^{-3} mole of NaOH provides 10^{-3} mole of hydroxyl ions.

The principles which were discussed under pH calculations of strong acids apply also to strong bases. To calculate the pH of a strong base it is easier first to determine the pOH and then to subtract this value from pK_w.

Example. Calculate the pH of a 10^{-3} molar solution of KOH.

Solution:

1. Observe that hydroxyl ion concentration is equal to $10^{-3}\ M + y$, where y is equal to the molar hydroxyl ion concentration contributed by the dissociation of water. The electroneutrality rule may be employed if desired, but for all practical purposes, the contribution of hydroxyl ion concentration by water is exceedingly small and is, therefore, disregarded in the calculation.

2. Hence,

$$[OH^-] = 10^{-3}\ M$$

$$pOH = \log \frac{1}{[OH^-]} = \log 10^3 = 3.00$$

$$pH = pK_w - pOH = 14.00 - 3.00 = 11.00$$

pH Calculation of a Weak Acid. Weak acids may be defined as proton donors whose chemical reaction with water does not proceed to completion and therefore produces a relatively small hydronium ion concentration. Weak acids may be uncharged chemical species, such as nicotinic acid or acetic acid, or they may be charged species, such as the ammonium ion, or the cationic form of many amine salts such as alkaloidal and antihistaminic salts ($R_3NH^+\ Cl^-$).

The following equilibrium reactions are involved in the calculation of hydronium ion:

$$2H_2O \rightleftharpoons H_3O^+ + OH^- \qquad K_w = [H^+][OH^-] \tag{14}$$

$$HA + H_2O \rightleftharpoons H_3O^+ + A^- \qquad K[H_2O] = K_a = \frac{[H_3O^+][A^-]}{[HA]} \tag{15}$$

where K = equilibrium constant

K_a = ionization constant of weak acid

Employing the electroneutrality rule, the hydronium ion concentration of a weak acid in an aqueous solution is found using the following relationship.

$$[H_3O^+] = [OH^-] + [A^-] \tag{16}$$

For all practical purposes the $[OH^-]$ contribution by water is small and may therefore be dropped from Eq. (16). Then

$$[H_3O^+] = [A^-]$$

Substituting $[H_3O^+]$ for $[A^-]$ into Eq. (15) and rearranging gives:

$$[H_3O^+] = \sqrt{(K_a)[HA]} \tag{17}$$

Let C equal the total titratable[4] hydrogen ion concentration of the weak acid.

$$C = [HA] + [H_3O^+]$$

Then

$$[HA] = C - [H_3O^+] \tag{18}$$

Substituting Eq. (18) into Eq. (17) gives:

$$[H_3O^+] = \sqrt{K_a(C - [H_3O^+])} \tag{19}$$

[4] Titratable hydrogen, as the name implies, refers to the hydrogen in a compound which can be titrated with base. This includes both the ionized and un-ionized forms of hydrogen.

When $[H_3O^+]$ is small as compared to C, it may be dropped. Equation (19) then reduces to:

$$[H_3O^+] = \sqrt{K_aC} \tag{20}$$

Taking the logarithm of both sides of Eq. (20) and changing signs of both sides, the following relation results:

$$-\log [H_3O^+] = \tfrac{1}{2}(-\log K_a - \log C) \tag{21}$$

Define pK_a as $-\log K_a$. Substitute pK_a and pH into Eq. (21) as indicated:

$$pH = \tfrac{1}{2}(pK_a - \log C) \tag{22}$$

Example. Calculate the pH of 0.002 M nicotinic acid solution. $K_a = 1.40 \times 10^{-5}$.

Solution:

1. Determine pK_a and $\log C$.

$$pK_a = \log \frac{1}{K_a} = \log \frac{1}{1.40 \times 10^{-5}} = \log \frac{10^5}{1.40}$$
$$5.00 - 0.15 = 4.85$$
$$\log C = \log 2 \times 10^{-3} = -3.00 + 0.30 = -2.70$$

2. Substituting into Eq. (22):

$$pH = \frac{1}{2}(4.85 - [-2.70]) = \frac{7.55}{2} = 3.78$$

pH Calculation of a Weak Base. Weak bases may be defined as chemical-proton acceptors whose reactions with water do not proceed to completion and therefore produce a relatively small hydroxyl ion concentration. Weak bases may be uncharged chemical species such as ammonia, atropine, morphine, chlorpromazine, etc., or they may be charged species such as acetate ion, barbiturate ion, or the anionic ion of salts of weak acids and strong bases. The derivation of the pH equation of a weak base will be illustrated, using ammonia as an example. The following equilibrium reactions must be considered.

$$2H_2O \rightleftharpoons H_3O^+ + OH^- \qquad K_w = [H_3O^+][OH^-] \tag{23}$$

$$NH_3 + H_2O \rightleftharpoons NH_4^+ + OH^- \qquad K[H_2O] = K_b = \frac{[NH_4^+][OH^-]}{[NH_3]} \tag{24}$$

where K = equilibrium constant

K_b = ionization constant of the weak base

As in the case of the strong base, it is easier to calculate first the $[OH^-]$ concentration.

Employing the electroneutrality rule:

$$[OH^-] = [H_3O^+] + [NH_4^+]$$

For all practical purposes the $[H_3O^+]$ concentration contributed by water is small and may therefore be dropped.

$$[OH^-] = [NH_4^+]$$

Substituting $[OH^-]$ for $[NH_4^+]$ into Eq. (24) and rearranging gives:

$$[OH^-] = \sqrt{K_b NH_3} \tag{25}$$

Let C equal total $[NH_3]$ and $[NH_4^+]$ concentration.

$$C = [NH_3] + [NH_4^+] \quad \text{and} \quad C = [NH_3] + [OH^-]$$

Then

$$[NH_3] = (C - [OH^-]) \tag{26}$$

Substituting Eq. (26) into Eq. (25) gives:

$$[OH^-] = \sqrt{K_b(C - [OH^-])} \tag{27}$$

When $[OH^-]$ is small compared to C, it may be dropped. Equation (27) then reduces to:

$$[OH^-] = \sqrt{K_b C} \tag{28}$$

Taking the logarithm of both sides of Eq. (28) and changing signs of both sides gives the following relation:

$$-\log [OH^-] = \tfrac{1}{2}(-\log K_b - \log C)$$
$$pOH = \tfrac{1}{2}(pK_b - \log C) \tag{29}$$

Substituting Eq. (29) into $pH = pK_w - pOH$ gives:

$$pH = pK_w - \tfrac{1}{2}pK_b + \tfrac{1}{2}\log C \tag{30}$$

Example. Calculate the pH of a 1.50×10^{-3} M atropine solution. $K_b = 4.50 \times 10^{-5}$.

Solution:

1. Determine the pK_b and $\log C$ values.

$$pK_b = \log \frac{1}{4.50 \times 10^{-5}} = \log \frac{10^5}{4.50} = 5.00 - 0.65$$

$$pK_b = 4.35$$
$$\log C = \log 1.50 \times 10^{-3} = -3.00 + 0.18 = -2.82$$

2. Substitute into Eq. (30):

$$pH = 14.00 - \tfrac{1}{2}(4.35) + \tfrac{1}{2}(-2.82) = 10.41$$

Salts. For purposes of this discussion salts will be defined as those chemical products, other than water, obtained from the neutralization of equivalent amounts of (1) a strong base and strong acid, (2) a strong base and weak acid, (3) a strong acid and weak base, and (4) a weak acid and weak base. It is important to point out (1) that the hydronium ion concentration of a salt solution obtained by dissolution is the same as that obtained during the neutralization reaction of chemically equivalent quantities of acid and base, and (2) that salts exist in solution as ions.

pH Calculation of a Salt Solution of a Strong Acid and Strong Base. Examples of this type of salt are sodium chloride and sodium sulfate. In dilute solutions salts of this kind do not affect the equilibrium reaction of water and therefore the pH may be considered the same as water, 7.00.

pH Calculation of a Salt Solution of a Weak Acid and a Strong Base. A few examples of this type of salt are sodium acetate, potassium penicillin, and sodium sulfamethazine. In deriving a pH equation for a salt of this nature, the anionic portion of the salt is considered the weak base, according to the Brönsted-Lowry definition; it reacts to some degree with water as shown in Eqs. (32) and (33). The cationic portion, however, does not react with water appreciably and therefore does not affect the pH of dilute solutions.

The derivation of the pH equation of this type of salt will be illustrated using sodium acetate as the example. The following equilibrium reactions are involved:

$$2H_2O \rightleftharpoons H_3O^+ + OH^- \qquad K_w = [H_3O^+][OH^-] \tag{31}$$

$$A^- + H_2O \rightleftharpoons HA + OH^- \qquad K_b = \frac{[HA][OH^-]}{[A^-]} \tag{32}$$

$$A^- + H_3O^+ \rightleftharpoons HA + H_2O \qquad K_a = \frac{[H_3O^+][A^-]}{[HA]} \tag{33}$$

To determine the hydronium ion concentration, the electroneutrality rule is employed. Hence,

$$[Na^+] + [H_3O^+] = [A^-] + [OH^-] \tag{34}$$

Solid sodium acetate salt contains an equal number of sodium and acetate ions; i.e., $[Na^+] = [A^-]$. However, in an aqueous solution of sodium acetate it can be seen from chemical reactions (32) and (33) that $[Na^+] = [A^-] + [HA]$. Substituting into Eq. (34) gives:

$$[A^-] + [HA] + [H_3O^+] = [A^-] + [OH^-]$$

Canceling $[A^-]$ on both sides of the equation gives:

$$[HA] + [H_3O^+] = [OH^-] \tag{35}$$

Now rearrange one or all of the Eqs. (31), (32), and (33) so that substitution into Eq. (35) eliminates [HA] and $[OH^-]$. *This is absolutely valid since* *any aqueous solution contains but one concentration of H_3O^+, OH^-, HA, and* A^-. Substituting as indicated gives:

$$\frac{[H_3O^+][A^-]}{K_a} + [H_3O^+] = \frac{K_w}{[H_3O^+]}$$

Clearing fractions:

$$[H_3O^+]^2[A^-] + [H_3O^+]^2K_a = K_wK_a$$

Rearranging and solving for $[H_3O^+]$:

$$[H_3O^+]^2([A^-] + K_a) = K_wK_a$$

$$[H_3O^+] = \sqrt{\frac{K_wK_a}{[A^-] + K_a}}$$

For all practical purposes this equation may be simplified by making two approximations:

1. Drop K_a in the *denominator* on the assumption that it does not contribute significantly to the sum in this part of the equation.
2. Assume that $[A^-]$ is equal to the original concentration C of the salt taken.

Then
$$[H_3O^+] = \sqrt{\frac{K_wK_a}{C}} \tag{36}$$

Taking logarithms of both sides of the equation and changing signs, the following equation results:

$$-\log [H_3O^+] = \tfrac{1}{2}(-\log K_w - \log K_a + \log C)$$

and
$$pH = \tfrac{1}{2}(pK_w + pK_a + \log C) \tag{37}$$

Example. Calculate the pH of a 2.00×10^{-3} M sodium benzoate solution. $K_a = 6.40 \times 10^{-5}$.

Solution:

1. Determine pK_a and $\log C$ values. Recall pK_w.

$$pK_w = 14.00 \qquad pK_a = -\log 6.40 \times 10^{-5} = 4.19$$
$$\log C = \log 2.00 \times 10^{-3} = -3.00 + 0.31 = -2.69$$

2. Substitute appropriate values into Eq. (37).

$$pH = \tfrac{1}{2}(14.00 + 4.19 - 2.69) = 7.75$$

pH Calculation of a Salt Solution of a Weak Base and a Strong Acid. A few examples of this type of salt are ammonium chloride, atropine hydrochloride, and chlorpromazine hydrochloride. In salts of this nature the cationic portion is considered the weak acid as defined by the Brönsted-Lowry definition and reacts with water according to the reactions given below. The anionic portion does not affect the pH of the solution. The derivation of the pH equation of this class of salts will be illustrated using ammonium chloride as the example. Note the similarity of the derivation of the pH equation between this type of salt and sodium acetate.

The following equilibrium reactions are involved:

$$2H_2O \rightleftharpoons H_3O^+ + OH^- \qquad K_w = [H_3O^+][OH^-] \tag{38}$$

$$NH_4^+ + H_2O \rightleftharpoons NH_3 + H_3O^+ \qquad K_a = \frac{[NH_3][H_3O^+]}{[NH_4^+]} \tag{39}$$

$$NH_4^+ + OH^- \rightleftharpoons NH_3 + H_2O \qquad K_b = \frac{[NH_4^+][OH^-]}{[NH_3]} \tag{40}$$

By the electroneutrality rule:

$$[NH_4^+] + [H_3O^+] = [OH^-] + [Cl^-] \tag{41}$$

In solid ammonium chloride there are an equal number of ammonium and chloride ions, i.e., $[Cl^-] = [NH_4^+]$. However, in a solution of this salt it can be seen from reactions (39) and (40) that $[Cl^-] = [NH_4^+] + [NH_3]$. Substituting into Eq. (41) gives:

$$[NH_4^+] + [H_3O^+] = [OH^-] + [NH_4^+] + [NH_3]$$

Canceling $[NH_4^+]$ on both sides of the equation gives:

$$[H_3O^+] = [OH^-] + [NH_3] \tag{42}$$

Rearranging one or all of the Eqs. (38), (39), and (40) so that substitution into Eq. (42) eliminates $[OH^-]$ and $[NH_3]$ gives:

$$[H_3O^+] = \frac{K_w}{[H_3O^+]} + \frac{[NH_4^+]K_w}{[H_3O^+]K_b}$$

Clearing fractions:

$$[H_3O^+]^2 = K_w + \frac{[NH_4^+]K_w}{K_b}$$

and solving for $[H_3O^+]$ gives:

$$[H_3O^+] = \sqrt{\frac{K_w[NH_4^+]}{K_b} + K_w} \tag{43}$$

For all practical purposes this equation may be simplified by making two approximations:

1. Drop the second K_w term in Eq. (43) on the assumption that it will not contribute significantly to the total.
2. Assume that $[NH_4^+]$ is equal to the original concentration C of the salt taken.

Then
$$[H_3O^+] = \sqrt{\frac{K_wC}{K_b}}$$

Taking logarithms of both sides of the equation and changing signs, the following equation results:

$$-\log [H_3O^+] = \tfrac{1}{2}(-\log K_w + \log K_b - \log C)$$
$$pH = \tfrac{1}{2}(pK_w - pK_b - \log C) \quad (44)$$

Example. Calculate the pH of a 2.30×10^{-3} M solution of atropine hydrochloride. $K_b = 4.50 \times 10^{-5}$.

Solution:

1. Determine pK_b and $\log C$ values. Recall pK_w.

$$pK_w = 14.00 \qquad pK_b = -\log 4.50 \times 10^{-5} = 4.35$$
$$\log C = \log 2.30 \times 10^{-3} = -3.00 + 0.36 = -2.64$$

2. Substitute appropriate values into Eq. (44) and solve.

$$pH = \tfrac{1}{2}(14.00 - 4.35 + 2.64) = 6.15$$

pH Calculation of a Salt of a Weak Acid and a Weak Base. Following the same approach as that used in the pH derivation of Eqs. (37) and (44), the following equation can be obtained.

$$[H_3O^+] = \sqrt{\frac{K_a K_w}{K_b}}$$

and

$$pH = \tfrac{1}{2}(pK_w + pK_a - pK_b) \quad (45)$$

Buffer Solutions. Buffer solutions have the property of resisting changes in pH when acids or bases are added to them. This property of resisting pH changes usually results from the presence of a pair of chemicals, often referred to as a "buffer pair," which consists of either a weak acid and some salt of the weak acid (acetic acid and sodium acetate) or a weak base and some salt of the weak base (ammonia and ammonium chloride). A discussion of the mechanism by which buffers act to resist changes in pH is presented in Chap. 3 and should be reviewed at this time.

The derivation of the pH equation of a buffer solution of a weak acid and some salt of the weak acid will be illustrated using acetic acid and sodium acetate. Consider the following reactions.

$$2H_2O \rightleftharpoons H_3O^+ + OH^- \qquad K_w = [H_3O^+][OH^-] \quad (46)$$

$$HA + H_2O \rightleftharpoons H_3O^+ + A^- \qquad K_a = \frac{[H_3O^+][A^-]}{[HA]} \quad (47)$$

$$A^- + H_2O \rightleftharpoons HA + OH^- \qquad K_b = \frac{[HA][OH^-]}{[A^-]} \quad (48)$$

In any aqueous solution in which two or more equilibrium reactions take place in the same flask, there is but one hydronium ion concentration which affects all the equilibrium reactions involving hydronium ions. This is true also of other ions in solution. It is well to recognize also that the salt in the buffer solution is the conjugate base of the acid, and a useful relationship can be observed when the Eqs. (47) and (48) are multiplied together and fractions cleared as illustrated.

$$K_aK_b = \frac{[H_3O^+][A^-]}{[HA]} \times \frac{[HA][OH^-]}{[A^-]} = [H_3O^+][OH^-] = K_w$$

$$K_aK_b = K_w \qquad (49)$$

Taking the logarithm of both sides of this equation and changing signs gives:

$$-\log K_a - \log K_b = -\log K_w$$

$$pK_a + pK_b = pK_w \qquad (50)$$

Equation (50) permits the calculation of the pK_b value of a conjugate base if the pK_a value of the acid is known, and vice versa. The usefulness of this relationship will become apparent in calculations of pH values of certain buffer solutions which will be discussed in the following paragraphs.

Although either Eq. (47) or (48) may be used to derive the pH equation for a buffer solution, it is obtained most simply by rearranging Eq. (47).

$$[H_3O^+] = K_a \frac{[HA]}{[A^-]}$$

Taking the logarithm of both sides of the equation, changing signs, and rearranging gives:

$$pH = pK_a + \log \frac{[A^-]}{[HA]} \qquad (51)$$

or, more general:

$$pH = pK_a + \log \frac{[\text{conj. base}]}{[\text{acid}]} \qquad (52)$$

By employing the electroneutrality rule it can be shown that the initial molar salt concentration, S, of sodium acetate for all practical purposes is equal to $[A^-]$ and the total molar titratable hydronium ion concentration, C, for all practical purposes is equal to [HA].

$$[Na^+] + [H_3O^+] = [A^-] + [OH^-]$$

and $$[A^-]^5 = [Na^+] + [H_3O^+] - [OH^-]$$

or $$[A^-] = [S] + [H_3O^+] - [OH^-]$$

As compared with S, both $[H_3O^+]$ and $[OH^-]$ are small and therefore, as a close approximation,

$$[A^-] \approx [S]$$

[5] Crystalline sodium acetate contains an equal number of sodium and acetate ions. Therefore

$$[S] = [Na^+A^-] = [Na^+] = [A^-]$$

From reactions (46), (47), and (48) it can be seen that

$$[HA] = C - [H_3O^+] + [OH^-]$$

and since $[H_3O^+]$ and $[OH^-]$ contribute negligibly to the sum, as a close approximation,

$$[HA] \approx C$$

Example. Calculate the pH of a solution which contains 2.0×10^{-3} mole of sodium acetate and 3.3×10^{-4} mole of acetic acid in 1 liter of solution. $pK_a = 4.74$.

Solution. Substitute into Eq. (51) or (52)

$$pH = 4.74 + \log \frac{2.0 \times 10^{-3}}{3.3 \times 10^{-4}}$$

$$pH = 4.74 + \log 2.0 \times 10^{-3} - \log 3.3 \times 10^{-4}$$
$$pH = 4.74 + [-3.00 + 0.30 + 4.00 - 0.52]$$
$$pH = 4.74 + 0.78 = 5.52$$

The derivation of the pH equation of a buffer solution of a weak base and one of its salts will be illustrated using ammonium chloride and ammonia.

$$2H_2O \rightleftharpoons H_3O^+ + OH^- \qquad K_w = [H_3O^+][OH^-] \qquad (53)$$

$$NH_4^+ + H_2O \rightleftharpoons NH_3 + H_3O^+ \qquad K_a = \frac{[NH_3][H_3O^+]}{[NH_4^+]} \qquad (54)$$

$$NH_3 + H_2O \rightleftharpoons NH_4^+ + OH^- \qquad K_b = \frac{[NH_4^+][OH^-]}{[NH_3]} \qquad (55)$$

Either Eq. (54) or (55) may be used to derive the pH equation. However, it is obtained most simply by rearranging Eq. (54).

$$[H_3O^+] = K_a \frac{[NH_4^+]}{[NH_3]}$$

Taking the logarithm of both sides of the above equation, changing signs, and rearranging gives:

$$pH = pK_a + \log \frac{[NH_3]}{[NH_4^+]} \qquad (56)$$

or

$$pH = pK_a + \log \frac{[\text{conj. base}]}{[\text{acid}]} \qquad (52)$$

Example. Calculate the pH of a solution which contains 3.7×10^{-3} mole of ammonium chloride and 4.8×10^{-2} mole of ammonia in 1 liter of solution. $pK_b = 4.74$.

Observe that in Eq. (54) the NH_4^+ ion acts as the acidic species and the NH_3 acts as the conjugate base. This does not affect the manner in which the pH of the buffer solution of this type is calculated, and the more general Eq. (52) is perfectly applicable.

Solution:

1. Determine the pK_a value using Eq. (50).

$$pK_a = pK_w - pK_b = 14.00 - 4.74 = 9.26$$

2. Substitute appropriate values in Eq. (52) or (56) and solve as shown in the previous example.

$$pH = 9.26 + \log \frac{4.8 \times 10^{-2}}{3.7 \times 10^{-2}} = 9.37$$

TITRATION CURVES

The manner in which the pH of a solution changes during a titration is best shown by preparing a graph in which the pH value is plotted on the Y axis and the volume of the titrant is plotted on the X axis. Titration curves may be prepared from pH data obtained experimentally, or the data for the graph may be calculated. For purposes of considering the principles underlying titrimetric methods based on neutralization reactions, data obtained from calculations are satisfactory. In actual practice, however, pH measurements are made experimentally and titration curves and their derivatives are used for the purpose of determining end points for titrimetric assays. In either case the equivalence point on the titration curve is that portion where the slope of the curve is steepest and is commonly referred to as the "break" in the titration curve. When an indicator is used to detect the equivalence point, it must give a sharp discernible color change at that point in the titration; otherwise the color change may lead to erroneous results. The transition ranges of phenolphthalein and methyl red are included on most of the titration curves discussed in this chapter for the purpose of showing graphically which indicator is suitable for a given titration.

The preparation of a titration curve is not difficult if one realizes that the calculations are based on only four types of points, which correspond to:

1. The pH of the solution being titrated but before any titrant is added
2. The pH of the solution after titrant is added but before the equivalence point is reached
3. The pH at the equivalence point
4. The pH beyond the equivalence point

Titration of a Strong Acid Using a Strong Base. In this type of titration a strong acid, such as HCl, is titrated with NaOH, a strong base. Although only four pH calculations will be illustrated, it can be seen from

TABLE 55. TITRATION-CURVE DATA*

ml of 0.1 *N* NaOH		Total ml of solution	pH
0.00	$pH = -\log [H_3O^+]$	100	1.70
5.00		105	1.85
10.00		110	2.04
15.00		115	2.36
17.00		117	2.59
19.00		119	3.08
19.50		119.5	3.38
19.90		119.9	4.08
19.95		119.95	4.38
20.00		120	7.00
20.05	$pH = pK_w - pOH$	120.05	9.62
20.10		120.1	9.92
20.50		120.5	10.62
21.00		121	10.92
23.00		123	11.39
29.00		129	11.84

* Calculated pH values for titration of 0.1 *N* HCl using 0.1 *N* NaOH as titrant. (20.00 ml of 0.1 *N* HCl tially taken and diluted to 100 ml.)

able 55 that more were required in order to prepare the titration curve ıown in Fig. 39. It must be clearly understood that these calculations pply only for strong acid–strong base titrations. As will be seen later, eak acid–strong base titrations and weak base–strong acid titrations quire different calculations at these four regions or points in the titration ırve. As can be seen in Fig. 39, either phenolphthalein or methyl red

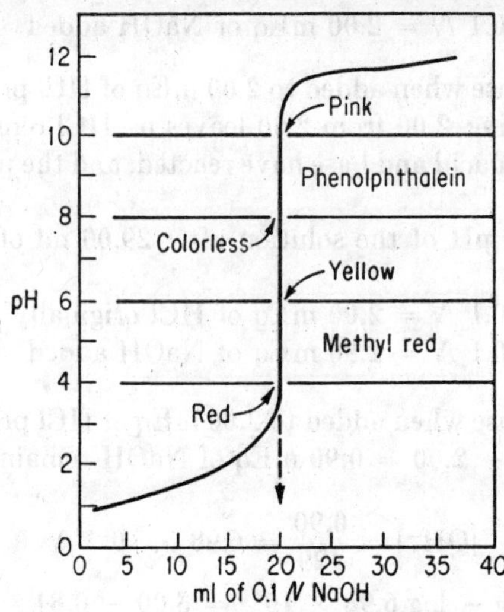

FIG. 39. Titration curve of HCl using data obtained from Table 55.

indicators are suitable. The pH equations used to calculate the pH of th solutions in the different regions of the titration curve are also shown i Table 55.

To illustrate, suppose that 20.00 ml of exactly 0.1 N HCl was taken diluted with sufficient distilled water to measure exactly 100 ml, an titrated with exactly 0.1 N NaOH solution.

1. Calculate the pH of the solution before any titrant is added:

$$20.00 \times 0.1\ N = 2.00 \text{ mEq of HCl in 100 ml solution}$$

$$[H_3O^+] = \frac{2.00 \text{ mEq}}{100 \text{ ml}} = 2.00 \times 10^{-2}\ N$$

$$\text{pH} = -\log 2.00 \times 10^{-2} = 2.00 - 0.30 = 1.70$$

2. Calculate the pH of the solution after 15.00 ml of 0.1 N NaOH added:

$$20.00 \times 0.1\ N = 2.00 \text{ mEq of HCl originally present}$$
$$15.00 \times 0.1\ N = 1.50 \text{ mEq of NaOH added}$$

This quantity of base when added to 2.00 mEq of HCl produces 1.50 mE of NaCl. Therefore, (2.00 − 1.50) or 0.50 mEq of HCl remains and:

$$[H_3O^+] = \frac{0.50}{115} = 4.35 \times 10^{-3}\ N$$

$$\text{pH} = -\log 4.35 \times 10^{-3} = 3.00 - 0.64 = 2.36$$

3. Calculate the pH of the solution after 20.00 ml of 0.1 N NaOH added:

$$20.00 \times 0.1\ N = 2.00 \text{ mEq of HCl originally present}$$
$$20.00 \times 0.1\ N = 2.00 \text{ mEq of NaOH added}$$

This quantity of base when added to 2.00 mEq of HCl produces 2.00 mE of NaCl; subtracting 2.00 from 2.00 leaves no HCl remaining. Stoichic metric quantities of acid and base have reacted, and the pH of the solutio is 7.00.

4. Calculate the pH of the solution after 29.00 ml of 0.1 N NaOH i added:

$$20.00 \times 0.1\ N = 2.00 \text{ mEq of HCl originally present}$$
$$29.00 \times 0.1\ N = 2.90 \text{ mEq of NaOH added}$$

This quantity of base when added to 2.00 mEq of HCl produces 2.00 mE of NaCl, and 2.90 − 2.00 = 0.90 mEq of NaOH remaining unreacted.

$$[OH^-] = \frac{0.90}{129} = 6.98 \times 10^{-3}\ N$$

$$\text{pOH} = -\log 6.98 \times 10^{-3} = 3.00 - 0.84 = 2.16$$

$$\text{pH} = \text{p}K_w - \text{pOH} = 14.00 - 2.16 = 11.84$$

TABLE 56. TITRATION-CURVE DATA*

ml of 0.1 N NaOH		Total ml of solution	pH
0.00	$pH = \frac{1}{2}(pK_a - \log C)$	100	3.22
5.00	$pH = pK_a + \log (A/HA)$	105	4.26
10.00		110	4.74
15.00		115	5.22
17.00		117	5.49
19.00		119	6.02
19.50		119.5	6.33
19.90		119.9	7.04
19.95		119.95	7.34
20.00	$pH = \frac{1}{2}(pK_w + pK_a + \log C)$	120	8.48
20.05	$pH = pK_w - pOH$	120.05	9.62
20.10		120.1	9.92
20.50		120.5	10.62
21.00		121	10.92
23.00		123	11.39
29.00		129	11.84

* Calculated pH values for titration of 0.1 N acetic acid using 0.1 N NaOH as titrant. (20.00 ml of 0.1 N HA initially taken and diluted to 100 ml.)

Titration of a Weak Acid Using a Strong Base. In this type of titration a weak acid such as acetic acid is titrated with NaOH, a strong base. Again only four pH calculations will be illustrated, but it must be clearly understood that more were required, as can be seen from Table 56 for the preparation of the titration curve shown in Fig. 40. Observe that

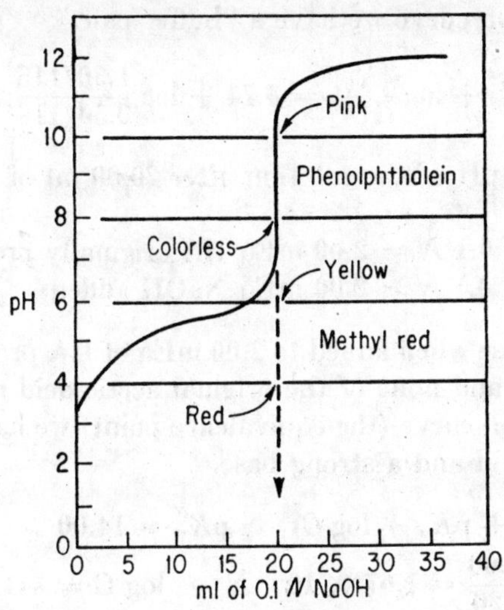

FIG. 40. Titration curve of acetic acid using data obtained from Table 56.

phenolphthalein is a suitable indicator for this titration, because a sharp color change is obtained at the equivalence point (20 ml of titrant). Methyl red, on the other hand, is unsuitable because the color change is too gradual, beginning at about 3 ml and finally changing to yellow at about 19 ml. The end point of the titration is neither sharp nor accurate, since a change in color at 19 ml would bring about an end point well in advance of the equivalence point. The equations used to calculate the pH of the solutions at the four different regions in the titration curve are also shown in Table 56.

To illustrate, assume that 20.00 ml of exactly 0.1 N acetic acid was taken, diluted with sufficient distilled water to measure exactly 100 ml, and titrated with exactly 0.1 N NaOH solution. pK_a (acetic acid) = 4.74.

1. Calculate the pH of the solution before any titrant is added.

$$20.00 \times 0.1\ N = 2.00 \text{ mEq of acetic acid originally present}$$

$$C \approx [\text{HA}] = \frac{2.00 \text{ mEq}}{100 \text{ ml}} = 2.00 \times 10^{-2}\ N$$

$$\text{pH} = \tfrac{1}{2}(\text{p}K_a - \log C) \qquad \log 2.00 \times 10^{-2} = -1.70$$

$$\text{pH} = \tfrac{1}{2}(4.74 + 1.70) = 3.22$$

2. Calculate the pH of the solution after 15.00 ml of 0.1 N NaOH is added

$$20.00 \times 0.1\ N = 2.00 \text{ mEq HA originally present}$$
$$15.00 \times 0.1\ N = 1.50 \text{ mEq NaOH added}$$

This quantity of base when added to 2.00 mEq of HA produces 1.50 mEq of sodium acetate, and 2.00 − 1.50 = 0.50 mEq of HA. Therefore, at this point in the titration curve we have a "buffer pair."

$$\text{pH} = \text{p}K_a + \log \frac{[\text{A}^-]}{[\text{HA}]} = 4.74 + \log \frac{1.50/115}{0.50/115} = 5.22$$

3. Calculate the pH of the solution after 20.00 ml of 0.1 N NaOH is added.

$$20.00 \times 0.1\ N = 2.00 \text{ mEq HA originally present}$$
$$20.00 \times 0.1\ N = 2.00 \text{ mEq NaOH added}$$

This quantity of base when added to 2.00 mEq of HA produces 2.00 mEq of sodium acetate, and none of the original acetic acid remains. At this point in the titration curve (the equivalence point) we have a solution of a salt of a weak acid and a strong base.

$$\text{pH} = \tfrac{1}{2}(\text{p}K_w + \text{p}K_a + \log C) \qquad \text{p}K_w = 14.00 \qquad \text{p}K_a = 4.74$$

$$C = \frac{2.00}{120} = 1.67 \times 10^{-2}\ N \qquad \log C = -1.78$$

$$\text{pH} = \tfrac{1}{2}(14.00 + 4.74 - 1.78) = 8.48$$

4. Calculate the pH of the solution after 29.00 ml of 0.1 N NaOH is added.

$$20.00 \times 0.1\ N = 2.00 \text{ mEq HA originally present}$$
$$29.00 \times 0.1\ N = 2.90 \text{ mEq NaOH added}$$

This quantity of base when added to 2.00 mEq of HA produces 2.00 mEq of sodium acetate, and 2.90 − 2.00 = 0.90 mEq of NaOH remains unreacted. Other sources of hydroxyl ions are those produced from the dissociation of water and from the hydrolysis of acetate ions. Since these contributions are negligible, the concentration of the unreacted NaOH only needs to be taken into account to calculate the pH of the solution at this concentration.

$$[OH^-] = \frac{0.90}{129} = 6.98 \times 10^{-3}\ N$$
$$pOH = -\log [OH^-] = -\log 6.98 \times 10^{-3} = 2.16$$
$$pH = pK_w - pOH = 14.00 - 2.16 = 11.84$$

Titration Curve of a Strong Base (NaOH) Using a Strong Acid (HCl). This type of titration curve is illustrated in Fig. 41. The pH calculations are identical with those used in the titration of a strong acid and a

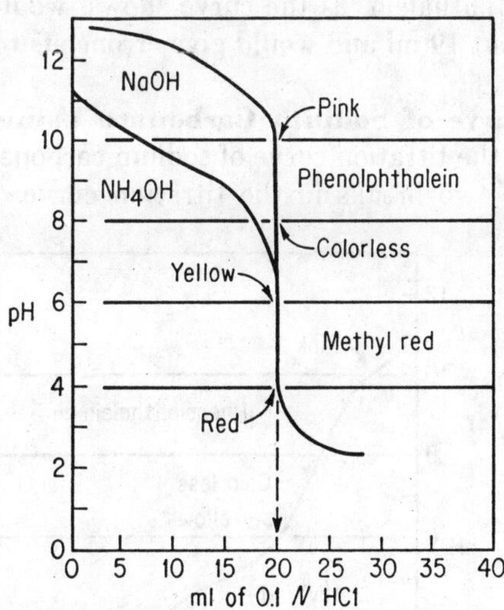

FIG. 41. Titration curve of strong and weak bases with a strong acid.

strong base, only the order is reversed. Again, either phenolphthalein or methyl red may be used as an indicator.

Titration Curve of a Weak Base (NH_4OH) Using a Strong Acid (HCl). This type of titration is also illustrated in Fig. 41. This curve demonstrates graphically the pH changes which take place during the titration of ammonia. Many medicinal agents, such as alkaloids, and other free amine compounds react similarly.

The pH calculations may be divided into four regions:

1. Before any titrant (HCl) is added to the weak base.

$$pH = pK_w - \frac{1}{2}pK_b + \frac{1}{2}\log C.$$

2. After titrant is added but before the stoichiometric end point is reached.

$$pH = pK_a + \log \frac{[NH_3]}{[NH_4^+]}$$

or

$$pH = pK_w - pK_b + \log \frac{[NH_3]}{[NH_4^+]}$$

3. At the equivalence point, we have a solution of a salt of a weak base and a strong acid ($NH_4^+Cl^-$). $pH = \frac{1}{2}(pK_w - pK_b - \log C)$.

4. Beyond the end point, the hydronium ion concentration of the unreacted HCl only needs to be considered when calculating the pH of the solution.

Figure 41 shows that methyl red is a suitable indicator for this type of titration. Phenolphthalein, as the curve shows, would give a premature end point at about 19 ml and would give erroneous results if used in this type of titration.

Titration Curve of Sodium Carbonate Using HCl as Titrant. Figure 42 shows the titration curve of sodium carbonate using HCl as the titrant. Observe two breaks in the titration curve. The first occurs at

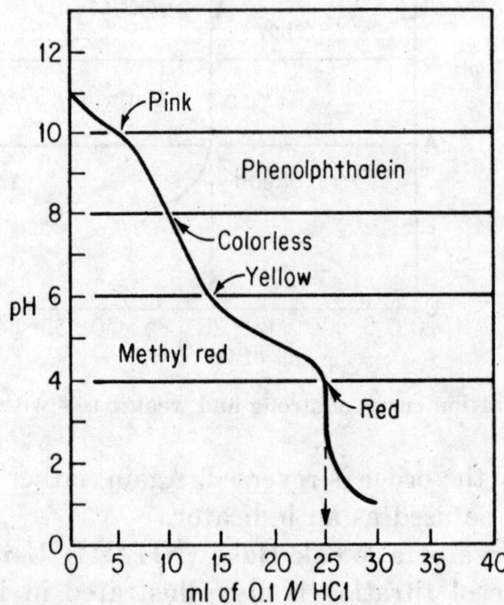

FIG. 42. Titration curve of carbonate with a strong acid.

about pH 8.3 for the reaction

$$CO_3^{=} + H_3O^+ \rightleftharpoons HCO_3^- + H_2O$$

and the second occurs at about pH 4 for the reaction

$$HCO_3^- + H_3O^+ \rightleftharpoons H_2CO_3 + H_2O$$

For this titration phenolphthalein, methyl red, or methyl orange may be used as the indicator. One equivalent of hydrogen is consumed at the phenolphthalein end point, and two equivalents of hydrogen are consumed at the methyl red and methyl orange end points.

Indicators. Indicators used in assays based on neutralization reactions may be classified chemically as weak organic acids and bases. In this discussion only the weak organic acid type of indicator will be considered. However, the principles may be applied equally well to indicators which are weak organic bases.

Three physical-chemical properties upon which the action of an indicator depends are (1) the color of the ionic species of an indicator is different from that of the undissociated species, (2) the color which the indicator imparts to a solution depends on the relative concentration of ionic and undissociated species in solution, and (3) the relative concentrations of the ionic and undissociated species in solution depend on the hydronium ion concentration. In this discussion, we will assume that the undissociated species, HIn, of the indicator is red, that the ionized species, In^-, is yellow, and that the K_a value equals 10^{-5}.

$$\underset{\text{(Red)}}{HIn} + H_2O \rightleftharpoons H_3O^+ + \underset{\text{(Yellow)}}{In^-} \qquad K_a = \frac{[H_3O^+][In^-]}{[HIn]}$$

We will further assume that the solution containing the indicator appears red to the observer when the ratio of $[In^-]/[HIn]$ is 1:10 or less, and yellow when the ratio is 10:1 or greater. (Experiments have shown this relationship to hold for many indicators.)

As previously stated, the pH of the solution determines the relative concentration of these species. As can be seen from the following calculation, the ratio $[In^-]/[HIn]$ is equal to 1:10 at pH 4; the solution will therefore appear red.

$$pH = pK_a + \log \frac{[In^-]}{[HIn]}$$

$$4.00 = 5.00 + \log \frac{[In^-]}{[HIn]}$$

$$\log \frac{[In^-]}{[HIn]} = -1 \qquad \frac{[In^-]}{[HIn]} = 10^{-1} = \frac{1}{10}$$

At lower pH values the ratio becomes smaller, indicating that the concentration of HIn (red species) increases and the concentration of the yellow species In^- decreases. Therefore, the intensity of the red color increases.

At a pH value of 6, the ratio $[In^-]/[HIn]$ is equal to 10:1 and the solutio appears yellow. At higher pH values the ratio increases, indicating tha the concentration of In^- increases and the concentration of HIn decrease The intensity of the yellow color therefore increases. Between these value (10:1 and 1:10) the ratio of concentration is such that both species con tribute significantly to the color, and the solution appears as a shade c orange.

An important conclusion which may be drawn from these consider ations and observations is that the minimum change in pH required fo an indicator to change from one perceptible color to another (red t yellow in this example) is about two (2) pH units. This statement may b expressed symbolically as $pH = pK_a \pm 1$.

A table of commonly used indicators (Table 7) showing the indicato range and other data is given in Chap. 3.

Questions

1. Refer to the current U.S.P. and N.F. and name five medicinal agents which ar salts of weak acids and a strong base, and five medicinal agents which are salts of weak base and a strong acid.

2. Calculate the pH of the following solutions:

(*a*) 0.0035 *M* HCl
(*b*) 0.00048 *M* H_2SO_4
(*c*) 0.0048 *N* H_2SO_4
(*d*) 0.0096 *M* acetylsalicylic acid ($K_a = 3.27 \times 10^{-4}$)
(*e*) 0.2 percent cocaine ($K_b = 2.6 \times 10^{-6}$

3. Calculate the hydronium ion concentration of the following (see Chap. 3):

(*a*) HCl solution, pH 3.41
(*b*) NaOH solution, pH 8.20
(*c*) Physostigmine solution pH 9.74
(*d*) Benzyl penicillin solution, pH 2.22

4. Forty-nine milliliters of 0.1 *M* NaOH solution is added to 50 ml of 0.11 *M* HC solution. Calculate the pH of the final solution.

5. Thirty grams of glacial acetic acid, 98.00 percent, is added to sufficient water t make 500 ml of final solution. Calculate the pH.

6. Assume that the K_a value of a weak acid is 4.89×10^{-3}. A solution is prepare by dissolving 10 mM of this acid (HA) and 20 mM of its sodium salt (A^-) in sufficien water to make 200 ml of solution. Calculate the hydronium ion concentration and pl of the solution.

7. Prepare a titration curve using the following data: Exactly 20 ml of 0.1 *I* NH_3 (K_b 1.80×10^{-5}) was placed in a beaker and diluted to 60 ml with water. Calcu late the pH of the solution after the following quantities (in milliliters) of 0.08 *N* HC solution were added to the original solution: (*a*) zero, (*b*) 5.00, (*c*) 10.00, (*d*) 14.00, (*e* 23.00, (*f*) 24.00, (*g*) 24.50, (*h*) 24.90, (*i*) 25.00, (e) 25.50, (*k*) 26.00, (*l*) 27.00.

EXPERIMENTAL DETERMINATION OF pH

Potentiometric Method. For purposes of measuring the pH o pharmaceutical preparations and solutions, the U.S.P. and N.F. hav adopted the modern operational scale which is defined by the equatio

$$pH = pH_s + \frac{E - E_s}{kT} \qquad (57$$

where pH_s = the pH value assigned to the standard buffer solution taken
E and E_s = the emf of the cell containing, respectively, the "unknown" solution and the standard buffer solution
k = a Nernst related constant
T = the temperature

The cell consists of an indicator electrode reversible to hydrogen ion (such as a glass electrode) and a reference electrode (such as the calomel electrode).

Equation (57) is seldom used as such to calculate pH values from emf measurements. pH values are usually obtained expediently, accurately,

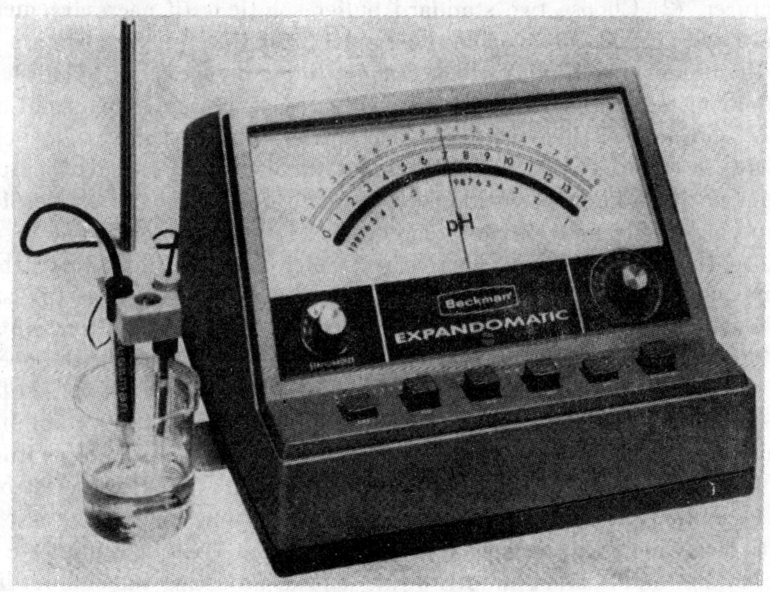

FIG. 43. Beckman Expandomatic pH meter. (*Courtesy of Beckman Instruments, Inc.*)

and reproducibly using a commercial laboratory model pH meter such as the Beckman Expandomatic, equipped with a glass and calomel electrode as shown in Fig. 43. However Eq. (57) provides manufacturers of pH meters with a fundamental operational definition of pH around which their instruments may be electronically designed. As an example, rearrangement of Eq. (57) to $(E - E_s)/(\text{pH} - \text{pH}_s) = kT$ shows that at a given temperature, one and only one potential value corresponds to a pH change of exactly 1 pH unit. At 25° the ratio $(E - E_s)/(\text{pH} - \text{pH}_s) =$ 59.16 millivolts per 1 pH unit, and at 60° this ratio has a value of 66.10 millivolts per 1 pH unit. Since a pH meter is essentially a voltmeter in which values of emf are converted to pH units, it is necessary that an adjustment control be available which will compensate for emf changes produced at different temperatures. Equation (57), along with suitable buffers used for standards, also provides a common bases for reproducible pH measurements, and it permits the use of the measured pH values as close approximations of the true value.

Potentiometric pH Measurements. As previously stated, pH values are usually obtained using a commercial model pH meter equipped with a glass and calomel electrode. No detailed instructions will be given for the operation of these instruments, as they are provided by the manufacturer. In general, however, the following procedure is recommended for pH determinations with a precision greater than 0.1 unit: (1) Allow the instrument to warm up in accordance with the instruction of the manufacturer. (2) Choose two standard buffer solutions (if necessary an exploratory pH measurement may be made to obtain the approximate pH of the unknown solution) which bracket the anticipated pH value of the unknown solution. (3) Allow the temperature of the buffer solutions, wash solution (distilled water in wash bottle), and glass and calomel electrodes to come within 2° of the temperature at which the pH measurement of the unknown solution is to be made. (4) Adjust the temperature knob of the pH meter to the temperature of the unknown solution. (5) Rinse the electrodes and sample container with distilled water three times and wipe carefully with clean absorbent tissue. (6) Place a buffer solution in the container and with the electrodes immersed in the solution rotate the standardization (asymmetry potential) knob until the needle of the pH meter reads the pH value corresponding to the buffer solution taken. Replace buffer solution, using fresh portions of the same buffer, until two successive pH readings come within ± 0.02 pH unit without changing the position of the standardization knob. (7) Rinse electrodes and sample container with distilled water three times and wipe carefully with clean absorbent tissue. Place second buffer in container, immerse the electrodes, and observe the pH value *without* adjusting the standardization knob. (8) Repeat step (7), using fresh portions of the second buffer, without changing the standardization knob until two successive readings agree within ± 0.02 pH unit. If the pH value of the second buffer comes within ± 0.02 pH unit of the first value, the pH meter is considered to be in satisfactory operating condition. (9) Finally, rinse electrodes and sample container with distilled water three times and wipe carefully with clean absorbent tissue. Place unknown solution in container, immerse electrodes in the solution, and note the pH reading. Using fresh portions of the unknown solution, again measure the pH until two successive readings agree within ± 0.02 unit. Record the pH value. The value so obtained is the pH of the unknown solution at the particular temperature at which the measurements were made. It is important to emphasize that pH measurements made at elevated temperatures must be made with pH meters and electrodes which have been standardized with buffer solutions at the same elevated temperature as the unknown solution.

The apparatus is calibrated with two buffers differing by at least two units to check the reliability of the pH meter and the electrodes used. The presence of a faulty glass electrode will be detected by failure to obtain a correct value for the pH of the second buffer when the meter has been standardized in terms of the first. If the pH of the solution to be tested is more alkaline than 9 it is necessary to use the type of glass electrode which has a negligible alkali error in the region of the test. Such electrodes are available from various commercial instrument firms. Some manufacturers provide a sodium ion correction monograph which may be used with glass electrodes to correct for the alkali error.

The pH of solutions which are only partially aqueous is approximated in the manner described above. Report the values so obtained as *apparent* pH values. It must be clearly recognized that, when a pH meter is standardized in terms of an aqueous buffer and then used to measure the "pH" of a nonaqueous solution or suspension, the ionization constant of the acid or base, the dielectric constant of the medium, the liquid-junction potential (which may give rise to errors of approximately 1 pH unit), and the hydronium ion response of the glass electrode are all changed. For these reasons the values so obtained must be reported as *apparent* pH values. This is particularly important for hydroalcoholic solutions.

Colorimetric Methods for Determining pH. The hydronium ion concentration of solutions can also be determined colorimetrically. The potentiometric method is more accurate, in general, and is to be preferred, even though it requires a more elaborate apparatus.

A number of colorimetric methods for the approximate determination of hydronium ion concentration have been developed. The most generally used of these methods involve the matching of the color obtained when a definite quantity of indicator is added to a definite volume of liquid being tested, with the color of one of a series of standard buffers differing by about 0.2 pH unit. The volume of the unknown and of the standard buffer solution must be the same, both of them must contain the same amount of indicator solution, and the colors must be matched by viewing them through the same depth or thickness of solution. Except in a few cases, the limit of accuracy attainable by the colorimetric method does not exceed 0.1 pH unit.

The colorimetric method depends on the assumption that an indicator in the same concentration will exhibit the same transition shade in different aqueous solutions if the hydrogen ion concentrations of the solutions are the same. When one has a series of buffer solutions of definite hydrogen ion concentrations, comparisons can be made between the shades of color produced in the buffer solutions and that produced by the same indicator in the solution to be tested.

In the determination of an unknown pH value, a suitable indicator must first be found. Only indicators that show an intermediate color between the extreme acid and alkaline colors can be used. Table 57 contains a list of indicators suitable for pH measurement. The indicator solutions are prepared by triturating 100 mg of the indicator with the volume of 0.05 *N* NaOH specified in Table 57 and diluting to 200 ml with carbon dioxide–free distilled water (0.05 percent indicator solution).

The first step in the choice of a suitable indicator is the determination of the approximate pH value of the solution under investigation. A few simple tests will usually supply the necessary information. Add a drop or two of phenolphthalein T.S. to a small portion of the solution. If the indicator remains colorless, the pH of the solution is less than 8.4. A second test is conducted in the same manner, using methyl orange T.S. as the indicator. If the solution assumes the alkaline color (yellow), the pH of the solution is greater than 4.4 and lies somewhere between 4.4 and 8.0. A few more tests with methyl red (pH interval 4.2 to 6.3), bromothymol

TABLE 57. pH INDICATORS

Indicator	pH range	Molecular weight	Color change	Solvent
Methyl yellow	2.9 to 4.0	225	Red–yellow	Alcohol
Bromophenol blue	3.0 to 4.6	669	Yellow–blue	3.0 ml of 0.05 *N* NaOH
Bromocresol green	3.8 to 5.4	699	Yellow–blue	2.8 ml of 0.05 *N* NaOH
Methyl red	4.2 to 6.3	269	Red–yellow	7.4 ml of 0.05 *N* NaOH
Bromocresol purple	5.2 to 6.8	540	Yellow–purple	3.7 ml of 0.05 *N* NaOH
Bromothymol blue	6.0 to 7.6	624	Yellow–blue	3.2 ml of 0.05 *N* NaOH
Phenol red	6.8 to 8.4	354	Yellow–red	5.7 ml of 0.05 *N* NaOH
Thymol blue	8.0 to 9.6	466	Yellow–blue	4.3 ml of 0.05 *N* NaOH
Thymolphthalein	9.3 to 10.5	430	Colorless–blue	Alcohol

blue (6.0 to 7.6), and phenol red (6.8 to 8.4) will give a close enough approximation of the pH value to show which indicator may be successfully used in the determination.

When the approximate pH value has been determined and a suitable indicator agreed upon, 2- to 10-ml portions of the unknown solution (depending upon the amount of the liquid available) are transferred to a hard, resistant-glass test tube approximately 15 cm long and with a 1.5-cm bore, and a measured amount of the indicator solution is added. As a rule, 0.10 to 0.20 ml of a 0.05 percent indicator solution, added from a 1-ml pipet, graduated to 0.01 ml per 10 ml of the solution being tested, constitutes a proper indicator concentration.

Transfer from four to six portions of buffer solutions (see Buffer Solutions, later in this chapter), the pH values of which overlap that of the unknown solution, to test tubes, and treat in exactly the same way as the solution being analyzed. The same amounts of indicator must be added to the unknown and to the buffer solutions. It is also essential that the test tubes used be of the quality and type already indicated. The color of the

TABLE 58. pH OF SOME OFFICIAL SUBSTANCES

Substance	pH
U.S.P.	
Iodinated I^{131} serum albumin	7.0 to 8.5
Aluminum acetate solution	3.6 to 4.4
Aluminum sulfate	n.l.t. 2.9 (20)
Aminosalicylic acid	3.0 to 3.7 (sat.)
Ammonium chloride	4.5 to 6.0 (20)
Ammonium chloride injection	4.0 to 6.0 (10)
Ascorbic acid injection	5.5 to 7.0
Atropine sulfate injection	4.0 to 6.0
Bemegride injection	4.5 to 6.5
Bentonite	9.0 to 10.0 (50)
Benzyl benzoate lotion	8.5 to 9.2
Betazole HCl injection	2.8 to 3.4
Bethanechol chloride injection	5.5 to 7.5
Caffeine and sodium benzoate injection	6.5 to 8.5
Calcium disodium edetate	6.5 to 8.0 (5)
Calcium disodium edetate injection	6.5 to 8.0
Carbachol ophthalmic solution	5.0 to 7.0
Cherry juice	3.0 to 4.0
Chlorocyclizine HCl	4.8 to 5.5 (100)
Chloroprocaine HCl injection	2.7 to 4.0
Codeine PO_4 injection	3.0 to 6.0
Corticotropin injection	3.0 to 7.0
Sterile corticotropin zinc hydroxide suspension	7.0 to 8.0
Cyanocobalamin Co^{57} solution	4.0 to 5.5
Cyanocobalamin Co^{60} solution	4.0 to 5.5
Cyclizine HCl	4.5 to 5.5 (50)
Cyclopentolate HCl	5.0 to 5.4 (100)
Cyclopentolate HCl ophthalmic solution	3.5 to 5.5
Dextrose injection	3.5 to 6.0
Dibucaine HCl injection	4.5 to 7.0
Disodium edetate	4.0 to 6.0 (20)
Disodium edetate injection	6.5 to 7.5
Edrophonium chloride injection	5.2 to 5.6
Emetine HCl injection	3.0 to 5.0
Epinephrine injection	2.5 to 5.0
Ergotamine tartrate injection	3.0 to 4.0
Evans blue injection	5.5 to 7.5
Folic acid injection	8.0 to 11.0
Glucagon for injection	2.5 to 3.0 (1,000)
Gold sodium thiomalate	5.8 to 6.5 (10)
Gold sodium thiomalate injection	5.8 to 6.5

Substance	pH
U.S.P.	
Homatropine HBr ophthalmic solution	6.0 to 6.5
Hydrocortisone acetate ophthalmic suspension	6.0 to 8.0
Sterile hydrocortisone acetate suspension	5.0 to 7.0
Hydrocortisone sodium succinate for injection	7.0 to 8.0 (20)
Hydroxyamphetamine HBr ophthalmic solution	4.2 to 6.0
Insulin injection	2.5 to 3.5
Isophane insulin injection	7.1 to 7.4
Insulin zinc suspension	7.1 to 7.5
Extended insulin zinc suspension	7.1 to 7.5
Prompt insulin zinc suspension	7.1 to 7.5
Protamine zinc insulin suspension	7.1 to 7.4
Iron dextran injection	5.2 to 6.0
Isoniazid	6.0 to 7.5 (10)
Isoniazid injection	6.0 to 7.0
Isoproterenol HCl injection	4.0 to 5.0
Levarterenol bitartrate injection	3.0 to 4.5
Lidocaine HCl injection	6.0 to 7.0
Lidocaine HCl injection with epinephrine	3.3 to 5.5
Lidocaine HCl jelly	6.0 to 7.0
Magnesium sulfate injection	5.5 to 7.0
Mechlorethamine HCl	3.0 to 5.0 (500)
Mechlorethamine HCl for injection	3.0 to 5.0 (50)
Sterile medroxyprogesterone acetate suspension	3.5 to 7.0
Meglumine diatrizoate injection	6.0 to 7.6
Meglumine iodipamide injection	6.5 to 7.7
Meglumine iothalamate injection	7.0 to 7.6
Menadiol sodium diphosphate injection	7.5 to 8.5
Meralluride injection	7.0 to 8.5
Metaraminol bitartrate	3.2 to 3.5 (20)
Metaraminol bitartrate injection	3.5 to 4.5
Methadone HCl	4.5 to 6.5 (100)
Methadone HCl injection	3.0 to 6.5
Methoxamine HCl injection	3.0 to 5.0
Methylene blue injection	3.0 to 4.5
Methylergonovine maleate	4.4 to 5.2 (5,000)
Morphine injection	3.0 to 6.0
Nalorphine HCl injection	6.0 to 7.5
Niacinamide injection	5.0 to 7.0
Nitrofurantoin oral suspension	About 4.5
Oxytocin injection	2.5 to 4.5
Parathyroid injection	2.5 to 3.0
Phenolsulfonphthalein injection	6.0 to 7.5

TABLE 58. pH OF SOME OFFICIAL SUBSTANCES (*Continued*)

Substance	pH
U.S.P.	
Phentolamine mesylate for injection	4.5 to 6.0 (1%)
Phenylephrine HCl injection	3.0 to 6.5
Phenylephrine HCl ophthalmic solution	4.0 to 7.5
Sterile phytonadione emulsion	5.0 to 7.0
Posterior pituitary injection	2.5 to 4.5
Sterile prednisolone acetate suspension	5.0 to 7.0
Prednisolone sodium phosphate	7.5 to 8.5 (100)
Prednisolone sodium phosphate injection	6.0 to 7.0
Prednisolone sodium phosphate ophthalmic solution	6.2 to 6.8
Primidone oral suspension	5.5 to 6.5
Procainamide HCl injection	4.0 to 6.0
Procaine HCl injection	3.3 to 5.5
Prochlorperazine edisylate injection	4.2 to 6.2
Promethazine HCl injection	4.0 to 5.5
Sterile propyliodone suspension	6.5 to 7.5
Protein hydrolysate injection	5.0 to 7.0
Pyridoxine HCl injection	2.0 to 3.8
Pyrvinium pamoate oral suspension	6.5 to 8.5
Raspberry juice	2.7 to 3.8
Reserpine injection	3.0 to 4.0
Riboflavin injection	4.5 to 7.0
Ringer's injection	5.0 to 7.5
Lactated ringer's injection	6.0 to 7.5
Rose bengal sodium I^{131} injection	7.0 to 8.5
Scopolamine HBr injection	4.0 to 6.0
Selenium sulfide detergent suspension	3.6 to 4.5
Silver nitrate ophthalmic solution	4.5 to 6.0
Sodium aminohippurate injection	7.0 to 7.6
Sodium aminosalicylate	6.5 to 8.5 (50)
Sodium amobarbital	9.6 to 10.4 (20)
Sodium chloride injection	4.5 to 7.0
Sodium chromate Cr^{51} injection	7.5 to 8.0
Anticoagulant sodium citrate solution	6.4 to 7.5
Sodium diaztrizoate injection	6.5 to 7.5
Sodium glucosulfone injection	5.0 to 6.5
Sodium heparin	5.0 to 7.5 (100)
Sodium indigotindisulfonate injection	2.8 to 3.8
Sodium iodide I^{131} solution	7.5 to 9.0
Sodium iodipamide injection	6.5 to 7.7
Sodium iodohippurate I^{131} injection	7.0 to 8.5
Sodium iothalamate injection	7.0 to 7.6
U.S.P.	
Sodium lactate injection	6.0 to 7.3
Sodium pentobarbital	9.8 to 11.0 (10)
Sodium pentobarbital for injection	10.0 to 10.5 (10)
Sodium pentobarbital injection	9.8 to 10.5
Sodium phenobarbital injection	8.5 to 10.0
Sodium phosphate P^{32} solution	5.0 to 6.0
Sodium secobarbital	9.7 to 10.5 (20)
Sodium sulfacetamide	8.0 to 9.5 (20)
Sodium sulfadiazine injection	8.5 to 10.5
Sodium sulfobromophthalein injection	5.0 to 6.5
Sodium suramin	5.5 to 7.0 (100)
Sodium thiopental for injection	10.0 to 11.0 (40)
Sodium thiosulfate injection	8.0 to 9.5
Sodium warfarin	7.2 to 8.3 (100)
Stibophen injection	5.0 to 7.0
Succinylcholine chloride injection	3.0 to 4.5
Tetracaine HCl injection	3.5 to 6.0
Thiamine HCl	2.7 to 3.4 (100)
Thiamine HCl injection	2.5 to 4.5
Thiamine mononitrate	6.0 to 7.5 (50)
Thiotepa for injection	7.0 to 8.2 (0.75%)
Tubocurarine chloride injection	3.0 to 5.0
Zinc sulfate ophthalmic solution	5.8 to 6.2
N.F.	
Acetaminophen	5.5 to 6.5 (sat.)
Acetyl sulfisoxazole oral suspension	5.0 to 5.5
Alphaprodine HCl injection	4.0 to 6.0
Aluminum phosphate gel	6.0 to 7.2
Amphetamine phosphate	4.0 to 5.0 (20)
Amphetamine sulfate injection	4.0 to 6.3
Anileridien HCl	2.5 to 3.0 (20)
Antazoline phosphate	4.0 to 5.0 (50)
Azacyclonol HCl	5.0 to 7.0 (200)
Benzypyrinium bromide	4.5 to 5.5 (100)
Benzpyrinium bromide injection	4.8 to 5.2
Blood-group specific substances A and B	6.0 to 6.8
Brompheniramine maleate	4.0 to 5.0 (100)

TABLE 58. pH OF SOME OFFICIAL SUBSTANCES (*Continued*)

Substance	pH
N.F.	
Brompheniramine maleate elixir	2.5 to 3.5
Butethamine HCl	About 5 (100)
Butethamine HCl and epinephrine injection	3.3 to 5.5
Calcium aminosalicylate	6.0 to 8.0 (50)
Calcium gluconate injection	6.0 to 8.2
Carbinoxamine maleate	4.6 to 5.1 (100)
Chlorotrianisene	5.0 to 7.0 (sat.)
Chlorpheniramine maleate injection	4.0 to 5.2
Choline dihydrogen citrate	3.5 to 4.5 (4)
Cobalamin concentrate	4.0 to 8.0 (200)
Sterile cortisone acetate suspension	5.0 to 7.0
Cyclopentamine HCl	About 6 (100)
Cyclophosphamide	4.3 to 4.7 (100)
Cycrimine HCl	5.2 to 5.8 (200)
Sterile desoxycorticosterone pivalate suspension	5.0 to 7.0
Dexamethasone sodium phosphate	7.5 to 10.5 (100)
Dexamethasone sodium phosphate ophthalmic solution	7.2 to 7.8
Dexchlorpheniramine maleate	4.0 to 5.0 (100)
Dextroamphetamine phosphate	4.0 to 5.0 (20)
Dextromethorphan HBr	5.2 to 6.5 (100)
Dihydroxyaluminum aminoacetate	6.5 to 7.5 (25)
Dihydroxyaluminum aminoacetate magma	6.5 to 7.5 (25)
Dihydroxyaluminum aminoacetate tablets	6.5 to 7.5 (25)
Dimethisoquin HCl	3.5 to 5.0 (100)
Diphemanil methylsulfate	4.0 to 6.0 (100)
Diphemanil methylsulfate injection	3.0 to 5.0
Ephedrine sulfate injection	4.5 to 7.0
Ethinamate	6.5 to 7.0 (sat.)
Ethoheptazine citrate	3.5 to 4.5 (100)
Fructose injection	3.0 to 6.0
Fructose and sodium chloride injection	3.0 to 6.0
N.F.	
Furazolidone and nifuroxime powder	3.0 to 4.5 (25)
Gallamine triethiodide injection	6.5 to 7.5
Gold Au^{198} solution	5.5 to 7.5
Halazone tablets*	n.l.t. 7.0
Hexylcaine HCl	4.0 to 6.0 (20)
Hexylcaine HCl injection	3.0 to 5.0
Hexylcaine HCl solution	3.0 to 5.0
Hydromorphone HCl injection	4.0 to 5.5
Impramine HCl injection	4.0 to 5.0
Isobucaine HCl	About 6 (50)
Isoproterenol sulfate	About 5 (100)
Levallorphan tartrate injection	4.0 to 4.5
Levorphanol tartrate injection	4.2 to 4.4
Mannitol injection	5.0 to 7.0
Menadione sodium bisulfite injection	2.0 to 4.0
Meperidine HCl injection	3.5 to 6.0
Mephentermine sulfate injection	4.0 to 6.5
Mepivacaine HCl	About 4.5 (50)
Mepivacaine HCl injection	4.5 to 6.8
Mepivacaine HCl and levonordefrin injection	3.3 to 5.5
Meprylcaine HCl	About 5.7 (50)
Mercurophylline injection	7.5 to 9.0
Metabutethamine HCl	About 6 (50)
Metabutoxycaine HCl	About 5.5 (50)
Metharbital	About 6 (sat.)
Methionine	5.6 to 6.1 (100)
Methscopolamine Br injection	4.5 to 6.0
Sterile methylprednisolone acetate suspension	3.5 to 7.0
Methylprednisolone sodium succinate for injection	7.0 to 8.0 (20)
Methyprylon elixir	4.6 to 5.0
Naphthazoline HCl	n.l.t. 6.1 (100)
Niacin injection	4.0 to 6.0
Nikethamide	6.5 to 7.5 (4)
Nitrofurazone	5.0 to 7.5 (100)
Noscapine HCl	2.5 to 3.5 (20)
Nylidrin HCl	4.5 to 6.5 (100)
Octylphenoxy polyethoxyethanol	7.0 to 9.0 (20)
Papaverine HCl injection	n.l.t. 3.0

TABLE 58. pH OF SOME OFFICIAL SUBSTANCES (*Continued*)

Substance	pH	Substance	pH
N.F.		**N.F.**	
Pentylenetetrazole injection....	6.0 to 8.5	Ringer's solution..............	5.0 to 7.5
Pheniramine maleate..........	4.5 to 5.5 (100)	Sodium butabarbital..........	9.0 to 10.2 (100)
Phenmetrazine HCl..........	4.5 to 5.5 (40)	Sodium dehydrocholate injection	8.5 to 9.5
Pipradol HCl................	5.0 to 7.0 (100)	Sodium hexobarbital.........	10.5 to 12.0 (10)
Polyethylene glycol 300........	4.0 to 7.0 (20)	Sodium methiodal injection....	5.0 to 8.0
Polyethylene glycol 1540........	4.0 to 7.0 (20)	Sodium methohexital for injection..........	10.4 to 11.4
Lactated potassic saline injection	6.5 to 7.5	Sodium sulfabromomethazine...	9.5 to 11.0 (10)
Potassium acetate............	7.5 to 8.5 (20)	Sodium sulfamerazine injection.	8.5 to 10.5
Potassium warfarin...........	7.2 to 8.3 (100)	Sodium thiamylal for injection..	10.5 to 11.0 (40)
Povidone-iodine solution.......	n.m.t. 6.0	Sodium vinbarbital injection...	8.0 to 10.0
Pramoxine HCl..............	About 4.5 (100)	Stannous fluoride*...........	2.8 to 3.5
Sterile progesterone suspension..	4.0 to 7.5	Sterile testosterone suspension..	4.0 to 7.5
Promazine HCl..............	4.2 to 5.2 (20)	Theophylline sodium glycinate..	8.5 to 9.5 (sat.)
Promazine HCl injection.......	4.0 to 5.5	Thimerosal.................	About 6.7 (100)
Propoxycaine HCl*...........	About 5.4	Thimerosal solution..........	9.8 to 10.3
Racephedrine HCl............	4.5 to 7.0 (20)	Trimethobenzamide HCl injection..........	4.8 to 5.2

* See U.S.P. or N.F. for specific directions.

Number in parentheses designates the final volume of solution prepared to make pH measurement. In each case one part of substance was diluted to the volume designated.

(sat.) = saturated solution; n.l.t. = not less than; n.m.t. = not more than.

unknown solution is then compared with the colors of the buffer solutions, and the pH value of the solution is thus determined.

Buffer Solutions. Buffer solutions which are required for calibrating pH meters and electrodes may be purchased from commercial sources, or they may be prepared from specified reagents and chemicals following the directions given in the U.S.P. XVII or N.F. XII. The five stock solutions needed to prepare the standard buffer solutions listed in the official compendiums are:

1. Hydrochloric acid, 0.2 *M*, and sodium hydroxide, 0.2 *M*. Prepare and standardize 1 liter of each solution following the general directions given in Exercises 9 and 11, respectively, in Chap. 3. The final concentration of these solutions must be at least 0.2000 ± 0.0005 *M*.

2. Potassium biphthalate, 0.2 *M*. Dissolve 40.846 Gm of potassium biphthalate in water and dilute to 1,000 ml.

3. Potassium phosphate, monobasic, 0.2 *M*. Dissolve 27.218 Gm of monobasic potassium phosphate in water and dilute to 1,000 ml.

4. Boric acid and potassium chloride, 0.2 *M*. Dissolve 12.366 Gm of boric acid and 14.911 Gm of KCl in water and dilute to 1,000 ml.

5. Potassium chloride, 0.2 *M*. Dissolve 14.911 Gm of KCl in water and dilute to 1,000 ml.

The directions and quantities of stock solution to be taken in the preparation of pH standard buffer solutions for ranges between pH 1.1 and 10.00 are given in Table 59, which is reproduced from the U.S.P. XVII and N.F. XII. The pH values given are considered reproducible within ±0.02 *M* at 25°. Since standard buffer solutions undergo changes

TABLE 59. COMPOSITION OF STANDARD BUFFER SOLUTIONS

Hydrochloric Acid Buffer

Place 50 ml of the potassium chloride solution in a 200-ml volumetric flask, add the specified volume of the hydrochloric acid solution, then add water to volume.

pH	1.2	1.3	1.4	1.5	1.6	1.7	1.8	1.9	2.0	2.1	2.2
HCl, ml	85.0	67.2	53.2	41.4	32.4	26.0	20.4	16.2	13.0	10.2	7.8

Acid Phthalate Buffer

Place 50 ml of the potassium biphthalate solution in a 200-ml volumetric flask, add the specified volume of the hydrochloric acid solution, then add water to volume.

pH	2.2	2.4	2.6	3.8	3.0	3.2	3.4	3.6	3.8	4.0
HCl, ml	49.5	42.2	35.4	28.9	22.3	15.7	10.4	6.3	2.9	0.1

Neutralized Phthalate Buffer

Place 50 ml of the potassium biphthalate solution in a 200-ml volumetric flask, add the specified volume of the sodium hydroxide solution, then add water to volume.

pH	4.2	4.4	4.6	4.8	5.0	5.2	5.4	5.6	5.8
NaOH, ml	3.0	6.6	11.1	16.5	22.6	28.8	34.1	38.8	42.3

Phosphate Buffer

Place 50 ml of the monobasic potassium phosphate solution in a 200-ml volumetric flask, add the specified volume of the sodium hydroxide solution, then add water to volume.

pH	5.8	6.0	6.2	6.4	6.6	6.8	7.0	7.2	7.4	7.6	7.8	8.0
NaOH, ml	3.6	5.6	8.1	11.6	16.4	22.4	29.1	34.7	39.1	42.4	44.5	46.1

Alkaline Borate Buffer

Place 50 ml of the boric acid and potassium chloride solution in a 200-ml volumetric flask, add the specified volume of the sodium hydroxide solution, then add water to volume.

pH	8.0	8.2	8.4	8.6	8.8	9.0	9.2	9.4	9.6	9.8	10.0
NaOH, ml	3.9	6.0	8.6	11.8	15.8	20.8	26.4	32.1	36.9	40.6	43.7

in pH with temperature, these effects must be taken into consideration when making pH measurements at elevated temperatures. The U.S.P. and N.F. have a table which lists the pH values for buffer solutions from 0 to 60°, which should be used when necessary. When commercial buffers are used to standardize the pH meter, pH-temperature data supplied by the manufacturer should be used.

Exercise 111

Object. To Determine the pH of Sodium Chloride Injection.

Materials Required. 10 or 15 ml of the above solution
A commercial-type pH meter equipped with a glass electrode and a reference electrode
A wash bottle
Standard buffer solutions, commercial or official

Procedure. Following the directions outlined earlier in this chapter under Experimental Determination of pH, determine the pH value of this pharmaceutical preparation potentiometrically and colorimetrically. Compare results.

Chapter 23

POLAROGRAPHY

Polarography is a method of analysis based on the measurement of current resulting from the electrolysis of an electroactive species at a given electrode potential under controlled conditions. Of the different polarographic methods which have been developed, only d-c (direct current) polarography, sometimes referred to as conventional polarography, will be considered. In this method, as in potentiometry, a reference electrode (SCE) and an indicator electrode are required. The reference electrode acts to maintain a constant potential throughout the measurement, and the indicator electrode assumes the potential impressed upon it from an external source. The apparatus is designed in such a manner that electrolysis of the electroactive species takes place at the indicator electrode, and the potential of the indicator electrode is always measured relative to the reference electrode. Figure 44 shows a schematic drawing of a simplified polarographic apparatus which will be used to explain the principles of polarography. Assume that the electrolysis cell of the apparatus contains a reducible species (abbreviated RS) such as Nitrofurantoin, U.S.P., and a supporting electrolyte which supports conduction of current but remains inert in the potential range of interest. When the movable contact M is placed at the extreme left position, the potential impressed across the electrodes as shown on the voltmeter V will be zero and the current measured in microamperes (μamp) on the ammeter A will also register zero. If the contact M is now moved to the right, say to position 0.5, a certain voltage will be impressed across the electrodes which may be read on the voltmeter V. Since the microelectrode is negative, ions (impurities in the salts and water) capable of picking up electrons at this low potential will

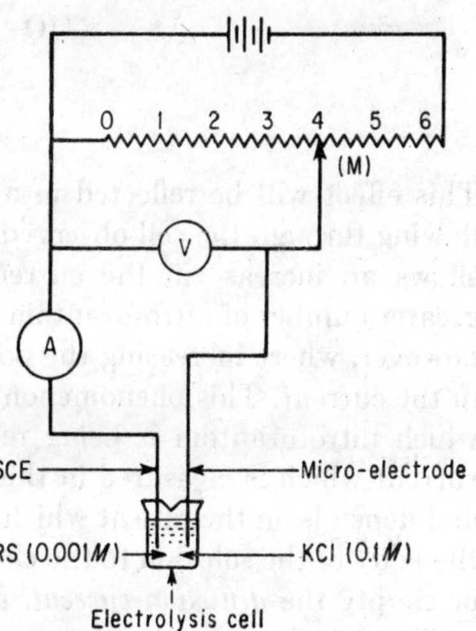

FIG. 44. Simplified polarographic apparatus.

be reduced and the current flowing through the cell will be registered on the ammeter. If the movable contact is advanced to position 1, a greater voltage will be impressed across the electrodes, more ions will be reduced, and an increase in current flowing through the cell will be observed. If we continue moving the contact, thereby increasing the voltage impressed upon the microelectrode, we will eventually reach a potential which will reduce nitrofurantoin according to the following equation.

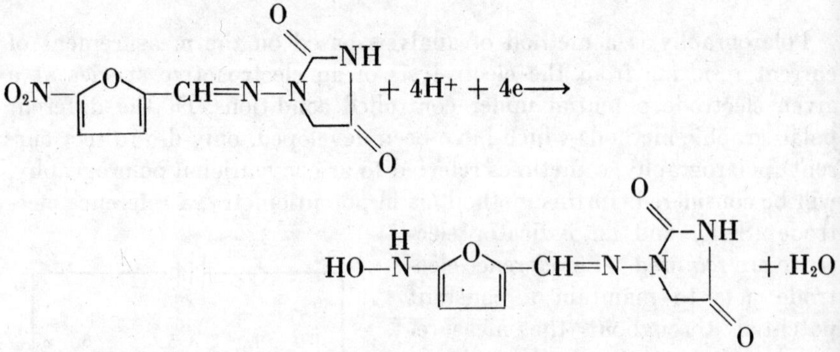

$$O_2N-\overset{O}{\square}-CH=N-N\overset{O}{\underset{O}{\square}}NH + 4H^+ + 4e \longrightarrow$$

$$HO-\overset{H}{N}-\overset{O}{\square}-CH=N-N\overset{O}{\underset{O}{\square}}NH + H_2O$$

This effect will be reflected in a relatively large increase in the current flowing through the cell observed on the ammeter. Increasing the voltage shows an increase in the current flowing through the cell, because a greater number of nitrofurantoin molecules are reduced. There is a region, however, where increasing the potential does not bring about an increase in the current. This phenomenon is explained by the fact that the rate at which nitrofurantoin is being reduced is constant. In other words, the current which is measured in this potential region is diffusion-controlled, and depends on the rate at which the nitrofurantoin molecules move from the body of the solution to the electrode. The diffusion-controlled current, or simply the *diffusion current, is a function of concentration.*

The data in polarography are best expressed graphically, as shown in Fig. 45, in which voltage versus SCE is plotted along the abscissa of the graph and current in microamperes is plotted along the ordinate. Such a graph is called a polarogram, and the trace is called a polarographic wave. The sawtooth effect seen on the polarogram indicates that the dropping mercury electrode (DME) was used as the indicator electrode. Polarographic data are obtained most easily from an automatic recording instrument (Fig. 46). These instruments are designed to change automatically the potential applied to the DME and to record simultaneously both potential and current on a strip chart recorder.

Polarogram. The polarogram is conveniently divided into three regions, designated as *A*, *B*, and *C* in Fig. 45. Region *A* results from the charging current of the electrode and reduction of traces of impurities

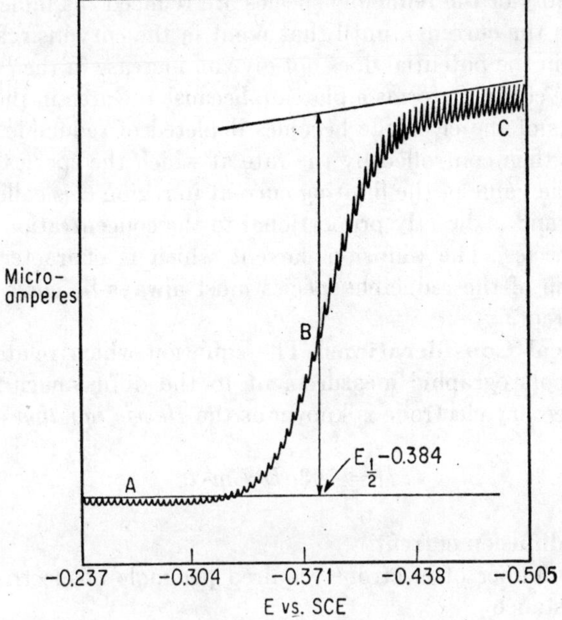

FIG. 45. Polarogram of an electroreducible species.

FIG. 46. S-29310 Sargent model XV polarograph.

present in the solution. This current is referred to as the *residual current*. Region *B* results from the reduction of the electroactive species in solution. The shape of the curve in this region indicates that there are more reducible species in the immediate space surrounding the electrode than can be reduced. As the potential of the electrode is increased, however, more and more of the reducible species are reduced (as indicated by the sharp rise in the current), until that point in the curve is reached where an increase in the potential does not give an increase in the current. This region of the curve, *C*, forms a plateau, because the area in the immediate surroundings of the electrode becomes depleted of reducible species and reduction is then controlled by the rate at which the species diffuse into this area. The value of the limiting current in region *C* is called the *diffusion current* and is directly proportional to the concentration of the electroactive species. The diffusion current which is characteristic of the concentration of the reducible species must always be corrected for the residual current.

Theoretical Considerations. The equation which relates the variables in a polarographic measurement to the diffusion current at the dropping mercury electrode is known as the *Ilkovic equation* and is given as follows:

$$i_d = 708nD^{1/2}Cm^{2/3}t^{1/6}$$

where i_d = diffusion current
n = number of electrons required per mole of electroactive substance
D = diffusion coefficient, cm² per sec
C = concentration, millimoles per liter
m = mass, mg of mercury flowing per second
t = drop time. Units: sec per drop
708 = a constant which depends in part on the units chosen for the variables, on the dimensions of the mercury drop, and on the instrumentation used

The Ilkovic equation clearly shows that the diffusion current i_d is a function of different variables. If during a particular analysis all variables except concentration are held constant, the Ilkovic equation reduces to

$$i_d = KC$$

where *K* indicates that all the variables are held constant during the measurement.

It is this relationship which is used during the polarographic analysis of the products official in the U.S.P. and N.F. The diffusion currents of a known concentration of reference standard are first determined, followed by the determination of the diffusion current (under identical conditions) of the solution of unknown concentration. The concentration of the unknown is then calculated using the following relationship:

$$\frac{(i_d)_S}{(i_d)_U} = \frac{C_S}{C_U} \quad \text{or} \quad \frac{(cm)_S}{(cm)_U} = \frac{C_S}{C_U} \qquad (1)$$

where S and U = reference standard and unknown, respectively
cm = height, cm, of the polarographic waves (see Measurement of Wave Height, below)

Measurement of Wave Height. Since most polarographic determinations are made using an automatic recording instrument, the peak heights obtained on the polarogram may conveniently be measured with a ruler and related directly to the concentrations of the solutions as shown in Eq. (1). It is not necessary to convert peak height values to diffusion current for quantitative purposes.

To determine the peak heights of a polarogram, both the residual and limiting currents are measured at the maximum of the oscillations by extending the straight lines, as shown in Fig. 45. A line is then drawn perpendicular to the abscissa, through the half-wave potential value and between the limiting and residual current lines. This vertical line represents the peak height of the curve.

Half-wave Potential. The half-wave potential is often used as a qualitative tool and serves as a means of identifying compounds. The half-wave potential ($E_{1/2}$) is independent of concentration and the capillary used. It is dependent on the solution composition and may vary with changes in pH.

Dissolved Oxygen. Dissolved oxygen present in solution gives rise to two reduction waves, as shown by the following equations.

Acid solution:

$$O_2 + 2H^+ + 2e \longrightarrow H_2O_2 \qquad \text{first wave}$$
$$H_2O_2 + 2H^+ + 2e \longrightarrow 2H_2O \qquad \text{second wave}$$

Alkaline solution:

$$O_2 + 2H_2O + 2e \longrightarrow H_2O_2 + 2OH^- \qquad \text{first wave}$$
$$H_2O_2 + 2e \longrightarrow 2OH^- \qquad \text{second wave}$$

Since these waves occur at the potentials which normally interfere with other waves, it is necessary to remove the oxygen. This is best done by bubbling nitrogen gas through the solution for 10 to 15 min before analyzing the sample.

Maximums. It has often been observed that polarographic waves after increasing, as an electroactive species is being reduced, fail to level off but rather fall again, thereby creating a maximum in the curve. This phenomenon is believed to result from the fact that some dissolved material in the solution absorbs to the surface of the mercury drop. The presence of a maximum makes it difficult to measure the peak height. It is best, therefore, to eliminate it. This can be accomplished by using 0.02 percent gelatin solution or some other surface-active agent.

For more complete information regarding the polarographic method of analysis, refer to the references given at the end of this chapter.

Official Products. The following official products are assayed by the polarographic method: (1) Acetazolamide, U.S.P., (2) Sodium Acetazolamide, U.S.P., (3) Sterile Sodium Acetazolamide, U.S.P., (4) Dichlorphenamide tablets, U.S.P., (5) Hydrochlorthiazide tablets, U.S.P., (6) Methazolamide tablets, U.S.P., (7) Nitrofurantoin tablets, U.S.P., (8) Nitrofurantoin Oral Suspension, U.S.P., (9) Dienestrol tablets, N.F.

Exercise 112

Object. Assay of Nitrofurantoin Oral Suspension.

Materials Required. Polarograph
Dropping Mercury Electrode
Saturated Calomel Electrode
Nitrofurantoin Reference Standard

Electrolyte Solution: "Transfer 67.5 ml. of stronger ammonia water to a 1000 ml. volumetric flask, and dilute to volume (A). Transfer 53.50 gm. of ammonium chloride to a 1000 ml. volumetric flask, and dilute to volume (B). Mix equal volumes of solutions A and B."

Gelatin Solution: "Dissolve 100 mg. of gelatin in 100 ml. of boiling water. Prepare this solution fresh weekly."

Procedure. "Transfer an accurately measured volume of Nitrofurantoin Oral Suspension, equivalent to about 50 mg. of nitrofurantoin, to a 100 ml. volumetric flask. Add 20 ml. of dimethylformamide, agitate for 5 minutes, and add *Electrolyte Solution* to volume. Filter the solution through a dry filter paper, discarding the first 25 to 30 ml. of the filtrate. Pipet 3 ml. of the clear filtrate into a 25 ml. volumetric flask, add 0.1 ml. of *Gelatin Solution*, and add water to volume. Transfer a portion of this solution to a polarographic cell that is immersed in a water bath regulated at 24.5° to 25.5°, and de-aerate by bubbling purified nitrogen through the solution for 10 minutes. Insert the dropping mercury electrode of a suitable polarograph, and record the polarogram from 0 volt to −1.0 volt, using a saturated calomel electrode as the reference electrode. Determine the height of the diffusion current at −0.60 volt. Calculate the quantity, in milligrams, of $C_8H_6N_4O_5$ in each milliliter of the Suspension taken by the formula $833(C/V)[(i_d)_U/(i_d)_S]$, in which V is the volume, in milliliters, of Suspension taken, and $(i_d)_U$ is the observed diffusion current of the unknown solution and $(i_d)_S$ is that determined similarly in a solution of U.S.P. Nitrofurantoin Reference Standard, the concentration of which is C mg. per ml. (about 60 mcg. per ml.)."

Before the polarograph can be used to assay this product it is necessary to study the operation manual published by the manufacturer of the instrument. Check all controls to make certain that they are functioning properly. A sensitivity setting on the instrument should be used which will give a peak height equal to at least half the width of the recording paper.

Sample Calculation. Exactly 10 ml of the oral suspension was taken for the assay and diluted as instructed. A polarogram of this dilution measured 12.45 cm. A polarogram of a reference standard solution containing 0.06 mg per ml measured 12.00 cm. Calculate the milligrams of nitrofurantoin in 100 ml of the oral suspension.

1. Determine the amount, in milligrams, of nitrofurantoin in the final dilution using Eq. (1).

$$C_U = C_S \frac{(cm)_U}{(cm)_S}$$

$$C_U = 0.06 \text{ mg/ml} \times \frac{12.45 \text{ cm}}{12.00 \text{ cm}} = 0.0622 \text{ mg/ml}$$

2. Determine the volume, in milliliters of the oral suspension (OS) in the final dilution.

$$\frac{10 \text{ ml (OS)}}{100 \text{ ml}} \times \frac{3 \text{ ml}}{25 \text{ ml}} = 0.012 \text{ ml (OS)/1 ml of dilution}$$

3. Divide (2) into (1) and multiply by 100 to convert to 100 ml.

$$\frac{0.0622 \text{ mg/ml}}{0.012 \text{ ml/1 ml}} \times 100 \text{ ml} = 518 \text{ mg}$$

the amount of nitrofurantoin in 100 ml of the oral suspension.

By combining all constants, the answer may be obtained using the U.S.P. equation.

$$833 \left(\frac{0.06}{10}\right)\left(\frac{12.45}{12.00}\right) = 5.18 \text{ mg/ml or } 518 \text{ mg/100 ml}$$

Karl Fischer Electrometric Titration. The determination of the water content of official drugs is very important in demonstrating compliance with the official standards. Water present in hydrates and that which is absorbed on the surface of drugs can be measured quantitatively in several ways. Certain of these methods are described in Chap. 11, Ash and Water Determinations. One of the important procedures for this determination, however, is the Karl Fischer electrometric titration method.

This method is the most rapid of the official methods, requires only a small sample, is specific for water, and is widely applicable in the determination of the water content of drug products.

The method involves the titration of the sample in reagent methanol with the Karl Fischer reagent which consists of iodine, sulfur dioxide, pyridine, and methanol.

Iodine is reduced by sulfur dioxide in the presence of pyridine and methanol as long as water is present according to the following equations:

$$I_2 + SO_2 + H_2O \rightleftharpoons 2HI + SO_3 \qquad (2)$$

$$SO_3 + C_5H_5N \longrightarrow \underset{\text{(Pyridine-sulfur trioxide)}}{C_5H_5N{\cdot}SO_3} \qquad (3)$$

$$C_5H_5N{\cdot}SO_3 + CH_3OH \longrightarrow C_5H_5N{\cdot}CH_3\text{—}O\text{—}SO_3H \qquad (4)$$

It is apparent that 1 mole of water is equivalent to 1 mole of iodine. The pyridine, by forming a complex with SO_3 Eq. (3) and by neutralizing the HI, prevents the oxidation of HI to I_2 [reversal of Eq. (2)]. This reverse reaction otherwise takes place as the concentration of the acidic products, HI and H_2SO_4 ($SO_3 + H_2O$), increases during the titration. Methanol reacts with the pyridine-sulfur trioxide, Eq. (4). In the absence of methanol, water would react with the complex, Eq. (5), and 2 moles of water would then be equivalent to 1 mole of iodine.

$$C_5H_5N{\cdot}SO_3 + H_2O \longrightarrow C_5H_5N{\cdot}H_2SO_4 \quad (5)$$

Pyridine also forms a complex with the SO_2 in the reagent.

The Karl Fischer reagent contains iodine, together with excess sulfur dioxide, pyridine, and methanol, in such concentration that each milliliter of the reagent is equivalent to approximately 5 mg of water. Since the reagent deteriorates gradually, it is standardized within an hour before use, or daily, if used continuously, by employing sodium tartrate, $Na_2C_4H_4O_6{\cdot}2H_2O$, as a primary standard or a water-methanol solution of known water concentration, as a secondary standard.

Any apparatus may be used which provides for adequate exclusion of atmospheric moisture and determination of the end point. Commercially available apparatus generally comprises a closed system consisting of one or two automatic burets and a tightly covered titration vessel fitted with the necessary electrodes and a magnetic stirrer. The air in the system is kept dry with a desiccant such as phosphorus pentoxide, anhydrous granular calcium chloride, or silica gel. The last may contain an indicator to reveal its state of hydration.

•Figure 47 gives a typical Karl Fischer titration assembly. The apparatus embodies a simple and inexpensive electrical circuit. The operation involves the setting of the movable contact of a potentiometer to that position which gives a current of 5 to 10 μamp through the circuit with the platinum electrodes immersed in the solution to be titrated. At this point in the titration, i.e., before titrant is added, the resistance of the cell is high and most current coming from the battery flows in the upper leg of the circuit. When the end point in the titration is reached, however, a slight excess of iodine is present in the solution. The iodine and the iodide ion (a product of the titration) permit a greater flow of current through the lower leg of the circuit, thereby causing an increase to be registered on the microammeter. The end point is taken when the current flow increases to between 50 and 100 μamp for 30 sec or longer, depending on the solution being titrated.•

Many commercial pH meters may also be adapted to detect the end point in a Karl Fischer titration. When using a pH meter for this purpose, a polarizing current of about 10 μamp is generated and maintained at a constant value throughout the entire titration. The potential developed between the electrodes is measured and is a function of the changes of the composition of the solution. At the end point a slight excess of iodine together with iodide produced during the titration brings about an abrupt drop in the potential of the electrodes, and the needle deflects from a high potential value to a low value. This method of end point detection is referred to as *potentiometry at constant current*.

For many substances the end point can be detected visually by observing the change from a brownish yellow to an amber color. However, for

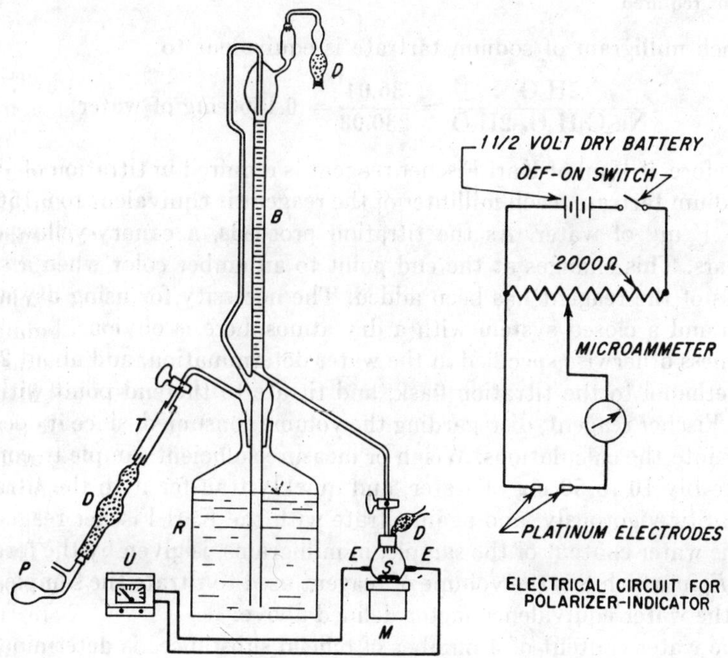

FIG. 47. Automatic buret and titration assembly for Karl Fischer water determination. *B*, buret; *D*, desiccant; *E*, platinum electrodes; *M*, electric magnetic stirring apparatus; *P*, pressure bulb; *R*, reservoir for Karl Fischer solution; *S*, plastic or glass-coated stirring bar; *T*, rubber tubing; *U*, polarizer-indicator.

colored substances or when the end point is not very sharply defined, the electrometric method is to be preferred since there is a definite sharp potential change at the end point.

Exercise 113

Object. Preparation and Standardization of Karl Fischer Reagent.

Materials Required.

About 150 ml of methanol

Sodium tartrate, $Na_2C_4H_4O_6 \cdot 2H_2O$

Karl Fischer Reagent may be purchased from commercial sources or it may be prepared following the directions given in the official compendiums. The solution may be standardized using the following directions.

Procedure. "Place about 36 ml. of methanol in the titration vessel and add sufficient of the *Karl Fischer Reagent* to give the characteristic end point. Quickly add 150 to 350 mg. of sodium tartrate ($Na_2C_4H_4O_6 \cdot 2H_2O$), accurately weighed by difference, and titrate to the end point. The water equivalence factor F, in milligrams of water per milliliter of reagent, is given by the formula $0.1566 \times W \div V$, in which W is the weight, in milligrams, of the sodium tartrate, and V is the volume, in milliliters, of the reagent required."

Each milligram of sodium tartrate is equivalent to

$$\frac{2H_2O}{Na_2C_4H_4O_6 \cdot 2H_2O} = \frac{36.04}{230.08} = 0.1566 \text{ mg of water}$$

Therefore, if V ml of Karl Fischer reagent is required in titration of W mg of sodium tartrate, each milliliter of the reagent is equivalent to $0.1566 \times W \div V$ mg of water. As the titration proceeds, a canary-yellow color appears. This changes at the end point to an amber color when a slight excess of the reagent has been added. The necessity for using dry apparatus and a closed system with a dry atmosphere is obvious.

Unless otherwise specified in the water determination, add about 25 ml of methanol to the titration flask, and titrate to the end point with the Karl Fischer reagent, disregarding the volume consumed, since it does not enter into the calculations. Weigh or measure sufficient sample to contain preferably 10 to 50 mg of water, and quickly transfer it to the titration flask. Stir vigorously, and again titrate with the Karl Fischer reagent.

The water content of the sample, in milligrams, is given by the formula $S \times F$, in which S is the volume of reagent used to titrate the sample, and F is the water equivalence factor defined above.

The water content of a number of official substances is determined by Karl Fischer titration.

Problems

1. Calculate the water equivalence factor F of Karl Fischer reagent if a 180-mg sample of sodium tartrate dihydrate required 15.00 ml of K.F. reagent.
2. Calculate the percent moisture in aminosalicylic acid if 9.00 ml of K.F. reagent, having a water equivalence factor of 4.10 were consumed by a 5.100-Gm sample.

References

1. Brezina, M., and P. Zuman, "Polarography in Medicine, Biochemistry and Pharmacy," Interscience Publishers, Inc., New York, 1958.
2. Ewing, G. W., "Instrumental Methods of Chemical Analysis," 2d ed., McGraw-Hill Book Company, New York, 1960.
3. Kolthoff, I. M., and J. J. Lingane, "Polarography," vols. I and II, Interscience Publishers, Inc., New York, 1952.

Chapter 24

ELECTROLYTIC METHODS

Quantitative analysis by means of electrolysis, or electroanalysis, as it is more generally called, is usually restricted to the determination of metals. This method of analysis is based upon the fact that an electrical current passed through a solution of the salt of a metal causes the deposition of the metal, usually in the elemental state, upon one of the electrodes. When deposition is complete the anode or cathode is removed, washed, dried, and weighed. The increase in weight over the tared electrode represents the weight of the substance deposited.

Theory. Those substances which conduct an electric current when dissolved in aqueous solution are termed *electrolytes.* Electrolytic determinations are based upon the fact that, when a direct current of electricity is passed through a solution of an electrolyte, some of the negatively charged ions are oxidized to the elemental state, and some of the positively charged ions are reduced to the elemental state. This decomposition of an electrolyte in solution is known as *electrolysis*, and when the decomposition of the electrolyte is carried out for purposes of analysis, the process is called *electroanalysis.* The conductor by which the current enters the solution of electrolyte is called the *anode*, and that by which it leaves the solution is known as the *cathode.* The anode and cathode are generally spoken of as positive and negative *electrodes*, respectively, or simply as electrodes. The source of current charges the positive electrode (anode) with positive electricity, and this anode attracts the negatively charged ions (*anions*), but it repels the positively charged ions (*cations*); the latter are attracted by the negatively charged cathode which repels the negatively charged anions, forcing them toward the anode. Since the positively charged ions gain an electron for each positive charge that they carry when they come in contact with the cathode, they are reduced; and since the negatively charged ions lose an electron for each negative charge that they carry when they come in contact with the anode, they are oxidized. The respective reductions and oxidations never take place in the mass of the solution of electrolyte but only at the electrodes, reduction occurring at the cathode and oxidation occurring at the anode. Simple metallic ions (cations), such as copper, mercury, silver, and zinc, travel to the cathode where they are reduced to the metallic state. The simple anions, such as chloride, bromide, or iodide, travel to the anode and are oxidized to the molecular condition, while solutions containing hydroxide, sulfate, or nitrate ions liberate oxygen at the anode. The halogen or oxygen liberated may escape from the solution or it may act directly or directly as an oxidizing agent.

The various anions are not all oxidized with the same ease, their relative order being sulfide, iodide, bromide, and chloride, of which the first named is the most readily oxidized. Nor are the different cations all reduced with the same ease, their relative order being auric, platinic, silver, mercurous, cupric, arsenous, bismuth, hydrogen, and lead, of which the first named is the most readily reduced. The electrolytic method of analysis is, therefore, a type of oxidation-reduction analysis, since the oxidation of one species of ion always takes place at the anode simultaneously with the reduction of another species of ion at the cathode.

Electrolytic methods of assay are not so simple as it might seem from the brief foregoing discussion, since numerous factors complicate conditions, *viz.*, resistance of the solution, polarization, overvoltage, reactions at the electrodes, the nature of the solutions used, the current density, the rate of stirring, the shape and size of the electrodes, and the temperature. Although numerous metals may be analyzed electrolytically and the method may be applied to a number of official substances, *viz.*, the salts of mercury, silver, copper, and zinc, there are only two official electrolytic assays. Thimersol Solution and Thimersol Tincture official in the N.F. are assayed by separating the mercury by electrolysis in a suitable electrodeposition apparatus, dissolving the mercury in nitric acid, and titrating the mercury with 0.1 *N* ammonium thiocyanate solution.

Electrolytic analysis is not commonly used in official methods since other methods of analysis, in general, have certain advantages over electroanalytic procedures. Therefore, a discussion of the above-mentioned factors and the various types of apparatus and procedures is beyond the scope of the intent of this text; those who desire a more comprehensive knowledge of this subject should consult the fourth edition of this book or the references listed below.

References

1. Ewing, G. L., "Instrumental Methods of Chemical Analysis," 2d ed., McGraw-Hill Book Company, New York, 1960.
2. Reichnitz, G. A., "Controlled-Potential Analysis," Pergamon Press, New York, 1963.

Chapter 25

CHROMATOGRAPHY

For the purposes of the United States Pharmacopeia, chromatography is defined as a procedure by which inert materials and drug principles encountered in pharmaceutical preparations are separated by fractional extraction, adsorption, or ion exchange on a porous solid. The drug principles so separated on a porous solid may be removed from the solid by means of flowing solvents or simple solvent extraction and assayed by any of the ordinary analytical methods appropriate for the specific drug.

The name *chromatography* resulted from the fact that the method was originally used for the separation of colored compounds on a suitable adsorbent. The method, however, is now applicable to the separation of colorless compounds as well.

Chromatography is particularly useful as a means of separating and purifying complex and closely related chemical substances which are difficult, if not impossible, to separate by classical methods based on differences in solubility and volatility. In addition, chromatographic techniques are more applicable, in general, in those instances where small quantities are involved in the assay or identification tests.

The principal objectives obtainable through the use of chromatography are:

1. Resolution of mixtures into constituent parts
2. Determination of homogeneity
3. Comparison of substances suspected of being identical
4. Purification
5. Concentration of substances from dilute solution
6. Identification and control of technical products
7. Quantitative separation from complex mixtures
8. Indication of molecular structure

Chromatography, depending upon the basic principles involved, may be divided into three basic types: (1) *adsorption column chromatography*, (2) *ion exchange column chromatography*, and (3) *paper partition chromatography*. The most important of these in the official assays and tests are adsorption column and paper partition chromatography.

Column Chromatography. The procedures by which adsorption and ion exchange column chromatography are carried out are essentially identical; therefore, the techniques described for adsorption chromatography apply to the use of ion exchange substances as well. The basic principles of separation, however, are somewhat different. Adsorption chromatography depends on the differential or selective affinities shown by the constituents of a mixture toward various adsorbents; whereas, in ion exchange chromatography, as the name implies, there is an exchange of ions between the bulk phase and the ions in solution. In adsorption chromatography no ion exchange occurs. In either case the constituents adsorbed or exchanged on the bulk phase of the column constitute what is called the *chromatogram.*

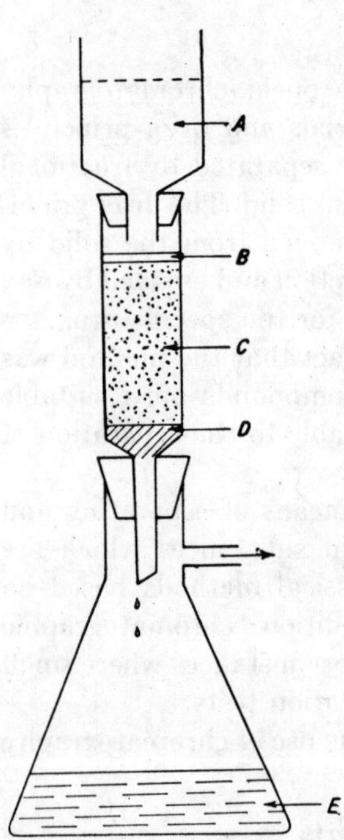

FIG. 48. Typical column chromatographic arrangement. *A*, solvent; *B*, filter paper or cotton on surface; *C*, adsorbent; *D*, cotton, perforated porcelain disk or sintered-glass disk; *E*, eluant.

The simplest type of chromatographic column consists of a suction flask and a cylindrical glass or quartz tube constricted at one end. The size of the column is determined by the quantity and adsorbability of the substances being separated. The material placed in the column to adsorb the drug is referred to as the *adsorbent.* (The U.S.P. and N.F. use the term *adsorbant* which is contrary to conventional use.) Purified siliceous earth, activated alumina, and calcium carbonate are examples of adsorbents commonly used. The adsorbent, as a dry solid or as a slurry, is uniformly and firmly packed into the tube, but it must not be so firm that the solution and subsequent solvents will not readily pass through. A rather concentrated solution of the drug in a suitable solvent (organic solvents are commonly used) is added to the top of the column and allowed to flow into the *adsorbent.* The drug principles are quantitatively removed from the solution by *adsorption* in a narrow band at the top of the column. An alternative procedure is to adsorb the constituents of the sample on a small amount of the adsorbent and add this to the top of the column. In either case, the column adsorbent is treated with fresh portions of solvent (referred to as *developer*), bringing about a separation of the components as each drug progresses down the column at a characteristic rate and, thus, resulting in the chromatogram.

If the separated compounds are colored or fluoresce under ultraviolet light, the adsorbent column can be forced, intact, from the tube and the fractions easily divided into segments with a knife. The appropriate segments may then be extracted with a suitable solvent to obtain the pure compounds, which may be analyzed by any appropriate method applicable to the drug in question. If the compounds are colorless, they may be visualized by painting or spraying the extruded column with a color-forming agent and proceeding as described above. If any of the components are radioactive, their position on the column may be determined by use of a Geiger-Müller counter or similar instruments.

Another procedure commonly used to develop chromatograms is known as "elution chromatography." The column is washed with suitable solvents, referred to as *eluants*, until each compound is successively obtained in the effluent liquid, known as the *eluate*. The separated drugs in different fractions of the eluate may then be determined by any appropriate method selected.

The U.S.P. specifies the use of purified siliceous earth as the adsorbent in the adsorption chromatographic separation of digitoxin in Digitoxin, Digitoxin Injection, and Digitoxin Tablets. Alcohol is used as the solvent, iso-octane as the developer, and benzene-chloroform mixture as the developer and eluant. The final analysis of the separated drug is accomplished by use of a spectrophotometric analysis at 495 mμ.

The thiamine assay of the U.S.P. requires that pharmacopeial articles containing thiamine be chromatographed using an artificially prepared base-exchange silicate for the purpose of removing interfering substances and concentration of the thiamine.

Paper Partition Chromatography. The basic principle of separation in partition chromatography is that of the differences in partition coefficients of substances between two immiscible liquids, one of which is a static phase supported on a solid adsorbent with the other mobile liquid phase flowing through it. If the solid adsorbent is filter paper (cellulose), the process is referred to as *paper partition chromatography*. In this process, the mobile phase, usually an organic solvent, moves slowly over the stationary phase, usually water, which is held in place by the fibers of the filter paper. Under such conditions, different substances move on the paper at different rates, depending upon the relative solubilities in the immiscible solvents, resulting in a separation by partition. Thus, if a solution of several different constituents is placed at one end of a piece of filter paper saturated with water under equilibrium conditions in a closed vessel, and an organic solvent is allowed to flow slowly through the paper, the different constituents (solutes) will move at different rates characteristic of each substance, and separation occurs. The ratio of the distance traveled on the paper sheets by a given compound to the distance traveled by the front of the mobile phase, from the point of application of the test substance, is designated as the R_f *value* of the compound.

$$R_f = \frac{\text{distance solute moves}}{\text{distance solvent moves}}$$

The ratio between the distances traveled by a given compound and a reference substance constitutes the *R_r value.* R_r as well as R_f values are characteristic constants for different compounds and may be used for identification purposes. Absolute R_f values, however, are difficult to establish, since slightly varying experimental conditions affect the R_f determination. Therefore, more definite identification is accomplished by using an authentic specimen of the compound in question as a reference substance. For this purpose three chromatograms are commonly prepared on the same sheet of paper: one from the substance to be identified, one from the authentic specimen, and one from a mixture of nearly equal amounts of the two substances. Each chromatogram should contain approximately the same quantity by weight of material. If the substance to be identified and the authentic specimen are identical, all three chromatograms will agree in color and R_r value, and the mixed chromatogram will show only one spot. In other words, the R_r value will equal one. Digoxin, U.S.P. is identified by paper partition chromatography using U.S.P. Digoxin Reference Standard as the authentic specimen. The R_f value is approximately 0.75 in this instance. The R_r value is 1.0.

There are three main methods or procedures for the preparation of paper partition chromatograms. These are (1) *descending chromatography*, which is accomplished by allowing the mobile phase to flow downward on the paper strip; (2) *ascending chromatography*, in which the mobile phase is allowed to rise upward on the paper by capillary attraction; and (3) *radial chromatography*, in which the mobile phase moves out in concentric circles from the center of a circular piece of paper. The latter type has not found as extensive use as the ascending and descending types. The equipment required for ascending and descending chromatographic procedures is essentially the same.

The assay for individual sulfonamides in a mixture is very difficult, if not impossible, by usual analytical methods; therefore, the U.S.P. and N.F. specify the use of paper partition chromatography for the separation of the drugs, followed by a spectrophotometric analysis of the pure compounds.

Exercise 114

Object. Assay of the Sulfadiazine, Sulfamerazine, and Sulfamethazine Content of Oral Trisulfapyrimidines Suspension.

Materials Required. 50 mg each of U.S.P. Sulfadiazine, Sulfamethazine, and Sulfamerazine Reference Standards

Oral Trisulfapyrimidines Suspension equivalent to about 1.5 Gm of total sulfapyrimidines

0.1 *N* sodium hydroxide

Whatman No. 1 filter paper, 4 × 57 cm

Stronger ammonia T.S.

n-Butanol

Ammonia solution (3 in 100)
Butyl nitrite
Acetic acid
N-(1-naphthyl) ethylenediamine dihydrochloride
2 *N* hydrochloric acid
0.12 *N* hydrochloric acid
Sodium nitrite solution (1 in 1,000)
Ammonium sulfamate (1 in 200)
Chromatographic chamber
Spectrophotometer

Procedure. 1. "*Standard curves.* Place 50 mg., accurately weighed, of U.S.P. Sulfadiazine Reference Standard in a 100-ml. volumetric flask, add 5 ml. of 0.1 *N* sodium hydroxide and, when solution is complete, add water to volume, and mix well. Each 0.01 ml. of this solution contains 5 mcg. of sulfadiazine."

The reference standard, an authentically pure sample of sulfadiazine, is dissolved in sodium hydroxide solution, forming the sodium salt which is readily soluble in water.

2. "Accurately measure, and place 0.01 ml., 0.02 ml., 0.03 ml., 0.04 ml., and 0.05 ml., respectively, as spots on individual chromatography strips, and proceed as directed below under *Procedure* [5, 6, 7], beginning with 'Suspend the strips until the spots have dried,' and ending with 'make any necessary correction.' From the absorbances observed for each of the five levels of sulfadiazine, plot the standard absorbance-concentration curve on arithmetic coordinate paper.

"Similarly, prepare standard absorbance-concentration curves using U.S.P. Sulfamerazine Reference Standard and U.S.P. Sulfamethazine Reference Standard, respectively."

The U.S.P. reference standards in varying quantities for each of the drugs are carried through the precise procedure used for the assay preparation, and the results are plotted on graph paper to provide a standard curve of absorbance versus concentration for the pure drugs. These curves are used to calculate the quantity of each of the sulfas in the assay preparation, using the absorbance values obtained from the assay procedure.

3. "*Assay Preparation.* Weigh accurately a quantity of well-shaken Oral Trisulfapyrimidines Suspension, equivalent to about 1.5 Gm. of total sulfapyrimidines and transfer the Suspension to a 250-ml. volumetric flask with the aid of a minimum volume of a mixture of 1 volume of stronger ammonia water and 9 volumes of water. Stopper the flask, and shake vigorously for 15 minutes, preferably on a mechanical shaker. Add the dilute ammonia to volume and mix thoroughly. Pipet 25 ml. of the solution to a 100-ml. volumetric flask, dilute to volume, and mix thoroughly. Each 0.01 ml. of the solution contains about 5 mcg. of each sulfapyrimidine."

The sulfapyrimidines suspension is dissolved in a dilute ammonia solution, forming the water-soluble ammonium salts. The solution is further diluted to make the sulfa concentrations comparable to the concentration of the U.S.P. reference standard solutions used in making the standard curves. Convenient volumes of solution, normally containing 1 to 20 μg of the compound, for chromatographic separation range from 0.01 to

0.1 ml. The concentration of about 5 μg of each sulfa in each 0.01 ml is satisfactory for the purposes of the chromatographic separation.

4. "*Procedure.* Cover the bottom of a chromatographic chamber suitable for descending chromatography with a butanol-ammonia emulsion freshly prepared by shaking together equal volumes of *n*-butanol and a solution of ammonia (3 in 100). For each paper strip to be used, place in the solvent trough 10 ml. of butanolammonia solution, prepared by allowing butanolammonia emulsion to separate and discarding the water layer. Cover the chamber, and allow sufficient time for the space within to become equilibrated with the vapor."

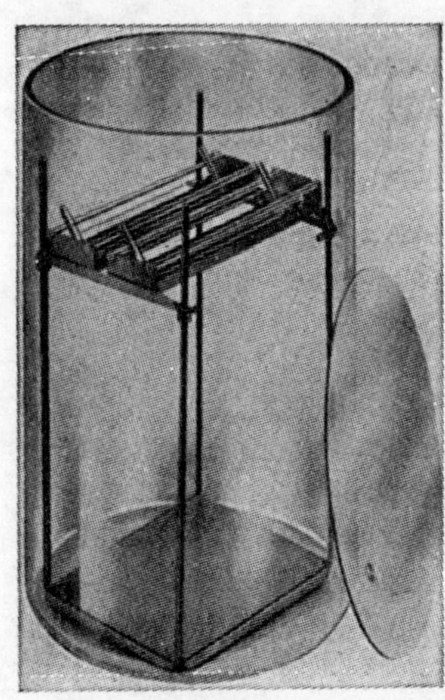

FIG. 49. A glass chromatographic chamber with the glass cover, metal rack, and glass antisiphoning rods used in descending paper partition chromatographic separations.

The essential equipment for descending chromatography consists of:

(*a*) A *vapor-tight chamber* provided with inlets for addition of solvent or for releasing internal pressure. The chamber should be constructed of nonabsorbent material inert to the solvents used, preferably made of glass, stainless steel, or porcelain. The chamber should also be designed to permit observation of the progress of the chromatographic run without opening the chamber. Tall glass cylinders are convenient if they may be made vapor-tight by sealing on glass covers with a sealing compound and providing two holes which may be closed.

(*b*) A *rack of corrosion-resistant material* about 5 cm shorter than the inside height of the chamber. The rack serves as a support for solvent troughs and for antisiphoning rods, which, in turn, hold up the chromatographic sheets.

(*c*) One or more *glass troughs* capable of holding a volume of solvent greater than that needed for one chromatographic run. The length of the trough must also be greater than the width of the chromatographic sheets.

(*d*) *Antisiphoning glass rods* supported by the rack and running outside of, parallel to, and slightly above the edge of the glass trough.

The bottom of the chamber is covered with butanol-ammonia emulsion to provide sufficient mobile solvent (butanol) and stationary phase (aqueous ammonium) to saturate the chamber completely with vapor after the chamber is sealed. Excess pressure in the chamber should be released as necessary. For large chambers, an equilibration time of 12 to 15 hr is necessary; small chambers are equilibrated after a period of 8 to 12 hr. The mobile solvent, butanol, is saturated with the stationary phase (aqueous ammonium) prior to being added to the solvent trough.

5. "Prepare the necessary number of chromatographic strips (Whatman No. 1 filter paper or equivalent) about 4 × 57 cm. in size, by drawing a light pencil line across each strip 45 mm. from one end and carefully creasing the paper on the line. With a micro-pipet or buret place 0.02 ml. of the *Assay preparation* as a spot on the pencil line of at least 3 of the strips. Suspend the strips until the spots have dried (about 15 minutes)."

Sheets of the prescribed type of filter paper are prepared in such a way that the length of the sheet is approximately that of the chromatographic chamber. The width of the sheet must be at least 2.5 cm but less than the length of the trough. A fine pencil line is drawn horizontally across the filter paper at a distance from one end such that, when the sheet is suspended from the antisiphoning rods with the upper end of the paper resting on the trough and the lower portion hanging free into the chamber, the line will be located a few centimeters below the rods. Care must always be taken to avoid contaminating the filter paper by excessive handling or by contact with dirty surfaces; otherwise, false color spots may appear later on the paper. If the total volume of solution applied to the paper is sufficient to produce a spot greater than 8 to 10 mm in diameter, it must be applied in separate portions to the same spot, allowing each portion to dry before the next is added.

6. "Keeping the folded ends uppermost, suspend the strips for descending chromatography for 18 hours or until the solvent has descended to approximately 25 mm. from the lower end of each strip. Remove the strips, and suspend in air until dry. Determine the position of the bands of adsorbed sulfapyrimidines on one of the strips by spraying lightly with a solution freshly prepared by mixing 25 ml. of *n*-butanol, 50 ml. of butyl nitrite, and 5 ml. of acetic acid. Allow to dry for 10 minutes, then spray lightly with a solution prepared by dissolving 100 mg. of *N*-(1-naphthyl) ethylenediamine dihydrochloride in 95 ml. of *n*-butanol and 5 ml. of 2 *N* hydrochloric acid. Violet-pink bands or spots appear. The spot which appears nearest the top is that of sulfadiazine, the intermediate spot is that of sulfamerazine, and the farthest spot is that of sulfamethazine. Using this strip as a guide, cut the replicate strips into segments representing, respectively, the sulfadiazine, sulfamerazine, and sulfamethazine, place them in separate 50-ml. Erlenmeyer flasks, and add to each flask exactly 10 ml. of 0.12 *N* hydrochloric acid. Swirl the flasks for 30 minutes, preferably on a mechanical device, transfer each solution to a centrifuge tube, centrifuge until the supernatant solution is clear, and pipet exactly 5 ml. of each solution into 25-ml. flasks."

After development and drying of the chromatograms, one of the strips is sprayed with diazotization and coupling solutions, which produce a color permitting the location of the positions of the three separate sulfas on the strip. Using this as a guide, the other strips are cut in such a way as to separate the individual sulfas. The sulfas are then individually extracted from the paper sections in hydrochloric acid.

7. "Into each flask pipet 1 ml. of freshly prepared sodium nitrite solution (1 in 1000), swirl the flask intermittently during 3 minutes, add by pipet 1 ml. of ammonium sulfamate (1 in 200), shake gently for 2 minutes, then add exactly 1 ml. of *N*-(1-naphthyl) ethylenediamine dihydrochloride (1 in 1000). Allow to stand, with

occasional shaking, for 15 minutes, then determine the absorbance at 545 mμ. Perform a blank extraction, diazotization, and coupling, with a segment of paper of the same dimensions and with the same qualities of the same reagents and make any necessary correction.

"Read from the appropriate standard absorbance-concentration curve the quantity of the individual sulfapyrimidine removed from each segment of paper. Average the results from three or more replicate strips, and calculate the quantity of each sulfapyrimidine in the portion of Oral Trisulfapyrimidines Suspension taken on the basis of the weight of 100 ml of suspension."

After the separation of the three sulfa drugs present in pure form by chromatographic procedures, the rest of the assay involves standard colorimetric procedures described in Chap. 20.

The radiochemical purity of Sodium Iodide I^{131} Solution is determined by the process of ascending chromatography, using a shielded Geiger-Müller counter to determine the location of iodide and iodate bands. In this instance the R_f values for the iodate and iodide bands must fall within ±5 percent of the values found for an iodine-131 sample of known purity when both are determined under parallel conditions.

Questions and Problems

1. Where is chromatography particularly useful in quantitative analytical chemistry?
2. Explain how chromatographic techniques can be used for the concentration of substances from dilute solutions.
3. How could you detect the presence of different compounds in different fractions of the eluate as the compounds are eluted from the column?
4. At the end of the development stage of a paper chromatogram, the solvent front was carefully marked and the paper dried. The solvent front had moved 22.5 mm from the initial point of drug application, and a component had moved 15.8 mm. What is the R_f value of the component? How could you identify this component?
5. Consult one of the references on paper chromatography at the end of the chapter, and explain what is meant by two-dimensional chromatography.
6. Why do R_f values differ for a given compound when separated under varying temperature conditions?
7. How could you detect the presence of uncolored phenol compounds on a paper chromatogram? How could you detect amino acids?

THIN-LAYER CHROMATOGRAPHY

The method of thin-layer chromatography (TLC) involves the spotting of a sample of a mixture of components at one end of an adsorbent-coated glass plate or other suitable support followed by passage of a solvent (developer) through the adsorbent for the purpose of separating the components of the sample. For many mixtures it offers the advantages over column and paper chromatography of (1) achieving separations in a relatively short time, 30 min or less, (2) accomplishing a complete analysis with as little as 20 nanograms[1] of material, and (3) providing a complete separation of components in complex mixtures (excellent resolution). In summary, the method is rapid and sensitive, and provides a high degree of resolution.

As in all chromatographic techniques, the physicochemical properties of the compounds to be separated will influence the choice of adsorbent and solvent used. Although a large variety of adsorbents are available from commercial sources, silica gel G and alumina are presently the most widely used. The letter "G" following "silica gel" stands for gypsum, which serves as a binder when added to this adsorbent. If the presence of the binder interferes with the separations, gypsum-free silica gel, or alumina, is used, or organic binders are incorporated into the adsorbent.

Adsorbents containing phosphors are also available. Depending on the nature of the phosphor incorporated, a white or colored background is observed when the developed chromatogram is examined under ultraviolet radiation. In most cases the presence of this fluorescent material renders the area of the spots on the chromatogram more distinct.

As implied previously, the solvent system used to develop the chromatogram greatly influences the degree of resolution which is obtained. If the substances being separated are unknown, benzene or chloroform containing 10 percent ethanol is the best solvent to use for an exploratory run. Examination of the developed chromatogram will then suggest which type of solvent to use. If the components migrate too rapidly, i.e., remain in the solvent front, a more weakly elutive solvent or solvent mixture is used. If the components remain at the base line, a more strongly elutive solvent or solvent mixture is used. The elutive properties of a mixture of ethanol in benzene increase with increasing amounts (2, 5, 10, 20 percent) of ethanol. A mixture in this series is recommended as a first step in the separation of components in a sample. If this fails, one of the solvents or solvent mixtures listed in the references at the end of this chapter is recommended.

For colored components such as dyes, it is not necessary to spray a special color-producing reagent in order to detect the location of the spot. Most components, however, are colorless and must therefore be treated with a reagent of some kind or examined under ultraviolet radiation in order to make the spots discernible. One general reagent which is often used in this technique and which cannot be used in paper chromatography for purposes of detecting organic components is concentrated sulfuric

[1] Microchemical mass units:

1 milligram (mg) = 0.001 Gm
1 microgram (μg or mcg) = 0.001 mg
1 nanogram (ng) = 0.001 μg
1 picogram (pg) = 0.001 ng
(pēkō)

acid. The organic components char after the chromatogram has been sprayed with sulfuric acid and heated. Another technique which is presently used involves the use of iodine vapor. In this method the chromatogram is placed in a closed container holding a few iodine crystals. The organic components react with the iodine vapor and form brown spots. It is interesting to note that in most cases the reaction between iodine and the organic compounds is physical rather than chemical and is reversible. When the chromatogram is placed in ordinary laboratory surroundings the iodine is released and the brown spots disappear. Use is also made of the fact that many organic compounds fluoresce when examined by ultraviolet radiation. Two ultraviolet light sources which are widely used and commercially available are ultraviolet short-wave (254 mμ) and long-wave (360 mμ) lamps.

Exercise 115

Object. To Determine the Limit of 5-Nitro-2-furfuraldazine in Nitrofurazone, N.F. XII.

Materials Required. Nitrofurazone, N.F.
Nitrofurazone, N.F., Reference Standard
5-Nitro-2-furfuraldazine, N.F., Reference Standard
Silica Gel G
Dimethylformamide
Methanol
Benzene
Acetone
Thin-layer chromatography apparatus

"*Reference Standard Solution*—Transfer 100.0 mg. of N.F. 5-Nitro-2-furfuraldazine Reference Standard to a 100 ml. volumetric flask. Add dimethylformamide to dissolve it and to make to volume. Pipet 1 ml. of the solution to a 50 ml. volumetric flask. Transfer 100.0 mg. of N.F. Nitrofurazone Reference Standard to the same 50 ml. volumetric flask, add 4 ml. of dimethylformamide to dissolve the powder, then add methanol to volume, and mix. Each milliliter of the final solution contains 20 mcg. of 5-nitro-2-furfuraldazine."

"*Test Solution*—Dissolve 100.0 mg. of Nitrofurazone in 5 ml. of dimethylformamide and dilute the solution to 50.0 ml. with methanol."

Procedure

1. "Perform thin-layer chromatography concurrently with the *Reference Standard Solution* and the *Test Solution*. Use a silica gel G plate, benzene-acetone solvent (8:2); spot 5 microliters (μl), and allow to run until the solvent front has moved about 10 cm."
2. "Observe the chromatogram near the solvent front under long ultraviolet light. Any spot from the *Test Solution* having the same R_f as the 5-nitro-2-furfuraldazine spot from the *Reference Standard Solution* should not exceed the Standard spot in size or intensity (1 percent)."

The usual dimensions of the plate which is used to prepare chromatograms are 20 × 20 cm (8 × 8 in.), and the layer thickness is approximately 250 μ. The glass plates used must be scrupulously clean. Dipping plates in chromic acid solution, followed by copious rinsings with water, is a recommended method of obtaining clean plates.

To prepare five glass plates, weigh 37 Gm of silica gel G and transfer to a mortar. Add 15 ml of water slowly with stirring. A slurry is formed in about 1 min. Tap the mortar lightly until all air bubbles have been removed (10 sec). Then transfer slurry to a thin-layer spreader; in one long stroke, the spreader delivers the adsorbent to the five glass plates at the desired thickness. Allow the plates to become firm before placing in a storage rack. Dry at 105° for 30 min and store plates in a desiccator until

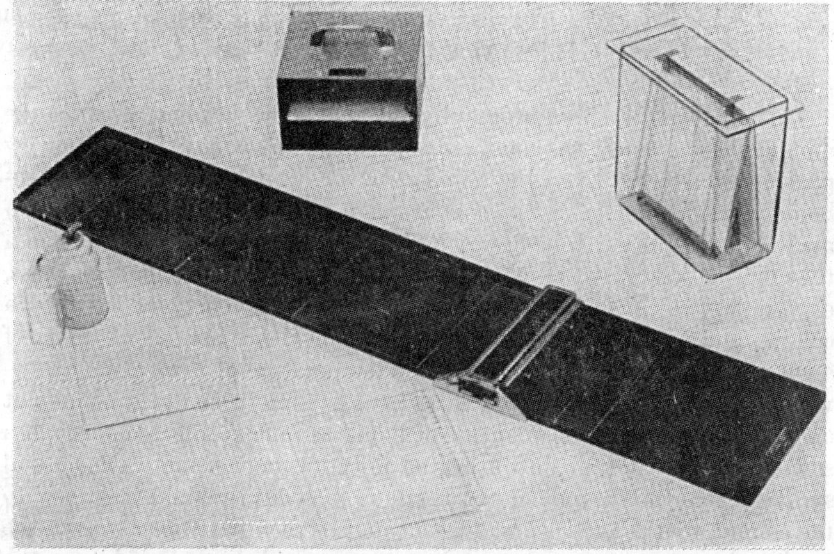

FIG. 50. Thin layer chromatography assembly. (*Courtesy of General Diagnostics Division, Warner-Chilcott Laboratory, Instruments Division.*)

needed. The essential equipment required to prepare and store plates for TLC is shown in Fig. 50.

Line the sides of a chromatographic chamber with a continuous strip of Whatman No. 1 chromatography paper and add sufficient solvent to the bottom of the chamber so that the filter paper becomes saturated. Permit chamber to stand overnight to ensure equilibration with the solvent mixture. The level of the solvent mixture on the bottom of the chamber should be sufficient to cover the lower edge of the adsorbent but must not touch the spot.

Apply 5 μl of the sample solution and standard solution at points about 1.5 cm apart and about 2 cm from the lower edge[2] of the plate. The use of

[2] "Lower edge" is defined as that part of the plate over which the spreader first moves during the application of the slurry.

a template will facilitate the application of the sample and reference solutions. After spots are dry, the plate is carefully placed in the chamber with the lower edge immersed in the solvent.

After the solvent front has moved about 10 cm, the plate is removed from the chamber and a light mark is made on the adsorbent to indicate the position of the solvent front. The plate is then allowed to dry.

The 1 percent limit placed on intensity and area in the N.F. is, of course, an approximation. If the spot approaches this limit it is advisable to purify the nitrofurazone.

GAS CHROMATOGRAPHY (GC)

This method of chromatography, as the name implies, uses as the mobile phase an inert gas commonly called the *carrier gas*. The stationary phase, also referred to as the *liquid substrate*, usually consists of a high-boiling liquid which is used to coat granular particles (No. 60 to 80 mesh) made of siliceous earth or firebrick. The coated particles are placed in a tube 0.6 cm × 2 m (¼ in. × 6 ft) which is usually made of copper, glass, or stainless steel. The tube containing the coated particles is called a column and is mounted in the constant-temperature heating chamber of a gas chromatography apparatus. The temperature of the column is elevated to a constant value; with carrier gas passing through the column at a constant rate, a small quantity of liquid sample is injected through a silicone rubber septum into a heated injection port using a needle and syringe. Here the components of the sample volatilize and are carried to the column by the carrier gas. The substances then partition between the liquid substrate and gas phase and are transported through and down the column at different rates, depending on the physicochemical properties of the substances, liquid substrate, solid support, and the operating conditions of the GC apparatus. When a component of the sample finally reaches the end of the column it passes through a differential detector connected through an electrical bridge circuit to a strip chart recorder. Here a peak is recorded on the chart paper. The position of the peak on the chromatogram (Fig. 51) is characteristic of the component and is measured in terms of its *retention* on the column. The terms most frequently used to describe retention are *retention time* and *retention volume*. Retention time t_R is defined as the time required by an average molecule of component to pass from the injection point through the column to the detector. It may be determined from a chromatogram by measuring the distance from the point of injection A (Fig. 51) to peak maximum C and multiplying this value by the reciprocal (minutes per millimeters) of the recorder speed. Retention volume V_R is defined as the volume of carrier gas necessary to carry an average molecule of the component from the point of injection to the detector. It is found by multiplying the retention time by the gas flow rate of the carrier gas. Usually the distance measured in centimeters from the point of injection to the peak maximum furnishes sufficient retention data for identification purposes. Because of the many operating variables of the GC apparatus which influence the retention data, most substances are best identified by determining relative retention α, which is defined by the equation

$$\alpha = \frac{x_2 - x_a}{x_1 - x_a}$$

where x_2, x_1, and x_a represent the retention from the point of injection

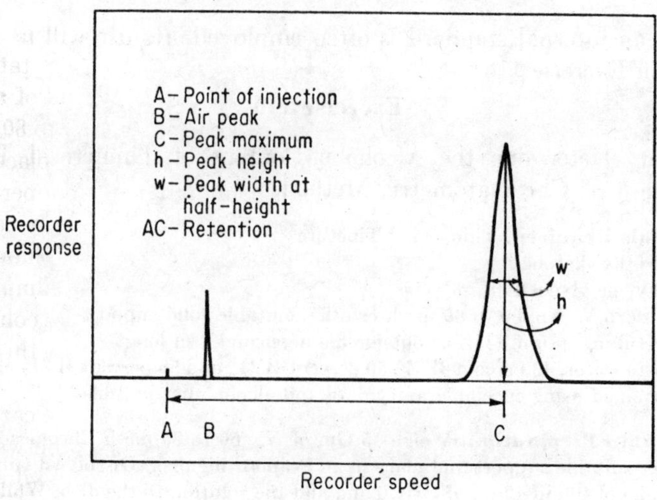

FIG. 51. Gas chromatogram.

measured in distance on the chromatogram to the peak maximums of the test drug, a standard such as *n*-pentane, and an inert compound such as air, respectively.

The essential parts of the gas chromatography apparatus, shown schematically in Fig. 52, consist of (1) inert gas, (2) detector, (3) injection port, (4) column, (5) exit port, and (6) recorder. No instructions will be given for the operation of a gas chromatography apparatus as they are supplied with the instruments.

The detector used in GC determines the shape of the chromatogram. When a differential detector, such as a thermal conductivity cell, is used a curve as shown in Fig. 51 is obtained. The peak area or peak height of the curve is proportional to the concentration of the component. For official gas chromatometric (denotes quantitative measurement) determi-

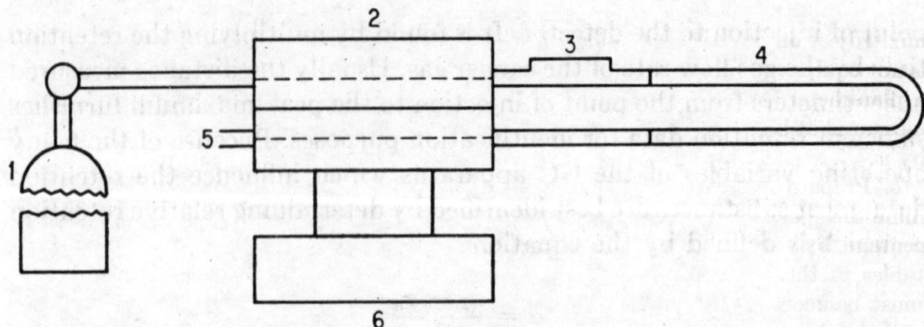

Fig. 52. Schematic diagram of gas chromatography (GC) apparatus. 1, tank of inert gas with regulator; 2, differential detector; 3, injection port; 4, column; 5, exit port; 6, strip chart recorder and associated electrical components.

nations, an internal standard is often employed. Its use will be demonstrated in Exercise 116.

Exercise 116

Object. Determine the Alcoholic Content of Thimerosal Tincture Using the Gas Chromatometric Method.

Materials Required. Thimerosal Tincture
Methyl ethyl ketone
Polyethylene glycol 400
Chromosorb W No. 60 to 80 mesh or other suitable solid support
Copper tubing, 6 mm (¼ in.) outside diameter and 2 m long
Alcohol in water, 43 percent W/V, 50 percent W/V, and 55 percent W/V, accurately determined using specific gravity data and alcoholimetric tables

1. Column Preparation. Weigh 15 Gm of No. 60 to 80 mesh Chromosorb W[3] or some other suitable support and place in an evaporating dish. Dissolve 3 Gm of polyethylene glycol 400 in 40 ml of $CHCl_3$ and add the solution to the dish. While stirring the solution, apply heat carefully (hot plate) in a hood until all the $CHCl_3$ has evaporated. (An alternate method of evaporating the solvent employs a rotating type of solvent evaporator fitted with a round-bottom flask.) Lightly pack one end of the copper tubing with glass wool. Then add the column material to the tube by aid of a funnel. Tap the tube on the floor while adding the packing material. After all the packing material has been added, a pledget of glass wool is placed in the tube and the column is bent to fit the gas chromatography apparatus. In order to prevent drifting base lines, the column is conditioned for several hours by heating to about 105°, with helium gas flowing slowly through the column (10 ml per min). When the GC apparatus is to be used, the column and exit port temperature are adjusted to 100° and the injection port temperature is adjusted to 160°. The regulator on the helium tank is adjusted so that the flow rate of the carrier gas is approximately 60 ml per min, and a suitable sensitivity value on the instrument is selected so that the peak heights can be measured to four significant figures in centimeters.

2. Procedure

a. Transfer 10.0 ml of the sample (Thimerosal Tincture) to a suitable glass-stoppered container, add 4.0 ml of the internal standard, methyl ethyl ketone (MEK), and

[3] Johns-Manville Corp., New York, N.Y.

mix. Inject an appropriate sample of the solution, (2 to 5 microliters). If the apparatus is not equipped with an integrator, the area of the peaks may be determined by using a planimeter or by multiplying the height of the peak by the corresponding width measured at half-height. In Fig. 51 the area of the peak would be equal to h multiplied by w.

b. Prepare three standard solutions by diluting alcohol with distilled water. Determine the specific gravity of the solutions at 25° using the results to ascertain the percentage, by volume, of C_2H_5OH contained in each solution, by using the alcoholimetric tables in the official compendiums. The alcoholic concentrations of these solutions must be accurately determined since they are used to prepare the calibration curve, and should contain about 43, 50, and 55 percent alcohol, by volume. Add 4.0 ml of accurately measured MEK to 10.0 ml of each solution and mix. Inject about 2 to 5 microliters of each mixture and determine the peak areas.

Calculations. The calculations will be illustrated by example:

1. Recall that the peak area (or height) is proportional to the concentration of the component. This statement may be expressed mathematically as

$$C_{\mu l/ml} = kA$$

where C = the concentration expressed on a V/V basis, rather than W/V
A = the area of the peak (or height)
k = the proportionality constant which is characteristic of the component for a given set of operating conditions

The value for k may vary from one analysis to another, depending on how well operating conditions of the instrument can be reproduced. Therefore it is necessary to calibrate the GC apparatus for each test component every time it is used for quantitative measurements. The calibration is usually accomplished using an internal standard since it is difficult to inject exactly the same volume of liquid each time.

2. Calibration of the GC instrument using known concentrations of test component (ethanol) and internal standard (MEK).

Place 10.0 ml of accurately analyzed aqueous-alcohol solution of about 43 percent V/V concentration in a glass-stoppered container. Add 4.0 ml of MEK and mix. Inject 2 to 5 microliters in the GC apparatus and determine the area of the alcohol and MEK peaks. Repeat using 50 and 55 percent aqueous-alcohol solutions. Assume subscripts 1 and 2 represent test component and internal standard respectively.

$$\frac{\mu l_1}{ml} = k_1A_1 \tag{1}$$

$$\frac{\mu l_2}{ml} = k_2A_2 \tag{2}$$

Divide Eq. (2) into Eq. (1) and cancel ml in each equation since these values will always be equal for any injection volume of the mixture taken. Then rearrange the equations to obtain:

$$\frac{\mu l_1}{\mu l_2} = \frac{k_1A_1}{k_2A_2}$$

k_1/k_2 represents the slope of a line which is obtained by plotting $\mu l_1/\mu l_2$ on the Y axis and A_1/A_2 on the X axis. The slope may change depending on the operating conditions used for the instrument and that is why it is necessary to recalibrate the instrument every time it is used.

3. Sample preparation of Thimerosal Tincture. "Place 10.0 ml. of Thimerosal Tincture in a glass-stoppered container. Add 4.0 ml. of MEK and mix. Inject 2 to 5 microliters in the GC apparatus and determine the areas of the alcohol and MEK peaks."

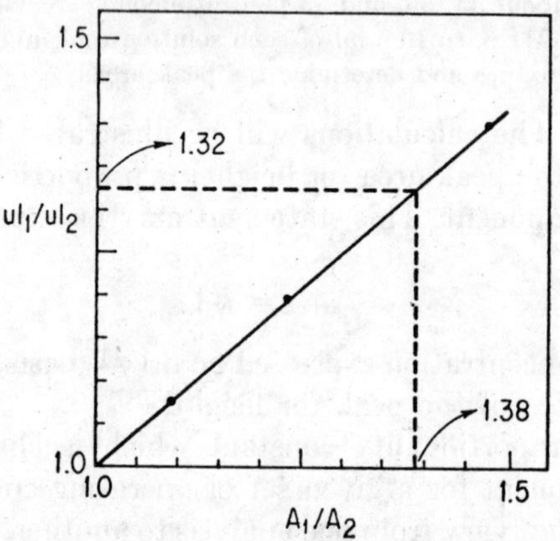

Fig. 53. Solid line: calibration curve for alcohol and methyl ethyl ketone (MEK). Dotted line: area ratios corresponding to volume ratios of alcohol and MEK.

Assume the data in Table 60 were obtained for the alcohol determination of Thimerosal. A plot of $\mu l_1/\mu l_2$ versus A_1/A_2 as obtained from the table is shown in Fig. 53 and is referred to as the calibration curve. The slope of the line is equal to k_1/k_2 and may be calculated and used in the determination of the alcoholic content of the tincture if desired. In this example, however, the calibration curve will be used. Observe from Table 60 that the areas of the alcohol and MEK peaks in the tincture are 12.4 and 9.0 cm², respectively. Dividing 12.4/9.0 gives a value of 1.38.

Table 60

Standard solution	Percent V/V alcohol	ml_1 alcohol	ml_2 MEK	A_1 alcohol, cm²	A_2 MEK, cm²	A_1/A_2	$\mu l_1/\mu l_2$
1	43.2	4.32	4.00	15.4	14.0	1.10	1.08
2	47.6	4.76	4.00	13.0	10.5	1.24	1.19
3	55.6	5.56	4.00	17.3	11.7	1.48	1.39
Tincture	....	....	4.00	12.4	9.0	1.38	...

This value is located on the abscissa of the calibration curve and is found to correspond to the value 1.32 on the ordinate. In words this value simply shows that the tincture contains 1.32 volumes of ethanol for each volume of MEK present. Since 4 ml. of MEK was taken, then

$$\frac{1.32\ \mu l_1}{1\ \mu l_2} \times 4.0\ \text{ml} = 5.28\ \text{ml ethanol}$$

the quantity present in the original volume of tincture taken. Then

$$\frac{5.28\ \text{ml ethanol}}{10.00\ \text{ml tincture}} \times 100 = 52.8\%\ \text{alcohol in the tincture}$$

References

1. Block, R. J., E. L. Durrum, and Gunter, Zweig, "A Manual of Paper Chromatography and Paper Electrophorese," Academic Press, Inc., New York, 1958.
2. Lederer, E., and M. Lederer, "Chromatography," 2d ed., Elsevier Publishing Company, Amsterdam, 1957.
3. Smith, Ivor, "Chromatographic Techniques, Clinical and Biochemical Applications," Interscience Publishers, Inc., New York, 1960.

Thin Layer Chromatography

1. Randerath, K., "Thin-layer Chromatography," Academic Press, Inc., New York, 1963.
2. Stahl, E., "Thin-layer Chromatography," Springer-Verlag OHG, Berlin, New York, 1965.

Gas Chromatography

1. Bayer, E., "Gas Chromatography," Elsevier Publishing Company, Amsterdam, 1961.
2. Burchfield, H. P., and E. E. Storrs, "Biochemical Applications of Gas Chromatography," Academic Press, Inc., New York, 1962.
3. Dal Nogare, S., and R. S. Junet, Jr., "Gas-Liquid Chromatography, Theory and Practice," Interscience Publishers, Inc., New York, 1962.

Chapter 26

RADIOACTIVITY

Radioactive substances are used in medical practice as diagnostic aids and as therapeutic agents.

The radioactive materials official in the U.S.P. XVII are (1) Sodium Iodide I^{131} Capsules and Solution, (2) Sodium Phosphate P^{32} Solution, (3) Sodium Chromate Cr^{51} Injection, (4) Iodinated I^{131} Serum Albumin, (5) Rose Bengal Sodium I^{131} Injection, (6) Cyanocobalamin Co^{57} Solution, (7) Cyanocobalamin Co^{60} Capsules and Solution, (8) Sodium Iodohippurate I^{131} Injection. Since these materials owe their usefulness to their radioactivity, exacting assay procedures for total radioactivity content and radiochemical purity are required. These assay and test procedures are necessarily unique and are, therefore, substantially different in principle from other official assay procedures.

The terminology used in describing the assay procedures is also somewhat different; therefore a brief discussion of the fundamentals and terminology of radioactivity is important. For more complete information the references listed at the end of this chapter should be consulted.

A nuclide is a species of atom characterized by the number of protons and neutrons present in its nucleus. Isotopes are nuclides of the same atomic number (belonging to the same element) but of different atomic mass numbers. For example, carbon-14, iodine-131, cobalt-57, etc., are all nuclides, but only carbon-14 is an isotope of carbon, since it belongs to the same element.

Depending on the stability of the nucleus, nuclides are classified as *stable* or *unstable*. The unstable nuclides are called *radioactive nuclides*, because as the nuclei change to a more stable form, *ionizing radiation* is emitted, giving rise to *radioactive properties*. The liberation of ionizing radiation from a radioactive nuclide, with the concurrent change to a more stable state, is often referred to as *radioactive decay* or *disintegration*. The radiations emanating from the radioactive decay process all have the ability to ionize the atoms of the material through which they pass.

There are three types of ionizing radiation emitted, singly or in combination, from radioactive isotopes, namely, *alpha* (α) *particles*, *beta* ($\beta^{\pm}$) *particles*, and *gamma* (γ) *rays*.

Alpha particles are the nucleus of the helium atom and consist of two protons and two neutrons in combination. Beta particles are either negatrons or positrons, and gamma rays are electromagnetic quanta or a wave form of energy in the same category as x-rays.

Individual radioactive isotopes differ from one another in (1) the type of ionizing radiation emitted (α, $\beta^{\pm}$, γ, or any combination of these three); (2) the energy with which the ionizing radiation is ejected, usually expressed in millions of electron volts (Mev); (3) the rate of radioactive decay, usually expressed in terms of the *half-life;* and (4) the method of decay process, spoken of as the *decay scheme.*

The rate of radioactive decay is constant and characteristic for each individual radioactive isotope. This rate of decay always follows an exponential course which permits expressing it in terms of the half-life. The *half-life period* ($T_{1/2}$) is the time in which the radioactive intensity decreases to one-half the original value. The exponential decay curve is described mathematically by the equation

$$N = N_o e^{-\lambda t}$$

where N = intensity of activity at a time t after initial intensity time
N_o = initial intensity
λ = disintegration constant characteristic of each individual radioisotope

The half-life period is related to the disintegration constant by the equation

$$T_{1/2} = \frac{0.693}{\lambda}$$

Radioactive decay corrections are calculated from the exponential equation, obtained from a decay curve plotted for the particular isotope involved, or obtained from previously prepared decay charts.

The quantity of radioactivity is expressed in absolute units (total number of atoms disintegrating per unit time) or comparative units (counts per unit time or divisions per unit time under specified conditions). The *curie* (C) and its metric subdivisions, *millicurie* (mc) and *microcurie* (μc), are the most commonly used absolute units. One curie equals 3.7×10^{10} disintegrating atoms per second and equals 10^3 mc and 10^6 μc.

Radioactivity may be detected and/or measured quantitatively by a number of different instruments, the operation of which depends on the collection and measurement of ion pairs, the measurement of fluorescence, or the measurement of the effect on a photographic emulsion.

The most frequently used instruments are *Geiger-Müller counters* and *scintillation counters,* attached to suitable electronic scaling devices (Fig. 54), of which many different kinds are available commercially. The Geiger-Müller counter detects and records individual ionizing radiations which strike its sensitive volume by the process of multiplicative ion collection. The scintillation counter depends upon the detection in a sensitive

FIG. 54. Scaling device. This scaling unit is typical of the basic counting instruments used for the routine counting of radiation. It may be used with Geiger or scintillation counters. (*Courtesy of Nuclear-Chicago Corp.*)

photoelectric cell of the discrete flashes of fluorescent light produced when ionizing radiation strikes fluorescent material.

RADIOACTIVITY MEASUREMENTS

Exercise 117

Object. Assay of Sodium Iodide I^{131} Solution.

Materials Required. A counter assembly and a suitable detector
A calibrated iodine-131 standard
Sodium Iodide I^{131} Solution

Procedure. "Determine the radioactivity, in counts per minute, of a calibrated iodine 131 standard by measurement in a suitable counting assembly, adjusting the concentration of the standard so as to obtain optimum counting accuracy with the particular assembly used. Place an accurately measured aliquot of Iodine 131 Solution, appropriately diluted, in a container identical to that used for the standard, and determine the radioactivity at approximately the same time and under the identical geometrical conditions used for the standard. Correct both determinations for background, and calculate the radioactivity, in microcuries or millicuries per milliliter, by the formula $SD(A/B)$, in which S is the microcurie or millicurie strength of the standard, D is the dilution factor, and A and B are the radioactivities expressed in counts per minute of the sample and the standard, respectively."

Iodine^{131} emits both beta particles and gamma rays when the radioactive atoms decay to a more stable state. Radioactivity is generally recorded on a counting assembly which consists of a sensing unit and an electronic scaling device. Either a Geiger-Müller tube or a scintillation detector may be used as the sensing unit.

The activity of the sample and the standard are determined at approximately the same time and under identical conditions in order to cancel out possible errors due to variation in the counting efficiency of the instrument from time to time, positioning of the sample, mass absorption, and scattering of the ionizing radiation.

In the Geiger-Müller counting of samples, correction must be made for coincidence when counting samples at high activity levels. Each individual counter and accompanying scaler has a characteristic resolving time. The *resolving time* is the time during which two or more ionizing radiations striking the tube are recorded as a single count. The phenomenon of two or more ionizing radiations striking the tube within the resolving time is termed *coincidence*. Coincidence correction may be made as follows: Prepare two radioactive sources on identical mounts, each having approximately the same activity in the range of 10,000 counts per minute. Carefully determine the activity of the sources individually and combined, being sure to retain identical geometry and back-scattering effects. Calculate the resolving time from the equation

$$\mathrm{T} = \mathrm{T}_1 \left[1 + \frac{\mathrm{T}_1}{2} (N_{1,2} - 3N_b) \right]$$

where T = resolving time

$$\mathrm{T}_1 = \frac{N_1 + N_2 - N_{1,2} - N_b}{2(N_1 - N_b)(N_2 - N_b)}$$

N_1, N_2, N_b = activities of samples 1 and 2 and the background, respectively

$N_{1,2}$ = activity of the combined samples

The activity corrected for coincidence may be obtained from the equation

$$N = \frac{n}{1 - n\mathrm{T}}$$

where N = activity corrected for coincidence

n = uncorrected activity

T = resolving time

Coincidence correction must always be made prior to the background correction.

Radioactivity due to the cosmic rays, spontaneous discharges in the atmosphere, and construction materials contributes what is known as the *background activity*. All sample radioactivity measurements must be corrected by subtracting background.

Sample Radioassay and Data

Resolving time of counter = 8.33×10^{-6} min
Background = 31 counts per minute
Dilution factor = 10^6
Sample activity = 2,560 counts per minute
Standard activity = 8,925 counts per minute
Millicurie strength of standard = 0.51×10^{-4} mc
Sample activity corrected for coincidence and background =

$$\frac{2{,}560}{1 - 2{,}560 \cdot 8.33 \times 10^{-6}} - 31 = 2{,}586$$

Standard activity corrected for coincidence and background =

$$\frac{8{,}925}{1 - 8{,}925 \cdot 8.33 \times 10^{-6}} - 31 = 9{,}599$$

$$\text{Activity per ml} = \frac{2{,}586}{9{,}599} \times 0.51 \times 10^{-4} \times 10^{6} = 13.7 \text{ mc}$$

The U.S.P. requires that Sodium Iodide I^{131} Solution contain not less than 95 percent and not more than 105 percent of the labeled amount of iodine^{131} as iodide at the time indicated in the labeling.

Questions and Problems

1. Define the following terms: (*a*) *isotopes*, (*b*) *α particle*, (*c*) *millicurie*, (*d*) *half-life*, (*e*) *cosmic rays*.

2. Explain the operation of a Geiger-Müller counter. (Consult one of the references at the end of this chapter.)

3. If a sample of iodine-131 has an activity of 10 mc per ml, what will be the activity 16 days later? The half-life of I^{131} is 8.0 days.

4. Consult the U.S.P. under "Limit of total iodide" for Sodium Iodide I^{131} Solution. Why is a negative test for iodide required?

5. Why should containers in which radioactive materials are placed be previously rinsed with a solution of a stable form of the element in the same chemical form as the radioelement?

6. Why is coincidence correction necessary with radioactive samples of higher activities?

References

1. CHASE, G. D., and J. L. RABINOWITZ, "Principles of Radioisotope Methodology," 2d ed., Burgess Publishing Company, Minneapolis, 1960.
2. FRANCIS, G. E., W. MULLIGAN, and A. WORMALL, "Isotopic Tracers," 2d ed., The Athlone Press, London, 1959.
3. OVERMAN, R. T., and H. M. CLARK, "Radioisotope Techniques," McGraw-Hill Book Company, New York, 1960.
4. PRICE, W. J., "Nuclear Radiation Detection," 2d ed., McGraw-Hill Book Company, New York, 1964.
5. QUIMBY, E. H., and S. FEITELBERG, "Radioactive Isotopes in Medicine and Biology," Lea & Febiger, Philadelphia, 1963.
6. WANG, C. H., and D. L. WILLIS, "Radiotracer Methodology in Biological Science," Prentice-Hall, Inc., Englewood Cliffs, N.J., 1965.

Chapter 27

ROTATORY POWER

Many organic substances, such as certain constituents of the volatile oils, alkaloids, and sugars, possess the power of rotating the plane of polarized light when the latter is passed through solutions containing them. Substances that possess this power are said to be *optically active* and are designated as dextrorotatory when the direction of the rotation is toward the right and as levorotatory when it is toward the left.

The optical activity of many substances is a function of their chemical constitution as well as of their concentration. Consequently, a determination of the rotatory power or lack of rotatory power of a substance may serve as a means of establishing both its identity and its purity. In some cases, the measurement of the optical activity of a substance may also give some indication of its therapeutic value; e.g., the levorotatory alkaloid hyoscyamine is much more active as a mydriatic than its optically inactive isomer atropine.

In a ray of *ordinary light*, the vibrations are transverse, i.e., they take place in a plane at right angles to the direction of propagation, but the vibration direction is constantly changing. In a ray of *plane-polarized light*, commonly termed *polarized light*, the vibrations are also transverse, but they take place in only one direction.

The polariscope and the polarizing microscope are used in the determination of the optical activities of liquid and solid substances, the polarizing microscope being employed more often for the examination of solid substances, while the polariscope is used primarily for the examination of liquids.

When a ray of ordinary light is passed through a specially constructed prism made of Iceland spar, such as the Nicol prism, the emergent light wave vibrations take place in one plane, and the light is said to be *plane-polarized.* If another Nicol prism is used to examine the plane-polarized light, it will be found that, on rotating the latter prism, the field of view appears alternately light and dark, the minimum brightness following the maximum as the prism is rotated through an angle of 90 deg. The prism by means of which the light is polarized is termed the *polarizer*, and the second prism, by means of which the light is examined, is called the *analyzer*.

If, when the field of view appears dark, as when the axes of the two prisms are at right angles to each other, a tube similar to that shown in Fig. 55 is filled with an optically active liquid and placed between the polarizer and the analyzer, the field will become light because of the rotation of the plane of polarization by the optically active substance. The extent to which the plane of polarization has been rotated can be determined by turning the analyzer through a certain angle until the field becomes dark again. When it is necessary to turn the analyzer to the right (clockwise), the optically active substance is said to be *dextrorotatory;* when the analyzer must be turned to the left (counterclockwise), it is said to be *levorotatory.*

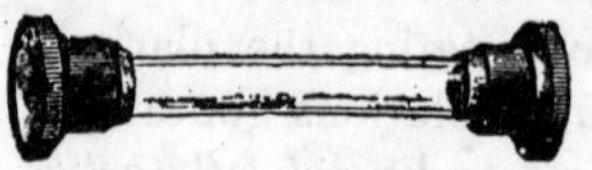

Fig. 55. Polariscope tube.

It is possible to obtain a position at which the field of view becomes dark by rotation of the analyzer to either the right or left, since in one complete rotation of the prism through 360 deg, there are two positions of the analyzer, 180 deg apart, at which the field becomes dark, and, likewise, two positions at which there is a maximum of light. In determining the sign of activity of a given substance, it is customary to take the direction in which the rotation required to reproduce the dark field is less than 90 deg.

The angle of rotation of the polarized light on passage through a substance varies directly as the length of the column of substance through which the light passes. In homogeneous mixtures or solutions, the angle of rotation depends upon the concentration of the optically active substances present. When more than one optically active substance is present in a solution, the resultant rotation will be the same as the sum of the several rotations of each separate constituent in the same volume of solution, provided that the substances do not react with each other. The angle of rotation also varies with the wavelength of the light employed (the shorter the wavelength the greater the angle of rotation), with the temperature, and with the nature of the substance. When all these factors are taken into account, the results may be calculated in terms of specific rotation or specific rotatory power.

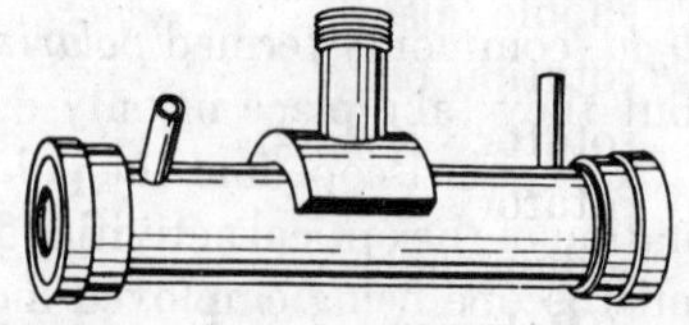

Fig. 56. Polariscope tube with jacket and thermometer well for constant temperature control.

The specific rotation $[\alpha]$ of a liquid is defined as *the angular rotation in degrees through which the plane of polarization of polarized monochromatic light is rotated by passage through* 1 *decimeter* (100 *mm*) *of the liquid, calculated to the basis of a specific gravity of* 1. *In the case of solutions of an optically active substance, the specific rotation is calculated to the basis of a concentration of* 1 *Gm of solute in* 1 *ml of solution.*

Since the specific gravity of liquids, as well as the optical activity of many substances, is influenced by the temperature, it is necessary to indicate the temperature at which the rotation and the specific gravity are determined. The temperature specified in the two official standards for the measurement of optical activity is generally 25°, except for sugar, which is determined at 20° by international agreement.

The calculation of the specific rotatory power of an optically active liquid or solution of an optically active solid may be made from the following general formulas:

1. For liquid substances $[\alpha]_D^t = \dfrac{a}{ld}$

2. For solutions $[\alpha]_D^t = \dfrac{100a}{lpd}$ or $[\alpha]_D^t = \dfrac{100a}{lc}$

For calculating the specific rotation $[\alpha]$ using these formulas, the determination of the following factors is necessary:

a = observed rotation in degrees of the liquid at temperature t using a sodium light

l = the length of the tube in decimeters

d = the specific gravity of the liquid or solution at the temperature of observation

p = concentration of solution expressed as the number of grams of active substance in 100 Gm of solution

c = concentration of solution expressed as the number of grams of active substance in 100 ml of solution

In expressing the specific rotation of a substance in solution, the concentration and solvent used must always be stated. The temperature should also be recorded; e.g., $[\alpha]_D^{20°} = -38°$ means that the specific rotation of the substance determined at 20° using sodium light is levorotatory 38 deg, and $[\alpha]_D^{20°} = +16°$ means that the substance is dextrorotatory 16 deg when the specific rotation is determined at 20°, using sodium light.

Polarimeters. The instruments used to measure the optical activity of liquid substances are called *polariscopes* or *polarimeters* and *saccharimeters.* Polarimeters permit the angle of rotation to be read off in degrees or fractions of a degree of a circle, and saccharimeters are usually constructed and set to give readings directly in terms of percentage of sugar. The so-called "half-shadow" instruments known as the *Laurent* and the *Schmidt and Hänsch* polarimeters are the most generally used in pharmaceutical work.

In the polarimeter of Laurent and that of Schmidt and Hänsch, the optical parts may be represented diagrammatically as in Fig. 57. Monochromatic light (sodium light) from the source L passes through the lens A, which renders the rays of light parallel, and then passes through the polarizing prism B, which renders the light ray plane-polarized. It then passes through the observation tube O, which contains the substance being examined, and thence through the analyzer prism *D*. The field of view is observed through the telescope *FE*. The circular opening *C* of the tube carrying the polarizer is one-half covered by a thin quartz plate as illustrated in *C′*, the thickness of which is so adjusted that the light in passing through the plate is altered in phase by half a wavelength but remains plane-polarized. In this way, two beams of polarized light are produced.

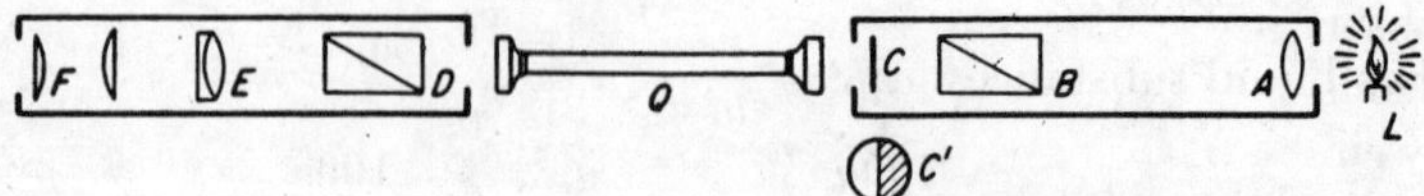

FIG. 57. Diagram of the optical parts of a polarimeter.

If the polarizer is rotated so that the plane of polarization forms an angle *X* with the quartz plate, the planes of polarization will also be inclined at an angle 2*X*. This is termed the *half-shadow angle*. On rotating the analyzer, a position will be found at which one beam of light will be completely, and the other only partly, extinguished. One half of the field of view will then appear dark while the other half will remain light. Upon rotating the analyzer farther through the angle 2*X*, a second position will be found at which the second beam of light will be extinguished while the first beam becomes visible. At this position of the analyzer, the half of the field which formerly appeared dark will be light, and that which was formerly light will be dark. When the analyzer occupies an intermediate position, a field of view of uniform brightness will appear, and this is the position at which the analyzer should be set.

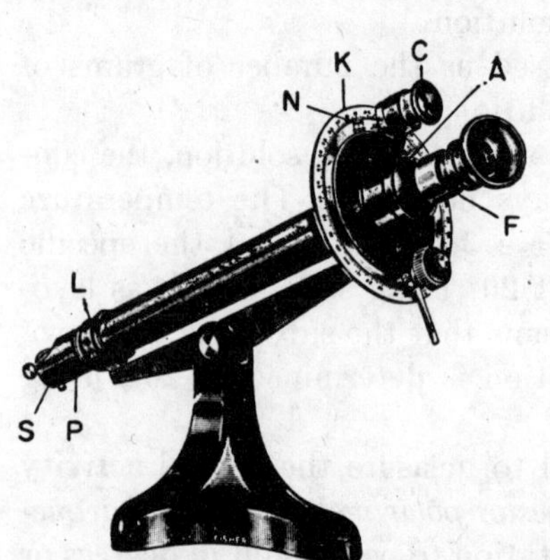

FIG. 58. Polarimeter.

The complete polarimeter is shown in Fig. 58. At the end *S*, which is directed toward the source of illumination, is the lens. The polarizing prism is at *P*, and it is connected with a lever *L*, by means of which it can be rotated to alter the angle of half-shadow. The observation tube is placed in the central part of the instrument and protected from extraneous light by a hinged cover. The analyzer is placed at *A*, and it can be rotated, independently of the graduated circle, by means of a screw. This permits the correction of the zero point of the instrument. *F* is a telescope with an eyepiece by means of which the field of view is observed. *K* is a graduated disk, which can be made to rotate with the analyzer and telescope past the fixed verniers *N* on the graduated circle. *C* is an eyepiece which magnifies the vernier scale reading. Most instruments are provided with one or more magnifying lenses to increase the accuracy of readings.

The official polarimetric determinations should be carried out in a darkroom using monochromatic light. Monochromatic light can be obtained easily by introducing a fused bead of sodium chloride on a platinum wire or gauze into the nonluminous flame of a gas burner.

The observation tube, in which the liquid to be examined is placed, usually consists of a tube of thick glass with accurately ground ends, closed by circular glass plates with parallel sides, which are pressed against the ends of the tube by means of screw caps. Other polarimeter tubes are constructed with a jacket through which water may be circulated to give a constant temperature. Since the unit of length in polarimetric measurements is 1 dcm, the observation tubes are generally of 1-dcm length or some multiple or fraction thereof, e.g., 2.0, 0.5, and 0.25 dcm. The length of the tube used in any definite determination usually depends upon the depth of color in the liquid and the magnitude of its rotatory power.

Readings made on a saccharimeter may be converted into the angular degrees of a polarimeter by means of the following factors:

One division of the Laurent sugar scale = 0.2167 deg angular rotation *D*.

One division of the Ventzke or Schmidt and Hänsch sugar scale = 0.3468 deg angular rotation D.

Exercise 118

Object. To Determine the Specific Rotation of Sucrose.

Materials Required. 26 Gm of cane sugar
A 100-ml volumetric flask calibrated at 20°
A polarimeter with a 200-mm observation tube

Procedure. 1. Adjust and determine the zero point of the polarimeter.

The polarimeter should be set up so that the polarizer end is in direct line with the bright sodium light and about 4 to 6 in. from it. Place a 200-mm observation tube filled with distilled water in the tube holder between the polarizer and the analyzer, and focus the telescope eyepiece on the line bisecting the field of view, rotating the analyzer if necessary in order to obtain unequal illumination of the two halves of the field. Now determine the zero point by rotating the analyzer until equal illumination of the two halves of the field is obtained. This position should be approached several times from either side of the zero point, readings being made from the verniers, and the mean of at least five readings taken as the zero point. Since the zero point is altered by the alteration of the angle of half-shadow, the lever that rotates the polarizer must be set before the

TABLE 61. SUBSTANCES ASSAYED BY THE MEASUREMENT OF ROTATORY POWER

Substance	Equivalent of 1° rotation, Gm	Official requirement, percent
U.S.P.		
Anticogulant acid citrate dextrose solution	1.0425	$C_6H_{12}O_6 \cdot H_2O$ = 95 to 105 l.a.
Dextrose injection	1.0425	$C_6H_{12}O_6 \cdot H_2O$ = 95 to 105 l.a.
Epinephrine inhalation	*	$C_9H_{13}NO_3$ = 0.9 to 1.15
Epinephrine injection	*	$C_9H_{13}NO_3$ = 0.09 to 0.115
Epinephrine solution	*	$C_9H_{13}NO_3$ = 0.09 to 0.115
Epinephrine suspension, sterile	*	$C_9H_{13}NO_3$ = 90 to 120 l.a.
Levarternol bitartrate injection	*	$C_8H_{11}NO_3 \cdot C_4H_6 \cdot H_2O$ = 90 to 115 l.a.
N.F.		
Epinephrine bitartrate ophthalmic ointment	*	$C_9H_{13}NO_3$ = 1
Fructose injection	0.5587	$C_6H_{12}O_6$ = 95 to 105
Fructose and sodium chloride injection	0.5587	$C_6H_{12}O_6$ = 95 to 105
Sodium chloride and dextrose tablets	1.0425	$C_6H_{12}O_6 \cdot H_2O$ = 92.5 to 107.5

* See U.S.P. XVII or N.F. XII for formula.
l.a. = labeled amount.

TABLE 62. SUBSTANCES WITH REQUIRED ANGULAR ROTATION

Substance	Official requirement,* angular rotation, °	Substance	Official requirement,* angular rotation, °
U.S.P.		**N.F.**	
Acacia	Slightly levorotatory	Caraway oil	+70 to +80
		Cardamom oil	+22 to +44
Anethole	n.m.t. 0.15	Eucalyptol	−0.3 to +0.3
Anise oil	−2 to 1	Lavender oil	−3 to −10
Cinnamon oil	−1 to 1	Myristica oil East Indian	+8 to +30
Clove oil	n.m.t. −1.5		
Coriander oil	+8 to +15	West Indian	+25 to +45
Eucalyptol	−0.3 to +0.3	Orange flower oil	+1.5 to +9.1
Fennel oil	+12 to +24	Pine needle oil	−5 to −15.5
Lemon oil	+57 to +65.6	Rose oil	−1 to −4
Orange oil	+94 to +99	Rosemary oil	−5 to −10
Peppermint oil	−18 to −32	Spearmint oil	−48 to −59
Tubocurarine chloride injection	+0.35 to +0.40	Thyme oil	Does not exceed −3
		dl-α-Tocopherol acetate	+0.25

* In 100-mm tube.
n.m.t. = not more than.

TABLE 63. SUBSTANCES WITH REQUIRED SPECIFIC ROTATION

Substance	Solvent used	Concentration in 100 ml, Gm	Specific rotation
U.S.P.			
Ascorbic acid	Water	10	+20.5 to +21.5
Aurothioglucose	Water	0.5	+65 to +75
Calcium pantothenate	Water	5	+25 to +27
Camphor	Alcohol	10	+41 to +43
Cholecalciferol	Alcohol	0.5	+105 to +112
Cholesterol	Dioxane	2	−34 to −38
Cocaine HCl	Water	2	−71 to −73
Colchicine	Water	1	−410 to −435
Desoxycorticosterone acetate	Dioxane	1	+168 to +176
Dextroamphetamine sulfate	Water	4	+20 to +23.5
Dextrose	Water	10	+52.5 to +53
Digoxin	Pyridine	10	+13.6 to +14.2
Dihydrotachysterol	Water	2	+100 to +103
Ephedrine sulfate	Water	5	−30.0 to −32.0
Epinephrine	0.5 *N* HCl	2	−50 to −53.5
Ergonovine maleate	Water	1	+50 to +55
Ergotamine tartrate	$CHCl_3$	0.35	−150 to −160
Estradiol benzoate	Dioxane	2.5	+41 to +47
Hydrocortisone	Dioxane	1	+150 to +156
Hydrocortisone acetate	Dioxane	1	+158 to +165
Hydrocortisone sodium succinate	Dioxane	1	+135 to +145
Lactose	Water	10	+52.2 to +52.5
Levarterenol bitartrate	Water	5	−10 to −12
Menthol (*l*-)	Alcohol	10	−45 to −51
Menthol (*dl*-)	Alcohol	10	−2 to +2
Medroxyprogesterone acetate	Dioxane	1	+45 to +51
Methamphetamine HCl	Water	2	+16 to +19
Morphine SO_4 (anhydrous)	Water	2	−107 to −109
Nalorphine HCl	Water	2	−122 to −125
Ouabain	Water	1	−31 to −32.5
Phenylephrine HCl	Water	2	−40 to −46
Physostigmine	Benzene	1	−119 to −121
Physostigmine salicylate	Water	1	−91 to −94
Physostigmine sulfate	Water	1	−113 to −116
Pilocarpine HCl	Water	2	+88.5 to +91.0
Pilocarpine nitrate	Water	2	+79.5 to +82.0
Prednisolone	Dioxane	1	+97 to +103
Prednisone	Dioxane	1	+167 to +175
Progesterone	Dioxane	2	+175 to +183
Propoxyphene HCl	Water	1	+52 to +57
Quinidine sulfate	Alcohol	1	+223 to +231
Riboflavin	0.1 *N* KOH in dilute alcohol	0.5	−112 to −122

TABLE 63. SUBSTANCES WITH REQUIRED SPECIFIC ROTATION (*Continued*)

Substance	Solvent used	Concentration in 100 ml, Gm	Specific rotation
U.S.P.			
Scopolamine HBr	Water	5	−24 to −26
Sodium ascorbate	Water	10	+103 to +108
Sodium levothyroxine	NaOH and alcohol	3	−5 to −6
Spironolactone	$CHCl_3$	1	−32 to −36
Sucrose	Water	26	n.l.t. +65.9
Testosterone cypionate	$CHCl_3$	2	+85 to +92
Testosterone propionate	Dioxane	1	+83 to +90
Tubocurarine chloride	Water	1	+208 to +218
N.F.			
Apomorphine HCl	0.02 *N* HCl	1.5	−49 to −51
Codeine sulfate	Water	2	−112.5 to −115
Cortisone acetate	Dioxane	1	+208 to +217
Dehydrochloric acid	Dioxane	2	+29 to +32.5
Deslanoside	Pyridine	2	+7 to +8.5
Desoxycorticosterone acetate pellets	Dioxane	1	+168 to +176
Desoxycorticosterone pivalate	Dioxane	1	+153 to +161
Dexamethasone	Dioxane	1	+72 to +80
Dexchlorpheniramine maleate	DMF solution	5	+39.5 to +43
Dextroamphetamine phosphate	Water	4	+15 to +19
Dextromethorphan HBr	Water	2	+26 to +28
Dimethyl tubocurarine iodide	Water	0.25	+148 to +158
Ephedrine (as HCl)	Water	5	−33 to −35.5
Ephedrine HCl	Water	5	−33 to −35.5
Estradiol	Dioxane	1	+76 to +83
Estradiol benzoate	Dioxane	1	+58 to +63
Estradiol cypionate	Dioxane	2	+39 to +44
Estradiol dipropionate	Dioxane	1	+37 to +41
Estrone	Dioxane	1	+158 to +165
Ethisterone	Pyridine	1	+29 to +33
Ethylmorphine HCl	Water	2	+102 to +105
Fructose	Water and NH_3	10	+89 to +91
Glutamic acid hydrochloride	HCl, 3.0 *N*	12.5	+23.5 to +25.5
Hyoscyamine sulfate	Water	5	n.l.t. 24
Lanatoside C	Methanol	2	+32 to +34.5
Levallorphane tartrate	Water	2	+37 to +39.2
Levonordefrin	HCl and water	5	+28 to +31
Levorphanol tartrate	Water	30	−13.5 to −15
Mannitol	Water	10	+23.3 to +24.3
Methscopolamine bromide	Water	5	−21 to −25
Methylprednisolone	Dioxane	1	+79 to +86
Methylprednisolone acetate	Dioxane	1	+97 to +105

TABLE 63. SUBSTANCES WITH REQUIRED SPECIFIC ROTATION (*Continued*)

Substance	Solvent used	Concentration in 100 ml, Gm	Specific rotation
N.F.			
Methylprednisolone sodium succinate	Alcohol	1	+96 to +104
Methyltestosterone	Benzene	1	+84 to +87
Progesterone	Dioxane	2	+175 to +183
Rescinnamine	Chloroform	1	−87 to −97
Sitosterols	Chloroform	2	−25 to −38
Syrosingopine	Chloroform	1	−93 to −109
Testosterone	Dioxane	1	+101 to +105
Testosterone enanthate	Dioxane	2	+77 to +82

n.l.t. = not less than.

zero point is determined. It is possible to rotate the analyzer in some instruments without rotating the graduated disk, and it is possible, in such cases, to adjust the instrument so that the zero-point correction is eliminated.

2. Dissolve 26 Gm of sucrose, accurately weighed and previously dried to constant weight at 105°, in sufficient distilled water to make 100 ml at 20°.

The sugar is dried to drive off moisture. The solution should be made up in a volumetric flask calibrated at 20°, and the final adjustment of the volume should be made at the same temperature.

3. Fill a clean dry observation tube with the solution. Determine the value of the angle of rotation. Calculate the specific rotation of the sample examined.

A polarimeter tube may be cleaned and dried easily by washing it successively with water, alcohol, and ether. In the determination of the angle of rotation, a number of successive readings should be taken as described under Procedure 1, and their mean taken as the angle of rotation. After correcting for the zero point of the instrument, the specific rotation can be calculated by the formulas given earlier in this chapter.

The specific rotation of the official sucrose must not be less than +65.9 deg when determined by the above method. Compare the specific rotation of the sample examined with the official requirement.

The measurement of rotatory power is used as a means of assaying several substances in the U.S.P. and N.F. These substances are listed in Table 61. Those substances for which the angular rotation is specified are given in Table 62. Those which have specific rotation requirements are listed in Table 63.

Questions and Problems

1. What is polarized light? How can it be obtained?

2. Show diagrammatically the essential parts of the polariscope used, and explain the function of each part.

3. What factors influence the rotatory power of a substance?

4. Define the term *specific rotatory power.*

5. Explain what the following terms mean: (1) $[\alpha]_D^{25°} = +4.210$; (2) $[\alpha]_D^{25°} = -10°$; (3) *dextrorotatory;* (4) *optically inactive.*

6. The optical rotation of a sample of oil of chenopodium determined in a 100-mm tube and having a density of 0.9750 is found to be −12.25 deg at 21°. Calculate the specific rotation of the oil.

7. A solution containing 10 Gm of sugar in 100 ml was found to exhibit an angular rotation of +11.5 deg when determined in a 100-mm observation tube. Calculate the specific rotation of the sucrose.

Chapter 28

REFRACTOMETRIC MEASUREMENTS

The index of refraction is a physical constant frequently made use of in the determination of the identity and purity of drug and food products. In many cases, it may be used to determine quantitatively the strength and purity of solutions or the proportions in which liquids are mixed; e.g., the percentage of sugar in syrup can be estimated directly from the refractivity of the solution, and the percentage of alcohol in water can also be determined in this way. Although the refractive index is a constant characteristic of many substances, such as fats, fatty oils, waxes, sugars, organic solvents, etc., it is applied almost exclusively in the official standards as a criterion of the purity of volatile oils.

Index of Refraction. When a ray of monochromatic light passes from one transparent substance into another of different optical density, it is deflected or refracted, except when it enters perpendicularly to the surface of contact between the substances. The extent and direction of the deflection are dependent upon the difference between the densities of the two substances. The angle between the ray in the first medium and a perpendicular to the dividing surface is termed the *angle of incidence* i, and the corresponding angle in the second medium is called the *angle of refraction* r. The sin of i and the sin of r are directly proportional to the velocities of light in the two mediums. The ratio $\sin i/\sin r$ is termed the *index of refraction*. Thus, the velocities of light in air and in water are in the ratio of about 4:3. The index of refraction of water with respect to air is therefore about 1.333. In Fig. 59 I is the less dense medium and II the more dense medium. A ray of light passing obliquely from I to II will be deflected so that the angle of refraction r will be less than the angle of incidence i, and according to the law of refraction, the relationship between these two angles will be $\sin i/\sin r = N/n$, where n is the index of refraction of the less dense and N the index of refraction of the more dense medium. When the angle i is increased, the angle r also is increased and attains its maximum value when the angle of incidence enters horizontally at 90 deg, termed *grazing incidence*, or at a right angle. In the latter case, since the sin of 90 deg = 1, the above equation becomes $1/\sin r = N/n$, or

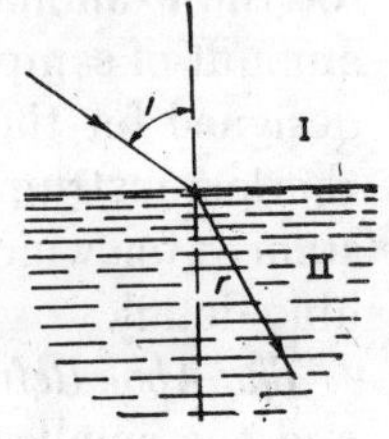

FIG. 59. Refraction of light.

$\sin r = n/N$. The critical angle of refraction, r, is defined as the angle of refraction when the angle of incidence is 90 deg. If the angle r is increased to a value at which i becomes equal to 90 deg, the beam of light will no longer pass from the first medium to the second but will travel through the first medium to the dividing surface at an angle of 90 deg with the perpendicular to the surface. If r is smaller than this particular value, however, the ray of light will pass on through the second medium, and, if greater, all light will be reflected from the surface back into the first medium. The critical ray is thus seen to make a boundary between the reflected and refracted light. This furnishes the basis for the observation of the refractive end point of a substance, and it is therefore used in many refractometers.

Refractometers. Indices of refraction may be determined by means of a polarizing microscope or refractometer. Refractometers are of several types. Some of them give figures on an arbitrary scale, the refractive index being found by reference to a table furnished with the instrument. The principal instruments in use are (1) the Pulfrich refractometer, an instrument adapted for use with liquids and solids of all degrees of refractive power, used chiefly for purely scientific investigations; (2) the Zeiss butyro-refractometer, designed especially for the examination of fixed oils and fats; (3) the Zeiss immersion refractometer, especially constructed for the examination of aqueous, oily, or alcoholic liquids, when a large amount of sample is available; (4) the Abbé refractometer, an instrument designed for the examination of liquids, solids, and plastic bodies. Since this last instrument is best suited for the examination of all the official substances for which refractive indices are given, it is described in considerable detail.

The Abbé Refractometer. Because of its simplicity, ease of manipulation, and the small amount of material required in its use, the Abbé refractometer is extensively used in pharmaceutical analysis. The wide range of indices measurable with this instrument and the fact that it can be used with ordinary white light make it capable of more general application than any of the other refractometers mentioned above. When measuring the refractive index of liquids, only a few drops, sufficient to form a film, need be placed between the prisms of the instrument. This is a distinct advantage over the other forms of refractometers, for frequently only a small quantity of the substance to be examined is available, and if the liquid is colored or turbid, the thinness of the film permits a determination of the index of refraction without difficulty.

The Abbé refractometer (Figs. 60 and 61) consists of four essential parts, viz.: (1) a telescope provided with an objective, an eye lens, and a cross-hair disk. The purpose of the telescope is to form an image of the border line of total reflection at the intersection of the cross hairs, where the image may be observed. (2) Two similar glass prisms (Abbé prisms) of

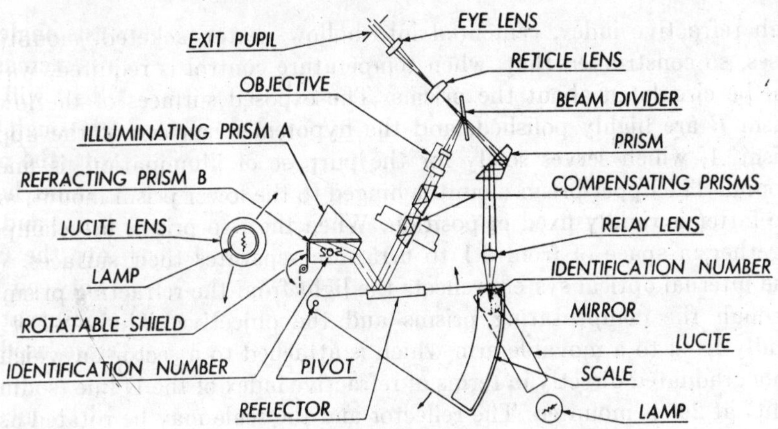

FIG. 60. Path of light, etc., through Abbé refractometer. (*Courtesy of Bausch & Lomb Optical Co.*)

FIG. 61. Precision refractometer and precision temperature controller. (*Courtesy of Bausch & Lomb Optical Co.*)

high refractive index, cemented into hollow, water-jacketed, mounting cases, so constructed that, when temperature control is required, water can be circulated about the prisms. The exposed surfaces of the lower prism *B* are highly polished, and the hypotenuse surface of the upper prism *A*, which serves solely for the purpose of illumination, is finely ground. The upper prism mount is hinged to the lower prism mount, and the latter is rigidly fixed in position. When the two prisms are clamped together, a space of from 0.1 to 0.15 mm separates their surfaces. (3) The internal optical system reflects the light from the refracting prism *B* through the compensating prisms and the objective. The *reflector* is rigidly fixed to a movable arm which is attached to a sector on which a scale graduated directly in terms of refractive index of the D line (sodium light) at 20° is mounted. The reflector and the scale may be rotated as a unit through the use of a large knob located on the side of the instrument. A fine adjustment knob in the center of the large knob assures high accuracy in making the final setting. The refractive index is read directly from the scale showing on the face of the instrument. The scale is illuminated by a suitable lamp housing. (4) The compensating prisms are used to correct for the dispersion of light. Thus, when white light is used in the place of sodium light, the beam is split up into colored beams, since short wavelengths of light are refracted to a greater extent than are long wavelengths. Unless this dispersion is compensated for in some way, the border line becomes a broad, varicolored band, which cannot be brought into accurate coincidence with the cross hairs of the telescope. The compensating prisms function to bring the various wavelengths of light together again. They are placed between the reflector and the objective and are rotated by means of the milled head located just below the refractive index scale.

Regulation of Temperature. Since the density of both the prisms and the substance under investigation varies with the temperature, it is necessary in refractivity measurements of any degree of accuracy to be able to maintain the prism and the substance at constant temperature. In the usual refractometer, this control of temperature is accomplished by circulating water, from a thermostat or other suitable constant-temperature device, about the jacketed prisms, the temperature at which the measurement is made being observed on a thermometer suitably located. A temperature control device such as that used with an Abbé refractometer is illustrated in Fig. 61.

Exercise 119

Object. To Determine the Refractive Index of Orange Oil.

Materials Required. About 1 ml of orange oil
A pipet
An Abbé refractometer

Procedure. 1. Attach the test plate to the refracting prism of the refractometer by means of the monobrom-naphthalene or other liquid provided with the test plate, by moistening the test plate with the liquid and pressing it against the prism *B*. Focus the light on the test plate, and adjust the instrument until the border line of total reflection coincides with the cross hairs in the telescope, and take the reading of the refractive index.

The refractive index of the test plate is etched upon it, and the reading of the instrument should correspond to that given on this plate. If the instrument is not in adjustment, a correction should be applied to the readings obtained, or adjustment should be made.

2. Remove the test plate, clean the prism with cotton saturated with alcohol, place 2 or 3 drops of the volatile oil on the prism, and clamp the prisms together firmly. Fix the light source so that the light is reflected through the prisms and adjust the instrument until the border line between the light and dark halves of the field of view exactly coincides with the cross hairs of the telescope, rotating the compensator prisms to obtain a sharp uncolored border line if necessary. Read the refractive index of the oil directly from the graduated scale to the fourth decimal place. Rotate the instrument out of adjustment and again determine the refractive index until three readings are obtained, taking the mean of the readings as the refractive index of the oil. Note the temperature at which the refractive index is read.

TABLE 64. SUBSTANCES WITH REQUIRED REFRACTIVE INDICES

Substance	Official requirement*	Substance	Official requirement*
U.S.P.		**U.S.P.**	
Anethole	1.5570 to 1.5610	Theobroma oil	1.4537 to 1.4585†
Anise oil	1.5530 to 1.5600	Triethanolamine	1.4810 to 1.4860
Cherry juice	n.l.t. 1.3500	**N.F.**	
Clove oil	1.5270 to 1.5350	Benzaldehyde	1.5440 to 1.5465
Coriander oil	1.4620 to 1.4720	Benzyl alcohol	1.5385 to 1.5405
Diethyltoluamide	1.5200 to 1.5235	Caraway oil	1.4840 to 1.4880
Ethohexadiol	1.4465 to 1.4515	Ethchlorvynol	1.4765 to 1.4800
Eugenol	1.5400 to 1.5420	Eucalyptol	1.4550 to 1.4600
Halothane	1.3695 to 1.3705	Eucalyptus oil	1.4580 to 1.4700
Iophendylate injection	1.5235 to 1.5255	Isopropyl alcohol	1.3763 to 1.3780
		Nikethamide	1.5220 to 1.5240
Lemon oil	1.4738 to 1.4755	Octylphenoxy polyethoxyethanol	About 1.4894
Methyl salicylate	1.5350 to 1.5380		
Orange oil	1.4720 to 1.4740	Phenylethyl alcohol	1.5310 to 1.5340
Paramethadione	1.4490 to 1.5010	Spearmint oil	1.4840 to 1.4910
Peanut oil	1.4625 to 1.4645†	*dl*-α-Tocopherol	1.5030 to 1.5070
Peppermint oil	1.4590 to 1.4650	*dl*-α-Tocopherol acetate	1.4940 to 1.4985
Phytonadione	1.5230 to 1.5252‡		
Raspberry juice	n.l.t. 1.3445	Undecylenic acid	1.4475 to 1.4485
Sorbitol solution	1.4550 to 1.4650	Zinc-eugenol cement	1.5280 to 1.5310

* Refractive index at 20° unless a different temperature is specified.

† 40°.

‡ 25°.

n.l.t. = not less than.

The refractive index of the oil should be expressed as $N_D^{20°} = 1.4725$ where $N_D^{20°}$ denotes the index of refraction for the D line (sodium light) measured at 20°.

Questions and Problems

1. Define the terms *index of refraction*, *critical ray*, and *angle of incidence*.
2. What advantages does the Abbé refractometer possess over the other types of refractometers?
3. A range is given for the refractive index of most of the official volatile oils. Why?
4. How do variations in temperature affect the indices of refraction of liquids?
5. Would you expect that orange oil would have a greater or smaller index of refraction at 25° than at 20°? Why?

Chapter 29

MELTING, CONGEALING, AND BOILING-POINT DETERMINATIONS

Melting Point. When an amorphous substance is heated, it gradually softens, eventually acquiring the properties characteristic of a liquid, but no definite point of transition from the solid to the liquid state occurs during the process of heating. When a crystalline solid is heated, there is a sharp change from the solid to the liquid state at a definite temperature, this temperature being known as the *melting point* of the substance.

The melting point of a compound may be defined as *the temperature at which the liquid and solid phases of the compound are in equilibrium.* Each pure fusible compound has a fixed and definite melting point, and this constant is often used to determine the purity of a solid, since the introduction of even a small amount of impurity lowers the melting point to a considerable extent. If appreciable quantities of impurities are present, the melting point is not sharp but ranges over a number of degrees and is usually low. In the official standards, the melting range or temperature of a substance is defined as *those points of temperature within which or the point at which the solid coalesces and is completely melted, when determined as directed.*

For the determination of the melting range, the official substances are divided into five classes, viz., (I) substances that can be readily reduced to a powder and which give reproducible melting ranges when heated at the rate of 3° per min starting approximately 30° below the expected melting range; (Ia) substances that can be readily reduced to a powder and which give reproducible melting points when heated at the rate of 3° per min starting at 10° below the expected melting range; (Ib) substances that must be precooled, dried in the capillary tube at reduced pressure, sealed in the capillary, and heated at the rate described in Class Ia to obtain reproducible results; (II) substances, such as fats, fatty acids, and waxes, that cannot be reduced to a powder readily; (III) substances requiring a specialized procedure, e.g., petrolatum.

The apparatus required to determine the melting range of substances in all classes except Class III consists of the following parts: (1) a glass container for a bath of colorless fluid; (2) a suitable stirring device; (3) an accurate thermometer covering the desired range of temperature and standardized for 76-mm stem immersion to compensate for the emergent mercury column (since the official melting-range thermometer is standardized to compensate for the error introduced by the emergent mercury column, no emergent stem corrections are necessary); and (4) a glass capillary tube about 10 cm long and 0.8 to 1.2 mm in internal diameter with walls 0.2 to 0.3 mm in thickness.

Any other apparatus or method capable of equal accuracy may be used. Six U.S.P. Melting Point Reference Standards are available and may be used for checking the accuracy of a method or apparatus.

The bath fluid is selected to secure the temperature required, but light paraffin or purified concentrated sulfuric acid is used generally for temperatures up to 200°. For higher temperatures, a pure grade of cottonseed oil (almost colorless) will serve, up to about 350°, for a limited number of determinations, and certain liquid silicones are well adapted to the higher temperature ranges.

Other, though less desirable, substitutes for use at high temperatures are (1) a pure grade of paraffin which has been freshly distilled; (2) clean, white, hydrogenated cottonseed oil. A very satisfactory bath is prepared by cautiously boiling together, for from 5 to 10 min under a hood, a mixture of 70 parts of sulfuric acid and 30 parts of potassium sulfate, stirring constantly until the potassium sulfate is completely dissolved.

The bath fluid is deep enough in the glass container to permit immersion of the thermometer to its specified immersion depth so that the bulb is still about 2 cm above the bottom of the bath. The heat may be supplied by an open flame or electrically.

The following exercises illustrate the methods used to determine the melting ranges of official substances.

Exercise 120

Object. To Determine the Melting Range of Salicylic Acid.

Materials Required. A melting-point apparatus, as described above
30 to 60 ml of light paraffin
About 0.05 Gm of salicylic acid

Procedure. 1. Reduce the salicylic acid to a fine powder and dry it at 100° for 2 hr.

The material is dried because some substances that contain water of crystallization or moisture must be rendered anhydrous to prevent liquefaction when heated through the solvent action of water, and because water as an impurity lowers the melting point.

2. Place sufficient light paraffin in the glass container for the bath and fill it to a depth to permit immersion of the thermometer to 76 mm (the specified immersion depth on the official thermometers) and still leave about 20 mm between the bulb of the thermometer and the bottom of the container.

This arrangement prevents unequal heating of the different parts of the thermometer bulb.

3. Introduce enough of the finely powdered, dry salicylic acid into the capillary tube to form a column about 0.3 cm in length.

The salicylic acid may be packed into the sealed end of the tube by tapping the open end of it into a little mound of the powder, then inverting the tube and tapping it gently on a solid surface, or by drawing the broad side of a triangular file across the surface of the tube just below the powder.

4. Heat the bath, with stirring, by means of a Bunsen flame until a temperature of about 130° is reached.

Attach the charged capillary tube to the thermometer by wetting both with the paraffin of the bath, and the tube will then be held in position by capillary attraction, or attach the capillary tube by means of a piece of platinum wire, in such a position that the column of salicylic acid is centrally located by the side of the thermometer bulb. Return the thermometer to the bath, continue the heating, with constant stirring, at a rate of rise approximately 3° per min until a temperature of 155° (3° below the expected melting point) is attained, then carefully regulate the rate to about 1 to 2° per min until melting is complete. Record the melting-interval temperature.

The temperature at which the column of salicylic acid is observed to collapse definitely in the capillary tube is regarded as the beginning of the melting; the temperature at which the salicylic acid becomes liquid throughout is regarded as the end of the melting; and the interval between the temperatures at which the melting begins and ends is taken as the melting range of salicylic acid.

The rate of rise in the temperature of the bath can be regulated easily if the burner is held in the hand so that more or less heat can be applied as required.

The melting range of a substance that cannot be readily reduced to a powder is determined in the same manner with the following exceptions: The substance is melted at as low a temperature as possible and drawn into a capillary tube, open at both ends, to a depth of about 1 cm. The charged tube is allowed to cool at 10° or at a lower temperature for 24 hr or in contact with ice for 2 hr. The melting range of the substance is then determined by attaching the charged tube to a standard thermometer as described in the above exercise and heating in a suitable bath so that the rate of rise of temperature does not exceed 0.5 to 1.0° per min for the last 5° before the expected melting range is reached. The temperature at which the substance is observed to rise in the capillary tube is taken as the melting range. Melting-range determinations by the above method are generally performed on substances that are mixtures and that do not give the sharp melting-range characteristic of pure organic compounds. Table 65 gives the melting ranges of some official substances.

Questions and Problems

1. Sketch a melting-point apparatus that meets the U.S.P. requirements.

2. List the characteristics of the thermometers that are listed as preferable types for use in determination of melting ranges. (See U.S.P.)

TABLE 65. MELTING RANGES OF SOME OFFICIAL SUBSTANCES

Substance	Melting range, °C	Substance	Melting range, °C
U.S.P.		**U.S.P.**	
Aminohippuric acid	197 to 199	Nalorphine HCl	260 to 263
Amobarbital	156 to 158.5	Neostigmine methylsulfate	142 to 145
Atropine sulfate	n.l.t. 188	Niacinamide	128 to 131
Bemegride	125 to 128	Noscopine	174 to 176
Benzoic acid	121 to 123	Pentylenetetrazol	57 to 59
Bishydroxycoumarin	287 to 293	Phenacetin	134 to 136
Busulfan	115 to 118	Phenobarbital	174 to 178
Caffeine	235 to 237.5	Phentolamine mesylate	175 to 180
Camphor	174 to 179	Phenylephrine HCl	140 to 145
Carbachol	201 to 205	Pilocarpine HCl	200 to 203
Chlorambucil	65 to 69	Pilocarpine nitrate	170 to 173
Chlorcyclizine HCl	222 to 227	Primidone	279 to 284
Chloroprocaine HCl	173 to 176	Probenecid	198 to 200
Chloropromazine HCl	195 to 198	Procainamide HCl	165 to 169
Chloroquine phosphate	193 to 195	Procaine HCl	153 to 156
	or 215 to 218	Propantheline bromide	155 to 160
Chlorpheniramine maleate	130 to 135	Propoxyphene HCl	162.5 to 168.5
Chlorpropamide	125 to 129	Propyliodone	187 to 190
Cholecalciferol	84 to 88	Propylparaben	95 to 98
Cholesterol	147 to 150	Propylthiouracil	218 to 221
Cyclopentolate HCl	137 to 141	Pyridostigmine bromide	154 to 157
Dapsone	175 to 181	Pyridoxine HCl	204 to 208
Desoxycorticosterone acetate	155 to 161	Pyrimethamine	238 to 241
Dibucaine HCl	95 to 100	Resorcinol	109 to 111
Dichlorphenamide	235 to 240	Saccharin	226 to 230
Dienestrol	227 to 231	Salicylic acid	158 to 161
Diethylcarbamazine citrate	135 to 138	Scopolamine HBr	195 to 199
Diethylstilbestrol	169 to 172	Spermaceti	42 to 50
Dihydrotachysterol	123.5 to 129	Spironolactone	198 to 207
Dimenhydrinate	102 to 107	Stearyl alcohol	56 to 60
Diphenhydramine HCl	166 to 170	Succinylcholine chloride	160 to 164
Echothiophate iodide	116 to 122	Sulfadiazine	252 to 256
Edrophonium chloride	169 to 174	Sulfamethazine	197 to 200
Epinephrine bitartrate	147 to 152	Sulfamethoxypyridazine	180 to 183
Ergocalciferol	115 to 118	Sulfapyridine	191 to 193
Estradiol valerate	143 to 150	Sulfisoxazole	192 to 195
Eucatropine HCl	183 to 186	Testosterone cypionate	98 to 104
Hexachlorophene	161 to 167	Testosterone propionate	118 to 123
Homatropine HBr	211 to 215	Tetracaine	41 to 46
Hydrocortisone	212 to 220	Thiotepa	52 to 57
Hydrocortisone acetate	216 to 222	Tolbutamide	126 to 132
Hydroxyamphetamine HBr	189 to 192	Trihexyphenidyl HCl	247 to 253
Iopanoic acid	152 to 158	Trimethadone	45 to 47
Isoniazid	170 to 173	Triplenennamine HCl	188 to 192
Isoproterenol HCl	167 to 172	Urea	132 to 134
Lanolin, anhydrous	36 to 42	Vanillin	81 to 83
Levarterenol bitratrate	100 to 106	Wax, white	62 to 65
Lidocaine	66 to 69	**N.F.**	
Lidocaine HCl	74 to 79	Acetaminophen	169 to 172
Lucanthone HCl	195 to 199	Acetyl sulfisoxazole	192 to 195
Medroxyprogesterone acetate		Acreoline HBr	170 to 175
	200 to 208	Alphaprodine HCl	218 to 220
Menadione	105 to 107	Aminobenzoic acid	186 to 189
Menthol (-*l*)	41 to 43	Antazoline PO_4	194 to 198
Methacholine chloride	170 to 173	Anthralin	175 to 181
Methamphetamine HCl	171 to 175	Antipyrine	110 to 112.5
Methenamine mandelate	127 to 130	Atropine	114 to 118
Methimazole	144 to 147	Benzestrol	161 to 163
Methoxamine HCl	212 to 216	Benzethonium chloride	160 to 165
Methylparaben	125 to 128	Benzocaine	88 to 92

TABLE 65. MELTING RANGES OF SOME OFFICIAL SUBSTANCES (*Continued*)

Substance	Melting range, °C
N.F.	
Benzpyrinium bromide	114 to 120
Benztropine mesylate	142 to 144
Bithionol	186 to 189
Brompheniramine maleate	130 to 135
Butethamine HCl	192 to 196
Butyl aminobenzoate	57 to 59
Carbetapentane citrate	90 to 95
Carbinoxamine maleate	116 to 121
Cetyl alcohol	45 to 50
Cetylpyridinium chloride	80 to 84
Chlordiazepoxide	212 to 218
Chlormerodrin	145 to 155
Chlorothen citrate	112 to 116 125 to 140
Chlorothymol	59 to 61
Chlorotrianisene	115 to 117
Codeine	154 to 158
Cyclopentamine HCl	113 to 116
Cycrimine HCl	241 to 244
Danthron	190 to 195
Dehydrochloic acid	231 to 242
Dexchlorpheniramine maleate	110 to 115
Dibucaine	62 to 65
Dienestrol	227 to 231
Diethylstilbestrol dipropionate	105 to 107
Dimethisoquin HCl	144 to 148
Diphemanil methylsulfate	189 to 196
Diphenadione	144 to 150
Diphenylhydantoin	292 to 299
Doxylamine succinate	100 to 104
Ephedrine HCl	217 to 220
Estradiol	173 to 179
Estradiol benzoate	190 to 196
Estradiol cypionate	149 to 153
Estradiol dipropionate	104 to 109
Estrone	256 to 262
Ethinamate	95 to 98
Ethisterone	267 to 275
Ethoheptazine citrate	135 to 145
Ethyl biscoumacetate	177 to 182
Ethyl vanillin	76 to 78
Glutethimide	86 to 89
Glyceryl monostearate	n.l.t. 55
Hexylcaine HCl	182 to 184
Hexylresorcinol	62 to 67
Homatropine methylbromide	190 to 198
Hydralazine HCl	270 to 280
Hydroxyzine HCl	196 to 204
Hyoscyamine sulfate	n.l.t. 200
Imipramine HCl	170 to 174
Inositol	224 to 227
Iodoalphionic acid	160 to 164
Iodoform	About 115
Isobucaine HCl	182 to 185
Levallorphan tartrate	174 to 177
Levonordefrin	205 to 215
Levorphanol tartrate	114 to 116
Mephenesin	70 to 73.5
N.F.	
Mephobarbital	176 to 181
Mepiridine HCl	186 to 189
Mepivacaine HCl	255 to 262
Meprobamate	103 to 107
Metabutethamine HCl	181 to 184
Metabutoxycaine HCl	117 to 120
Methacholine bromide	147 to 150
Methacholine chloride	170 to 173
Methantheline bromide	171 to 177
Methapyrilene HCl	161 to 165
Metharbital	151 to 155
Methprylon	74 to 77
Methscopolamine bromide	222 to 228
Methsuximide	50 to 56
Methylbenzethonium chloride	162 to 163
Methyltestosterone	163 to 168
Monbenzone	115 to 118
Naphazoline HCl	253 to 258
Niacin	234 to 237
Nordefrin HCl	175 to 180
Penthienate bromide	122 to 128
Pentylenetetrazol	58 to 61
Petrolatum	38 to 60
Phenacaine HCl	n.l.t. 190
Phenindamine tartrate	160 to 162
Pheniramine maleate	104 to 108
Phenmetrazine HCl	172 to 182
Phenolphthalein	n.l.t. 258
Phensuximide	68 to 74
Phentolamine HCl	238 to 242
Phenylbutazone	103 to 106
Phenylmercuric nitrate	175 to 185
Pipradrol HCl	285 to 295
Pramoxine HCl	169 to 172
Promazine HCl	172 to 182
Propoxycaine HCl	146 to 151
Pyrilamine maleate	99 to 103
Pyrrobutamine phosphate	127 to 131
Salicylamide	139 to 142
Salicylanilide	136 to 138
Stannous fluoride	212 to 214
Sulfacetamide	181 to 184
Sulfadimethoxine	197 to 202
Sulfamethizole	209 to 212
Testosterone	153 to 157
Testosterone enanthate	34 to 39
Thenyldiamine HCl	167 to 171
Theophylline	270 to 274
Thonzylamine HCl	173 to 176
Thymol	48 to 51
d-α-Tocopheryl acid succinate	73 to 78
Tribromoethanol	79 to 82
Tridihexethyl chloride	198 to 202
Trimethobenzamide HCl	187 to 190
Urethan	48 to 50
Vinbarbital	160 to 163
Yellow wax	62 to 65

n.l.t. = not less than.

3. How may a thermometer and the melting-range apparatus in which it is used be checked for accuracy?

4. Why must thermometers be standardized at a given immersion depth in the melting-range bath?

5. Given two samples that show the same melting range, how might you determine whether or not they were the same substance?

Congealing Temperature. The congealing, or solidifying, temperature is a physical constant, characteristic of certain liquids, such as anise oil, eucalyptol, and the fatty acids of fixed oils, that may serve as a criterion of purity and quality just as the melting range of a solid does, since the presence of impurities also causes a depression of the congealing temperature. The congealing temperature of a solid is frequently referred to in commerce as the *melting temperature*, although its determination involves solidification or congelation from the molten state upon controlled cooling. With respect to fatty acids, it is spoken of as the *solidification temperature* or the *titer*. The method for the determination of congealing temperatures is as follows (Fig. 62):

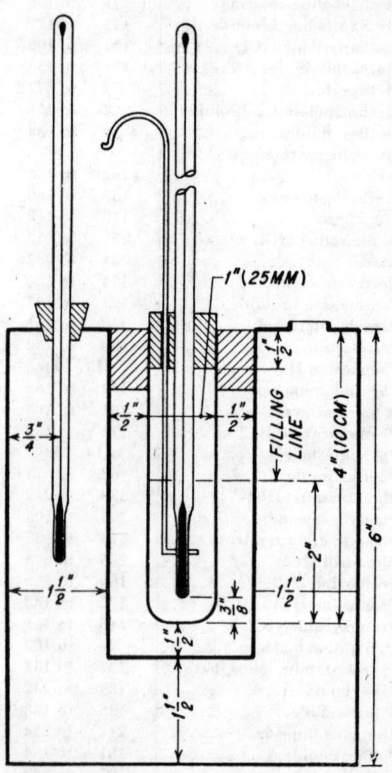

FIG. 62. Congealing-temperature apparatus.

Assemble an apparatus in which the container for the substance is a 25- × 100-mm test tube. This is provided with a Type III thermometer suspended in the center, and a wire stirrer, about 30 cm long, bent at its lower end into a horizontal loop around the thermometer.

The sample container is supported, by means of a cork, in a suitable water-tight cylinder about 50 mm in inside diameter and 11 cm in length. The cylinder, in turn, is supported in a suitable water bath sufficient to provide at least a 37-mm layer of water surrounding the sides and bottom of the cylinder. The outside water bath is provided with a Type I thermometer.

Melt the substance, if a solid, at a temperature not more than 20° above its expected congealing point, and pour it into the test tube to a height of 50 mm. Assemble the apparatus with the bulb of the test-tube thermometer immersed halfway between the top and bottom of the sample in the test tube. Fill the water bath to within 12 mm of the top with water at a temperature 4 to 5° below the expected congealing point.

When the sample has cooled to about 10° above its expected congealing point, adjust the water bath to a temperature 15 to 20° below the expected congealing point. Stir the sample continuously during the remainder of the test by moving the loop up and down throughout the entire depth of the sample at a regular rate of 100 complete cycles per minute.

Record the reading of the test-tube thermometer every 30 sec. The temperature will be expected to fall gradually at first, then become almost constant or rise slightly before becoming constant, and again fall gradually. Pronounced supercooling may cause deviation from this normal pattern of temperature changes. If the latter occurs, repeat the test, introducing small particles of the material under test in solid form at 1° intervals as the temperature approaches the expected congealing point.

Continue recording the temperature in the test tube every 30 sec for at least 3 min after the temperature again begins to fall after remaining constant. The average of not less than four consecutive readings that lie within a range of 0.2° constitutes the congealing point.

In case the substance is a liquid at room temperature, carry out the determination in a bath temperature 10 to 15° below the expected congealing point.

Congelation frequently may be induced by rubbing the inner walls of the test tube with the thermometer, or by introducing a small fragment of the previously congealed substance. Discontinue the stirring, and note the rise in temperature every 5 to 10 sec. The highest temperature remaining constant for about 1 min during the congelation is defined as the congealing temperature.

The official procedure for the determination of the solidification of fatty acids is illustrated in the following exercise.

Exercise 121

Object. Determination of the Solidification Temperature of the Fatty Acids of Cottonseed Oil.

Materials Required. 50 ml of cottonseed oil
25 Gm of potassium hydroxide
100 ml of glycerin
15 ml of alcohol
A standard thermometer for determining the titer of fatty acids (see U.S.P.)

Procedure. 1. "Heat 75 ml. of glycerin-potassium hydroxide solution (made by dissolving 25 Gm. of potassium hydroxide in 100 ml. of glycerin) to 150° in an 800-ml. beaker, and add 50 ml. of the clarified fat, melted if necessary. Heat the mixture for 15 min. with frequent stirring, but do not allow the temperature to rise above 150°. When saponification is complete, the mixture is homogeneous, with no particles clinging to the beaker at the meniscus."

The oil is saponified by the KOH with the formation of the potassium salts of the fatty acids and glycerin. See Saponification Value (Chap. 13) for the reactions that occur.

2. "Pour the contents of the beaker into 500 ml. of nearly boiling water in an 800-ml. beaker or casserole, add slowly 50 ml. of dilute sulfuric acid (made by adding 1 volume of sulfuric acid to 3 volumes of water), and heat the solution, with frequent stirring, until the fatty acids separate cleanly as a transparent layer. Wash the acids with boiling water until free from sulfuric acid, collect them in a small beaker, and place on a boiling water bath or steam bath until the water has settled and the fatty acids are clear, filter into a dry beaker while hot, and dry for 20 minutes at 100°.

"Place 3 ml. of the dry acids in a test tube and add 15 ml. of alcohol. Heat the solution to boiling and add an equal volume of ammonia T.S. A clear solution should result."

The fatty acids are liberated from their potassium soaps by the sulfuric acid. They are washed to remove potassium sulfate, excess sulfuric acid, and glycerin, and dried to remove the water. If saponification is incomplete, the acids will be contaminated with unsaponified oil, which would alter the solidification temperature of the acids. A turbidity in the alcohol solution indicates incomplete saponification, since the oil is insoluble in the diluted alcohol and the ammonium soap is soluble.

3. "Cool the dry, filtered acids to from 15 to 20 degrees above the expected reading, and transfer to a glass tube 25 mm. in diameter and 100 mm. in length, the glass being 1 mm. in thickness. By means of a perforated cork fasten the tube in a wide-mouth bottle of clear glass, approximately 70 mm. in diameter and 150 mm. in height. Suspend a Type V thermometer [see U.S.P.] in the melted acids so that it will serve as a stirrer, cool if necessary, and stir the mass slowly until the mercury remains stationary for 30 seconds. Then allow the thermometer to hang quietly, with the bulb in the center of the acids, and observe the rise of the mercury column. The highest point to which it rises is the Solidification Temperature of the fatty acids."

TABLE 66. SUBSTANCES WITH CONGEALING TEMPERATURE REQUIREMENTS

Substance	Congealing temperature, °C	Substance	Congealing temperature, °C
U.S.P.		**U.S.P.**	
Acetic acid, glacial	n.l.t. 15.6	Fennel oil	n.l.t. 3
Anethole	n.l.t. 20	Menthol (*dl*-)	30.5 to 32
Anise oil	n.l.t. 15	Oleic acid	n.m.t. 10
Benzyl benzoate	n.l.t. 18	Paraldehyde	n.l.t. 11
Chlorophenothane	n.l.t. 89	Phenol	n.l.t. 39
Fatty acids from:		Polyoxyl 40 stearate	39 to 44
Corn oil	14 to 20	Stearic acid	n.l.t. 54
Cottonseed oil	31 to 35	Undecylenic acid	n.l.t. 21
Olive oil	17 to 26	**N.F.**	
Peanut oil	26 to 32	Eucalyptol	n.l.t. 0
Sesame oil	20 to 25	Eucalyptus oil	n.l.t. −15.4
Theobroma oil	45 to 50	Fatty acids from lard	36 to 42
		p-Chlorophenol	42 to 44

When the melted acids solidify, the "latent heat of fusion" is liberated, and a rise in temperature occurs, which remains constant for a short time. The fatty acids give a distinct rise in temperature when they solidify, whereas the fats and fatty oils are usually characterized by the fact that they yield a constant temperature, which for some is variable when they solidify.

Questions and Problems

1. The U.S.P. requires that the congealing temperature of phenol should not be below 39°. Why?

2. Define *melting temperature* and *titer* as they are used in connection with congealing range or temperature.

3. Why is there a rise in temperature which remains constant for a short time when fatty acids solidify?

Boiling and Distilling Ranges. One of the characteristic properties of a liquid is its boiling point. If a substance consists of a single compound, it is generally considered to be pure when it distills at a constant boiling point, under constant pressure. In like manner, the purity of a mixture of substances, such as a volatile oil, may be judged in part by the range of temperature at which the substance distills.

The boiling point of any liquid may be defined as *that temperature at which the vapor pressure of the liquid is equal to the atmospheric pressure.* Where a given temperature is stated as the boiling point of a liquid, the inference is that the corresponding pressure is standard atmospheric pressure, 760 mm, unless otherwise given.

The boiling or distillation range or temperature is *the range of temperatures between which all of a given liquid or a specified percentage of the liquid will distill.*

To determine the temperatures between which an official liquid may boil or the percentage of the material which distills between two specified temperatures, use Method I or Method II as directed. The lower of the two temperatures is the thermometer reading when the first 5 drops of distillate have been collected, and the upper temperature is the reading when the last liquid evaporates from the bottom of the flask (dry point) or when the proportion specified has been collected.

Method I

Use this method for liquids for which the permissible range in boiling temperature is 5° or less, unless otherwise specified in the individual monograph. In order to have the entire volatile portion distill over at the prescribed rate, it is best to make a preliminary distillation of a separate portion of 25 ml of the liquid, during which the source of heat is regulated until the distillation proceeds at the prescribed rate. Stop the distillation, clean the flask, recharge it with a fresh portion of 25 ml of the liquid, and conduct a new distillation as directed.

Apparatus. Use a distilling flask having a bulb of 50- to 60-ml capacity and a neck 10 to 12 cm long and 14 to 16 mm in internal diameter. Attached about midway on the neck is an outlet tube 10 to 12 cm long and 4 to 5 mm in internal diameter, which forms an angle of 70 to 75° with the lower portion of the neck.

Use a straight glass condenser with a water jacket 40 to 60 cm long, the distance from the upper end of the jacket to the neck of the flask being 18 to 25 cm.

Provide an asbestos board 12 to 15 cm square and 3 to 5 mm thick and having a circular perforation, located centrally, to hold the flask. The edge of the perforation should fit the flask closely when the latter is set into it. The size of the

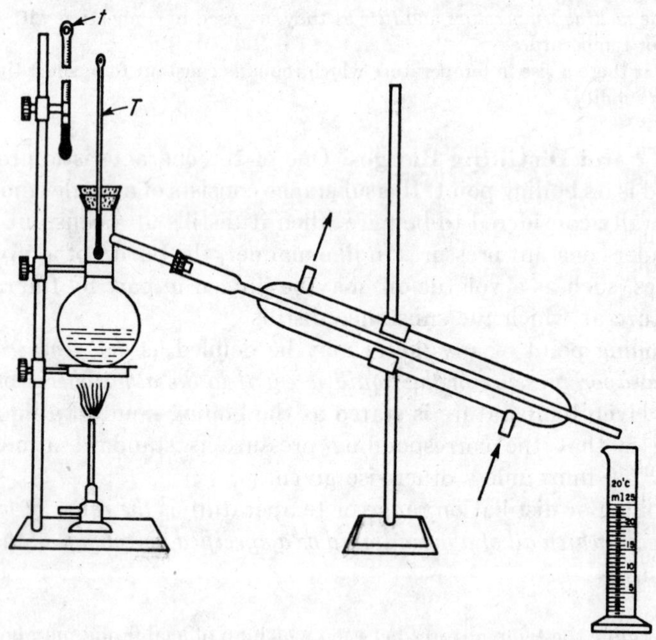

FIG. 63. Apparatus for boiling- and distilling-point determinations.

perforation should be such that when the flask is set into it the portion of the flask below the upper surface of the asbestos has a capacity of 3 to 4 ml.

In order to avoid the necessity for an emergent-stem correction, an accurately standardized thermometer of the Anschütz type may be employed. When placed in position, the top of the bulb of the thermometer is level with the center of the opening of the outlet tube.

When a thermometer of Type VI or VII is employed, correct for the temperature of the emergent stem by the formula $0.00015 \times N(T - t)$, in which N represents the number of degrees of emergent stem from the bottom of the stopper, T, the observed temperature of distillation, and t, the temperature registered by an auxiliary thermometer the bulb of which is placed midway of the emergent stem, adding the correction to the observed readings of the main thermometer (Fig. 63).

Procedure. Place the asbestos board on a tripod or other suitable support. Place in the distilling flask 25 ml of the liquid to be tested, insert the thermometer, stand the flask in an upright position in the perforation of the asbestos board, and connect it with the condenser. Then distill the liquid by the application of heat, from a suitable source, at the rate of 4 to 5 ml per min, noting the temperature as soon as 5 drops of the liquid have distilled into the receiver, and when the last liquid evaporates from the bottom of the flask or when the specified percentage has distilled over. Correct the observed temperature readings for any variation in the barometric pressure from the normal (760 mm) by allowing 0.1° for each 2.7 mm of variation, adding if the pressure is lower, or subtracting if higher than 760 mm, and apply emergent-stem correction when necessary.

Method II

Use this method with liquids for which the permissible range in boiling temperature exceeds 5°.

Apparatus. Use apparatus similar to that specified for Method I, except that the bulb of the distilling flask is of 200-ml capacity, and the internal diameter of the neck is 18 to 24 mm and that of the outlet tube is 5 to 6 mm. The perforation in the asbestos board is 50 mm in diameter.

For liquids that distill below 80°, connect the condenser and the receiver by means of a suitable adapter fitted loosely enough to allow air to escape readily from the receiver. Arrange to keep the receiver immersed in ice to within 2.5 cm of its height during the distillation.

Procedure. Place the distilling flask in an upright position in the perforation in the asbestos board and connect it with the condenser. The outlet tube is to extend 25 to 35 mm into the condenser beyond the connecting stopper.

Measure 100 ml of the liquid to be tested, using a cylinder having 1-ml graduations. Note the temperature of the liquid, and transfer it as completely as possible to the distilling flask. Use this cylinder as the receiver for the distillate without rinsing out any of the adhering liquid. Insert the thermometer, connect the distilling flask with the condenser as described in Method I, and distill by the application of heat, at the rate of from 4 to 5 ml per min, collecting the distillate coming over between the temperatures specified in the text. Bring the distillate to the same temperature as that at which the liquid was originally measured, and note its volume. Correct the temperature reading for barometric pressure, and if necessary, for the emergent stem of the thermometer.

Cool all liquids which begin to distill below 80° to between 10 and 15° before measuring the 100 ml for the test.

Exercise 122

Object. To Determine the Boiling Range of Ethyl Acetate.

Materials Required. 25 ml of ethyl acetate
A distillation apparatus as illustrated

Procedure. Follow the procedure described under Method I above. Report the boiling range found and compare with the official limits as given in the N.F.

The temperature obtained when a thermometer is immersed in a boiling liquid is usually greater than the true boiling point of the liquid, because the vapor bubbles form below the surface of the liquid and consequently are at a pressure greater than atmospheric. There is also an appreciable error due, in this case, to superheating of the liquid. When the thermometer is suspended in the vapor of the liquid, however, the bulb becomes coated with a thin film of liquid in contact with its vapor, and superheating is avoided.

It is often best to run a preliminary distillation, of a 25-ml portion of liquid, prior to the boiling-point determination, in order to ascertain the regulation of heating necessary to cause the entire portion of volatile liquid to distill at the prescribed rate. Superheating of the liquid and consequent bumping can be prevented by placing a few small pieces of porous plate in the distillation flask. The correction for the barometric pressure may be made simply, as illustrated in the following example: If the range of temperature within which 95 percent of glacial acetic acid distills is 116.5 to 116.9° at 733 mm, the corrected temperatures would be

$$116.5 + 0.1\left(\frac{760 - 733}{2.7}\right) = 117.5$$

and

$$116.9 + 0.1\left(\frac{760 - 733}{2.7}\right) = 117.9$$

The boiling point of the acid should then be reported as $\text{b.p.}_{733} = 117.5$ to 117.9° cor., the subscript 733 indicating the barometric pressure at which the determination was made, and the abbreviation "cor." indicating that the boiling point as reported has been corrected to standard pressure.

If an accurately standardized thermometer of the Anschütz type is not used in the determination, a further correction for the emergent stem must be made, as described under Method I. Correct for the temperature of the emergent stem by the formula $0.00015 \times N(T - t)$, in which N represents the number of degrees of emergent stem from the bottom of the stopper to the distilling temperature, T, the observed temperatures of distillation, and t, the temperature registered by an auxiliary thermometer the bulb of which is placed outside the distillation flask, midway of the emergent stem. The correction is added to the observed reading of the main thermometer; i.e., if in the determination the main thermometer is placed in the stopper to the 30° graduation mark, the distilling range is found to be 172 to 174°, and the temperature registered on the auxiliary thermometer is 44°, the correction for the distilling temperature would be

$$0.00015 \times 142(172 - 44) = 2.81°$$

and the corrected distilling range becomes

$$172 + 2.7 = 174.7° \text{ to } 174 + 2.7 = 176.7°$$

When boiling points are given with a range, the figures are to be regarded as inclusive. The approximate boiling points of such substances as glacial acetic acid, ether, ethyl oxide, ethyl chloride, and alcohol are given in the monographs under the "description" heading and should be regarded as informative rather than as criteria of identity and purity.

Questions and Problems

1. Define the terms *boiling range* and *distilling range.*

2. How is the distillation range of eucalyptol directed to be determined in the N.F.?

3. If ether contained alcohol as an impurity, would the boiling point of the ether be raised or lowered?

4. A sample of carbon tetrachloride is found to distill between 76.8 and 77.9° at 664 mm. Calculate the corrected distilling range of the carbon tetrachloride.

5. Consult the appended table and list those substances the boiling points of which should be determined by Method II of the U.S.P.

6. Which is more pure as indicated by the required boiling-point ranges, ethyl acetate of the N.F. or ethyl acetate reagent of the U.S.P.?

Chapter 30

SPECIFIC GRAVITY AND DENSITY

The *absolute density* of a solid or liquid is the mass of unit volume of the substance. Generally, however, in using the term *density*, we refer to the *relative density* or the density at a given temperature relative to the density of distilled water, either at the same temperature, represented by d, or at 4°, represented by $d_{4^\circ}{}^{t^\circ}$. When, in the latter case, the density is corrected for the buoyancy of the air, the number obtained represents the *true specific gravity* of the substance. In common practice, however, and in all except very exact work, the specific gravity of a substance is taken as *the ratio of the weight of the substance to the weight of an equal volume of distilled water at the same temperature.*

In the official standards, the specific gravity of a substance has been established as *the ratio of the apparent weight of the substance in air at* 25° *to that of an equal volume of distilled water at the same temperature*, or $d_{25^\circ}{}^{25^\circ}$, except when otherwise stated and for alcohol, which should be determined at 15.56° and compared with water at the same temperature. The latter is the standard temperature established by the U.S. Internal Revenue Service for the measurement of the specific gravity of alcohol.

A considerable number of methods have been developed by which the specific gravities of various substances can be determined. Only those methods of practical importance most commonly used in pharmaceutical analysis to establish the identity and purity of medicinal substances will be considered in the following pages. These methods may be classified as follows:

A. Methods used to determine the specific gravity of liquids
  1. By the use of a pycnometer
  2. By the use of the Westphal balance
  3. By weighing a solid of known specific gravity and weight in the liquid
  4. By the use of hydrometers

B. Methods used to determine the specific gravity of solids
  1. Solids heavier than and insoluble in water
     a. By weighing in air and then in water
     b. By determining the weight of water displaced

2. Solids lighter than and insoluble in water
   a. By weighing immersed in water
   b. By flotation
3. Solids heavier than and soluble in water: By weighing in air and then in some liquid of known specific gravity and in which the solid substance is insoluble
4. Solids lighter than and soluble in water: By weighing immersed in some liquid of known specific gravity and in which the substance is insoluble

METHODS USED TO DETERMINE THE SPECIFIC GRAVITY OF LIQUIDS

1. By the Use of a Pycnometer. The specific gravity of liquids can be letermined very accurately by means of some form of pycnometer or

FIG. 64. Specific-gravity bottle.

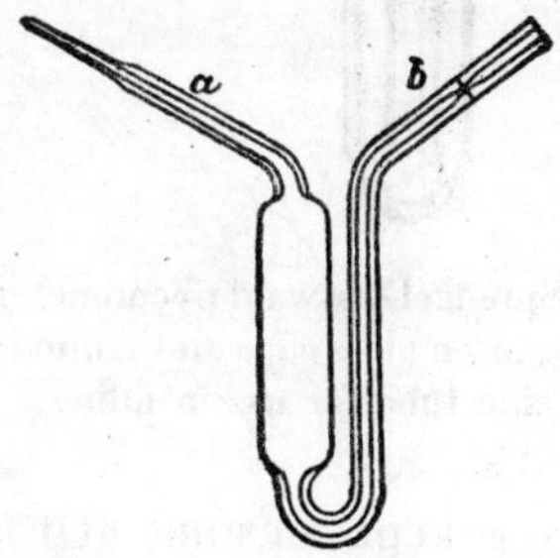

FIG. 65. Sprengel-Ostwald pycnometer (modified form).

pecific-gravity bottle, such as that shown in Fig. 64. Although a number f different forms of pycnometers have been developed by investigators, hose described may be used for practically all the determinations of the pecific gravity of liquids required in pharmaceutical analysis.

The Sprengel-Ostwald Pycnometer. One of the simplest and most generlly useful forms of pycnometers is that known as the Ostwald modificaion of the Sprengel tube (Fig. 65). This form of pycnometer, which can be asily made by the student, consists of a thin glass tube, 10 to 15 mm in liameter, to which narrow tubes *a* and *b* are sealed and bent as shown in he figure. At about the middle of tube *b*, the bore should be slightly contricted and a graduation mark scratched or etched on the glass. Tube *a* hould be drawn out to a point and fire-polished, so that the bore is contricted. Some of the commercially available pycnometers of this type Fig. 66) are provided with ground-on glass caps which fit over the ends f the tubes *A* and *B*. In all except very accurate work, however, these aps are unnecessary.

The volume of the pycnometer should be about 5 to 15 ml, thus permitting an accuracy of about one unit in the fourth decimal place. When only a very small volume of liquid is available for a determination, however, a tube of small bore, bent in the shape indicated in Fig. 66, and of a capacity of 0.5 ml, may be used in approximate specific-gravity determinations.

In carrying out a determination with this type of pycnometer, it should be cleaned by washing successively with water, alcohol, and ether and then dried by drawing a current of warm air through the tube. The pycnometer, cleaned and dried, is first weighed empty. To do this, it is suspended from the end of a balance beam by means of a double hook made from either platinum or copper wire, preferably the former. The pycnometer is then filled with distilled water, which has been previously boiled to expel dissolved carbon dioxide and allowed to cool, by applying gentle suction through a rubber tube attached to the tube *b* while the tip of the tube *a* dips into the water. The introduction of air bubbles into the pycnometer must be carefully avoided. The filled pycnometer is then suspended in a beaker of water, along with a thermometer, and the temperature is kept constant at the desired level.

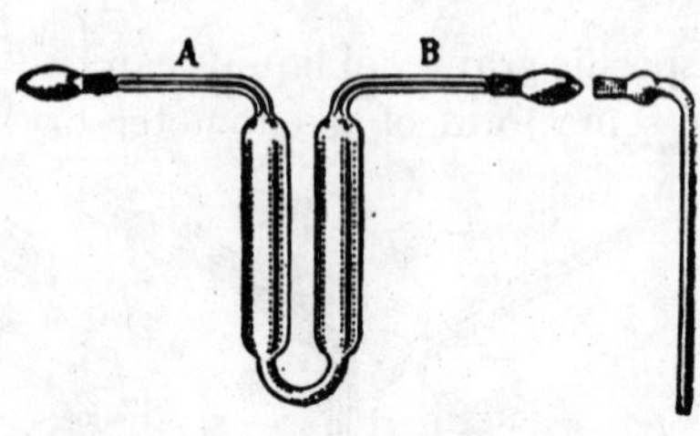

FIG. 66. Sprengel-Ostwald pycnometer with ground-on glass caps and ground-on glass side tube for use in filling.

When the pycnometer and its contents have acquired the same temperature as the bath, usually in about 20 min, the amount of water must be so adjusted that it exactly fills the pycnometer from the point of the tube *a* to the graduation mark on *b*. If there is too much water in the pycnometer, a piece of filter paper is carefully touched to the tip of the tube *a*, thereby drawing water from the pycnometer until the meniscus corresponds to the mark on *b*. If the pycnometer is not completely filled, a rod or tube carrying a drop of water is touched against the end of the tube *a*, and water is drawn into the tube by capillarity.

The pycnometer is removed from the bath and the outside is dried with a cloth, using care to prevent the expulsion of water from the tube by expansion due to the heat of the hands, and weighed.

After the pycnometer filled with water has been weighed, it is emptied, dried once more, and filled with the liquid the specific gravity of which is to be determined. It is placed in the constant-temperature bath, the volume of liquid is adjusted to the mark, and the pycnometer is dried with a cloth in the same manner as before and weighed.

The Geissler pycnometer or specific-gravity bottle (Fig. 67) is a type of pycnometer well suited to accurate determinations and can be procured in sizes varying from 25- to 100-ml capacity. Pycnometers of this type are provided with a thermometer stopper and a capillary overflow tube fitted with a ground-glass cap to prevent evaporation. The cap should have a very small perforation in its top to permit the liquid to expand without forcing the cap loose. The thermometers provided with these pycnometers are not always accurate and should be tested by comparison with one of known accuracy whenever exact determinations are made.

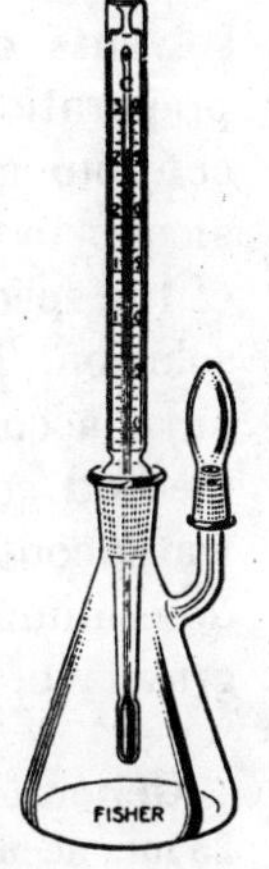

FIG. 67. Geissler pycnometer.

In ascertaining the specific gravity of a liquid with this pycnometer, the thermometer and cap are removed, the previously cleaned bottle is filled with the liquid, which has been cooled to from 1 to 3° below the temperature at which the specific gravity is to be determined. The thermometer is then fitted to the bottle, using due precautions, so that no air bubbles remain in the bottle. The pycnometer with its contents is warmed to the required temperature, any excess liquid that exudes from the capillary tube is carefully wiped off, the cap is fitted to the side tube, and the whole is weighed.

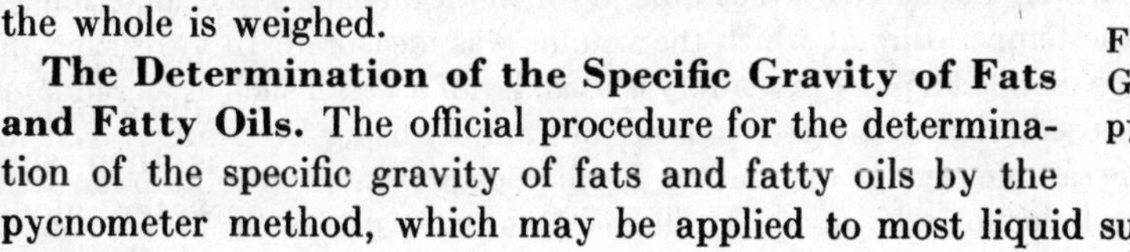

The Determination of the Specific Gravity of Fats and Fatty Oils. The official procedure for the determination of the specific gravity of fats and fatty oils by the pycnometer method, which may be applied to most liquid substances, is as follows:

"Determine the specific gravity of a fat or oil at 25°, except when the substance is a solid at that temperature, in which case determine the specific gravity at the temperature directed in the respective monograph and referred to water at 25°.

Procedure. "Clean a pycnometer (use a Sprengel or other suitable pycnometer with a well-fitted capillary stopper) by filling it with chromic acid cleansing mixture and allowing it to stand for at least 4 hours. Empty the pycnometer, and rinse it thoroughly with water. Then fill it with recently boiled water previously cooled to about 20°, and place in a constant temperature bath at 25°. At the end of 30 minutes adjust the level of the water to the proper point on the pycnometer, insert the perforated cap or stopper, remove from the bath, wipe dry with a clean cloth free from lint, and, after allowing to stand for 30 minutes, weigh. Empty the pycnometer, rinse several times with alcohol and then with ether, allow it to become perfectly dry, remove any ether vapor, and weigh. Ascertain the weight of the contained water at 25° by subtracting the weight of the pycnometer from its weight when full.

"Fill the clean, dry pycnometer with the oil at a temperature below that at which the determination is to be made. Place it in a constant temperature bath at the specified temperature for 30 minutes, adjust the level of the oil to the mark on the pycnometer, insert the cap or stopper, wipe dry, allow to stand for 30 minutes, and weigh. Subtract the weight of the empty pycnometer from its weight when filled with oil, and divide the difference by the weight of water contained at

25°. The quotient is the specific gravity at the temperature of observation, referred to water at 25°."

The Determination of Alcohol Content Using a Pycnometer. It is very difficult to effect a quantitative separation of alcohol from the solvents ordinarily employed in the manufacture of pharmaceutical preparations, but it is comparatively easy to prepare an aqueous distillate containing all the alcohol in the original mixture and to estimate the alcohol in the distillate by some physical means, such as the determination of the specific gravity, the index of refraction, or the boiling point of the solution. The specific-gravity method is official, and it is probably the most accurate and satisfactory of the methods mentioned above. This method consists of separating the alcohol from substances other than water contained in the preparation by distillation or other procedures and determining the percent of alcohol in the preparation from the specific gravity of the distillate.

General Procedure. "Transfer to a suitable distilling apparatus not less than 25 ml., accurately measured, of the liquid in which the alcohol is to be determined, and note the temperature at which the volume was measured. In measuring the sample to be distilled a pipet preferably is used, but if a vessel such as a graduated cylinder is used, rinse the latter free of the sample, using the water taken for diluting the sample prior to distillation. For liquids presumed to contain less than 30 percent of alcohol, add an equal volume of water. Distil, and collect a volume of distillate about 2 ml. less than the volume taken of the original test liquid, adjust to the temperature at which the original test liquid was measured, add sufficient water to measure exactly the original volume of the test liquid, and mix thoroughly. The distillate is clear or not more than slightly cloudy, and does not contain more than traces of volatile substances other than alcohol and water. Determine the specific gravity of the liquid at 25°, using this result to ascertain the percentage, by volume, of alcohol contained in the liquid examined by reference to the *Alcoholometric Table* [see U.S.P.].

"For liquids presumed to contain more than 30 percent of alcohol, proceed as directed above, except: dilute the sample with about twice its volume of water and collect a volume of distillate about 2 ml. less than twice the volume of the original test liquid, bring to the temperature at which the original liquid was measured, add sufficient water to measure exactly twice the original volume of the test liquid, and determine its specific gravity. The proportion of alcohol, by volume, in this distillate, as ascertained from its specific gravity, equals one-half that in the liquid examined.

"This general method is suitable for examining most fluidextracts and tinctures, provided the capacity of the distilling flask is sufficient (commonly two to four times the volume of the liquid to be distilled) and the rate of distillation is such that clear distillates are produced. Cloudy distillates may be clarified by agitation with talc, or with precipitated calcium carbonate, and filtered, after which the temperature of the filtrate is adjusted and the alcohol determined from the specific gravity. During all manipulations take precautions to minimize the loss of alcohol by evaporation.

"**Frothing.** Liquids that froth to a troublesome extent during distillation may be strongly acidified with phosphoric or sulfuric acid, or treated with a slight excess of calcium chloride solution, or with a little paraffin or yellow wax before distillation.

"**Bumping.** Liquids that tend to bump when heated (particularly resinous solutions) may be made alkaline with magnesia magma, or pieces of pumice, glass beads, or similar materials may be placed in the distilling flask with the liquid before distillation.

"**Glycerin.** Dilute liquids that contain glycerin with sufficient water so that the residue, after distillation, contains at least 50 percent of water.

"**Iodine.** Treat all solutions containing free iodine before distilling with powdered zinc or decolorize with just sufficient solution of sodium thiosulfate followed by a few drops of sodium hydroxide T.S. to fix volatile sulfur compounds.

"**Volatile Substances.** Treat spirits, elixirs, tinctures, etc., that contain appreciable proportions of volatile materials other than alcohol and water, such as volatile oils, chloroform, ether, camphor, etc., as follows: Mix the accurately measured liquid with about an equal volume of saturated solution of sodium chloride in a separator, then add a volume of petroleum benzin equal to that of the sample, and shake the mixture to extract the interfering volatile ingredients. Draw off the separated lower layer, and extract the benzin solution with two successive portions of a saturated solution of sodium chloride, using about one-half as much each time as was used in the first extraction mixture. Combine the saline solutions, and distil the mixture in the usual way, collecting a volume of distillate having a simple ratio to the volume of the original liquid.

"If a troublesome emulsion develops in the liquid mixture when shaken with the petroleum benzin, dilute a fresh portion of the original liquid with water and distil it as directed under *General Procedure.* Then treat this distillate as directed above, using petroleum benzin and sodium chloride solution, and distil the saline solution so produced to obtain a distillate which is free from volatile substances other than alcohol and water.

"In preparing collodion for distillation, use water in place of the saturated solution of sodium chloride directed above.

"If volatile oils are present in small proportions only, and a cloudy distillate is obtained, the benzin treatment not having been employed, the distillate may be clarified and rendered suitable for the specific gravity determination by shaking it with about one-fifth its volume of petroleum benzin, or by filtering it through a thin layer of talc.

"**Other Preparations Requiring Special Treatment.** Preparations containing free ammonia, such as aromatic ammonia spirit, must be rendered slightly acid with sulfuric acid before being distilled. If volatile acids are present, the preparation must be rendered slightly alkaline with sodium hydroxide."

Exercise 123

Object. To Determine the Alcohol Content of Belladonna Tincture.

Materials Required. 25 ml of belladonna tincture
A distilling apparatus
A flask graduated to hold 50 ml
A pycnometer

Procedure. 1. Place 25 ml of the preparation, measured at 25°, in a distilling flask of about 200-ml capacity, provided with a high side tube connected with a suitable condenser; add 50 ml of distilled water, distill into a 50-ml graduated flask, at such a rate that about 48 ml of distillate will be received in ½ hr; if volatile products other than alcohol are known to be absent, bring to the original temperature and dilute to 50 ml with water of the same temperature.

In very accurate work, it is best that the volume of the sample and of the distillate be measured at the same temperature, i.e., 15.56°. When the temperatures differ, the volume of the sample used and that of the distillate collected are not the same, since the coefficient of expansion of alcohol is much greater than that of water. In the volume percentages for alcohol, it is assumed that the volumes are measured at 15.56°.

2. Determine the specific gravity of the distillate at 25° by means of a pycnometer, and find the percent of absolute alcohol by volume by reference to the alcoholometric table in the U.S.P., page 1108, and multiply by the dilution factor to find the percent of alcohol in the original preparation.

The specific gravity of the distillate may be determined by means of either the Sprengel-Ostwald or the Geissler pycnometer, as previously described. The latter instrument is the more satisfactory, since the temperature of the alcoholic solution can be determined while the specific-gravity determination is being made.

The method of recording the data and calculating the results is illustrated in the example:

Volume of sample at 25° = 25 ml
Volume of distillate at 25° = 50 ml
Dilution factor = $^{50}/_{25}$ = 2

The specific gravity of the distillate at 25°/25° found = 0.9561.

By referring to the alcoholometric table, the percentage by volume of alcohol in the distillate is found to be 34 + 0.4 = 34.4. This percentage is based on the apparent specific gravity of alcohol at 15.56°/15.56°. The percent of absolute alcohol by volume contained in the sample tested is found by multiplying the percent found in the distillate by the dilution factor, or 34.4 × 2 = 68.8 percent. The corresponding percent of absolute alcohol by weight is found by consulting the table on page 1108 of the U.S.P.XVII; e.g., 68 and 69 percent of alcohol by volume are found to correspond to 60.33 and 61.38 percent of alcohol by weight, respectively. Therefore, 68.8 percent of alcohol by volume corresponds to 61.09 percent by weight.

The amount of alcohol present in various preparations such as fluidextracts and tinctures is subject to considerable variation because of the varying proportions of moisture and extractive contained in the drug used and the methods of extraction and concentration employed in their preparation. The official standards, therefore, generally give a range of alcohol content for preparations of this type; e.g., belladonna tincture is required to contain 65 to 70 percent of alcohol by volume.

2. By the Use of the Westphal Balance. The specific gravity of liquids may be determined quickly and with a fair degree of accuracy by means of the apparatus known as the *Westphal balance.* The mechanism of this device is based upon the generally known law of physics that a body immersed in a liquid is buoyed up by a force equal to the weight of the

FIG. 68. Plummet.

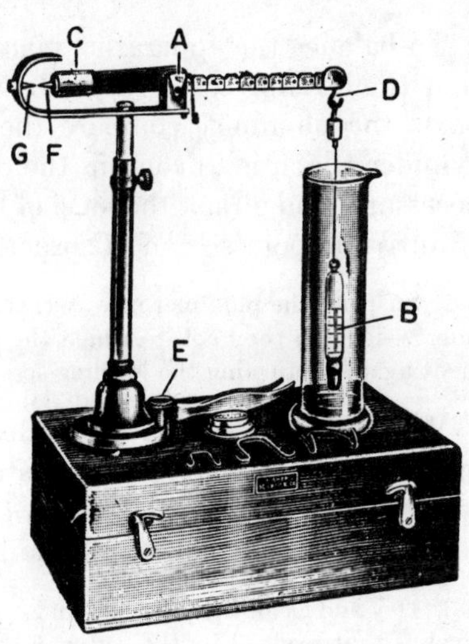

FIG. 69. Westphal specific-gravity balance.

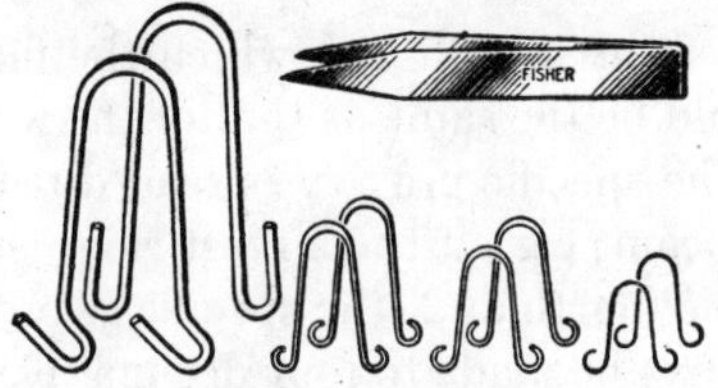

FIG. 70. Riders.

liquid displaced. This apparatus (Figs. 68, 69, and 70) consists of a beam balanced on a knife-edge *A* with a sinker or plummet *B* suspended from one end and counterpoised by a fixed brass weight *C* at the other. The distance between the knife-edge *A* and the plummet support *D* is graduated into 10 divisions of equal length. The plummet, which is made of glass and provided with a thermometer, is made of such size that it will displace exactly 5 Gm of water at a given temperature, usually 15.56°. The plummet and the platinum wire by which it is suspended are usually made of a definite weight, so that they are interchangeable on different balances.

Exercise 124

Object. To Determine the Specific Gravity of a Volatile Oil.

Materials Required. A Westphal balance
A 50-ml cylindrical graduate
50 ml of volatile oil

Procedure. 1. Balance the apparatus in air.

To balance the apparatus, place it upon a firm, level table and be sure that the leveling screw in the base is directly beneath the arm that supports the plummet, compare the height of the beam with that of the cylinder which is to contain the oil, lengthen or shorten the standard, if necessary, and adjust the level of the beam by means of the leveling screw *E* until the points *F* and *G* exactly coincide.

2. Suspend the plummet in water, contained in the glass cylinder, place the largest rider weight on the hook to which the plummet is attached, and balance the instrument again by turning the leveling screw.

When determinations are made at the exact temperature for which the instrument is designed, no adjustment should be necessary at this point, but in taking the specific gravity at 25°, a slight adjustment is required to correct for the expansion of the liquid at the higher temperature.

3. Dry and clean the plummet and cylinder, fill the cylinder with the oil to the same level as that reached by the water in balancing the apparatus, lift the plummet, and place the cylinder beneath the end of the beam. Replace the plummet so that it hangs free in the oil without touching the cylinder wall, and place the riders on the beam until balance is restored.

The temperature of the oil (25°) at which the final adjustment of the balance is made should be the same as that of the water used to standardize the apparatus. The specific gravity is read directly from the position of the riders on the beam; e.g., if the largest rider is at 9, the second at 7, the third at 5, and the fourth at 2, the specific gravity of the oil is 0.9752. Readings should always be made to four decimal places; i.e., all four sizes of riders should be used.

3. By Weighing a Solid in Air, in Water, and in the Liquid. Plummets or sinkers of various forms made of sealed glass tubes partly filled with mercury or of some metal such as platinum have been devised for specific-gravity determinations of this type, but a brass weight will suffice for the determination of the specific gravity of liquids that do not attack the brass. For example, if a piece of brass weighs 10 Gm in air, 8.824 Gm in water, and 9 Gm in a liquid of unknown specific gravity, the specific gravity of the last can be calculated from a simple proportion as follows:

$$1.176:1::1':X \qquad X = 0.85$$

where 1.176 is the loss of weight in water ($10 - 8.824 = 1.176$) or the weight of the water displaced; 1 is the loss of weight in the liquid; 1′, the specific gravity of water; and X, the specific gravity of the unknown liquid.

The principles involved and the method of making the determination are similar to those given in detail under the determination of the specific gravity of solids heavier than, and insoluble in, water (see Solids Heavier Than and Insoluble in Water, later in this chapter).

4. By the Use of Hydrometers. Hydrometers are instruments used to determine quickly the approximate specific gravity or density of liquids by flotation. They usually consist of a sealed glass tube weighted with mercury or shot and frequently are provided with a thermometer, so that the temperature at which the specific gravity is determined may be read at the same time. They are variously called *alcoholometers*, *lactometers*, *urinometers*, etc., according to the use for which they are designed. They are either constructed to read specific gravity directly, as, for example, the hydrometers for general use and lactometers, or they are graduated according to some arbitrary scale of degrees, as in the case of the Baumé and Twaddell hydrometers. Specific-gravity hydrometers are usually made in sets of four or six, each instrument being calibrated to cover a different range in specific gravities, so that the complete set will suffice to determine the specific gravity of almost any liquid; e.g., the different instruments in the set of four are calibrated to read from 0.600 to 1.000, from 1.000 to 1.400, from 1.400 to 1.800, and from 1.800 to 2.200.

When making a specific-gravity determination with a hydrometer, the thoroughly cleaned instrument should be allowed to sink in the liquid gradually until it is at rest; it is then depressed about 1 in. to moisten a portion of the stem, and allowed to come to equilibrium, when the reading is taken.

Exercise 125

Object. To Prepare 200 ml of Sulfuric Acid Containing 10 Gm of H_2SO_4 per 100 ml.

Materials Required. 20 ml of concentrated H_2SO_4
A hydrometer calibrated from specific gravity 1.000 to 1.400 at 25°

Procedure. Dilute 20 ml of pure concentrated sulfuric acid to 100 ml with distilled water, allow the resulting solution to acquire a temperature of 25°, and take its specific gravity as accurately as possible with the hydrometer.

If the specific gravity of the solution is found to be 1.2603, it contains 35 percent of H_2SO_4 as found by referring to the table, page 804 of the U.S.P. XII. The weight of the prepared acid solution to be diluted will then be $\frac{10 \times 2 \times 100}{35} = 57.1430$ Gm. This weight of solution containing

35 percent of H_2SO_4 when diluted to 200 ml will form a solution containing 10 Gm of H_2SO_4 per 100 ml.

Solutions prepared as directed above are subject to slight inaccuracies of concentration, since specific-gravity determinations by means of hydrometers are not exact, but if the specific gravity is determined accurately by means of a pycnometer, the concentration of a solution prepared in this manner will be sufficiently accurate for all ordinary qualitative purposes.

METHODS USED TO DETERMINE THE SPECIFIC GRAVITY OF SOLIDS

1. Solids Heavier Than and Insoluble in Water. *a.* Weigh the substance in air and then in water.

According to the law of Archimedes, a body suspended in a fluid is buoyed up by a force equal to the weight of the fluid displaced. The weight of a solid in air, therefore, less its weight in water, gives the weight of an equal volume of water, since the volume of water displaced is equal to the volume of the solid. If the weight in air is divided by the loss of weight in water, the result is the specific gravity of the substance at the temperature at which the determination is made. For example: Suppose that a piece of brass weighs 10.0040 Gm in air at 25° and 8.8010 Gm immersed in water at 25°. The weight in air divided by the loss of weight, 1.2030 Gm, gives the specific gravity of the brass at 25°/25° or 10.0040/1.2030 = 8.32.

b. If the substance is in small pieces or is a powder, the specific gravity may be determined as illustrated in the following exercise.

Exercise 126

Object. To Determine the Specific Gravity of Powdered Barium Sulfate.

Materials Required. 7 to 10 Gm of powdered barium sulfate
A small volumetric flask (25 or 50 ml)

Procedure. 1. Weigh about 10 Gm of the powdered barium sulfate accurately.

2. Accurately weigh the small volumetric flask filled exactly to the graduation mark with recently boiled distilled water at 25°.

3. Introduce the powdered barium sulfate into the flask quantitatively, remove any air occluded in the powder by agitating the mixture, draw off the water with a pipet until the level of the liquid is again exactly at the graduation mark when the liquid is at 25°, and weigh the flask and its contents accurately.

It is evident that a volume of water equal to that of the barium sulfate has been displaced and removed from the flask. The specific gravity of the barium sulfate may then be calculated, e.g.:

Weight of powdered barium sulfate = 8.0 Gm
Weight of flask filled with water = 80 Gm
Weight of flask, barium sulfate, and water = 85 Gm

80 + 8 − 85 = 3 Gm, the weight of the volume of water displaced by the barium sulfate; and 8/3 = 2.6666, the specific gravity of the barium sulfate.

2. Solids Lighter Than and Insoluble in Water. *a. By Weighing in Water.* If a substance is lighter than water, it will float, and its volume cannot be obtained by weighing it in water unless it is completely submerged by means of a weight or sinker heavier than water. The method of determining and calculating the specific gravity of a substance of this type is illustrated in the following exercise.

Exercise 127

Object. To Determine the Specific Gravity of Camphor.

Materials Required. A piece of camphor (2 to 4 Gm)
A suitable sinker such as a 2-Gm weight

Procedure. 1. Weigh the piece of camphor in air.
2. Weigh the sinker immersed in water.
3. Attach the sinker to the piece of camphor and find their combined weight when immersed in water.

Since the camphor is lighter than water, the weight of the sinker and camphor is less than the weight of the sinker alone. The difference between the weight of the sinker and camphor in water and the weight of the sinker alone in water plus the weight of the camphor in air is a measure of the water displaced by the camphor and also a measure of its buoyant power, e.g.:

Weight of the camphor in air = 4 Gm
Weight of the sinker in water = 1.6 Gm
Weight of the camphor and sinker in water = 1.5 Gm

The difference between the weight of the sinker in water and the weight of the sinker and camphor in water = 1.6 − 1.5 = 0.1 Gm.

The weight of the camphor in air plus this difference of 0.1 Gm is the weight of a volume of water equal to the weight of the piece of camphor or 4 + 0.1 = 4.1 Gm of water displaced.

Since the specific gravity of a substance is found by dividing its weight by the weight of an equal volume of water, the specific gravity of the camphor is 4/4.1 = 0.9756.

b. By the Flotation Method. This method consists of bringing the substance into equilibrium with a liquid of known specific gravity. It is used in the determination of the specific gravity of such substances as white and yellow wax and spermaceti, and by food and drug analysts to determine the specific gravities of certain fats and oils.

Exercise 128

Object. To Determine the Specific Gravity of Yellow Wax.

Materials Required. About 0.25 Gm of yellow wax
25 ml of alcohol
A pycnometer

Procedure. 1. Melt the wax at a low temperature and allow it to fall into separated drops from just above the surface into alcohol that has been warmed to from 45 to 50°. Allow the globules to remain in the alcohol until it has cooled spontaneously to room temperature (20 to 25°); then remove the wax and keep it at room temperature for 24 hr.

The melted wax separates in globules free from occluded air, which solidify when the mixture is cooled. The wax, which has a high coefficient of expansion, is kept at room temperature for 24 hr to permit the globules to acquire their normal volume.

2. Prepare a mixture of four volumes of alcohol and enough distilled water to make ten volumes, and allow it to stand until free from air bubbles. Moisten the globules of wax with distilled water by means of a brush, and place them by means of forceps in the alcohol solution just prepared and contained in a beaker. Then add alcohol or air-free distilled water, as required, to the mixture kept at 25°, until the globules of the wax float and rest indifferently in the liquid. Finally, determine the specific gravity of the alcohol-water mixture. The figure thus obtained represents the specific gravity of the wax examined.

The wax globules are moistened with water, so that the water-alcohol mixture will come in contact with all portions of the globules without occluding air at the interface. When the globules are in equilibrium in the alcoholic solution, their weight must be equal to the weight of liquid that they displace; otherwise they would rise to the top of the liquid and float, or sink to the bottom of the beaker. Consequently, when such equilibrium is attained, the specific gravity of the wax and that of the liquid are equal, and a determination of the specific gravity of the liquid is a measure of the specific gravity of the wax.

3. Solids Heavier Than and Soluble in Water. The specific gravity of solids heavier than and soluble in water may be determined by weighing the solid in air and then in some liquid of known specific gravity in which the solid substance is insoluble, such as oil of turpentine, chloroform, benzene, etc. Thus, the weight of a piece of alum in air divided by its loss in weight when weighed immersed in oil of turpentine gives the density of the alum relative to the oil of turpentine, and if the specific gravity of the oil of turpentine is known, that of the alum can be ca'culated, e.g.:

Weight of the alum in air = 10 Gm
Weight of the alum in oil of turpentine = 6.5 Gm
Specific gravity of the oil of turpentine = 0.860

10 − 6.5 = 3.5 Gm, loss in weight of the alum in the oil of turpentine equal to the weight of an equal volume of the latter.

$$3.5:X::0.86:1 \qquad X = 4.070 \text{ Gm}$$

where X is the weight of a volume of water equal to the volume of the oil of turpentine displaced.

$$10/4.070 = 2.4570 \qquad \text{specific gravity of the alum}$$

4. Solids Lighter Than and Soluble in Water. The specific gravity of substances lighter than and soluble in water may be determined by weighing the substance with a sinker in some liquid of known specific gravity in which the substance is not soluble. For example, the specific gravity of lithium may be determined from the data below as follows:

Weight of the lithium in air = 2.0 Gm
Weight of the sinker in air = 5.0 Gm
Specific gravity of the sinker (brass) = 8.4
Specific gravity of liquid (kerosene) = 0.830
Weight of lithium and sinker in kerosene = 3.8 Gm

The density X of the sinker referred to kerosene as a standard is

$$0.830:1::8.4:X \qquad X = 10.1205$$

The weight Y of the sinker in kerosene may be calculated,

$$\frac{5.0}{5.0 - Y} = 10.1205 \qquad Y = 4.5095$$

and the density of lithium compared to that of kerosene is

$$\frac{2}{2 + 4.5060 - 3.8} = 0.7391$$

The specific gravity Z of the lithium is, therefore,

$$0.830:1::Z:0.7391$$
$$Z = 0.6135$$

Questions and Problems

1. Define the terms *specific gravity, apparent specific gravity*, and *true specific gravity*.
2. By what different methods might the specific gravity of alcohol be determined?
3. A 10-Gm weight of brass, specific gravity 8.4, submerged in glycerin weighed 8.5135 Gm. What is the specific gravity of the glycerin?

TABLE 67. SUBSTANCES WITH SPECIFIC GRAVITIES

Substances	Specific gravity, 25°
U.S.P.	
Alcohol	n.m.t. 0.816*
Alcohol, diluted	0.935 to 0.937*
Anethole	0.983 to 0.988
Anise oil	0.978 to 0.988
Benzyl benzoate	1.116 to 1.120
Castor oil	0.945 to 0.965
Cherry juice	1.045 to 1.075
Cinnamon oil	1.045 to 1.063
Clove oil	1.038 to 1.060
Collodion	0.765 to 0.775
Coriander oil	0.863 to 0.875
Cottonseed oil	0.915 to 0.921
Diethyltoluamide	0.9965 to 1.002
Dimercaprol	1.238 to 1.240
Ether	0.713 to 0.716
Eugenol	1.064 to 1.070
Fennel oil	0.953 to 0.973
Glycerin	n.l.t. 1.249
Halothane	1.872 to 1.877
Iophendylate injection	1.245 to 1.260
Lemon oil	0.849 to 0.855
Methyl salicylate (natural)	1.176 to 1.182
Methyl salicylate (synthetic)	1.180 to 1.185
Mineral oil	0.860 to 0.905
Olive oil	0.910 to 0.915
Orange oil	0.842 to 0.846
Peanut oil	0.912 to 0.920
Peppermint oil	0.896 to 0.908
Peruvian balsam	1.150 to 1.170
Petrolatum, white	0.815 to 0.880†
Polyethylene glycol 400	1.110 to 1.140
Polysorbate 80	1.06 to 1.10
Propylene glycol	1.035 to 1.037
Raspberry juice	1.025 to 1.045
Sesame oil	0.916 to 0.921
Sorbital solution	n.l.t. 1.285
Syrup	About 1.313
Tetrachlorethylene	1.603 to 1.615
Theobroma oil	0.858 to 0.864‡
Trichloroethylene	1.458 to 1.463
Triethanolamine	1.1204 to 1.1284
N.F.	
Acetone	n.m.t. 0.789
Alcohol, dehydrated	n.m.t. 0.798*
Alcohol rubbing compound	0.8797 to 0.8874*¶ 0.8691 to 0.8771[a]
Almond oil	
Amyl nitrite	0.871 to 0.875
Amylene hydrate	0.803 to 0.807
N.F.	
Benzaldehyde	1.041 to 1.046
Benzyl alcohol	1.040 to 1.050
Caramel	m.l.t. 1.30
Caraway oil	0.900 to 0.910
Cardamom oil	0.917 to 0.947
Creosote	n.l.t. 1.076
Ethyl acetate	0.894 to 0.898
Eucalyptus oil	0.905 to 0.925
Honey	n.l.t. 1.099[b]
Isopropyl alcohol	0.7828 to 0.7869
Isopropyl alcohol, rubbing compound	0.8690 to 0.8790
Kaolin mixture with pectin	1.10 to 1.15
Lead subacetate solution	1.25
Merbromin solution	1.010 to 1.015
Merbromin solution, surgical	0.9110 to 0.9140
Nikethamide	1.058 to 1.066
Nitric acid	About 1.41
Nitromersol solution	1.005 to 1.010
Nitromersol tincture	0.9050 to 0.9150
Octylphenoxy polyethoxyethanol	1.029 to 1.033
Orange flower oil	0.863 to 0.880
Petrolatum, light, liquid	0.828 to 0.880
Phenylethyl alcohol	1.017 to 1.020
Phosphoric acid	About 1.71
Phosphoric acid, diluted	About 1.057
Pine needle oil (dwarf)	0.853 to 0.871
Pine oil	0.927 to 0.940
Polyethylene glycol 300	1.124 to 1.130¶
Potassium iodide solution	About 1.700
Propylhexedrine	0.848 to 0.852
Resorcinol monoacetate	1.203 to 1.207
Rosemary oil	0.894 to 0.912
Rosin	1.07 to 1.09
Silver nitrate solution, ammoniacal	About 1.48
Sodium hypochlorite solution, diluted	n.m.t. 1.025
Sodium phosphate solution	About 1.385
Spearmint oil	0.917 to 0.934
Sulfuric acid	About 1.84
Sulfuric acid, diluted	About 1.067
Thyme oil	0.910 to 0.935
dl-α-Tocopherol	0.947 to 0.958
Tocopherols, mixed, concentrate	0.932 to 0.955
d-α-Tocopheryl acetate	0.950 to 0.964
dl-α-Tocopheryl acetate	0.950 to 0.964
Vinyl ether	0.767 to 0.771

* 15.56°.
† 60°.
‡ 100°.
§ 20°.
¶ Made with specially denatured alcohol, Formula 23-G.
[a] Made with specially denatured alcohol, Formula 23-H.
[b] Diluted with 2 volumes of water.
n.l.t. = not less than; n.m.t. = not more than.

4. Given the specific gravity of sulfuric acid at 15° as 1.8422 compared with water at 15°, what is its specific gravity referred to water at 21°? (See U.S.P. XII.)

5. How much nitric acid, specific gravity 1.45, and how much water must be mixed to form 1 liter of nitric acid, specific gravity 1.20?

6. Calculate the specific gravity of a piece of sandalwood from the following data:

Weight in air = 10.100 Gm
Weight of sinker in water = 12.200 Gm
Weight of sinker and sandalwood in water = 11.900 Gm

7. How many grams of water will a pycnometer calibrated to hold 100 Gm of water at 25°/25° and 760 mm pressure contain at 19° and 740 mm pressure? (See U.S.P. XII.)

8. Give the specific gravity of alcohol compared with water at 15.56° and 760 mm pressure in a solution containing 16.5 percent of alcohol by volume. (See U.S.P.)

9. Briefly state how the specific gravity of each of the following might be determined with comparative accuracy: olive oil, benzene, powdered borax, granular zinc, and cacao butter.

The approximate specific gravities for many official substances, such as acetone, acids, and numerous preparations, are given in the U.S. Pharmacopeia and the National Formulary simply as information for the conversion of volume to weight. When the minimum or maximum specific gravity or a range of specific gravity is given, it is a definite constant, of value in determining the identity, purity, and concentration of the substance. Some of the substances for which definite specific gravity limits are given in the official standards are shown in Table 67.

Chapter 31

SOLUBILITY

The statements of the solubilities of official substances are given primarily for the information of physicians and pharmacists for use in the preparation and dispensing of medicines. They should not always be regarded as exact physical constants, upon which the identification or the determination of the purity of a substance may be based. Only when a special solubility test is given, and is designated as such, is it a test for purity. Frequently, the official standards are desc[illegible] terms to indicate the approximate solubility of substances, viz.: (1) "Very soluble," 1 part of the substance is soluble in less than 1 part of the solvent; (2) "freely soluble," 1 part of the substance is soluble in from 1 to 10 parts of solvent; (3) "soluble," 1 part of the substance is soluble in from 10 to 30 parts of solvent; (4) "sparingly soluble," 1 part of the substance is soluble in from 30 to 100 parts of solvent; (5) "slightly soluble," 1 part of the substance is soluble in from 100 to 1,000 parts of solvent; (6) "very slightly soluble," 1 part of the substance is soluble in from 1,000 to 10,000 parts of solvent; (7) "practically insoluble" or "insoluble," more than 10,000 parts of solvent are required to dissolve 1 part of the substance.

There are nine possible classes of solutions, viz.: (1) a gas in a gas; (2) a gas in a liquid; (3) a gas in a solid; (4) a liquid in a gas; (5) a liquid in a liquid; (6) a liquid in a solid; (7) a solid in a gas; (8) a solid in a liquid; (9) a solid in a solid. Although examples of all these different types of solutions are known, the only classes of pharmaceutical importance are solutions of gases in liquids, of liquids in liquids, and of solids in liquids.

The official determination of the solubility of gases in water, such as HCl and NH_3, is generally effected by chemical titration methods, while the solubility of liquids in liquids and of solids in liquids may be determined by physical methods, such as specific gravity, optical rotation, refractive index, etc., as well as by usual titrimetric or gravimetric analytical methods.

Numerous factors affect the rate and the extent of the solubility of a substance in a given solvent, as follows:

1. The solubility of most of the official substances is increased by a rise in the temperature at which solution is effected. There are, however, numerous exceptions to this generality; e.g., gases, calcium salts, and ether are less soluble in hot than in cold water. Since the solubility of sub-

stances varies markedly with slight changes in temperature, it is very important that a constant temperature be accurately maintained throughout a solubility determination.

2. Substances in a fine state of division dissolve more rapidly than large crystals or particles because the total surface area exposed to the action of the solvent is much greater when the substance is powdered.

3. The purity of the substance and of the solvent must be assured in all solubility determinations, since slight amounts of impurity in either may cause considerable variation in the results. In the official statements of solubility, it is assumed that both the substance and the solvent employed conform to the official tests for purity.

4. The position of the solute in the solvent affects the rate of solution. If the solute is allowed to lie on the bottom of the vessel, it becomes surrounded by a layer of concentrated solution, which prevents the access of fresh portions of the solvent to the surface of the solute.

The determination of the solubility of a solid in a liquid necessitates the preparation of a saturated solution at a carefully controlled temperature of the substance. This is followed by ascertaining, by some type of analysis, the amount of the substance present in a definite quantity of the solution. Complete saturation can be attained most readily by agitation of an excess of finely divided solid material in the appropriate solvent contained in a closed vessel. The agitation may be accomplished by means of a mechanical stirrer or by shaking. At periodic intervals the undissolved material is allowed to settle and an accurately measured sample of the solution is removed for analysis. When two consecutive results agree, saturation of the solution is known to be complete. Such determinations must be made in a temperature-controlled atmosphere, such as through the use of a constant-temperature water bath.

Exercise 129

Object. To Determine the Solubility of Boric Acid in Water at 25°.

Materials Required. 1.5 Gm of boric acid
A solubility apparatus and thermostat

Procedure. 1. In a suitable container, place about 1.5 Gm of finely powdered boric acid and 25 ml of distilled water. After fitting it with the stirrer, place it in the thermostat so that it is immersed in the water up to the level of the stopper, and stir the solution vigorously for about 3 hr.

The boric acid is powdered finely because substances in fine particles dissolve more rapidly than do large crystals; subdivision increases the surface area exposed to the action of the solvent.

The container is immersed in the thermostat bath to maintain a constant temperature throughout the mixture, since the solubility of boric acid varies appreciably with slight changes in temperature. The temperature in the thermostat should not vary by more than 0.1°, above or below 25°, during the determination. The mixture is stirred continuously to bring fresh portions of solvent in contact with the undissolved particles of boric acid and to ensure equal concentration throughout the solution.

2. Stop the stirrer and remove it from the tube, stopper the latter securely, allow the undissolved crystals to settle, remove about 5 ml of the clear solution from the solubility tube by means of a pipet and weigh accurately. Evaporate this portion of the solution to dryness on a water bath at 80° and finally to constant weight at 90° in an oven, and weigh the residue.

The weight of the boric acid contained in the residue, subtracted from the weight of the solution analyzed, gives the weight of the water in which the boric acid was dissolved.

3. Attach the stirrer and allow the stirring to continue for about 1 hr. Again remove a 5-ml portion of the solution, weigh it accurately, and determine the amount of boric acid it contains, in the same manner as before.

If this result agrees with the former determination, it shows that saturation of the solution is complete, but if the amount of dissolved solid is greater in the second case, the stirring must be continued until the concentration of the solution becomes constant.

4. Express the results as grams of boric acid soluble in 100 Gm of water.

If a 5.5-Gm portion of a saturated boric acid solution yielded 0.3000 Gm of boric acid upon drying to constant weight, the 0.3 Gm of boric acid was dissolved in $5.5 - 0.3 = 5.2$ Gm of water. The amount of boric acid soluble in 100 Gm of water at 25° may therefore be calculated from the simple proportion

$$0.3:5.2::X:100$$

in which X = the number of grams of boric acid soluble in 100 Gm of water at the temperature at which the determination was carried out.

Calculate the solubility of boric acid in water at 25° from the results of the above determination.

In addition to the general tests for solubility described in the official standards, a solubility test for hypodermic tablets is given. This test applies generally to all tablets of pharmacopeial drugs intended for the preparation of solutions for hypodermic administration except where the labeling specifies dissolving the tablet at a temperature higher than 25° or in a volume greater than 2.5 ml.

To carry out the test, place 1 tablet of the sample to be tested in each of six test tubes about 15 mm in diameter and containing 2.5 ml of purified water. Shake the tubes with a moderate sidewise motion for 2 min, and observe the tablets: each tablet has dissolved completely without effervescence to give a clear solution substantially free of undissolved particles.

Questions and Problems

1. What factors influence the rate of solution of boric acid in water?

2. Why is the accurate control of temperature throughout a solubility determination essential?

3. Why is it necessary to carry out a solubility determination in a closed container?

Chapter 32

VISCOSITY MEASUREMENTS

The viscosity of a fluid substance is constant for any given temperature and is a measurable characteristic of the substance. It is used chiefly as an index of the composition and lubricating value of oils. The viscosities of solutions and liquid mixtures often vary with their concentration and composition; this property, however, may be used, in many cases, as a rapid means of analysis. The U.S.P. utilizes the viscosity as a means of standardizing mineral oil, which is employed therapeutically as an intestinal lubricant, and also specifies the viscosity of methylcellulose.

Fluidity is the reciprocal of viscosity; i.e., fluidity = 1/viscosity. Fluidity may be regarded as a measure of the tendency of a liquid to flow, whereas viscosity is a measure of the resistance that a liquid exerts against the tendency to flow. Viscosity may be defined as *the force of friction that tends to retard movement in a fluid body*. The unit of absolute viscosity is defined as *the tangential force required to move a unit area of plane surface within the liquid with unit velocity relative to another parallel unit area of plane surface one unit distant from the former surface, where the unit of force is the dyne (the force necessary to produce an acceleration of 1 cm per second per second on a mass of 1 Gm), the unit of velocity is the centimeter per second, the unit of area is the square centimeter, and the unit of distance is the centimeter*. This unit of absolute viscosity is known as the *poise*, and it is generally expressed as dyne-second per square centimeter. The poise is a relatively large unit, and the one-hundredth part of the poise, or a *centipoise*, is commonly employed as the unit of absolute viscosity.

Instead of giving results in terms of absolute viscosity, most methods of determination give the relative viscosity, i.e., the viscosity of the liquid compared with that of some standard of known viscosity, such as water under similar conditions. As a convenience, the *kinematic viscosity*, which is *the ratio of the absolute viscosity in poises to the density*, has been adopted in the U.S.P. as the method of expressing viscosity. The unit of kinematic viscosity is the *stoke*, which is equivalent to 100 centistokes. The approximate viscosity in centistokes at room temperature of ether is 0.2, of water, 1; of kerosene, 2.5; of mineral oil, 20 to 70; and of honey, 10,000. As the viscosity of a liquid changes very rapidly with the temperature, decreasing about 2 percent for each degree rise in temperature, it is important to know the exact temperature of the liquid at the time of measurement and to use an instrument provided with some means whereby the sample under examination may be maintained at constant temperature.

Apparatus. In the official determination any suitable viscometer may be employed, and the results obtained with these instruments may be converted into the kinematic viscosity by the use of a conversion table. Viscosity can be determined by any method that will measure the resistance to shear offered by the liquid. For ordinary liquids, it is customary to determine the time required for a given sample of the liquid to flow, at a regulated temperature, through a small capillary tube, and to compare this time with that required by the reference liquid, such as water. If the viscosity of the reference liquid is known, the kinematic viscosity of the unknown liquid may be calculated simply. Many capillary tube viscometers have been devised, and nearly all are modifications of the Ostwald type. Several types are described, with directions for their use, by the American Society for Testing Materials.

Viscosity for commercial products is often expressed on arbitrary scales and measured by empirical methods in special instruments. Thus semisolids are often examined by so-called "rotary-instruments," which measure the resistance to rotation of a cylinder, placed concentrically within another, the annular space being filled with the sample. The viscosity of oils is expressed on arbitrary scales, which vary from one country to another, there being several corresponding instruments. The most widely used are the Redwood No. I and No. II, the Engler, the Saybolt Universal, and the Saybolt Furol. The Redwood No. II and the Saybolt Furol viscometers are designed for extremely viscous oils, in that they have a wider capillary for the measurement. The Saybolt Furol readings are approximately one-tenth of those obtained on the same oil with the Saybolt Universal. Results for all the instruments are expressed in seconds of time required for outflow of a standard amount of liquid.

When the Engler instrument is used, the viscosity is expressed in terms of *Engler degrees*, obtained by dividing the time of discharge for the liquid being tested by the time of discharge of the same volume of water at 20°. In this country, the Saybolt Universal viscometer is most commonly employed in the determination of the viscosity of lubricating oils. The results obtained with this instrument are expressed as the time in seconds required for the discharge of the liquid being tested.

The *Saybolt Universal viscometer* (Fig. 71) consists of a metallic cup of about 80-ml capacity to contain the liquid under examination, with an outlet in the bottom closed by a cork stopper. This cup is surrounded by a jacket which serves as a constant-temperature bath. Most instruments are provided with devices for stirring and heating the bath, with a thermometer to measure the temperature, and with a receiving flask graduated to contain 60 ml at 20°; a stop watch with which the time of dis-

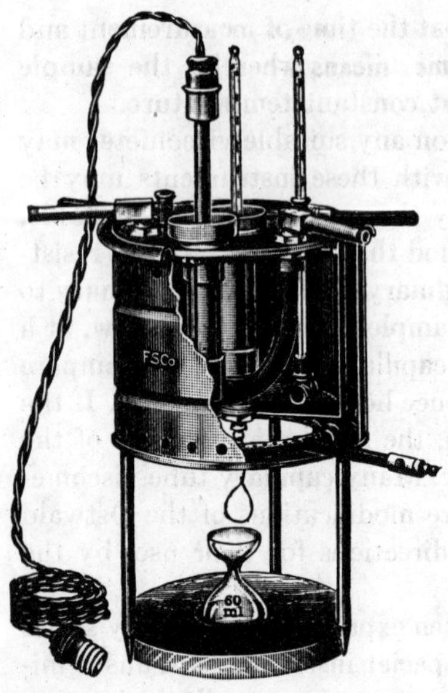

FIG. 71. Saybolt Universal viscometer.

charge of the liquid may be measured is usually a part of the standard equipment supplied with the instrument, but any stop watch may be used.

Exercise 130

Object. To Determine the Kinematic Viscosity of Mineral Oil.

Materials Required. A Saybolt viscometer
A stop watch
100 ml of mineral oil

Procedure. 1. Assemble the viscometer, fill the constant-temperature bath with water, and bring the temperature of the latter to about 40°. Tightly stopper the outlet tube, and pour the sample of mineral oil into the oil cup until the level is above the rim of the main cup. When the temperature of the oil sample is 37.8°, pipet the excess oil from the overflow cup until the level in it is below the rim of the main cup.

The water in the temperature bath should be stirred continually to ensure an even distribution of temperature throughout the sample of oil.

2. When the thermometer in the oil cup registers exactly 37.8°, remove the cork from the outlet tube, starting the stop watch at the same instant, and collect the effluent oil in the flask graduated to hold 60 ml. When the level of the oil in the flask reaches the 60-ml graduation mark, stop the watch.

The time in seconds required for the efflux of 60 ml of the mineral oil is called the *Saybolt Universal viscosity*.

The determination of kinematic viscosity with modern instruments is more rapid and convenient than with the use of the Saybolt and similar equipment. The Ostwald pipet type of viscometer is therefore displacing the metal capillary variety, although the Saybolt, Redwood, and Engler scales have been retained.

To calibrate the Ostwald type of viscometer, determine the viscometer constant k for each viscometer by the use of an oil of known viscosity. Place an excess of the oil (adjusted to 20° ± 0.1°) in the filling tube, and transfer it to the orifice tube by gentle suction, taking care to prevent bubble formation in the liquid by keeping the air vent tube closed. Adjust the meniscus of the column of liquid in the orifice tube to the level of the top graduation line. Open both the vent and orifice tubes in order to permit the liquid to flow into the reservoir against atmospheric pressure. (Failure to open the vent tube before releasing the orifice tube will yield false values.) Record the time in seconds for the liquid to flow from the upper mark to the lower mark in the orifice tube.

Calculate the viscometer constant k from the equation $k = v/dt$, in which v is the known viscosity of the liquid in centipoises, d is the specific gravity of the liquid tested at 20°/20°, and t is the time in seconds for the liquid to pass from the upper mark to the lower mark.

If a viscometer is repaired, it must be recalibrated, since even minor repairs frequently cause significant changes in the value of its constant k.

Measurement of the viscosity of solutions of the high-viscosity types of methylcellulose is a special case, since they are too viscous for the commonly available viscometers. The Ubbelohde viscometer may be adapted to the measurement of the ranges of viscosity encountered in methylcellulose solutions. The principle of measurement is essentially the same as that of the Saybolt viscometer except that larger orifice tubes are used and the design is somewhat different.

Kinematic viscosities determined in the Ostwald and Ubbelohde pipets are convertible to Saybolt and similar values by means of equations and conversion charts obtainable from the American Society for Testing Materials.

Consistency measurements of ointments and ointment bases such as White Petrolatum U.S.P. are made using a device called a *penetrometer*. As the name implies, a penetrometer penetrates the ointment, by means of a truncated cone-shaped metal plunger, the depth of which is taken as a measure of consistency.

Questions and Problems

1. Consult the viscosity conversion table (U.S.P. XIV) and express the results obtained in the above exercise in terms of kinematic viscosity.
2. If the viscosity of a sample of liquid petrolatum is found to be less than the official requirement, what does the determination indicate relative to the composition of the oil?
3. Would viscosity requirements be of value as a constant for (*a*) glycerin; (*b*) castor oil; (*c*) simple syrup?
4. Why is the Ubbelohde viscometer used in the determination of the viscosity of methylcellulose?
5. Define "truncated" cone-shaped metal plunger.
6. What is meant by consistency with regard to ointments and ointment bases?

REFERENCE TABLES

Logarithms of Numbers*

Natural numbers	0	1	2	3	4	5	6	7	8	9	Proportional parts								
											1	2	3	4	5	6	7	8	9
10	0000	0043	0086	0128	0170	0212	0253	0294	0334	0374	4	8	12	17	21	25	29	33	37
11	0414	0453	0492	0531	0569	0607	0645	0682	0719	0755	4	8	11	15	19	23	26	30	34
12	0792	0828	0864	0899	0934	0969	1004	1038	1072	1106	3	7	10	14	17	21	24	28	31
13	1139	1173	1206	1239	1271	1303	1335	1367	1399	1430	3	6	10	13	16	19	23	26	29
14	1461	1492	1523	1553	1584	1614	1644	1673	1703	1732	3	6	9	12	15	18	21	24	27
15	1761	1790	1818	1847	1875	1903	1931	1959	1987	2014	3	6	8	11	14	17	20	22	25
16	2041	2068	2095	2122	2148	2175	2201	2227	2253	2279	3	5	8	11	13	16	18	21	24
17	2304	2330	2355	2380	2405	2430	2455	2480	2504	2529	2	5	7	10	12	15	17	20	22
18	2553	2577	2601	2625	2648	2672	2695	2718	2742	2765	2	5	7	9	12	14	16	19	21
19	2788	2810	2833	2856	2878	2900	2923	2945	2967	2989	2	4	7	9	11	13	16	18	20
20	3010	3032	3054	3075	3096	3118	3139	3160	3181	3201	2	4	6	8	11	13	15	17	19
21	3222	3243	3263	3284	3304	3324	3345	3365	3385	3404	2	4	6	8	10	12	14	16	18
22	3424	3444	3464	3483	3502	3522	3541	3560	3579	3598	2	4	6	8	10	12	14	15	17
23	3617	3636	3655	3674	3692	3711	3729	3747	3766	3784	2	4	6	7	9	11	13	15	17
24	3802	3820	3838	3856	3874	3892	3909	3927	3945	3962	2	4	5	7	9	11	12	14	16
25	3979	3997	4014	4031	4048	4065	4082	4099	4116	4133	2	3	5	7	9	10	12	14	15
26	4150	4166	4183	4200	4216	4232	4249	4265	4281	4298	2	3	5	7	8	10	11	13	15
27	4314	4330	4346	4362	4378	4393	4409	4425	4440	4456	2	3	5	6	8	9	11	13	14
28	4472	4487	4502	4518	4533	4548	4564	4579	4594	4609	2	3	5	6	8	9	11	12	14
29	4624	4639	4654	4669	4683	4698	4713	4728	4742	4757	1	3	4	6	7	9	10	12	13
30	4771	4786	4800	4814	4829	4843	4857	4871	4886	4900	1	3	4	6	7	9	10	11	13
31	4914	4928	4942	4955	4969	4983	4997	5011	5024	5038	1	3	4	6	7	8	10	11	12
32	5051	5065	5079	5092	5105	5119	5132	5145	5159	5172	1	3	4	5	7	8	9	11	12
33	5185	5198	5211	5224	5237	5250	5263	5276	5289	5302	1	3	4	5	6	8	9	10	12
34	5315	5328	5340	5353	5366	5378	5391	5403	5416	5428	1	3	4	5	6	8	9	10	11
35	5441	5453	5465	5478	5490	5502	5514	5527	5539	5551	1	2	4	5	6	7	9	10	11
36	5563	5575	5587	5599	5611	5623	5635	5647	5658	5670	1	2	4	5	6	7	8	10	11
37	5682	5694	5705	5717	5729	5740	5752	5763	5775	5786	1	2	3	5	6	7	8	9	10
38	5798	5809	5821	5832	5843	5855	5866	5877	5888	5899	1	2	3	5	6	7	8	9	10
39	5911	5922	5933	5944	5955	5966	5977	5988	5999	6010	1	2	3	4	5	7	8	9	10
40	6021	6031	6042	6053	6064	6075	6085	6096	6107	6117	1	2	3	4	5	6	8	9	10
41	6128	6138	6149	6160	6170	6180	6191	6201	6212	6222	1	2	3	4	5	6	7	8	9
42	6232	6243	6253	6263	6274	6284	6294	6304	6314	6325	1	2	3	4	5	6	7	8	9
43	6335	6345	6355	6365	6375	6385	6395	6405	6415	6425	1	2	3	4	5	6	7	8	9
44	6435	6444	6454	6464	6474	6484	6493	6503	6513	6522	1	2	3	4	5	6	7	8	9
45	6532	6542	6551	6561	6571	6580	6590	6599	6609	6618	1	2	3	4	5	6	7	8	9
46	6628	6637	6646	6656	6665	6675	6684	6693	6702	6712	1	2	3	4	5	6	7	7	8
47	6721	6730	6739	6749	6758	6767	6776	6785	6794	6803	1	2	3	4	5	5	6	7	8
48	6812	6821	6830	6839	6848	6857	6866	6875	6884	6893	1	2	3	4	4	5	6	7	8
49	6902	6911	6920	6928	6937	6946	6955	6964	6972	6981	1	2	3	4	4	5	6	7	8
50	6990	6998	7007	7016	7024	7033	7042	7050	7059	7067	1	2	3	3	4	5	6	7	8
51	7076	7084	7093	7101	7110	7118	7126	7135	7143	7152	1	2	3	3	4	5	6	7	8
52	7160	7168	7177	7185	7193	7202	7210	7218	7226	7235	1	2	2	3	4	5	6	7	7
53	7243	7251	7259	7267	7275	7284	7292	7300	7308	7316	1	2	2	3	4	5	6	6	7
54	7324	7332	7340	7348	7356	7364	7372	7380	7388	7396	1	2	2	3	4	5	6	6	7

*See page 12 for examples illustrating the use of logarithmic tables in calculations.

Logarithms of Numbers (*Continued*)

Natural numbers	0	1	2	3	4	5	6	7	8	9	Proportional parts								
											1	2	3	4	5	6	7	8	9
55	7404	7412	7419	7427	7435	7443	7451	7459	7466	7474	1	2	2	3	4	5	5	6	7
56	7482	7490	7497	7505	7513	7520	7528	7536	7543	7551	1	2	2	3	4	5	5	6	7
57	7559	7566	7574	7582	7589	7597	9604	7612	7619	7627	1	2	2	3	4	5	5	6	7
58	7634	7642	7649	7657	7664	7672	7679	7686	7694	7701	1	1	2	3	4	4	5	6	7
59	7709	7716	7723	7731	7738	7745	7752	7760	7767	7774	1	1	2	3	4	4	5	6	7
60	7782	7789	7796	7803	7810	7818	7825	7832	7839	7846	1	1	2	3	4	4	5	6	6
61	7853	7860	7868	7875	7882	7889	7896	7903	7910	7917	1	1	2	3	4	4	5	6	6
62	7924	7931	7938	7945	7952	7959	7966	7973	7980	7987	1	1	2	3	3	4	5	6	6
63	7993	8000	8007	8014	8021	8028	8035	8041	8048	8055	1	1	2	3	3	4	5	5	6
64	8062	8069	8075	8082	8089	8096	8102	8109	8116	8122	1	1	2	3	3	4	5	5	6
65	8129	8136	8142	8149	8156	8162	8169	8176	8182	8189	1	1	2	3	3	4	5	5	6
66	8195	8202	8209	8215	8222	8228	8235	8241	8248	8254	1	1	2	3	3	4	5	5	6
67	8261	8267	8274	8280	8287	8293	8299	8306	8312	8319	1	1	2	3	3	4	5	5	6
68	8325	8331	8338	8344	8351	8357	8363	8370	8376	8382	1	1	2	3	3	4	4	5	6
69	8388	8395	8401	8407	8414	8420	8426	8432	8439	8445	1	1	2	2	3	4	4	5	6
70	8451	8457	8463	8470	8476	8482	8488	8494	8500	8506	1	1	2	2	3	4	4	5	6
71	8513	8519	8525	8531	8537	8543	8549	8555	8561	8567	1	1	2	2	3	4	4	5	5
72	8573	8579	8585	8591	8597	8603	8609	8615	8621	8627	1	1	2	2	3	4	4	5	5
73	8633	8639	8645	8651	8657	8663	8669	8675	8681	8686	1	1	2	2	3	4	4	5	5
74	8692	8698	8704	8710	8716	8722	8727	8733	8739	8745	1	1	2	2	3	4	4	5	5
75	8751	8756	8762	8768	8774	8779	8785	8791	8797	8802	1	1	2	2	3	3	4	5	5
76	8808	8814	8820	8825	8831	8837	8842	8848	8854	8859	1	1	2	2	3	3	4	5	5
77	8865	8871	8876	8882	8887	8893	8899	8904	8910	8915	1	1	2	2	3	3	4	4	5
78	8921	8927	8932	8938	8943	8949	8954	8960	8965	8971	1	1	2	2	3	3	4	4	5
79	8976	8982	8987	8993	8998	9004	9009	9015	9020	9026	1	1	2	2	3	3	4	4	5
80	9031	9036	9042	9047	9053	9058	9063	9069	9074	9079	1	1	2	2	3	3	4	4	5
81	9085	9090	9096	9101	9106	9112	9117	9122	9128	9133	1	1	2	2	3	3	4	4	5
82	9138	9143	9149	9154	9159	9165	9170	9175	9180	9186	1	1	2	2	3	3	4	4	5
83	9191	9196	9201	9206	9212	9217	9222	9227	9232	9238	1	1	2	2	3	3	4	4	5
84	9243	9248	9253	9258	9263	9269	9274	9279	9284	9289	1	1	2	2	3	3	4	4	5
85	9294	9299	9304	9309	9315	9320	9325	9330	9335	9340	1	1	2	2	3	3	4	4	5
86	9345	9350	9355	9360	9365	9370	9375	9380	9385	9390	1	1	2	2	3	3	4	4	5
87	9395	9400	9405	9410	9415	9420	9425	9430	9435	9440	0	1	1	2	2	3	3	4	4
88	9445	9450	9455	9460	9465	9469	9474	9479	9484	9489	0	1	1	2	2	3	3	4	4
89	9494	9499	9504	9509	9513	9518	9523	9528	9533	9538	0	1	1	2	2	3	3	4	4
90	9542	9547	9552	9557	9562	9566	9571	9576	9581	9586	0	1	1	2	2	3	3	4	4
91	9590	9595	9600	9605	9609	9614	9619	9624	9628	9633	0	1	1	2	2	3	3	4	4
92	9638	9643	9647	9652	9657	9661	9666	9671	9675	9680	0	1	1	2	2	3	3	4	4
93	9685	9689	9694	9699	9703	9708	9713	9717	9722	9727	0	1	1	2	2	3	3	4	4
94	9731	9736	9741	9745	9750	9754	9759	9763	9768	9773	0	1	1	2	2	3	3	4	4
95	9777	9782	9786	9791	9795	9800	9805	9809	9814	9818	0	1	1	2	2	3	3	4	4
96	9823	9827	9832	9836	9841	9845	9850	9854	9859	9863	0	1	1	2	2	3	3	4	4
97	9868	9872	9877	9881	9886	9890	9894	9899	9903	9908	0	1	1	2	2	3	3	4	4
98	9912	9917	9921	9926	9930	9934	9939	9943	9948	9952	0	1	1	2	2	3	3	4	4
99	9956	9961	9965	9969	9974	9978	9983	9987	9991	9996	0	1	1	2	2	3	3	3	4

ANTILOGARITHMS

Logarithms	0	1	2	3	4	5	6	7	8	9	Proportional parts								
											1	2	3	4	5	6	7	8	9
0.00	1000	1002	1005	1007	1009	1012	1014	1016	1019	1021	0	0	1	1	1	1	2	2	2
0.01	1023	1026	1028	1030	1033	1035	1038	1040	1042	1045	0	0	1	1	1	1	2	2	2
0.02	1047	1050	1052	1054	1057	1059	1062	1064	1067	1069	0	0	1	1	1	1	2	2	2
0.03	1072	1074	1076	1079	1081	1084	1086	1089	1091	1094	0	0	1	1	1	1	2	2	2
0.04	1096	1099	1102	1104	1107	1109	1112	1114	1117	1119	0	1	1	1	1	2	2	2	2
0.05	1122	1125	1127	1130	1132	1135	1138	1140	1143	1146	0	1	1	1	1	2	2	2	2
0.06	1148	1151	1153	1156	1159	1161	1164	1167	1169	1172	0	1	1	1	1	2	2	2	2
0.07	1175	1178	1180	1183	1186	1189	1191	1194	1197	1199	0	1	1	1	1	2	2	2	2
0.08	1202	1205	1208	1211	1213	1216	1219	1222	1225	1227	0	1	1	1	1	2	2	2	3
0.09	1230	1233	1236	1239	1242	1245	1247	1250	1253	1256	0	1	1	1	1	2	2	2	3
0.10	1259	1262	1265	1268	1271	1274	1276	1279	1282	1285	0	1	1	1	1	2	2	2	3
0.11	1288	1291	1294	1297	1300	1303	1306	1309	1312	1315	0	1	1	1	2	2	2	2	3
0.12	1318	1321	1324	1327	1330	1334	1337	1340	1343	1346	0	1	1	1	2	2	2	2	3
0.13	1349	1352	1355	1358	1361	1365	1368	1371	1374	1377	0	1	1	1	2	2	2	3	3
0.14	1380	1384	1387	1390	1393	1396	1400	1403	1406	1409	0	1	1	1	2	2	2	3	3
0.15	1413	1416	1419	1422	1426	1429	1432	1435	1439	1442	0	1	1	1	2	2	2	3	3
0.16	1445	1449	1452	1455	1459	1462	1466	1469	1472	1476	0	1	1	1	2	2	2	3	3
0.17	1479	1483	1486	1489	1493	1496	1500	1503	1507	1510	0	1	1	1	2	2	2	3	3
0.18	1514	1517	1521	1524	1528	1531	1535	1538	1542	1545	0	1	1	1	2	2	2	3	3
0.19	1549	1552	1556	1560	1563	1567	1570	1574	1578	1581	0	1	1	1	2	2	3	3	3
0.20	1585	1589	1592	1596	1600	1603	1607	1611	1614	1618	0	1	1	1	2	2	3	3	3
0.21	1622	1626	1629	1633	1637	1641	1644	1648	1652	1656	0	1	1	2	2	2	3	3	3
0.22	1660	1663	1667	1671	1675	1679	1683	1687	1690	1694	0	1	1	2	2	2	3	3	3
0.23	1698	1702	1706	1710	1714	1718	1722	1726	1730	1734	0	1	1	2	2	2	3	3	4
0.24	1738	1742	1746	1750	1754	1758	1762	1766	1770	1774	0	1	1	2	2	2	3	3	4
0.25	1778	1782	1786	1791	1795	1799	1803	1807	1811	1816	0	1	1	2	2	2	3	3	4
0.26	1820	1824	1828	1832	1837	1841	1845	1849	1854	1858	0	1	1	2	2	3	3	3	4
0.27	1862	1866	1871	1875	1879	1884	1888	1892	1897	1901	0	1	1	2	2	3	3	3	4
0.28	1905	1910	1914	1919	1923	1928	1932	1936	1941	1945	0	1	1	2	2	3	3	4	4
0.29	1950	1954	1959	1963	1968	1972	1977	1982	1986	1991	0	1	1	2	2	3	3	4	4
0.30	1995	2000	2004	2009	2014	2018	2023	2028	2032	2037	0	1	1	2	2	3	3	4	4
0.31	2042	2046	2051	2056	2061	2065	2070	2075	2080	2084	0	1	1	2	2	3	3	4	4
0.32	2089	2094	2099	2104	2109	2113	2118	2123	2128	2133	0	1	1	2	2	3	3	4	4
0.33	2138	2143	2148	2153	2158	2163	2168	2173	2178	2183	0	1	1	2	2	3	3	4	4
0.34	2188	2193	2198	2203	2208	2213	2218	2223	2228	2234	1	1	2	2	3	3	4	4	5
0.35	2239	2244	2249	2254	2259	2265	2270	2275	2280	2286	1	1	2	2	3	3	4	4	5
0.36	2291	2296	2301	2307	2312	2317	2323	2328	2333	2339	1	1	2	2	3	3	4	4	5
0.37	2344	2350	2355	2360	2366	2371	2377	2382	2388	2393	1	1	2	2	3	3	4	4	5
0.38	2399	2404	2410	2415	2421	2427	2432	2438	2443	2449	1	1	2	2	3	3	4	4	5
0.39	2455	2460	2466	2472	2477	2483	2489	2495	2500	2506	1	1	2	2	3	3	4	5	5
0.40	2512	2518	2523	2529	2535	2541	2547	2553	2559	2564	1	1	2	2	3	4	4	5	5
0.41	2570	2576	2582	2588	2594	2600	2606	2612	2618	2624	1	1	2	2	3	4	4	5	5
0.42	2630	2636	2642	2649	2655	2661	2667	2673	2679	2685	1	1	2	2	3	4	4	5	6
0.43	2692	2698	2704	2710	2716	2723	2729	2735	2742	2748	1	1	2	3	3	4	4	5	6
0.44	2754	2761	2767	2773	2780	2786	2793	2799	2805	2812	1	1	2	3	3	4	4	5	6
0.45	2818	2825	2831	2838	2844	2851	2858	2864	2871	2877	1	1	2	3	3	4	5	5	6
0.46	2884	2891	2897	2904	2911	2917	2924	2931	2938	2944	1	1	2	3	3	4	5	5	6
0.47	2951	2958	2965	2972	2979	2985	2992	2999	3006	3013	1	1	2	3	3	4	5	5	6
0.48	3020	3027	3034	3041	3048	3055	3062	3069	3076	3083	1	1	2	3	4	4	5	6	6
0.49	3090	3097	3105	3112	3119	3126	3133	3141	3148	3155	1	1	2	3	4	4	5	6	6

ANTILOGARITHMS (*Continued*)

Logarithms	0	1	2	3	4	5	6	7	8	9	Proportional parts 1	2	3	4	5	6	7	8	9
0.50	3162	3170	3177	3184	3192	3199	3206	3214	3221	3228	1	1	2	3	4	4	5	6	7
0.51	3236	3243	3251	3258	3266	3273	3281	3289	3296	3304	1	2	2	3	4	5	5	6	7
0.52	3311	3319	3327	3334	3342	3350	3357	3365	3373	3381	1	2	2	3	4	5	5	6	7
0.53	3388	3396	3404	3412	3420	3428	3436	3443	3451	3459	1	2	2	3	4	5	6	6	7
0.54	3467	3475	3483	3491	3499	3508	3516	3524	3532	3540	1	2	2	3	4	5	6	6	7
0.55	3548	3556	3565	3573	3581	3589	3597	3606	3614	3622	1	2	2	3	4	5	6	7	7
0.56	3631	3639	3648	3656	3664	3673	3681	3690	3698	3707	1	2	3	3	4	5	6	7	8
0.57	3715	3724	3733	3741	3750	3758	3767	3776	3784	3793	1	2	3	3	4	5	6	7	8
0.58	3802	3811	3819	3828	3837	3846	3855	3864	3873	3882	1	2	3	4	4	5	6	7	8
0.59	3890	3899	3908	3917	3926	3936	3945	3954	3963	3972	1	2	3	4	5	5	6	7	8
0.60	3981	3990	3999	4009	4018	4027	4036	4046	4055	4064	1	2	3	4	5	6	6	7	8
0.61	4074	4083	4093	4102	4111	4121	4130	4140	4150	4159	1	2	3	4	5	6	7	8	9
0.62	4169	4178	4188	4198	4207	4217	4227	4236	4246	4256	1	2	3	4	5	6	7	8	9
0.63	4266	4276	4285	4295	4305	4315	4325	4335	4345	4355	1	2	3	4	5	6	7	8	9
0.64	4365	4375	4385	4395	4406	4416	4426	4436	4446	4457	1	2	3	4	5	6	7	8	9
0.65	4467	4477	4487	4498	4508	4519	4529	4539	4550	4560	1	2	3	4	5	6	7	8	9
0.66	4571	4581	4592	4603	4613	4624	4634	4645	4656	4667	1	2	3	4	5	6	7	9	10
0.67	4677	4688	4699	4710	4721	4732	4742	4753	4764	4775	1	2	3	4	5	7	8	9	10
0.68	4786	4797	4808	4819	4831	4842	4853	4864	4875	4887	1	2	3	4	6	7	8	9	10
0.69	4898	4909	4920	4932	4943	4955	4966	4977	4989	5000	1	2	3	5	6	7	8	9	10
0.70	5012	5023	5035	5047	5058	5070	5082	5093	5105	5117	1	2	4	5	6	7	8	9	11
0.71	5129	5140	5152	5164	5176	5188	5200	5212	5224	5236	1	2	4	5	6	7	8	10	11
0.72	5248	5260	5272	5284	5297	5309	5321	5333	5346	5358	1	2	4	5	6	7	9	10	11
0.73	5370	5383	5395	5408	5420	5433	5445	5458	5470	5483	1	3	4	5	6	8	9	10	11
0.74	5495	5508	5521	5534	5546	5559	5572	5585	5598	5610	1	3	4	5	6	8	9	10	12
0.75	5623	5636	5649	5662	5675	5689	5702	5715	5728	5741	1	3	4	5	7	8	9	10	12
0.76	5754	5768	5781	5794	5808	5821	5834	5848	5861	5875	1	3	4	5	7	8	9	11	12
0.77	5888	5902	5916	5929	5943	5957	5970	5984	5998	6012	1	3	4	5	7	8	10	11	12
0.78	6026	6039	6053	6067	6081	6095	6109	6124	6138	6152	1	3	4	6	7	8	10	11	13
0.79	6166	6180	6194	6209	6223	6237	6252	6266	6281	6295	1	3	4	6	7	9	10	11	13
0.80	6310	6324	6339	6353	6368	6383	6397	6412	6427	6442	1	3	4	6	7	9	10	12	13
0.81	6457	6471	6486	6501	6516	6531	6546	6561	6577	6592	2	3	5	6	8	9	11	12	14
0.82	6607	6622	6637	6653	6668	6683	6699	6714	6730	6745	2	3	5	6	8	9	11	12	14
0.83	6761	6776	6792	6808	6823	6839	6855	6871	6887	6902	2	3	5	6	8	9	11	13	14
0.84	6918	6934	6950	6966	6982	6998	7015	7031	7047	7063	2	3	5	6	8	10	11	13	15
0.85	7079	7096	7112	7129	7145	7161	7178	7194	7211	7228	2	3	5	7	8	10	12	13	15
0.86	7244	7261	7278	7295	7311	7328	7345	7362	7379	7396	2	3	5	7	8	10	12	13	15
0.87	7413	7430	7447	7464	7482	7499	7516	7534	7551	7568	2	3	5	7	9	10	12	14	16
0.88	7586	7603	7621	7638	7656	7674	7691	7709	7727	7745	2	4	5	7	9	11	12	14	16
0.89	7762	7780	7798	7816	7834	7852	7870	7889	7907	7925	2	4	5	7	9	11	13	14	16
0.90	7943	7962	7980	7998	8017	8035	8054	8072	8091	8110	2	4	6	7	9	11	13	15	17
0.91	8128	8147	8166	8185	8204	8222	8241	8260	8279	8299	2	4	6	8	9	11	13	15	17
0.92	8318	8337	8356	8375	8395	8414	8433	8453	8472	8492	2	4	6	8	10	12	14	15	17
0.93	8511	8531	8551	8570	8590	8610	8630	8650	8670	8690	2	4	6	8	10	12	14	16	18
0.94	8710	8730	8750	8770	8790	8810	8831	8851	8872	8892	2	4	6	8	10	12	14	16	18
0.95	8913	8933	8954	8974	8995	9016	9036	9057	9078	9099	2	4	6	8	10	12	15	17	19
0.96	9120	9141	9162	9183	9204	9226	9247	9268	9290	9311	2	4	6	8	11	13	15	17	19
0.97	9333	9354	9376	9397	9419	9441	9462	9484	9506	9528	2	4	7	9	11	13	15	17	20
0.98	9550	9572	9594	9616	9638	9661	9683	9705	9727	9750	2	4	7	9	11	13	16	18	20
0.99	9772	9795	9817	9840	9863	9886	9908	9931	9954	9977	2	5	7	9	11	14	16	18	20

Atomic Weights

Adopted by the International Union of Pure and Applied Physics (1960) and by the International Union of Pure and Applied Chemistry (1961)

Name	Symbol	Atomic Number	Atomic Weight*	Name	Symbol	Atomic Number	Atomic Weight*
Actinium	Ac	89	(227)	Mercury	Hg	80	200.59
Aluminum	Al	13	26.9815	Molybdenum	Mo	42	95.94
Americium	Am	95	(243)	Neodymium	Nd	60	144.24
Antimony	Sb	51	121.75	Neon	Ne	10	20.183
Argon	Ar	18	39.948	Neptunium	Np	93	(237)
Arsenic	As	33	74.9216	Nickel	Ni	28	58.71
Astatine	At	85	(210)	Niobium	Nb	41	92.906
Barium	Ba	56	137.34	Nitrogen	N	7	14.0067
Berkelium	Bk	97	(247)	Nobelium	No	102	—
Beryllium	Be	4	9.0122	Osmium	Os	76	190.2
Bismuth	Bi	83	208.980	Oxygen	O	8	15.9994†
Boron	B	5	10.811†	Palladium	Pd	46	106.4
Bromine	Br	35	79.909‡	Phosphorus	P	15	30.9738
Cadmium	Cd	48	112.40	Platinum	Pt	78	195.09
Calcium	Ca	20	40.08	Plutonium	Pu	94	(242)
Californium	Cf	98	(249)	Polonium	Po	84	[210]
Carbon	C	6	12.01115†	Potassium	K	19	39.102
Cerium	Ce	58	140.12	Praseodymium	Pr	59	140.907
Cesium	Cs	55	132.905	Promethium	Pm	61	[147]
Chlorine	Cl	17	35.453‡	Protactinium	Pa	91	(231)
Chromium	Cr	24	51.996‡	Radium	Ra	88	(226)
Cobalt	Co	27	58.9332	Radon	Rn	86	(222)
Copper	Cu	29	63.54	Rhenium	Re	75	186.2
Curium	Cm	96	(247)	Rhodium	Rh	45	102.905
Dysprosium	Dy	66	162.50	Rubidium	Rb	37	85.47
Einsteinium	Es	99	(254)	Ruthenium	Ru	44	101.07
Erbium	Er	68	167.26	Samarium	Sm	62	150.35
Europium	Eu	63	151.96	Scandium	Sc	21	44.956
Fermium	Fm	100	(253)	Selenium	Se	34	78.96
Fluorine	F	9	18.9984	Silicon	Si	14	28.086†
Francium	Fr	87	(223)	Silver	Ag	47	107.870‡
Gadolinium	Gd	64	157.25	Sodium	Na	11	22.9898
Gallium	Ga	31	69.72	Strontium	Sr	38	87.62
Germanium	Ge	32	72.59	Sulfur	S	16	32.064†
Gold	Au	79	196.967	Tantalum	Ta	73	180.948
Hafnium	Hf	72	178.49	Technetium	Tc	43	[99]
Helium	He	2	4.0026	Tellurium	Te	52	127.60
Holmium	Ho	67	164.930	Terbium	Tb	65	158.924
Hydrogen	H	1	1.00797†	Thallium	Tl	81	204.37
Indium	In	49	114.82	Thorium	Th	90	232.038
Iodine	I	53	126.9044	Thulium	Tm	69	168.934
Iridium	Ir	77	192.2	Tin	Sn	50	118.69
Iron	Fe	26	55.847‡	Titanium	Ti	22	47.90
Krypton	Kr	36	83.80	Tungsten	W	74	183.85
Lanthanum	La	57	138.91	Uranium	U	92	238.03
Lead	Pb	82	207.19	Vanadium	V	23	50.942
Lithium	Li	3	6.939	Xenon	Xe	54	131.30
Lutetium	Lu	71	174.97	Ytterbium	Yb	70	173.04
Magnesium	Mg	12	24.312	Yttrium	Y	39	88.905
Manganese	Mn	25	54.9380	Zinc	Zn	30	65.37
Mendelevium	Md	101	(256)	Zirconium	Zr	40	91.22

* Values in parentheses or brackets denote mass numbers of selected radioactive isotopes. Those in parentheses are for the isotopes of longest known half-life; those in brackets are for isotopes that are better known than the corresponding isotopes of the longest half-life. All atomic weight values are based on the atomic mass of $^{12}C = 12$.

† The atomic weight varies because of natural variations in the isotopic composition of the element. The observed ranges are: boron, ±0.003; carbon, ±0.00005; hydrogen, ±0.00001; oxygen, ±0.0001; silicon, ±0.001; sulfur, ±0.003.

‡ The atomic weight is believed to have an experimental uncertainty of the following magnitude: bromine, ±0.002; chlorine, ±0.001; chromium, ±0.001; iron, ±0.003; silver, ±0.003. For other elements, the last digit given is believed reliable to ±0.5.

Index